O ye Gentlemen

Arabic Studies on Science and Literary Culture

Islamic Philosophy, Theology and Science

Texts and Studies

Edited by
H. Daiber

VOLUME LXXIV

Remke Kruk

O ye Gentlemen
Arabic Studies on Science and Literary Culture

In Honour of Remke Kruk

Edited by

Arnoud Vrolijk and Jan P. Hogendijk

BRILL

LEIDEN • BOSTON
2007

This book is printed on acid-free paper.

A C.I.P. record for this book is available from the Library of Congress.

ISSN: 0169-8729
ISBN: 978 90 04 15794 1

PRINTED IN THE NETHERLANDS

CONTENTS

SCIENCE & PHILOSOPHY

MEDICINE & MAGIC

HISTORY

LITERATURE

LANGUAGE

ART & MUSIC

LAW & RELIGION

FOREWORD

O ye Gentlemen explores two vital strands in Arabic culture: the Greek tradition in science and philosophy and the literary tradition. They are permanent and, though drawing on Islam as a dominant religion, they are by no means dependent on it. That the strands freely interweave within the broader scope of *Schrifttum* is shown by essays on subjects as varied as the social organisation of bees, spontaneous generation in the Shiʿite tradition, astronomy in the *Arabian Nights*, the benefits of sex, precious stones in a literary text, the virtue of women in Judaeo-Arabic stories, animals in Middle Eastern music and the transmission of Arabic science and philosophy to the medieval West.

The choice of subjects reflects the scholarly interests and principles of the dedicatee of this volume, Remke Kruk (*b.* Apeldoorn, 1942), who, after an initial career at the University of Utrecht, held the Leiden chair of Arabic from 1990 until her retirement in 2007. It is no doubt an exceptional career, for at the University of Leiden, an organisation firmly in the grip of bureaucrats in suits and ties, a female professor may not be as rare as a unicorn, but it is still very much a *rara avis*.

Remke's thesis on the Arabic version of Aristotle's *De partibus animalium*, defended at the University of Amsterdam in 1978 under the supervision of the eminent philologist H.J. Drossaart Lulofs, revealed a profound interest in the history of life sciences in Arabic-Islamic culture, with a discerning eye for the extraordinary, the magical and the wondrous. Precisely these elements led her to the study of the Arabic popular epic, the *sīra*, and the role of the leading ladies in these romances, like princess Dhāt al-Himma and Qannāṣa bint Muzāḥim.

Remke has always pursued her studies with a complete disregard for the artificial boundaries between Arabic *belles lettres* and the larger corpus of classical and post-classical Arabic texts on history, philosophy, religion and science. The basic notion of *Schrifttum* may have lost ground, just as the whole philological tradition seems to have given way to social sciences and Middle East studies, but to Remke – and to many of those who have contributed to this volume – it remains a sensible and rewarding approach to Arabic culture.

From her relatively marginal position as a female professor, breaking down barriers must have come naturally to Remke. When the Arabic department of Utrecht University found itself in heavy weather in the 1980s, Wim Gerritsen and his fellow mediaevalists generously gave her shelter, helping her to increase her awareness of the value of interdisciplinarity. In 2000 she was elected a member of the Royal Netherlands Academy of Sciences, a platform where eminent scholars from all disciplines meet on an equal footing. In the Leiden faculty of Arts, Remke and her colleague Sjef Houppermans have for years been the moving force behind the 'Conventie & Originaliteit' working group, where representatives from 'occidental' and 'oriental' studies meet to discuss and to publish thought-provoking books intended for a general readership. In the same spirit Remke has reached out to the public through her translations of classics like Ibn Ḥazm's *Ṭawq al-ḥamāma* (together with Jan Just Witkam, 1977), Ibn Ṭufayl's *Ḥayy b. Yaqẓān* (1985, rev. ed. 2005) and most recently Ibn Rushd's *Faṣl al-maqāl* (2 editions, 2006). The results of her work may be gleaned from the list of publications included in this volume.

More important still, however, is the fact that Remke has consistently refused to isolate her scholarly work from the rest of her life. Her students from the 1970s still recall how Remke used to breastfeed her baby during class because it was the natural thing to do. Her knowledge of Nature has enabled her to create one of the loveliest gardens in Leiden, with the soft cackle of her hens in the distance. Her passion for food, which she shares with colleagues like David Waines and Manuela Marín, has always been an 'alimentary' principle in her life. Yet there is no matronly dowdiness in Remke's character. She is a fiercely independent spirit with a corrosive sense of humour, mitigated only by her understanding of the weaknesses of man.

This volume is offered to Remke Kruk by colleagues, friends and students as a token of our appreciation. We wish her well in her future career, no longer burdened with managerial duties and academic red tape.

The title *O ye Gentlemen* is a translation of 'Yā Sāda', the cry used by the narrator of the Arabic folk epic to arrest the attention of his audience. The masculine form of the expression was chosen to amuse, but also to vex the dedicatee.

PUBLICATIONS OF REMKE KRUK, 1976–2006

Monographs

Aristoteles Semitico-Latinus. The Arabic version of Aristotle's Parts of Animals. Book XI–XIV of the Kitāb al-Ḥayawān. A critical edition with introduction and selected glossary, by Remke Kruk. Verhandelingen der Koninklijke Nederlandse Akademie van Wetenschappen, afd. Letterkunde, Nieuwe reeks; 97 (Amsterdam [etc.], 1979).

Een kaart van de kat! Tussen Arabistiek en Mediëvistiek. Rede uitgesproken bij de aanvaarding van het ambt van gewoon hoogleraar in het Arabisch en de Arabische cultuur aan de Rijksuniversiteit te Leiden op vrijdag 22 februari 1991 (Leiden, 1991).

Edited volumes

G. Endress & R. Kruk (eds.), *The ancient tradition in Christian and Islamic Hellenism. Studies on the transmission of Greek philosophy and sciences dedicated to H.J. Drossaart Lulofs on his ninetieth birthday* (Leiden, 1997).

S. Houppermans, W.L. Idema & R. Kruk (eds.). *Op avontuur! Aspecten van avonturenverhalen in Oost en West* (Zutphen, 1998).

S. Houppermans, R. Kruk & Henk Maier (eds.), *Van gene zijde. Verhalen over hemel en hel* (Rijswijk, 2000).

S. Houppermans, R. Kruk & Henk Maier (eds.), *Rapsoden en rebellen; Literatuur en politiek in verschillende culturen* (Amsterdam, 2003).

R. Kruk & S. Houppermans (eds.), *Een vis in een fles raki. Literatuur en drank in verschillende culturen* (Amsterdam, 2005).

Articles

'Pseudo-Aristotle. An Arabic version of *Problemata physica* X,' *Isis* 67/237 (June 1976), pp. 251–256.

'Arabische poezie. Enkele thema's en motieven,' *De Gids* 143/9–10 (1980), pp. 639–647.

'Arabische courtisanes (*qiyān*),' in R.E.V. Stuip & C. Vellekoop (eds.), *Middeleeuwers over vrouwen.* Utrechtse bijdragen tot de mediëvistiek; 3 (Utrecht, 1985), I, pp. 95–110.

'Hedgehogs and their "chicks". A case history of the Aristotelian reception in Arabic zoology,' *Zeitschrift für Geschichte der arabisch-islamischen Wissenschaften* 2 (1985), pp. 205–234.

'Aristoteles, Avicenna, Albertus, en de locusta maris,' in A.M.J. van Buuren, O.S.H. Lie [et al.] (eds.), *Tussentijds. Bundel studies aangeboden aan W.P. Gerritsen ter gelegenheid van zijn vijftigste verjaardag.* Utrechtse bijdragen tot de mediëvistiek; 5 (Utrecht, 1985), pp. 147–156.

'Some late mediaeval zoological texts and their sources,' in *Actas del XII Congreso de la U.E.A.I., Malaga 1984* (Madrid, 1986) pp. 423–429.

'Arabische bronnen over de gewone mens. Het curieuze en het alledaagse,' in R.E.V. Stuip & C. Vellekoop (eds.), *Gewone mensen in de Middeleeuwen. Bundel studies aangeboden aan F.W.N. Hugenholtz ter gelegenheid van zijn afscheid.* Utrechtse bijdragen tot de mediëvistiek; 7 (Utrecht, 1987), pp. 254–268.

'An 18th-century descendant of Ḥayy Ibn Yaqẓān and Robinson Crusoe. Don Antonio de Trezzanio,' *Arabica* 34 (1987), pp. 357–365.
'Pregnancy and its social consequences in mediaeval and traditional Arab society,' *Quaderni di studi arabi* 5–6 (1987–88) = *Atti del XIII congresso dell'Union européenne d'arabisants et d'islamisants*, pp. 418–430.
'A frothy bubble. Spontaneous generation in the medieval Islamic tradition,' *Journal of Semitic studies* 35 (1990), pp. 265–282.
'Neoplatonists and after. From Ibn Ṭufayl to Ibn an-Nafīs,' in A. Vanderjagt & D. Pätzold (eds.), *The neoplatonic tradition. Jewish, Christian and Islamic themes* (Köln, 1991), pp. 75–85.
'Nabāt,' in *Encyclopaedia of Islam*. New ed., VI (Leiden, 1991), pp. 831–834.
'De bloei der tuinen. Tuinen en tijd in de middeleeuwse Arabische literatuur,' in. R.E.V. Stuip & C. Vellekoop (eds.), *Tuinen in de Middeleeuwen*. Utrechtse bijdragen tot de mediëvistiek; 11 (Hilversum, 1992), pp. 71–89.
'De geschiedenis van koning 'Umar an-Nu'mân. Over Arabische ridderromans,' in R.E.V. Stuip & C. Vellekoop (eds.), *Oraliteit en Schriftcultuur*. Utrechtse bijdragen tot de mediëvistiek; 12 (Hilversum, 1993), pp. 171–187.
'De onbaatzuchtige wormen,' in W.L. Idema [et al.] (eds.), *Mijn naam is haas. Dierenverhalen in verschillende culturen* (Baarn, 1993), pp. 149–158.
'Zār amulets,' in C. Huygens, F. Ros [et al.] (eds.), *Dreaming of paradise. Islamic art from the collection of the Museum of Ethnology Rotterdam* (Rotterdam, 1993), pp. 146–149.
'Warrior women in Arabic popular romance. Qannâṣa bint Muzâḥim and other valiant ladies,' Pt. 1, *Journal of Arabic literature* 24/3 (1993), pp. 213–30; Pt. 2, *ibid.* 25/1 (1994), pp. 16–33.
'Lichaam en ziel. Verzwegen en geafficheerde pijn. Aspecten van de Arabische cultuur,' in. H.F.J. Horstmanshoff (ed.) *Pijn en balsem, troost en smart. Pijnbeleving en pijnbestrijding in de oudheid*. Nieuwe Nederlandse bijdragen tot de geschiedenis der geneeskunde en der natuurwetenschappen; 47 (Rotterdam, 1994), pp. 95–108.
'Clipped wings. Medieval Arabic adaptations of the Amazon myth,' *Harvard Middle Eastern and Islamic review* 1/2 (1994), pp. 132–154.
'Rāqid,' in *Encyclopaedia of Islam*. New ed., VIII (Leiden, 1995), p. 407.
'Traditional Islamic views of apes and monkeys,' in R. Corbey & B. Theunissen (eds.), *Ape, man, apeman. Changing views since 1600. Evaluative proceedings of the Symposium Ape, Man, Apeman [...], Leiden, The Netherlands, 28 June–1 July, 1993* (Leiden, 1995), pp. 29–42.
'Knopen en ontknopen. Magie in de Arabische letterkunde,' in M. Schipper & P. Schrijvers (eds.), *Bezweren en betoveren. Magie en literatuur wereldwijd* (Baarn, 1995), pp. 24–34.
'History and apocalypse. Ibn an-Nafīs' justification of Mamluk rule,' *Der Islam* 72/2 (1995), pp. 324–337.
'Ibn Baṭṭūṭa. Travel, family life, and chronology. How seriously do we take a father?,' *Al-Qantara* 16/2 (1995), pp. 369–385.
'Ibn Ṭufayl. A medieval scholar's views on nature,' in L.I. Conrad (ed.), *The world of Ibn Ṭufayl. Interdisciplinary perspectives on Ḥayy ibn Yaqẓān*. Islamic philosophy, theology and science, Texts and studies; 24 (Leiden, 1996), pp. 69–89.
'Ibn Bājja's commentary on Aristotle's *De Animalibus*,' in G. Endress & R. Kruk, *The ancient tradition in Christian and Islamic Hellenism* (Leiden, 1997), pp. 165–180.
'Sidr,' in *Encyclopaedia of Islam*. New ed., IX (Leiden, 1997), pp. 549–550.
'Back to the boudoir. Arabic versions of the *Sîrat al-amîr Ḥamza*, warrior princesses, and the Sîra's literary unity,' in. L. Jongen & S. Onderdelinden (eds.), *'Der muoz mir süezer worte jehen'. Liber amicorum für Norbert Voorwinden*. Amsterdamer Beiträge zur älteren Germanistik; 48 (Amsterdam [etc.], 1997), pp. 129–148.

'Islamitische visies op de rechten van het dier,' in R. van den Bos (ed.), *Welzijn van dieren en dierenwelzijnsbeleid. Essays over doelstellingen, instrumenten en evaluatie van het dierenwelzijnsbeleid in Nederland* (Tilburg, 1997), pp. 83–97.
'Aldebaran, alchimie en algebra. Wereldbeeld en wetenschap in de Arabische wereld tussen 1150 en 1400,' in D.E.H. de Boer (ed.), *Kennis op kamelen. Europa en de buiten-Europese wereld (1150–1350)* (Amsterdam, 1998), pp. 176–210.
'The Bold and the Beautiful. Women and *fitna* in the *Sīrat Dhāt al-Himma*. The story of Nūrā,' in G.R. Hambly (ed.), *Women in the medieval Islamic World. Power, patronage and piety* (Basingstoke, 1998), pp. 99–116.
'Popular Arabic literature as a mirror of social attitudes and preoccupations,' in R. Deguilhem, *Individu et Société dans le monde méditerranéen musulman. Questions et sources* (Aix-en-Provence, 1998), pp. 99–112.
'Het artsenberoep in het middeleeuwse Arabische discours,' *Gewina* 21/3 (1998), pp. 94–107.
''Antar komt zichzelf tegen en andere avonturen van 'Antar ibn Shaddâd,' in J. Houppermans [et al.] (eds.), *Op Avontuur! Aspecten van avonturenverhalen in Oost en West* (Zutphen, 1998), pp. 204–223.
'De goede arts. Ideaalbeeld en werkelijkheid in de middeleeuws-arabische wereld,' *Hermeneus* 71/2 (april 1999), pp. 140–149.
'On Animals. Excerpts of Aristotle and Ibn Sīnā in Marwazī's *Ṭabāʾiʿ al-ḥayawān*,' in C. Steel [et al.] (eds.), *Aristotle's animals in the Middle Ages and Renaissance*. Mediaevalia Lovaniensia, Series 1, Studia; 27 (Leuven, 1999), pp. 96–125.
'"In the Popular Manner". Sīra-recitation in Marrakesh anno 1997,' with C. Ott, *Edebiyat* N.S. 10/2 (1999), pp. 183–198.
'De stad van de snavelhakkers,' *Madoc* 13/4 (1999), pp. 279–286.
'Takwīn,' in *Encyclopaedia of Islam*. New ed., X (Leiden, 2000), pp. 147–148.
'Tawallud,' in *Encyclopaedia of Islam*. New ed., X (Leiden, 2000), pp. 378–379.
'Een dame, een paard en een toren vóór' = 'A lead of queen, knight and rook,' in H. Reerink (ed.), *Dame aan zet. Vrouwen en schaken door de eeuwen heen = Queen's move. Women and chess through the ages* (Den Haag, 2000), pp. 13–34. (Cat. exhibition 'Dame aan zet', Koninklijke Bibliotheek, 28 August–17 November 2000).
'Kiespijn en kastelen. Voorstellingen van het hiernamaals in de islamitische traditie,' with Mariëtte van Beek, in. S. Houppermans [et al.] (eds.), *Van gene zijde. Verhalen over hemel en hel* (Rijswijk, 2000), pp. 142–156.
'Ibn Abī l-Ashʿath's *Kitāb al-ḥayawān*. A scientific approach to anthropology, dietetics and zoological systematics,' *Zeitschrift für Geschichte der arabisch-islamischen Wissenschaften* 14 (2001), pp. 119–168.
'Is schaken wel een spel voor heren? Over het schaakspel in de middeleeuwse Arabische cultuur,' in W.S. van Egmond & M. Mostert, *Spelen in de Middeleeuwen. Over schaken, dammen en dobbelen en kaarten*. Utrechtse bijdragen tot de mediëvistiek; 17 (Hilversum, 2001), 105–122.
'Of Rukhs and rooks, camels and castles,' *Oriens* 36 (2001), pp. 288–298.
'Sharaf az-Zamān Ṭāhir Marwazī (fl. ca. 1100 A.D.) on Zoroaster, Mānī, Mazdak, and other pseudo-prophets,' *Persica* 17 (2001), pp. 51–68.
'Timotheus of Gaza's *On Animals* in the Arabic tradition,' *Le Muséon* 114/3-4 (2001), pp. 355–387.
'Ibn Sīnā on animals. Between the First Teacher and the Physician,' in J. Janssens & D. de Smet (eds.), *Avicenna and his heritage. Acts of the international colloquium, Leuven-Louvain-la-Neuve September 8–September 11, 1999*. Ancient and medieval philosophy, Ser. 1; 28 (Leuven, 2002), pp. 325–341.
'Click of needles. Polygamy as an issue in Arabic popular epic,' in M. Marín & R. Deguilhem, *Writing the feminine. Women in Arab sources* (London [etc.], 2002), pp. 3–24.

Introduction to: Jamal Sleem Nuweihed, *Abu Jmeel's Daughter and other stories. Arab folk tales from Palestine and Lebanon. Retold by Jamal Sleem Nuweihed, translated by members of her family with Christopher Tingley* (New York, 2002), pp. xii–xix.

'Sharaf ul-Insān. An Arabist's questions,' in J. Schmidt (ed.), *Essays in honour of Barbara Flemming =. Journal of Turkish studies* 26/2 (2002), pp. 61–68.

'Waḥam,' in *Encyclopaedia of Islam*. New ed., XI (Leiden, 2002), p. 32.

'Strijders voor het geloof? Bouillon in Mekka en Baybars in Rome,' in S. Houppermans [et al.] (eds.), *Rapsoden en rebellen. Literatuur en politiek in verschillende culturen* (Amsterdam, 2003), pp. 61–80.

'De treurige zanger,' in A.A. Seyed-Gohrab [et al.] (eds.), *De band met u, mijn vriend, verbreek ik niet. Liber amicorum voor Kamil Banak* (Leiden, 2003), pp. 85–92.

'La zoologie aristotélicienne. Tradition arabe,' in R. Goulet (ed.), *Dictionnaire des philosophes antiques*. Supplément (Paris, 2003), pp. 329–334.

'The princess Maymūnah. Maiden, mother, monster,' *Oriente moderno* 83 = N.S. 22 (2004), pp. 425–442.

'The *Arabian Nights* and the popular epics,' in U. Marzolph & R. van Leeuwen (eds.), *The Arabian Nights encyclopedia* (Santa Barbara [etc.], 2004), I, pp. 34–38.

'Ibn Ḥazm's tadpoles. A Ẓāhirite reads the Book of Nature,' in R. Arnzen & J. Thielmann (eds.), *Words, texts and concepts cruising the Mediterranean Sea*. Orientalia Lovaniensia analecta; 139 (Leuven, 2004), pp. 401–419.

'Harry Potter in the Gulf. Contemporary Islam and the occult,' *British Journal of Middle Eastern Studies* 32/1 (2005), pp. 47–74.

'Suckling lamb and free range chicken. Ibn Abī l-Ashʿath on domestic animals and dietetics,' in M. Marín & Cristina de la Puente (eds.), *El banquete de las palabras. La alimentación en los textos árabes*. Estudios Árabes e Islámicos, Monografías; 10 (Madrid, 2005), pp. 89–103.

'Shaken, not stirred? Drank en de vrouwelijke detective,' in R. Kruk & S. Houppermans (eds.), *Een vis in een fles raki. Literatuur en drank in verschillende culturen* (Amsterdam, 2005), pp. 131–150.

'Sīrat ʿAntar ibn Shaddād,' in R. Allen & D.S. Richards (eds.), *Arabic literature in the post-classical period*. The Cambridge history of Arabic literature; 6 (Cambridge, 2006), pp. 292–306.

Translations

Ibn Hazm, *De ring van de duif. Over minnaars en liefde. Vertaald uit het Arabisch en ingeleid door Remke Kruk en J.J. Witkam*. De oosterse bibliotheek; 6 (Amsterdam, 1977).

Abu Bakr Muhammad Ibn Tufayl, *Wat geen oog heeft gezien, geen oor heeft gehoord en in geen mensenhart is opgekomen. De geschiedenis van Hayy ibn Yaqzan. Uit het Arabisch vertaald en ingeleid door Remke Kruk*. De oosterse bibliotheek; 23 (Amsterdam, 1985).

ʿAli ibn Ahmad Ibn Hazm (994–1064), 'De ring van de duif. Fragmenten,' in A. Vrolijk (ed.), *De taal der engelen. 1250 Jaar klassiek Arabisch proza* (Amsterdam, 2002), pp. 381–397.

Ikhwan al-Safaʾ (2e helft 10e eeuw), 'Verhandelingen van de Broeders der Zuiverheid. De zaak van de dieren tegen de mensen,' in A. Vrolijk (ed.), *De taal der engelen. 1250 Jaar klassiek Arabisch proza* (Amsterdam, 2002), pp. 283–299, 594–595.

Sharaf al-Din Tahir al-Marwazi (ca. 1100), 'Over de naturen van de levende wezens. Reusachtige zeedieren. De zeeslang,' trans. Remke Kruk, in A. Vrolijk (ed.), *De taal der engelen. 1250 Jaar klassiek Arabisch proza* (Amsterdam, 2002), pp. 319–324, 600.

'Sirat al-amira Dhat al-Himma. De geschiedenis van Qannasa, dochter van Muzahim,' in A. Vrolijk (ed.), *De taal der engelen. 1250 Jaar klassiek Arabisch proza* (Amsterdam, 2002), pp. 536–553, 602–603.

ʿAbd al-Salām Bālī, *The cutting sword. Challenging the evil sorcerers, trans. by Remke Kruk* (Leiden, 2005). http://www.tcmo.leidenuniv.nl/index.php3?m=140&c=259, accessed 20 September 2006.

Abu Bakr Muhammad Ibn Tufayl, *Hayy ibn Yaqzan. Een filosofische allegorie uit Moors Spanje. Uit het Arabisch vert. en ingel. door Remke Kruk* (Amsterdam, 2005). (rev. ed. of *Wat geen oog heeft gezien*... (1985)).

Averroes, *Geloof en wetenschap in de islam. Averroes' 'Het beslissende Woord'. Ingel., vert. en geann. door Remke Kruk* (Kampen, 2006).

LIST OF PLATES

PLATES

Plate 1. Engraving by Hans Holbein the Younger in the corner of a world map by Sebastian Munster (1532), depicting Vartoman as a traveller and as a prisoner at Aden.

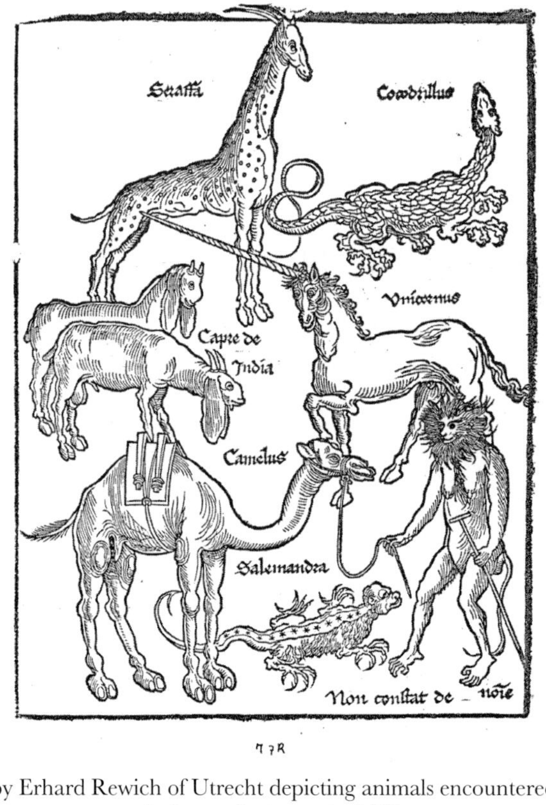

Plate 2. Woodcut by Erhard Rewich of Utrecht depicting animals encountered in the Holy Land during a pilgrimage in 1483.

Plate 3. Woodcut by Jörg Breu the Elder (1515), depicting the unicorns Varthema saw at Mecca.

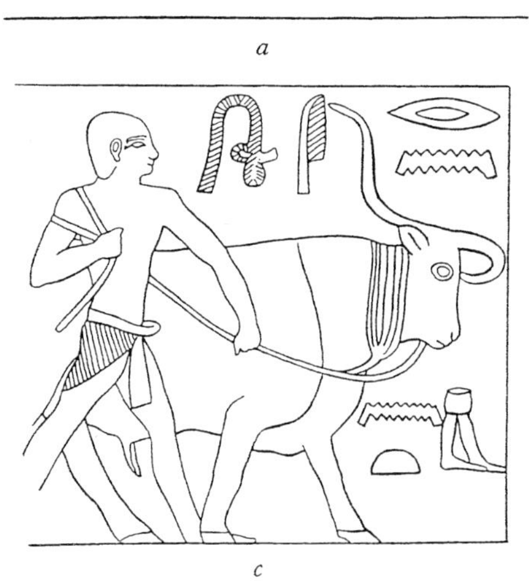

Plate 4a. Ox with asymmetric horns on the Egyptian tomb of Manufer (c. 2700 BC).

Plate 4b. Dinka ox sketched around 1930.

Plate 5. André Rouveyre, 'Docteur J.-C. Mardrus,' *Mercure de France* 80 (1909), p. 405.

Plate 6. Willem Frederik Dupont, 'Kees van Dongen,' charcoal, 1930. Leiden University Library, Prints & Drawings collection, inv. no. PK-T-AW-5470.

Plate 7. © Kees van Dongen, Shahrazād and Sultan Shahriyar, 1918 c/o Beeldrecht Amsterdam 2006. *Hassan Badreddine El Bassraoui. Conte des 1001 Nuits* (Paris, 1918), p. 203.

Plate 8. Testas, pencil on tracing paper, 8.5 × 16 cm; Private collection, inv. no. TTS 095.

Plate 9. Litho of the interior of the *qibla* wall of the funerary complex of Sultan Qaitbay, Prisse d'Avennes 1877 I, pl. 21.

Plate 10. Testas, pencil on tracing paper, 8.5 × 16 cm; Private collection, inv. no. TTS 006.

Plate 11. Testas, pencil on tracing paper, 13.3 × 14.5 cm, Private collection, inv. no. TTS 088.

Plate 12. Testas, 'Saïs ou palefrenier avant coureur Egyptien,' watercolour, 29.4 × 39.8 cm, signed W. Testas 69; Haarlem, Teylers Museum, inv. no. DD26.

Plate 13. Testas, 'A street scene in el-Kab,' watercolour, 33.3 × 24 cm; Private collection, inv. no. TTS 029.

Plate 14. Testas, 'Courtyard of a house in Cairo,' oilpainting, 84 × 62 cm, signed W. Testas; Amsterdam, Rijksmuseum, inv. no. SK-A-1184.

Plate 16. Testas, 'Porte de maison avec ornaments en bronze,' pencil on tracing paper, 12 × 7.2 cm; Private collection, inv. no. TTS 010.

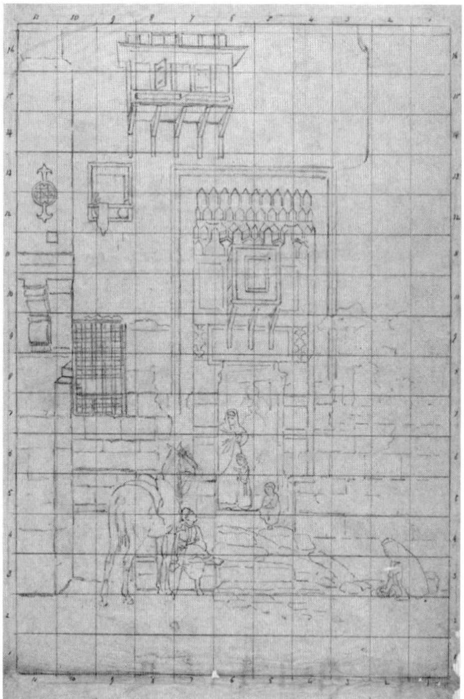

Plate 15. Testas, pencil on tracing paper, 51.5 × 33.7 cm; Haarlem, Teylers Museum, inv. no. KT 1710.

Plate 17. Abuna Arsanious, Priest of the Coptic Church of St Mary's in Amsterdam, wearing the *qalansuwa* and standing in front of the gallery with icons that decorate the entrance to the altar.

Plate 18. Scene from the flight to Egypt painted on the walls of the church built in a cave in the Muqattam hills.

Plate 19. Representation of the Virgin Mary as she appeared in Zeitoun (1968 & 1986).

Plate 20. Mary, a Coptic laywomen, holding a poster of Children portrayed as angels surrounding the Holy Family on their flight to Egypt.

Plate 21. The Convent of St Mercurius (Abu Saifein) in Old Cairo, restored by Mother Irini.

Plate 22. The farm built by the Convent of St Mercurius in Sidi Krir.

Plate 23. Books on women saints from the Coptic heritage and books on Russian women saints.

Plate 24. A nun in the Convent of St Dimyanah painting an icon in the convent's workshop.

Plate 25. Twenty-four elders of the Acopalypse, depicted inside the altar of a newly built church in Nasr City.

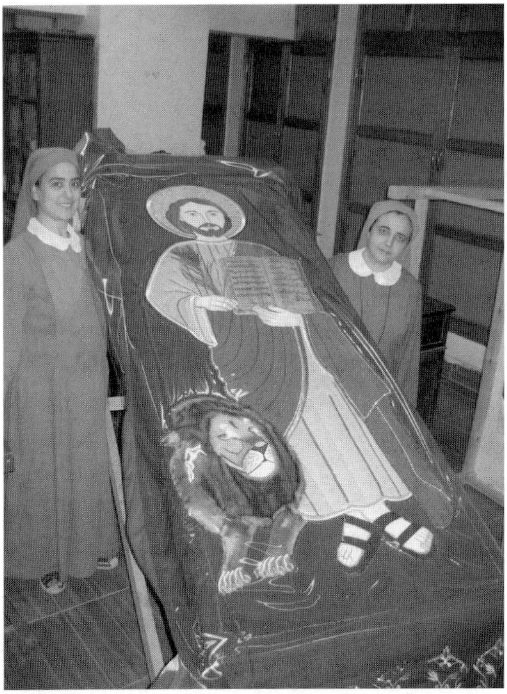

Plate 26. Nuns of the Convent of the Sisters of St Mary showing a newly manufactured altar curtain.

Plate 27. An altar curtain portraying the Virgin Mary *Theotokos*, designed and made by the Nuns of the Sisters of St Mary.

Plate 28. Mary and Joseph bathing the child Jesus under the holy tree at Matariya.

Plate 29. St Barbara.

Plate 30. St Marina.

Plate 31. Icon of patron saint Abu Saifein, paired with a rendition of St Mary appearing in Zeitoun.

Plate 32. Saints Irini, Marina and Dimyanah and her forty virgins.

Plate 33. St Dolagi and her four children.

Plate 34. St Perpetua.

Plate 35. St Mahra'il.

Plate 36. St Butamina.

Plate 37. St Efrusina (Euphrosyne).

Plate 38. Booklet with the picture of abbess Mother Aghapi holding the relics of St George.

Plate 39. Photographs of Sister Hannah and groups of nuns surrounding a portrait of Bishop Athanasius.

SCIENCE & PHILOSOPHY

LUDOVICO DE VARTHEMA AND THE UNICORNS IN MECCA (1504)

Willem Pieter Gerritsen

Ludovico de Varthema (or Barthema), a native of Bologna, was one of the first Christians to visit Mecca. The year was probably 1504, possibly 1503. Disguised as a Mamluk soldier, he walked around in the town, observing the holy places and the dense throngs of pilgrims. Behind the great mosque, he chanced upon a grated enclosure in which he perceived two unicorns. This is how he described the animals in the account of his travels which was first published in 1510 and soon turned out to be a European bestseller.

> They [the unicorns] were shown to the people as creatures astounding on account of their rarity and strange nature, and this not without reason. One of them, which is much higher than the other, is similar to a foal of thirty months of age. On its forehead grows a single horn, which is straight and four cubits long. The other animal is much younger, about a year of age, and like a young foal; this one's horn being of a length of four hand-spans. The colour of this animal is like that of a dark horse; its head resembles that of a stag, but it has no long neck, its thin mane hanging down on one side. Their legs are thin and slender, like that of deer. The hoofs of the fore-legs are cloven, like the feet of a goat. The back of their hind legs is thickly covered with hair. This animal certainly seems to be very wild and fierce, although its wildness is mitigated by a certain comeliness. These unicorns were presented to the Sultan of Mecca as a most precious and rare gift. They were sent to him from Ethiopia by a king of that country who wished to propitiate the Sultan towards him.[1]

---

[1] Ludovico de Varthema, *Reisen im Orient*, transl. by F. Reichert (Sigmaringen, 1996), pp. 77–79 (my translation of Reichert's German version, which is based on Ramusio's edition of Varthema's work). Earlier editions of Varthema's works: *The Travels of Ludovico di Varthema in Egypt, Syria, Arabia Deserta and Arabia Felix, in Persia, India, and Ethiopia, A.D. 1503 to 1508.* Translated from the original Italian edition of 1510, with a preface by J.W. Jones, and edited, with notes and an introduction, by G.P. Badger. Works issued by the Hakluyt Society I; 32 (London, 1863), and *The itinerary of Ludovico di Varthema of Bologna from 1502 to 1508, as translated by John Winter Jones in 1863 for the Hakluyt Society. With a discourse on Varthema and his travels in Southern Asia*, by R.C. Temple (London, 1928).

Born in all probability in the early seventies of the fifteenth century, Ludovico de Varthema was the son of a Bolognese physician. According to his own account, he was trained as a builder of cannon, but in reality he may only have obtained some experience in the soldier's craft. Around the year 1500, spurred by curiosity 'to see the diversity of the realms of the Earth,' he travelled via Alexandria, Cairo, Beirut and Aleppo to Damascus, which he described as a wonderful city of great wealth. Here, he came in contact with a group of Mamluks. The Mamluks of that time were an elite consisting of former slaves who had been captured, usually at a tender age, in Turkey, in the Balkans, in the Caucasus or in Russia. After several years of rigorous indoctrination and military training they obtained their freedom and served henceforth as a military aristocracy. Among them were also numerous former Christians who had forsaken their faith. Learning that a caravan was about to depart for the *Hajj* to Mecca, Varthema struck up friendship with the leader of the armed contingent of Mamluks which was to escort the pilgrims. In return for a handsome reward, this man, a renegade Christian, permitted him to join the escort, dressed up as a Mamluk. Setting out in April – he gives the year as 1503, but it was probably 1504 – the caravan reached Mecca towards the end of May, after a dangerous journey of about forty days. Varthema describes the city and the throngs of pilgrims flocking in from all over the Muslim world. He watched the enormous crowds of pilgrims walking seven times round the Ka'ba and receiving the ritual ablutions. After the sacrificial slaughtering of sheep he took part in a festive meal, and recollects that he and his companions threw the rinds of melons they had eaten to the poor people waiting outside their tent. The ritual 'stoning of the devil' by throwing pebbles at a wall[2] near the town of Minā was cut short by the menace of the caravan being attacked and plundered by a band of Bedouins.[3] Back in Mecca, Varthema was struck by the vast numbers of pigeons flying unhindered everywhere in the city and were not prevented from soiling the monuments. He was told that killing the pigeons would entail the downfall of the state.

The next chapter of Varthema's travel-story is his above-quoted description of the two unicorns he saw in an enclosure behind the city's

[2] Actually, at the columns in front of the wall, but Varthema may not have realised this.

[3] Note that the sequence of the Hajj rituals as related by Varthema differs from current practice, see for instance A.J. Wensinck, 'Ḥadjdj,' in *Encyclopaedia of Islam* (New ed.), III, p. 37.

great mosque. Before discussing this chapter in greater detail, it will be useful to provide a short account of Varthema's further adventures and the success of his book.

Instead of travelling back to Damascus, Varthema joined a caravan departing for India via Aden, which at that time was the main port for the maritime spice trade in the Islamic world. In Aden, however, he was recognized as a Christian and arrested. After an imprisonment of several months he succeeded in regaining his freedom, which he owed to the intercession of the Sultan's wife who had taken a liking to him, whereupon he embarked on a ship bound for India by way of Persia. Calling in at Muscat (in present-day Oman) and at Hormuz at the entrance of the Persian Gulf, the ship reached the coastal city of Cambay (now in the Indian federal state of Gujarat). From there, they travelled south along the Malabar coast, visiting cities like Bidar (state of Karnataka), Cannanore (state of Kerala) and Calicut (present-day Kozhikode on the Kerala coast). Here they entered the sphere of influence of the Portuguese.

Vasco da Gama had reached Calicut in 1498 on his quest 'for Christians and spices', but had been forced to set out on the return journey without gaining a firm foothold on the Indian subcontinent. Two years later, Pedro Álvares Cabral, commanding a squadron of thirteen ships with more than 1200 men, had been more successful. On Cabral's insistence, the rajah of Calicut had been forced to consent to the foundation of a Portuguese trading station. Likewise, the rajah of Cannanore had to accept a treaty with the Portuguese. After another two years of partly failed, partly successful attempts to eliminate the Muslim trade on India, King Manuel I of Portugal decided to institute an 'Estado da Índia' with the task of administrating the Portuguese possessions in the East Indies. Francisco de Almeida was appointed as viceroy. When Varthema arrived in Cannanore, he decided to leave his companions and to defect to the Portuguese. He was taken to the Portuguese fort where Lourenço de Almeida, the viceroy's son, received him with enthusiasm and sent him on to his father's headquarters. After joining the Portuguese ranks, Varthema took part in several battles, of which he gave a lively account. On the 4th of December 1507, the viceroy personally conferred a knighthood upon him in reward of his dedication to the Portuguese cause.

Shortly afterwards, Varthema boarded a ship bound for Portugal via the Cape of Good Hope. After his arrival in Lisbon, in July of the next year, he was received by King Manuel. The king kept him at his palace for several days, listening with great interest to the story of his adventures in the Orient. King Manuel's endorsement of the charter conferring a knighthood to Ludovico de Varthema is preserved in the Portuguese national archives. Back in his native Italy, Varthema lectured with great success about his experiences to a Venetian audience. He composed, in Italian, a written account of his travels which he dedicated to the Duchess of Tagliacozzo. The Duchess's daughter Vittoria was presented a copy written by the famous calligrapher Ludovico degli Arrighi. In November 1510 a first edition of Varthema's account appeared in Rome, printed by Stephano Guillireti and Hercule de' Nani.[4] The next year, a Latin edition, in which Varthema's name was latinised as Vartomannus, was published in Milan. This inaugurated the European success of the book. A German edition, illustrated with a series of 46 woodcuts by Jörg Breu the Elder, was printed at Augsburg in 1515, followed by editions in Spanish (1520), Dutch (1544), French (1556), and English (1576). Until the middle of the seventeenth century, more than 30 editions of the book have been recorded. Moreover, Varthema's text was included, in full or in extracts, in several of the great geographical collections of the time, like John Huttich's *Novus orbis regionum ac insularum veteribus incognitarum* (Basel, 1532) and the *Navigationi e viaggi* compiled by Giovanni Battista Ramusio, first published at Venice, 1550. Several sixteenth-century cartographers took note of Varthema's experiences. In a small engraving enlivening the right-hand bottom corner of a map of the world published in 1532, Hans Holbein the Younger depicted two scenes from Varthema's book: we see the restless traveller roaming through the world with forceful footsteps and, while detained in Aden, playing the fool under the eyes of the Sultan's wife (Plate 1).[5]

Not all that Varthema tells his readers, however, is necessarily true. As early as the second half of the sixteenth century doubts began to be expressed as to the credibility of some of his stories. Nowadays, the account of his travels in the interior of Persia (he even claims to have visited Samarkand) is thought to be largely fictitious, as well as what he tells about his travels to Ceylon, Burma and Indonesia. In all likelihood, Varthema has never penetrated farther east than the western coast of the Indian subcontinent. Like most of his predecessors – Sir John de Mandeville and Marco Polo are notorious examples – he appar-

[4] Varthema, *Itinerario* (Rome, 1510), facsimile edition by E. Musacchio (Bologna, 1991).

[5] Reproduced in Varthema, *Reisen im Orient*, transl. Reichert, p. 27.

ently made no bones about embellishing his account with a variety of tall stories that circulated in eastern ports and trade centres. Nor did he have any scruples about appropriating what he heard from local informants and presenting it as part of his own experiences. This is a literary device found in many medieval and early modern travel accounts. On the other hand, much of what he tells about his travels in Syria, Arabia and along the Malabar coast clearly rings true and undoubtedly rests on personal observation. Varthema was one of the very first Europeans to have visited the Muslim holy cities of Medina and Mecca, and his detailed descriptions of these cities are for the most part in agreement with those of later travellers. There is, in short, no reason to suspect that Varthema would have invented his report of having seen two unicorns in an enclosure behind the great mosque at Mecca.

What kind of animals did Varthema see? His description of them is sufficiently detailed and precise to enable us to visualize their build. It evokes the image of a dark-furred, straight-horned antelope-like animal, not unlike a kind of Oryx (*Oryx gazella*?) or a Thomson's Gazelle (*Gazella thomsoni*). Varthema's description is conspicuously at variance with the traditional representation of unicorns in medieval bestiaries and works of art, where they are usually portrayed as white-furred, horse-like animals, whether or not with cloven hoofs. Some twenty years earlier, a unicorn had been sighted in the Sinai desert. This was recorded by Felix Fabri, a Swiss Dominican friar who travelled in 1483 with a large group of pilgrims to the monastery of St Catherine's in the southern part of the Sinai peninsula. Once upon a day, the pilgrims perceived from afar a big animal standing on the summit of a mountain.[6] They took it for a camel, but were assured by the dragoman and his colleagues that what they saw was indeed a unicorn. To the same group of pilgrims, of which the most important member was Bernhardt Breydenbach, a wealthy canon from Mainz, also belonged a Dutch artist and printer, Erhard Rewich of Utrecht, who in 1488 published in Mainz an account of the pilgrimage, illustrated with woodcuts of his own invention. One of these depicts the animals which the pilgrims had encountered during their voyage.[7]

---

[6] Félix Fabri, *Voyage en Egypte* (Le Caire, 1975), I, pp. 159–62; cf. III, pp. 830–31.

[7] Bernhardt van Breydenbach, *Die heylighe bevarden tot dat heylighe grafft in Iherusalem* (Mainz: Eerhaert Rewich van Utrecht, 1488), f. 158a. The woodcut is also reproduced by Odell Shepard, *The lore of the unicorn* (London, 1930, [repr. 1967]), after p. 192. In accordance with the medieval iconographical tradition, the unicorn's horn is depicted as having spiralled grooves. As a native of Utrecht, Erhard Rewich may have

Among these animals is a unicorn which resembles a horse with cloven hoofs and a very long straight, spiralled horn protruding from its brow. This vivid portrait, allegedly drawn from life, answers in every detail to the traditional description of the unicorn in medieval western art (Plate 2).

It is interesting to compare Rewich's woodcut with the one by Jörg Breu the Elder, the illustrator of the German edition of Varthema's work, printed at Augsburg in 1515. He situated the two animals Varthema had seen in Mecca in a cage with a tiled floor, surrounded by a strong fence.[8] Onlookers are peering through the openings in the fence. Having in all likelihood no more information about the 'unicorns' than what he could deduce from Varthema's verbal description, Breu portrayed them in his woodcut as robust goats whose necks are covered with a thick mane. The top side of their slightly bowed frontal horn is notched like a saw and points more or less horizontally forward (Plate 3).

If we assume that Varthema did not invent the story of his visit to the Meccan unicorns and that his observation of the animals was on the whole correct, he must have seen two animals having a single horn protruding from the brow. Now modern zoology is adamant in its assurance that no mammal of that description exists or has ever existed in Nature. A way out of this apparent paradox, however, is provided by two important details in his report. The first of these is that the animals were kept in captivity. This may imply that they were raised under human surveillance. The second point is Varthema's statement that the unicorns were presented to the ruler of Mecca as a gift from an Ethiopian king.

Before further exploring these issues, it is not without interest to note that the unicorns Varthema saw at Mecca were not the only single-horned animals he encountered in the course of his travels. Having set out from Aden on its way to the Indian Ocean, his ship was driven off course by a gale and ended up in the port of Saylac (present-day Zeila) in northern Somalia, near Djibouti. There he saw kinds of cattle he had never seen before. Among these were cows carrying antlers like stags, which the sultan of the city had received as a gift, but he also observed

> a different sort of cow which carried a single horn on its brow, of a length of one and a half hand-span, and this horn was pointing more towards

seen the famous 'unicorn horns' [in reality: narwhal teeth] belonging to the treasure of the Utrecht church of St Mary's.

[8] Breu's woodcut is reproduced in Varthema, *Reisen im Orient*, transl. Reichert, p. 78.

> the back of the animal than forwards. Their colour is red, while the ones I described above were black.[9]

Although Varthema does not tell us explicitly where he saw these animals, it seems plausible to assume that this must have been in some kind of menagerie belonging to the ruler of the city, a location not unlike the one he describes having seen in Mecca. Moreover, in both cases reference is made to peculiar animals which had been presented as a gift to the ruler in question.

In the sixteenth century, Ethiopia was believed to be a country full of marvels. It was thought to be the realm of a Christian sovereign, the legendary Prester John. An Elizabethan adventurer, Edward Webbe, claims to have visited the country and to have seen

> in a place like a Park adioyning vnto prester Iohns Court, three score and seuen-teene [76], Vnicornes and Eliphants all aliue at one time, and they were so tame that I haue played with them as one would play with young Lambes.[10]

The report of another early traveller to Ethiopia sounds more credible than that of Edward Webbe. The Portuguese Jesuit Lobo (1593–1678) worked for several years in Abyssinia as a missionary until the Jesuit fathers were banished from the country in 1632. He left a description of Abyssinia which was published in a French translation. The French edition was translated into English by no less an author than Samuel Johnson who later was to acquire fame as a lexicographer. Here is Johnson's version of Lobo's account of the unicorn:

> In the province of Agaus, has been seen the unicorn, that beast so much talk'd of, and so little known; the prodigious swiftness with which this creature runs from one wood into another has given me no opportunity of examining it particularly, yet I have had so near a sight of it as to be able to give some description of it. The shape is the same as that of a beautiful horse, exact and nicely proportion'd, of a bay colour, with a black tail, which in some provinces is long, in others very short; some have long manes hanging to the ground. They are so timerous that they never feed but surrounded with other beasts that defend them.[11]

---

[9] Varthema, *Reisen im Orient*, transl. Reichert, p. 110.

[10] Edward Webbe, *Chief Master Gunner. His travailes, 1590*, ed. E. Arber. English reprints, large paper ed. (London, 1869), p. 25. The passage is quoted by Shepard, *The lore of the unicorn*, p. 195.

[11] S. Johnson, *A voyage to Abyssinia. (Translated from the French)*, ed. J.J. Gold. The Yale edition of the works of Samuel Johnson; 15 (New Haven [etc.], 1985), p. 46; see also pp. 188–89.

In a footnote, Joel J. Gold, the editor of Samuel Johnson's translation, points out that 'This first-person account produces an effect very different from that of the French: "Comme cet animal passe vite d'un bois à un autre, on n'a pas eu le temps de l'examiner; on l'a néanmoins assez bien considéré pour pouvoir le décrire."'

The African unicorn continued to fascinate the European public. In its March issue of the year 1844 the *Journal asiatique* published a letter by Fulg. Fresnel, a consular agent at Jeddah, on the subject of a single-horned animal to be found in Dar-Bargou, a region north-west of Darfur in present-day Sudan. This '*abou-karn*' (Ar. *abū qarn*), as it was known locally, had the build of a bull, but its legs and feet were like that of an elephant. It was a pachyderm with a hairless body, but emphatically was not a rhinoceros. Sprouting not from the nose but from between the eyes, its single horn, grey with a scarlet tip, was movable. While walking calmly, the animal let the horn swing to left and right, but when aroused it could fix it like a bayonet for the attack. Fresnel had to admit not having seen the *abou-karn* himself; he relied upon the testimony of some Arab informants, whom he maintains could be trusted without restriction. They had assured him having more than once hunted the animal and having killed several specimens. In Fresnel's opinion, the *abou-karn* was without doubt the same animal as the *re'em* referred to in the Hebrew Bible.[12]

It would be easy to adduce more reports about unicorn sightings in East-Africa, but these would not inspire more trust than the three ones quoted above. Since the middle of the nineteenth century, the age-old search for a four-footed animal which a single horn sprouting from it brow has been given up. After a scientific debate which had lasted for several centuries, in which important scholars and scientists like Samuel Bochart, Ambroise Paré, Thomas Bartholinus, Gottfried Wilhelm Leibniz, and Georges Cuvier had taken part, the unicorn was definitively banished to the ranks of the fabulous animals. It did not exist and never had existed.

Do we have to conclude, then, that Varthema lied about the unicorns he saw in Mecca?

---

[12] F. Fresnel, 'Extraits d'une lettre de M. Fulg. Fresnel, agent-consulaire de France à Djeddah, à M. Jomard, membre de l'Institut de France, sur certains quadrupèdes réputés fabuleux,' *Journal asiatique* (Mars 1844), pp. 129–59, spec. 131–54.

The answer, I think, must be no.[13] Since Antiquity, it has been known that it is possible to manipulate the budding horns of a new-born ruminant. The practice was referred to as early as 77 C.E. by the elder Pliny in his great *Naturalis historia:*

> Farmers heal the horns of their oxen when worn by greasing the horn of the hoof with fat; and the substance of horn is so ductile that even the horns of living cattle can be bent with boiling wax, and they can be slit at birth and twisted in opposite directions, so as to produce four horns on one head.[14]

A more recent confirmation of Pliny's statement is to be found in the travel book of the François Le Vaillant who in the years 1780 to 1785 explored the interior of southern Africa. Le Vaillant relates that the Zulus have a custom of transforming the horns of their cattle by twisting them into various shapes. By this operation, he maintains, they were able to multiply the number of horns on a cow's head, and also to model the animal's horns into all forms prompted by their fantasy.[15]

The next piece of evidence brings us nearer to the allegedly Ethiopian origin of the unicorns Varthema saw in Mecca. The Dinka and the Nuer are tribes of Nilotic pastoralists living near the White Nile in the South of Sudan. Anthropologists report that the men of both tribes cultivate an emotional relationship with their cattle. In their great work on the pagan tribes of the Nilotic Sudan, C.G. Seligman and his wife Brenda Z. Seligman comment on the so-called 'cattle-names' current among the Dinka: a Dinka man can refer to himself by a name which he shares with his pet ox.

> Such oxen as these have their horns trained so that one grows forward and the other backward, and are known as *muor cien*; they act as leaders of the herd. It is difficult to describe their importance to their masters or the love and care the latter have for their beasts, but it is certainly no exaggeration to say that it amounts to what psychologists would term "identification".[16]

---

[13] As far as I am aware, Odell Shepard has been the first to point to the possibility of 'unicorns' having been produced artificially. See *The lore of the unicorn*, pp. 227–29.

[14] Pliny, *Naturalis historia*, Book XI, chap. XLVII (130).

[15] François Le Vaillant, *Reize in de Binnenlanden van Afrika, langs de Kaap de Goede Hoop in de jaaren 1780 tot 1785 gedaan door den Heer Le Vaillant.* Uit het Fransch door J.D. Pasteur (Leyden [etc.]: Honkoop & Allart, 1791–1798), II, pp. 166–67.

[16] C.G. Seligman & B.Z. Seligman, *Pagan tribes of the Nilotic Sudan* (London, 1932), p. 169.

Elsewhere in their work, the Seligmans point out that the practice of artificial deformation of the horns of cattle must have reached the Nilotic tribes from ancient Egypt, where an oxen whose horns are trained forward and backward in exactly the same way is pictured on the tomb of Manufer of the fifth dynasty (around 2700 B.C.) (Plate 4a–b).[17]

The practice of artificially twisting the horns of cattle may be traceable back to Egyptian culture of the third millennium B.C., and may have spread from there over a considerable part of the African continent. This does not, of course, prove that the same method has also been in use in order to make the two horns of an animal coalesce into one, in other words: in order to produce, artificially, a 'unicorn'. That this can be done was demonstrated in 1933 by W. Franklin Dove, a biologist of the University of Maine. Dr Dove succeeded in artificially producing a 'unicorn' by transplanting pedicled flaps of horn-forming tissues on the head of a calf from the sides of the skull to the centre. In his report to *The Scientific Monthly*, he reports as follows:

> The experiment was successful. The animal, now two and a half years old, bears upon the forehead the stamp of the once fabulous unicorn. The two buds have coalesced and have formed one exceptionally large and long horn molded into the skull bones of the forehead for support. [...] The horn curves slightly upward toward the tip and gracefully extends the curve of the back and neck when the animal stands at attention.[18]

In my view, it is highly likely that the unicorns Varthema saw in Mecca were artificially produced specimens. This would mean that the Ethiopian king who, according to Varthema, had presented them to the ruler of Mecca, had had the new-born animals – possibly some kind of antelopes – fashioned into 'unicorns' by having their horns coalesce into a single one. That artificial deformation of the horns of cattle was practised by East-African herdsmen is clear from the examples quoted above, although it should be admitted that I have been unable to find evidence as to artificial formation of a 'unicorn' out of a normal antelope (or, for that matter, out of a goat or sheep). And this is by no means

[17] Seligman & Seligman, pp. 35–6. See also C.G. Seligman, 'Egyptian Influence in Negro Africa,' in R. Mond (ed.), *Studies presented to Francis Llewellyn Griffith* (Oxford, 1932), pp. 457–62 and Plate 74. Striking photographs of oxen with asymmetric horns are to be found in J. Ryle, *Warriors of the White Nile. The Dinka* (Amsterdam, n.d.), pp. 94–95.

[18] See W.F. Dove, 'Artificial production of the fabulous unicorn', *The scientific monthly* 42 (1936), pp. 431–36.

the only question which I have to leave open. It would be enlightening to know which thoughts lay behind the presenting of these 'unicorns' to the Meccan ruler. Were they, in the eyes of the giver, no more than curious, exceptional animals and, as such, a fitting royal present for an important political relation across the Red Sea? Or did they represent, like the unicorns in Christian and Islamic traditions, a 'mythology', in the sense that their properties were associated with abstract, magical or religious, concepts?[19] Asking questions like these, I feel like the man in Jörg Breu's woodcut who is peering curiously through the openings in the fence. The past will never give up all its secrets.

[19] For the Christian tradition see J.W. Einhorn: *Spiritalis Unicornis. Das Einhorn als Bedeutungsträger in Literatur und Kunst des Mittelalters*. 2nd ed. (München, 1998). Odell Shepard's *The lore of the unicorn* remains the most general survey of the unicorn tradition. An overview of the Muslim tradition is given by R. Ettinghausen, *The unicorn*. Studies in Muslim iconography; 1 (Washington, 1950). See also A. Contadini, 'A bestiary tale. Text and image of the unicorn in the *kitāb Naʿt al-ḥayawān* (British Library, Or. 2784),' *Muqarnas* 20 (2003), pp. 17–33.

AL-MAQRĪZĪ'S TREATISE ON BEES

Giovanni Canova

Information regarding bees and honey is to be found in a large number of Arabic sources, ranging from pre-Islamic poetry to the Qur'an and its commentaries, the hagiographic literature, the philosophic, zoological, and medical treatises, the great lexicons and the *adab* works.[1] Honey bees are considered in Islamic culture as extraordinary insects, that received divine revelation (Qur. 16:68).[2] If the Ikhwān al-Ṣafā highly admire the bees' body structure and the perfect hexagonal architecture of the cells, Avicenna and Abū Bakr al-Rāzī praise the extraordinary healing virtues of their honey. The hive is ruled by a just king, the *yaʿsūb*, who assigns equitably the various tasks to the bees and takes care that they are carried out. Bees represent an ideal model for the believers (al-Ghazzālī) and the perfect organization of their colony is a metaphor for a well-ordered human society.

The knowledge of the bees by the Arabs increased considerably when they came into contact with the Greek scientific heritage, in particular the zoological writings of Aristotle, the translation of which is attributed to Yaḥyā [Yuḥannā] Ibn al-Biṭrīq (c. 200/815),[3] translator of works by Plato, Aristotle, Galen and Hippocrates. Also in the description of bees Arab works often refer to the writings of Aristotle, 'Ṣāḥib al-manṭiq'. It seems, however, that quite often these references are just literary citations, without the support of a personal experience. And yet, the careful

---

[1] For an analysis of the sources and bibliographycal references, see our essay 'Api e miele tra sapere empirico, tradizione e conoscenza scientifica nel mondo arabo-islamico,' in G. Canova (ed.), *Scienza e Islam*. Quaderni di studi Arabi, Studi e testi; 3 (Roma [etc.], 1999), pp. 69–92.

[2] 'And thy Lord revealed unto the bees, saying: "Take unto yourselves, on the mountains, houses, and of the trees, and of what they are building. Then eat of all manner of fruit, and follow the ways of your Lord easy to go upon."' (Arberry's translation).

[3] See Arisṭūṭālīs, *Ṭibāʿ al-ḥayawān*, ed. ʿAbd al-Raḥmān Badawī (Kuwayt, 1977); *Ajzāʾ al-ḥayawān*, ed. ʿAbd al-Raḥmān Badawī (Kuwayt, 1977); *Fī aʿḍāʾ al-ḥayawān* = *The Arabic version of Aristotle's Parts of Animals*, ed. R. Kruk (Amsterdam [etc.], 1979); *Fī kawn al-ḥayawān* = *Generation of Animals*, ed. J. Brugman and H.J. Drossaart Lulofs (Leiden, 1971).

observations of the bees' behaviour by the Bedouins, found in particular in the poetry of the Banū Hudhayl, are remarkable.[4]

In spite of the fact that some authors, such as Abū Ḥanīfa al-Dīnawarī[5] or Ibn ʿAwwām,[6] write extensively about bees and honey, it seems to us that a true Arabic treatise on bee-keeping in the Arab world is missing. An exception is a short monograph of the great Egyptian historian al-Maqrīzī.

Taqī al-Dīn Abū 'l-ʿAbbās Aḥmad b. ʿAlī al-Maqrīzī (d. 845/1442) is the author of the well known *Kitāb al-Mawāʿiẓ wa 'l-iʿtibār fī dhikr al-khiṭaṭ wa 'l-āthār* and of dozens of other works. He also wrote a brief treatise about bees with the significant title *Kitāb Naḥl ʿibar al-naḥl* ('the book of the gift of the bees' moral teachings'), where the double meaning of the term *naḥl* is emphasized. The lexicographer al-Zajjājī reminds us that bees are called *naḥl* because God donated (*naḥala*) honey to man.

The work is structured into eleven chapters of various length which give a clear idea of the content: 1. The bee and its names; 2. The kinds, colours, sizes, and peculiarities of bees; 3. Wax, the building of combs, the hives; 4. Honey, its lexicon, its colours, and its varieties; 5. Information from Aristotle and other scholars, honey's nature and origin; 6. Qurʾan, sūra of the Bee (16:68–69), other Qurʾanic references and various exegetic comments; honey as a 'healing for men'; 7–8. The Prophet's traditions; 9. various opinions about what bees eat; the *zakāt* on honey; 10. Anecdotes about wax; 11. Poems about wax.

Often the author seems to have a personal knowledge of the world of bees, even if there is no sufficient proof that he was a bee-keeper. His experience as a public servant sometimes brings him to comparing the bees' frugality to men's wastefulness. In his treatise, he borrows much information from Avicenna, al-Qazwīnī, al-Damīrī, and from the 'learned men of the past' (i.e. Aristotle). In spite of his personal observations, Maqrīzī's knowledge about the world of bees with its 'kings' (actually, 'queens') and their subjects, or about the true process of producing

---

[4] See our essay 'Cacciatori di miele. Dalla poesia huḏaylita alle pratiche tradizionali nel Dhofar (Oman),' *Quaderni di studi Arabi* 20–21 (2002–2003), pp. 185–206.

[5] Abū Ḥanīfa al-Dīnawarī, *Kitāb al-nabāt = The Book of Plants. Part of the Monograph Section*, ed. B. Lewin (Wiesbaden, 1974), pp. 275–294; Id., *Kitāb al-ʿasal wa 'l-naḥl*, attributed to Abū ʿUmar al-Zāhid, ed. Muḥammad Jabbār al-Muʿaybid, *al-Mawrid* 3/1 (1974), pp. 113–142.

[6] Ibn ʿAwwām, *Kitāb al-Filāḥa*, ed. J.A. Banqueri (Madrid, 1802 [repr. 1988]), II, pp. 717–730.

honey – from the nectar elaborated by forager bees in their crops, i.e. their honey stomachs –, remains essentially that of his time, based on the beliefs of the Ancients. (The situation in Europe was not dissimilar until the discoveries on the biology of bees in the following centuries, thanks to such men of science as L. Mendes de Torres, C. Butler, R. Remnant, F. Cesi and F. Stelluti, R.-A. Ferchault de Réaumur, or J. Swammerdam, and the invention of the microscope.)[7]

As far as we know, there are six manuscripts of the *K. Naḥl ʿibar al-naḥl*, which are kept in the libraries of Leiden (Cod. Or. 560 (6), XVth century), Cambridge (Add 746[3], dated 1112/1701; Qq. 141[2], dated 1232/1817), Paris (BNF, ar. 4657, XVIIIth century), Damietta (83.65, dated 1229/1814), Istanbul (Nuruosmaniye 4937[4]). The Leiden library has the privilege of owning a miscellaneous manuscript of al-Maqrīzī, with a dozen of short treatises, including the *K. Naḥl ʿibar al-naḥl* revised by the author himself.[8] Signing his revision, al-Maqrīzī honestly acknowledges to be the *jāmiʿ* (collector), besides being the work's author. But he is not very happy with the copyist's work, because he remarks that he revised the text to 'the extent possible, in spite of the great weakness (*suqm*) of the copy' (f. 61b). In fact there are many additions in the margins of the pages in his own hand.

Jamāl al-Dīn al-Shayyāl, editor of various writings by al-Maqrīzī, published a first edition of the *K. Naḥl ʿibar al-naḥl* in 1946 (Cairo: Maktabat al-Khānjī), based on the Damietta manuscript; the numerous missing parts, due to the poor state of the codex, were reconstructed with passages taken from the *Ḥayāt al-ḥayawān al-kubrā* of al-Damīrī (d. 905/1405),[9] a work certainly known to the author even if he does not quote it. A second edition is by ʿAbd al-Majīd Diyāb (Cairo: Dār al-Faḍīla, 1997), this time based on the Leiden Cod. Or. 560 (6); the edition is provided with lexical notes and rich indexes. The Paris MS was used by T. Fahd in his essay about bees in Islam.[10]

---

[7] See Canova 1999, p. 85.

[8] Described by R.P.A. Dozy, *Notices sur quelques manuscrits arabes* (Leyde, 1847–1851), pp. 17–28; and M.J. de Goeje [et al.], *Catalogus codicum orientalium Bibliothecae Academiae Lugduno-Batavorum*, V (Lugd. Bat., 1873), pp. 100–101, where the number is given as Cod. 560 (5). Cf., however, P. Voorhoeve's *Handlist of Arabic manuscripts [...]*. 2nd enl. ed. (Leiden, 1980), p. 247, where the MS is listed as Cod. Or. 560 (6).

[9] Al-Damīrī, *Ḥayāt al-ḥayawān al-kubrā* (Damascus, 2005), s.v. *naḥl*, III, pp. 29–52.

[10] See T. Fahd, 'L'abeille en Islam', in R. Chauvin (ed.), *Traité de biologie de l'abeille*, V, *Histoire, ethnographie et folklore* (Paris, 1968), pp. 61–83.

Below we present the translation of the second chapter of the *K. Naḥl ʿibar al-naḥl*, the most interesting in our opinion. The numbers of the folios of the Leiden manuscript[11] and those of the pages of Diyāb's edition are given here.

The Book of the Gift of the Bees' Moral Teachings

Chapter 2[12]

[f. 44a; p. 49] There are nine species (*aṣnāf*) of bees; six of these live in community, sharing their tasks: some produce the wax, others bring the honey and place it in the comb's cells (*abyāt al-shuhd*), still others bring water and mix it with the honey. As far as their colour is concerned, they are of three kinds: dust-coloured, the smallest ones; black, of medium size; yellow, the larger ones. Bees and ants are the most productive and hardest-working animals. The 'noble' bee (*al-naḥla al-karīma*, i.e. 'the good working-bee'] is small, its body is round and of various colours; the long bee is neither noble nor industrious, and is not suited for working. The small bees throw out of the hive and chase away the larger ones; if the larger ones are the winners, it is the end of the hive's well-being (*karam*). The small bees are the workers (*ʿummāl*), their colour is dark, burnt-like. Bees are pure and uncontaminated: they seem to be busily working virgins. They expel the lazy ones, that do not care about the honey. [p. 50] The bees that [f. 45a] forage (*tasraḥu*) in the mountains are smaller and better workers than those of the plains.

God the Highest gave the bees a king to whom they obey, called *yaʿsūb*. He is the heir of his ancestors' kingdom, because the *yaʿsūb* begets only *yaʿsūb*s. They are the kings of the bees and their leaders. The bees form a community around them and carry out their orders; they move where the king moves and settle where the king settles. The *yaʿsūb* is like the prince that must be obeyed. It is marvellous that the *yaʿsūb* does not leave the hive (*kuwar*, pl.) and does not go out to forage (*raʿy*); if he left the hive, all the bees would leave with him and the work would stop.

[11] I wish to thank Dr Arnoud Vrolijk, who kindly provided me with copies of the Leiden manuscript.

[12] Cf. Aristotle, *Historia animalium*, 553a26–554b21; 623b5–627b23; Arisṭūṭālīs, *Ṭibāʿ*, pp. 429–441; Ibn Sīnā, *al-Shifāʾ. al-Ṭabīʿāt*, ed. Ibrāhīm Madkūr [et al.], VIII, *al-Ḥayawān* (Cairo, 1970), pp. 131–136.

When a *yaʿsūb* is not able to fly, he is carried by the bees. If he dies, the bees in the hive (*khaliyya*) stop building the combs and producing honey: they are in mourning and fly with their heads bent to the ground. From this one infers that [p. 51] the *yaʿsūb* is dead. Then the bees look for another one, they take him and place him in the hive. After that, they start working again. If there is no *yaʿsūb,* [the bees] die very rapidly. His body is as big as that of two bees. He orders them (*yaʾmuruhum*) to work and assigns each of them the more suitable [task]. He orders some of them to build their home and some others to make honey; he chases away from the hive those that do not perform their tasks well and forbids them to be an example of laziness to the others. He places guardians (*bawwāb*, sg.) on the entrance of the hive to prevent impurities to be brought in. When the *yaʿsūb* plans to get out of the hive [to swarm], he starts to hum (*ṭanna*) one or two days in advance, to let the young bees know of his intentions, so that they can get ready.

The kinds (*ajnās*) of bees are numerous. As far as the *yaʿsūb* is concerned, there are two kinds: the first is reddish in colour and he is the best one; the second is blackish and variegated. Some of them have bodies as big as that of four bees. The *yaʿsūb* has a sting (*ḥuma*); it is black in its front side and red in its back side. In each hive there is only one *yaʿsūb*, yet at times there may be more than one if the hive is very large. If there is more than one, each one has his own colony (*ṭāʾifa*) of bees. If one of the *yaʿsūb*s gets out of the hive, all his bees follow it. The big *yaʿsūb* is called *jaḥl* ('king of the bees'). The kings of the bees do not sting (*taldhaʿu*), nor get enraged, because the *yaʿsūb* is very gentle. [p. 52] In fact, in this quality there is a lesson to be learned (*ʿibra*): were it present in one of the wise men, the best among all the living beings, it would be an extraordinary occurrence (*ʿajab*). In this respect, God the Highest told us what He revealed to this insect in spite of its weakness; He said: 'Surely therein is a sign for people who understand' (Qur. 16:69), in order that [men] should think about the extraordinary ability and the subtle understanding that was inspired in bees, in spite of their weak body. Thinking of this, a learned man of the past said: 'Bees are the animals that are most similar to man in their ability to organize themselves' and also 'The order that governs them is similar to the order of the person that governs populous cities'.[13]

---

[13] Cf. ed. Diyāb: *amruhunna shabīhun bi-amīrin yasūsu ʾl-madāʾin*; MS: *amruhunna shabīhun bi-amri man yasūsu ʾl-madāʾin.*

Bees build for their kings separate lodgings where they live, and they do the same for their drones (the male bees). Some people say, however, that the drones build their own lodgings. But they do not do any work. The female bees are the workers: they feed their kings and their males. The bees have no other food beside honey. The males hardly ever get out: they do this only when they want [f. 45b] to move their bodies in order to relieve themselves. Then they get out together, they rise in the air and hum; then they turn around and get back into the hive. If there is a drought and there is shortage of honey, the bees kill the drones, but if the drones realize that, they often fly away. One can see battles at the sides of the hive, on the outside. This is a proof, as already mentioned, that the bees protect their honey, care about it, do their best to economize it, and do not waver in facing any threat; this in spite of their good nature and their docility during quiet moments and times of good foraging (*kasb*).

[p. 53] Truly bees are marvellous creatures and have a very fine intelligence. We recall how they chase the idle and lazy ones, those that exploit the work and the supplies of the other bees. For us to act in a similar manner towards our lazy companions would be more wise and more useful. Among the proofs that they stock honey in their hives gathering food for themselves, and not for others, there is their zeal in economizing their honey, in saving and protecting it, their frenzy when it is threatened and their boldness in facing dangers: they fight against everything that could threaten their supplies. They keep away foreign bees and stand firm in their hive. At times some bees want to sack another colony and the two groups fight until one of them has slaughtered or defeated the other one; the losers flee, leaving behind their possessions. Avicenna said: 'The bees fight the foreign bees that try to seize their hive. If a bee-keeper [*rajul*] is watching his bees [while they are fighting the robbers] he is never stung by his own bees.'[14] If they are stronger [than the intruders], they kill them or they chase them away. It is for this reason that those who gather honey confuse the bees with smoke to keep them away and get to the honey.

[14] Cf. Ibn Sīnā, *al-Ḥayawān*, p. 135. Aristotle says that 'Before now, when a hive had been unhealthy, some of the bees attacked a hive belonging to others, and as they were winning the battle began carrying out the honey; but when the bee-keeper started to kill them, the others came out to attack them and began repelling them, and refrained from stinging the man.' Aristotle, *History of Animals*, trans. D.M. Balme (Cambridge-London, 1991), III, p. 359 [626b11 ff.].

Avicenna says in his book *al-Shifāʾ*: 'If [p. 54] a bee stings (*ladaghat*) an animal, leaving its sting behind, it will die. However, at times it is the bee that kills its victim: one time a bee even killed a horse.' He adds: 'I heard of a village where there were some hives. Once, the Kurds were attacking it and were on the winning hand, but the villagers incited the bees against them: the bees defeated them, stinging them and their horses.'[15] When the bees sting someone, they leave behind the sting (*ḥuma*), not being able to pull it back; they get free [, but trying to extract] their sting they die. *Al-ḥumāh* is a 'hair' in their back with which they sting. They can put it out or hide it at will. *Al-ḥuma* in Arabic means 'poison', but peoples call that hair *ḥumāh*. Avicenna said: 'It is not wrong to say that the bee's sting, although it is a weapon, is useful in transferring a humid part into the honey, giving it some strength.[16] If the bees are enveloped by smoke, they realize that the honey is going to be taken from the hive and they hurry in eating it frantically. They eat it even if they do not taste it.

Among the drones there is a kind that tricks the bees, getting into their hives and eating [p. 55] their honey. They are called 'robbers' (*luṣūṣ*). If the bees defeat them inside their hives, they kill them. They do not abandon their hive to forage without leaving someone behind to defend it.

When bees are 'noble', they do not allow insects into their hive that will damage the comb: they kill them or chase them away. Instead, bees that are not 'noble' languish, are lazy, and neglect their work; [f. 46a] [the colony] decays and dies. If the bees are not industrious and neglect the hive, it exhumes a putrid smell and decays.

The bee is the most delicate of all the animal species, and therefore it abhors any stinking and foul-smelling foraging area. Bees do not come close to stinking and dirty [persons]. They abhor evil-smelling and fat [persons], even if they use perfumes: if they come close, bees sting them. Bees like pleasant and melodious sounds; they do not harm people's way of life.

Bees love wild thyme (*ṣaʿṭar*), the best thyme is the white one. They seek protection from the wind, they drink only pure water, after spewing their saliva. [p. 56] When they go foraging, people say *jarasat tajrisu jarsan*, whether they collect wax or honey from flowers, both [actions] are called *jars*. Bees carry the wax on their limbs: one can see them burdened,

---

[15] Ibn Sīnā, *al-Ḥayawān*, p. 134.

[16] *Ibid.*, p. 133.

while the wax swings under them. Some people tried to observe the bees when they collect wax. Some maintain that the wax is attached inside the flowers as some kind of dust: a sticky substance (*luzūja*), such as one finds in flowers. Bees are seen carrying the wax on their forelegs, then scraping it with their middle legs; if there is any left they scrape it with their hind legs.

As far as honey is concerned, it is found inside the flowers (*anwār*) and it is a delicate nourishment for plants. When it is mature, it is sweet and of a pleasing taste. The bees insert their tongues inside the flowers (*nuwwār*) sucking that 'harvest' (*janāh*). Whoever tried it, knows it; we have licked many flowers and we found inside them that sugary substance. The bees' sucking means gathering honey. Their tongue is hollow, long and sharp, and it is specialised for this purpose and not for producing sounds, because bees do not emit them. The flies do not emit them, either; and also the bee is a fly (*dhubāba*). [p. 57] With their tongue they remove harmful [insects] (*adhiyya*) and at the same time they convey food into their bellies, because their nourishment consists of liquid substances only. They lick them and then they convey the food into the receptacles of their bellies. They are called tongues; in fact they are neither tongues nor proboscides, but they look more like tongues. When the bees suck this sweet substance from the flowers (*azhār*) they collect it in their breasts, they go to the comb and they 'vomit' it (*fa-atāʿathu*), that is they pour it into the cells (*nakhārīt* = *nakhārīb*). These cells are holes of a regular shape, with waxy walls, similar to the openings of the lodgings of wasps. If a bee lands on a flower of a certain kind, and what it gathered is not enough, it moves over to a flower of the same kind, and then comes back to the hive and vomits (*fa-tamujju*) everything it has collected. Then it goes back to forage. When the comb's cells are full of honey, bees cap them with a thin layer of wax, so that the wax covers [f. 46b] all sides – just as the mouth of a vase is sealed with paper –, to let the honey mature. If the bees did not act this way, the comb would putrefy, and a worm, called *ʿankabūt*, would emerge. If the bees succeed in keeping the comb clean, it stays intact, otherwise everything would putrefy.

When the grass flowers, the bees produce wax, and for this reason it is better to gather [p. 58] wax in that season. If there is shortage of wax, the bees replace it immediately [as much as is needed]. Bees produce honey in spring and autumn, but spring honey is better and more plentiful. Bees bring into the hive also another substance, which is neither wax nor honey, but a dry and somewhat elastic mixture. If it is pressed, it cracks; it is not very sweet, and it does not taste good. The Ancients compared its taste to the taste of figs. The bees carry this substance on the middle and the lower part of their legs, just as they carry the wax. The Arabs call it *ʾikbir*, that is the *mūm* ('wax', in Persian); it is also called *ʿikbir*. One can see the flying bees with this attached *ʿikbir*; they placed it in the combs' cells instead of the honey, but they collect it in abundance only in years of drought. They get it mostly from the Zizyphus trees (*sidr*). People eat it like bread and they are satiated. They carry it in their bags when they travel, but it gets spoiled more quickly than honey. Bees eat it only if they do not find other food.

[p. 59] Bees drink only water that is pure and has a good flavour and they look for it only in places where pure water is found. They eat only as much honey as they need; if there is shortage of honey in their hive, they mix it with water to dilute it, being afraid that it may run out. Bees defecate, and often they expel their excrements while flying, because they stink, and bees hate anything that stinks. If they defecate in their hive, they do it in a far corner, so that their lodgings are not contaminated and their honey does not get spoiled. When the cells are full of honey, the bees cap them, they seal also their brood (*firākh*, i.e their eggs) with a thin layer of wax. 'Sealing' (*khatm*) means to close the cells openings with a thin layer of wax, so that the wax seals the honey on every side. After being emptied, the cells are often smeared with a very black substance, very pungent in smell, that looks like wax [i.e. *propolis*]. This substance is one of the best remedies for swellings and wounds, in Persian it is called *miyāy*; it is very precious and scarce. Among its peculiarities, it helps in extracting thorns and arrowheads. As the saying goes: 'Who takes it as his companion, will leave to it every pain, and will be distracted from all [frivolous] dreams'. Cold and rain disturb bees; an indication of this is their insulating their hives.

There is something marvellous in the gift of sensitiveness of many animals. In this gift there is a warning for reasoning people: 'So blessed be God, the fairest of creators!' (Qur. 23:14).

THE ARABIC TRANSMISSION OF THE *HISTORIA ANIMALIUM* OF ARISTOTLE

Lou Filius

Ibn al-Nadīm is the most important source on the transmission of the Arabic translation of Aristotle's Book of Animals. In his *Kitāb al-Fihrist*[1] he writes:

> وهو تسع عشرة مقالة. نقله ابن البطريق. وقد يوجد سرياني نقلا قديما اجود من العربي. وله جوامع قديمة، كذا قرأت بخط يحيى بن عدى في فهرست كتبه. ولنيقولاوس اختصار لهذا الكتاب. من خط يحيى بن عدى، وقد ابتدأ ابو علي بن زرعة بنقله الى العربي وتصحيحه.

The translation of this passage reads: 'This book contains 19 maqālāt (*i.e.* in Arabic, the Book of Animals or *Kitāb al-Ḥayawān* consists of *De Generatione Animalium*,[2] *De Partibus Animalium*[3] and *Historia Animalium*).[4] Ibn al-Biṭrīq translated the book. There may be a Syriac <version> in an ancient translation, better than the Arabic one. This <Syriac version> existed in an old miscellany. Thus I read in a manuscript of Yaḥyā b. ʿAdī in the catalogue of his books. Nicolaus wrote a summary of this book, <as is handwritten> by Yaḥyā b. ʿAdī. Abū ʿAlī Ibn Zurʿa had started to translate this book into Arabic and to correct it'.

Therefore there may have been a Syriac translation and it is possible that this one was superior to the later Arabic translation by Ibn al-Biṭrīq. However, we know hardly anything about this Syriac translation; even if it had existed at all it is now lost. It is also unclear what the relation was between the Syriac translation and the Arabic. Was the Arabic translation based on this supposed Syriac translation, or was the Arabic translation directly rendered from Greek, but influenced by Syriac as the mother tongue of the Christian translator? H.J. Drossaart Lulofs made the assumption that the Arabic version of *De Generatione Animalium*

[1] Ibn al-Nadīm, *Kitab al-Fihrist*, ed. G. Flügel (Leipzig, 1871–1872), p. 251; ed. Riḍā Tajaddud (Tehran, 1391/1971), p. 312.

[2] Aristotle, *Generation of animals. The Arabic translation commonly ascribed to Yahya ibn Bitriq*, ed. J. Brugman & H.J. Drossaart Lulofs (Leiden, 1971).

[3] R. Kruk (ed.), *The Arabic version of Aristotle's Parts of Animals Book XI–XIV of the Kitāb al-Ḥayawān* (Amsterdam [etc.], 1979).

[4] An edition is currently being prepared by R. Kruk, J. den Heijer and L.S. Filius.

was translated from Syriac. He illustrated his supposition[5] with many examples, like the beginning of a sentence with وايضا, which could be traced to the Syriac starting expression ܛܘܒ (*tob*) and the rendering of the Greek ὁ αὐτός with هو وهو, going back to the Syriac ܗܘ ܟܕ ܗܘ (*hū kad hū*). Nevertheless, these phenomena can also be considered as nothing more than syriacisms, because all translators were Christians who used Syriac as their mother tongue, but this does not demonstrate, however, that a Syriac translation was an intermediary between the Greek origin and the Arabic translation.

In this respect R. Kruk comes closer to the truth by saying: 'Aristotle's Zoology was translated from Greek into Arabic, possibly with a Syriac translation in between,'[6] but later[7] on she even decided: 'Whether there was any relation between this Syriac translation, of which nothing further is known, and the Arabic one which has come to us, cannot be ascertained. There is not even any proof that the Arabic text was translated from the Syriac, either from this old translation or from a translation specially made by the translator as a preparation for the translation in Arabic (as was often done). Several syriacisms may be pointed out in the translation, but these only prove that the translator's language was strongly influenced by Syriac.' It is supposed that the same goes for the *Historia Animalium* too, and perhaps even more so than for the *De Partibus Animalium*. The Arabic version of the *Historia Animalium* has been translated directly from Greek into Arabic by Christians who had Syriac as their native language. Thus there are many syriacisms in the *Historia Animalium*,e.g. تنور for the Greek word θώραξ from the Syriac ܬܢܘܪܐ (*tanūrā*) and الآذان for the Greek word βράγχια (e.g. 533b4) from the Syriac ܐܕܢܐ (*ednā*), which also means *ear* and *gill*,[8] but these words do not prove the use of a Syriac intermediary, because these are only isolated words.[9] The Arabic transcription of the Greek words seems to

[5] See Brugman & Drossaart Lulofs, pp. 5–10 for the identity of the translator from Syriac into Arabic. In that connexion the editors did not mention these examples as evidence of the Syriac origin, for that was their starting-point, but to demonstrate the difference between the anonymous translator and Usṭāt. But also chapter 2 (pp. 11–17) does not give a decisive answer about the original language of the Arabic translation.

[6] See Kruk, p. 14.

[7] See Kruk, p. 22.

[8] See H.J. den Heijer, 'Syriacisms in the Arabic version of Aristotle's Historia Animalium,' *ARAM* 3 (1991), pp. 105–106.

[9] In this respect the usage of a Syriac intermediary for the translation of Aristotle's *De Caelo* is obvious, see G. Endress, *Die arabischen Übersetzungen von Aristoteles' Schrift De Caelo* (Frankfurt a/M., 1966), pp. 32–35.

point in the same direction through the rendering of various case endings of Greek words from the Greek text.

The Arabic version of the *Historia Animalium* I–X has been transmitted in two manuscripts, a London manuscript (L)[10] and a manuscript from Tehran (T).[11] The London manuscript contains corrections by two or more hands, of which mainly the first hand is correct. The relation between both manuscripts is straightforward: The Tehran manuscript has been copied from the London manuscript, as is evident from e.g. 487b15, where L[1] reads يعوم against L[2] يعوص and T يغوص, whereas the Greek text reads νευστικά. Another example is 487b30: L اخر became اجزاء in T, whereas Scotus read *ultimo*, which is missing in Greek, but it must be correct in Arabic in L. Compare also حس in L against جنس in T, where Scotus read *sensum* in 491a23 and in many other places. Many cases can be mentioned where other words are concerned. Also the lacunas in T match with the text of L. Therefore L is the most important manuscript, but T is complete, so that T is important when L is defective.

The Hebrew tradition seems to be a very long tradition. Steinschneider[12] mentioned a Hebrew transmission of the *Historia Animalium*, which appeared to be, however, a translation from a Latin translation. Therefore this translation is irrelevant for the Arabic transmission. But according to M. Zonta[13] there may have been a long tradition among Jewish scholars to study the whole corpus of Aristotle, including the *Historia Animalium*, but only a compendium of it has been transmitted. On account of this the importance for the Arabic tradition seems to be very limited.

Going back to the Arabic transmission we ought to consider the Latin translation from Arabic by Michael Scotus. If his translation was made from our Arabic text, it could serve to reconstruct the Arabic text. When we read the text of Michael Scotus, it seems a very accurate translation

---

[10] London, British Library, MS Add. 7511 (12th century), contains only book I, II and a part of book III and at the end a part of book IX and book X. The notice in F. Sezgin, *Geschichte des arabischen Schrifttums (GAS)*, III (Leiden, 1970), pp. 350–351 is incorrect.

[11] Tehran, Majlis Library, coll. Ṭabāṭabāʾī, no. 1143. This manuscript is complete, 17th or 18th century.

[12] M. Steinschneider, *Die hebraeischen Uebersetzungen des Mittelalters und die Juden als Dolmetscher. Ein Beitrag zur Literaturgeschichte des Mittelalters, meist nach handschriftlichen Quellen* (Berlin 1893, repr. Graz 1956), pp. 479–483, see also F. Sezgin, *GAS*, III, p. 351.

[13] See M. Zonta, *La filosofia antica nel Medioevo ebraico. Le traduzioni ebraiche medievali dei testi filosofici antichi* (Brescia, 1996), pp. 153–154 and 160–161.

of the Arabic text. He translated the whole Book of Animals, although we have no evidence that he did it for his later patron Frederick II.[14] He had a profound knowledge of Arabic and a deep insight in the texts he translated. 'Apparently, his aim was to provide a clear and precise text, as faithful as possible to the Arabic version, at the same time avoiding superabundant or obscure Arabic periphrasis as much as he could,' as A.M.I. van Oppenraay characterises him.[15] Owing to the quality of his Latin translation his text must be of eminent importance for the constitution of the Arabic text. But was the Arabic text of Michael Scotus the same text that we have now or another one, e.g. the text corrected by Ibn Zurʿa, who started to make a new translation and correct the old one, possibly that of Ibn al-Biṭrīq, or some other unknown translation? At this moment it is assumed 'mit einiger Wahrscheinlichkeit' that the text of the London and Tehran manuscripts both contain the translation by Yaḥyā Ibn al-Biṭrīq.[16]

It is therefore necessary to make a comparison between both texts in order to decide whether the Arabic text of Michael Scotus was the same one as our text or not, to give the text of Scotus its correct appraisal, and to decide to what extent this translation is decisive for the constitution of the Arabic text.

The beginning of Book 4

وقد ذكرنا فيما سلف حال الحيوان الدمي ووصفنا جميع اعضائه العامية والخاصية التي لكل جنس من الاجناس ولخصنا ايضا اعضاءها التي اجزاؤها لا يشبه بعضها بعضا وكل ما كان منها في ظاهر الجسد وكل ما كان في باطنه.

The translation of Scotus

> *Iam diximus superius dispositionem animalis habentis sanguinem et narravimus omnia eius membra communia et propria, et pertransivimus etiam membra eius consimilia manifesta et intrinseca.*[17]

[14] See C.H. Haskins, *Studies in the history of mediaeval science* (Cambridge, Mass., 1924), pp. 274–275.

[15] A.M.I. van Oppenraay, *Aristotle, De Animalibus. Michael Scot's Arabic Latin translation.* Pt. 2, Books XI–XIV, *Parts of Animals.* Aristoteles Semitico-Latinus; 5 (Leiden 1998), p. xii.

[16] See M. Ullmann, *Wörterbuch zu den griechisch-arabischen Übersetzungen des 9. Jahrhunderts* (Wiesbaden 2002), pp. 57–58. However, Endress, pp. 113–115 is less convinced of it.

[17] The Latin text is a provisional text, quoted here with the kind permission of A.M.I. van Oppenraay.

The omission of التي لكل جنس من الاجناس is remarkable: the Greek text does not have it and perhaps it is an example of the Arabic periphrasis and therefore omitted by Michael Scotus. The rendering of ولخصنا is somewhat strange, because normally it has a translation[18] different from *pertransivimus*. Yet there is no indication that Scotus used a different text from our transmitted text. Other examples confirm this. It is evident that the Greek text will help to decipher the Arabic manuscripts, especially with regard to improbable letter combinations, as in the same part 523b17, where T has this version: ومن هذا الصنف ما هو محزز الجسد وليس له جناح. T incorrectly reads جسد instead of جناح, which conforms with the Greek ἄπτερα. When correcting difficult letter combinations or graphemes the starting point must always be the Greek text, since it was the origin of the Arabic text. For example, نياله in T 523b7, written without any punctuation, provides the conjecture يتك, based on the Greek θλαστόν. In the same manner one could assume in 525b25 that πόδες, which is in T ورجل, must be slightly altered into وارجل in order to conform with the Greek text.

But what to do with e.g. 525b23, where Scotus reads according to the Greek text: *et sequuntur ipsos alii tres subtiles in utraque pars*, εἶτ' ἄλλους ἐχομένους τρεῖς ἐφ' ἑκάτερα, but in T this part is missing. Is it to be added, if this addition is based on the Greek text and on Scotus' translation?

As said, the Tehran manuscript is not always reliable, because it is a copy from the London manuscript, but also L has these problems.

L in 487a27

وبعض الحيوان الذي يأوي في الماء بحري وبعضه نهري <...> وبعضه نقاعي مثل الضفدع والحيوان الذي ....

Scotus translated

> *Et quaedam animalia quae manent in aqua marina, et quaedam fluminea, <et quaedam stagnaea,> et quaedam paludosa, sicut ranae, et animal quod......*

[18] Cf. Van Oppenraay, p. 557, s.v.

Greek text

> Τῶν δ' ἐνύδρων τὰ μέν ἐστι θαλάττια, τὰ δὲ ποτάμια, <τὰ δὲ λιμναῖα,> τὰ δὲ τελματαῖα, οἷον βάτραχος καὶ...

The part within <...> is missing in Arabic in both L and T. In some cases it is implicated in the whole text that sometimes words have to be added, because elsewhere one will find the same combination, e.g. 507a22: Scotus translated *quadrupedibus* and the Greek text has τῶν τετραπόδων, so that we can add here in Arabic the necessary words الحيوان <الذي له اربعة ارجل>. But is this also correct in 487a27? Till 514b16 we possess the text of the London manuscript and we can limit our additions to some places where an addition is necessary. Normally the text of the London manuscript does not seem to differ too much from the text of Scotus. But after 514b16 we only possess the Tehran manuscript.

In 514b25 an addition is obvious: من <العرق> العظيم for this type of missing words is frequent and also the Greek text has ἀπὸ τῆς μεγάλης φλεβός and Scotus *a vena maiori*.

An addition like the one in 518b16: مختلف الوان <الشعر فمختلف الوان> الجلد, based on Scotus' text *in pilis et etiam diversi coloris*, supported by the Greek text ὅσα δὲ ποικίλα τῶν ζῴων κατὰ τὰς τρίχας, τούτοις καὶ ἐν τῷ δέρματι προυπάρχει ἡ ποικιλία, is easily to explain as a so-called parablepsis[19] from مختلف to الوان.

On the other hand, an addition like one finds in 519a19–20, where the Greek text is τὸν Ὅμηρόν φασιν ἀντὶ Σκαμάνδρου Ξάνθον προσαγορεύειν αὐτόν and Scotus reads *Homerus vocabat illud flumen rubeum*, would be at least odd from the perspective of the Arabic text in general. Firstly, because the name of Homer is not very frequent in the transmitted version, only in 574b33 and 618b25 against nine occurrences in the Greek text and probably somewhat less frequent in Scotus, but also because quotations from poets do not normally appear in the Arabic text. Moreover, such an omission does not cause any vagueness in the text. On the other hand, the evident lacuna in 536a10 in the Tehran manuscript has obviously to be added from Scotus' *cannae* with <الانبوبة>, based on the Greek τὸν φάρυγγα. However, it clearly shows the difficulty of making a decision, because we have no certainty about the text of Scotus on account of the indifferent quality of the Tehran manuscript.

---

19 M.L. West calls this phenomenon *saut du même au même* in *Textual criticism and editorial technique* (Stuttgart, 1973), pp. 24–25.

From book VI Scotus translated more frequently parts of the text that are missing in the transmitted Arabic text. Often they concern Greek terms which are not mentioned in Arabic, but which are present in Scotus, e.g. 561b32 *Et in isto tempore tela absolvens intestina quae assimilatur chorion*, and 562a3 *et apparet etiam alia tela similis chorion*, where the Greek text has χοριοειδὴς: καὶ φανερὸς κατὰ τοῦτον τὸν χρόνον ὅ τε χοριοειδὴς ὑμὴν…(561b32) and καὶ ὁ ἕτερος ὑμὴν χοριοειδὴς ὤν…(562a3). In Arabic it is in 562a3: وفي هذا الزمان يظهر الصفاق المحدق بالرطوبة <…>. And in 562a3: ويظهر الصفاق الآخر <…>

Also 566a18: Scotus reads: *et incipit coire in mense qui dicitur mamatireon* for the Greek ἄρχονται δ΄ ὀχεύεσθαι μηνὸς Μαιμακτηριῶνος. This part is completely absent in the Arabic text. More and more these typically Greek terms, regularly mentioned by Scotus, are omitted from the Arabic text and have not been transmitted to us.

On the whole Scotus exactly follows the Greek text, whereas our Arabic text omits many parts of it. It is tempting to complete the Arabic text with the text of Scotus, but we have to do with a bad Arabic text, like the one that Ibn al-Nadīm mentioned, whereas Scotus used a corrected and 'newly' translated Arabic text, possibly that of Ibn Zurʿa.

In 623b10 the writing of the words *vespe et hakihokeuz* is strange in comparison with the Arabic text.

Scotus 623b10

> *Et sex ex eis congregantur sicut greges et sunt apes et kankin quod est semper cum apibus et vespe et hakihokeuz et iribi et kataneridez.*

Greek

> Τούτων δ'ἐστι γένη ἐννέα, ὧν τὰ μὲν ἓξ ἀγελαῖα, μέλιττα, [βασιλεῖς τῶν μελιττῶν], κηφὴν ὁ ἐν ταῖς μελίτταις, σφήξ, ὁ ἐπέτειος, ἔτι δ'ἀνθρήνη καὶ τενθρηδών.[20]

---

[20] Aristotle, *Historia Animalium*, I, Books I–X, *Text*, ed. by D.M. Balme, prepared for publication by A. Gotthelf. Cambridge classical texts and commentaries; 38 (Cambridge, 2002).

Arabic

وهي تسعة اصناف منها ستة اصناف يأوي بعضها مع بعض اعني النحل والذكورة التي تكون في النحل والدبر الذي يأوي علي وجه الارض والدبر الصغير الاصفر والدبر الاسود المستطيل.

The Greek text has σφήξ ὁ ἐπέτειος which seems to have been corrupted in Scotus' text to *vespe et hakihokeuz*, words that cannot be a translation of the Arabic text we have: والدبر الذي يأوي على الارض.

A second case that deserves mention here is 609a1:

> Διὰ τὰς ὠφελείας γὰρ ἡμεροῦται, οἷον ἐνιαχοῦ τὸ τῶν κροκοδείλων γένος πρὸς τὸν ἱερέα διὰ τὴν ἐπιμέλειαν τῆς τροφῆς.

Scotus

> *Et propter hoc vivunt simul et sunt quasi domestica. Genus autem karokokiloz et etiam ardon habet pacem cum kehhin et domesticatur cum illo quando cogitat de suo cibo.*

Arabic

وما كان من اصناف الحيوان صعب الخلق نزقا بريا يكون هناك آنس لحال كثرة المنفعة كما يعرض في بعض المواضع التي يكون فيها جنس الجراذين يأوي في قرب البزاة لحال كثرة الطعم
....

Here Scotus reads *habet pacem cum kehhin*, which must be a translation of the Arabic كهان, the plural of كاهن, not our Arabic: البزاة, which should be in Greek ἱεράκα. The conclusion must be that Scotus did not use our Arabic text, but another translation, a better one, which may be the one of Ibn Zurʿa, because the text of Scotus is closer to the Greek text. Especially from book VI there are many deviations, of which many are not reducible to the existing Arabic translation. It is possible that Scotus used our translation for the first five books, generally assumed to be translated by Ibn al-Biṭrīq, and that from book VI he used the 'new' translation by Ibn Zurʿa, who started his own translation according to Ibn al-Nadīm, and possibly he began with book VI, but these last facts are only hypothetical and further investigation of the Arabic text, and especially of the text of Scotus, is necessary. Because the Syriac translation was a good translation according to Ibn al-Nadīm, it is to be supposed that the transmitted Arabic text is not the translation of Ibn al-Biṭrīq, because he must have known the Syriac translation.

It is possible to suppose that the London manuscript has another version from this point, but because of the difference in graphema the

London manuscript is supposed to have the same version in the case of 609a1, because T is a copy from L. And when we compare the London manuscript with the Tehran manuscript with regard to book IX[21] and X, one can discover the same type of translation, many parts being omitted without many Greek terms and omissions not being indicated.

Therefore it is obvious that Scotus used a different translation from the one we know, and for this reason we can use Scotus' translation for the constitution of the Arabic text only to a limited degree: to correct a single word, to fill up a lacuna or to complete a formula. In any case, when the editions, Greek, Arabic and Latin are ready, it will be very interesting to possess the whole tradition of this text of Aristotle.

Further study of the history of this text in particular will possibly throw new light on the study of Aristotle in general.

[21] I.e. Book IX as it is found in the Arabic Book of Animals and in Scotus, in the arrangement adopted by D.M. Balme in his Loeb edition, *Historia Animalium*, part III (Cambridge, Mass. [etc.], 1991), and in the edition prepared by A. Gotthelf (Cambridge, 2002). The usual order is from I. Bekker, *Aristotelis Opera*, vol. I–V (Berlin, 1831–1870) and P. Louis, Aristote, *Histoire des Animaux*, t. I–III (Paris, 1964–1969): BG: VI 558b–576a (BL: VI 558b–581a), BG: IX (BL: VII 581a–588a), BG: VII (BL: VIII–588a–608a), BG: VIII (BL: IX 608a–633b), BG: X (BL: X 633b–638b).

SOME RECENT FINDINGS IN MICHAEL SCOT'S ARABIC-LATIN TRANSLATION OF ARISTOTLE'S *HISTORY OF ANIMALS*

Aafke M.I. van Oppenraay

During my work on the edition[1] of Michael Scot's 13th cent. Arabic-Latin translation of Aristotle's *History of Animals* it appeared that Scot had at his disposal an Arabic translation which clearly diverged from the text that has come down to us, partly in two manuscripts from London and Leiden, in its entirety in a single manuscript from Tehran.[2]

For instance, names of animals within this translation tradition are either translated or transcribed, if it was impossible to find an equivalent. But I observed in a number of cases that the Latin text contains transcriptions which obviously agree with what once must have been Arabic transcriptions that were more closely related to the forms found in the Greek text, instead of giving a Latin translation, or a transcription, of the corresponding Arabic words as we find them in our preserved text. See for instance example 1:

1. HA 7.589b26 ἕν δὲ μόνον νῦν ὦπται τοιοῦτον, ὁ καλούμενος <u>κορδύλος</u>. Only one such has been seen at present, the so-called <u>cordylus</u>. وقد ظهر حيوان واحد الذى يسمّى <u>تمساحا</u>. Et non fuit visum animal talis dispositionis nisi unus modus tantum, qui dicitur *codiloz*.

Here, Scot's 'codiloz' clearly transcribes an Arabic transcription of the Greek κορδύλος ('water-newt', 'salamander'), whereas the extant Arabic

---

[1] The edition of this text is now being prepared by Aafke M.I. van Oppenraay at the Huygens Instituut in The Hague, whereas a critical edition of the Arabic translation is prepared by L.S. Filius, J. den Heyer and J.N. Mattock (Leiden). This project is supervised by Remke Kruk. Scot's translations of *De partibus animalium* and *De generatione animalium* have been edited: A.M.I. van Oppenraay (ed.), Aristotle, *De animalibus. Michael Scot's Arabic-Latin Translation*. Part two, books XI–XIV, *Parts of Animals*. Aristoteles Semitico-Latinus; 5/2 (Leiden [etc.], 1998); idem, Part three, books XV–XIX, *Generation of Animals*. Aristoteles Semitico-Latinus; 5/3 (Leiden [etc.], 1992).

[2] MS London, British Museum, Or. Add. 7511; MS Leiden, University Library, Or. 166; MS Tehran, Majles Library 1143.

text offers a translation of the Greek κροκόδειλος ('crocodile'), which by the way is the reading here of some of the Greek MSS.[3]

Likewise, we do find equivalents of combinations of words, or even of entire parts of sentences, from the original Greek text in Scot's translation, whereas these are not found in the extant Arabic text. See examples 2–5 (the words in question from Scot's text have been put in italics).

2. HA 1.487a26–8 τῶν δ' ἐνύδρων τὰ μέν ἐστι θαλάττια, τὰ δὲ ποτάμια, τὰ δὲ λιμναῖα, τὰ δὲ τελματιαῖα, οἷον βάτραχος καὶ κορδύλος. Some water-animals live in the sea, some in rivers, some in lakes, some in marshes, for instance the frog and the newt. وبعض الحيوان الذى يأوى فى الماء بحرى وبعضه نهرى وبعضه نقاعى مثل الضفدع والحيوان الذى يسمّى باليونانية قردولوس. Et quaedam animalia quae manent in aqua sunt marina, et quaedam fluminea, et quaedam stagnaea, *et quaedam paludosa* sicut ranae et animal quod nominatur Graece cardoloquios.

3. HA 1.489b1 ζῳοτόκα μὲν οἷον ἄνθρωπος καὶ ἵππος καὶ φώκη καὶ τὰ .ἄλλα ὅσα ἔχει τρίχας Examples of viviparous animals are man, horse, seal and all other hairy animals. فاما الذى يلد حيوانا فمثل الانسان والفرس وجميع ما له شعر. Illa vero quae generant animalia sunt sicut homo et equus *et koki* et omne pilosum.

4. HA I 491b18…ὑφ' αἷς ὀφθαλμοί. οὗτοι κατὰ φύσιν δύο. τούτων μέρη ἑκατέρου βλέφαρον τὸ ἄνω καὶ κάτω. τούτου τρίχες αἱ ἔσχαται βλεφαρίδες. Below the eyebrows are the eyes, of which the natural number is two. Parts of them are as follows: each eye has an upper and lower eyelid; the hairs on their edges are eyelashes. وتحت الحاجبين العينان واجزاؤهما الشفر الاعلى والاسفل. Et sub superciliis sunt oculi. Et partes eorum sunt palpebrae, superior et inferior, *et pili in inferiori et superiori dicuntur cilia.*

Note that in the Greek text the usual number of eyes is specified, which the Arabic translator rendered by a mere dual form, whereas Scot ignores the signal, probably taking the number of eyes for granted here.

[3] The MSS group α; see D.M. Balme & A. Gotthelf (eds.), Aristotle, *Historia animalium*, I (Cambridge, 2002), p. 50. The English translation in this article is from A.L. Peck (ed.), *Historia animalium*, Books I–VI (Cambridge, MA, 1964, repr. 1970) and D.M. Balme (ed.), *Aristotle, History of Animals*, Books VII–X (Cambridge, MA [etc.], 1991).

5. HA II 504a6 ἔτι δὲ πολυσχιδεῖς τρόπον τινὰ πάντες · τῶν μὲν γὰρ πλείστων διῄρηνται οἱ δάκτυλοι, τὰ δὲ πλωτὰ στεγανόποδά ἐστι, διηρθρωμένους δ’ ἔχει καὶ χωριστοὺς τοὺς δακτύλους. All (birds) in one way or another have numerous toes; in most birds the toes are quite distinct, and the swimmers, although web-footed, have separately articulated toes. وجميعها مشقوقة الرجلين وشقوق رجلي بعضها مفَصلة بينة. فاما ما يعوم منها ويأوي فى الماء ففيما بين اصابع رجليه جلد قوى متصل. Et omnes sunt fissi pedis et fissurae digiti quorundam sunt distinctae, quae vero sunt natatilia habent corium continuans digitos *et divisionem in extremitatibus digitorum.*

This phenomenon of parts of sentences in the Latin text agreeing with the original Greek text, whereas no equivalent is found in our Arabic text, occurs from the very first book onwards. We see here that in the first two books these parts of sentences are often only found in the text or in the margin of the Vatican manuscript,[4] usually wholly or partly supported in this by marginal glosses in the Cambridge manuscript, or, sometimes, inserted in the text, and occasionally in another manuscript, but ignored by most of the others. From the third book onwards, however, such parts of sentences have been seamlessly integrated into the running text of all the manuscripts.[5]

Following on from this observation, I would like to suggest – obviously with caution and reservations – that Michael Scot, when starting on his translation of Aristotle's biological books, may have used an Arabic text which rendered the original Greek text less literally, but shortly afterwards, in any case when the first two books had been completed, somehow obtained a copy of an Arabic text which contained a better, i.e. more literal, translation. With this newly acquired translation in hand, he then added some passages in the first two books – some existing lacuna's in the Vatican MS were filled in later, too – after which he continued his translation of the later books using this text as his basis.

---

[4] MS Vaticanus Chigi E. VIII. 251; MS Cambridge, Gonville and Caius College Library 109/178; MS Berlin, Staatsbibliothek Preussischer Kulturbesitz Hdschr. 194.

[5] In an article from 1999 I cited other examples; these are mostly from the later books, because at the time I thought that the phenomenon occurred from the sixth book onwards; see A.M.I. van Oppenraay, 'Michael Scot's Arabic-Latin translation of Aristotle's Books on Animals. Some remarks concerning the relation between the translation and its Arabic and Greek sources,' in C. Steel, G. Guldentops [et al.] (eds.), *Aristotle's Animals in the Middle Ages and Renaissance*. Mediaevalia Lovaniensia. Series I, Studia; 27 (Leuven, 1999), pp. 31–43.

This 'better text' may go back to an Arabic translation which was produced after the Greek source text had first been put into Syriac as an intermediate translation, by way of an aid, either in the mind of the translator or actually committed to paper. This is well known to be a translation technique used by the best translators; we find it described by Ḥunayn b. Isḥāq, and later it was also adopted by his son Isḥāq. It may be possible to indirectly confirm at least the one-time existence of a better Arabic version than the extant one – whether including a Syriac intermediate phase or not – by the recent findings in Michael Scot's Arabic-Latin translation.

It will probably be impossible to find out exactly how many versions of the Arabic translation played a role in the transmission of the copy used by Scot, or who their possible translators or revisers were. In the *Introduction* to her edition of the Arabic version of *De partibus animalium*,[6] Remke Kruk offers a survey both of testimonies to the transmission within the Arabic tradition itself and of the many names connected with it. Besides the name of Yaḥyā Ibn al-Biṭrīq, to whom the current translation was, wrongly, attributed in the Arabic tradition, other great names that circulated were Ḥunayn b. Isḥāq, who was said to be the author of a 'better version' of the translation (*iṣlāḥ*), and Ibn Zurʿa, a well-known corrector of primitive earlier translations. Also, the statement by Ibn al-Nadīm in his *Fihrist* that there once was a Syriac translation of the biological books which was *older and better* than the Arabic one seems valuable in this connection.[7] It it is highly regrettable, therefore, that neither of the reported 'better versions' appear to have been passed down.

[6] R. Kruk (ed.), Aristotle, *The Arabic version of Aristotle's Parts of Animals. Books XI–XIV of the Kitāb al-Ḥayawān*. Aristoteles Semitico-Latinus; *2 (Amsterdam [etc.], 1979), p. 15ff.

[7] B. Dodge (ed.), *The Fihrist of al-Nadīm*. 2 vols. (New York [etc.], 1970), II, p. 605.

SCARABÉES, SCORPIONS, CLOPORTES ET CORPS CAMPHRÉS. MÉTAMORPHOSE, RÉINCARNATION ET GÉNÉRATION SPONTANÉE DANS L'HÉTÉRODOXIE CHIITE

Daniel De Smet

Dans une de ses nombreuses contributions à l'étude des sciences naturelles en Islam médiéval, Remke Kruk relève quatre notions qui constituent le contexte philosophique de la théorie de la génération spontanée chez les auteurs arabes: (1) la conception aristotélicienne de la génération; (2) l'influence du climat et du milieu naturel sur la formation des êtres vivants, thèse que l'auteur met en rapport avec le galénisme; (3) le thème néoplatonicien de la *scala naturae*, la continuité et la stricte hiérarchisation des règnes de la nature, dont les êtres représentent autant de maillons d'une chaîne ininterrompue; (4) la doctrine alchimique d'une matière inorganique 'vivante', car susceptible de croître et d'être purifiée, transmuée et sublimée en une matière plus noble.[1]

La littérature chiite de tradition ismaélienne et druze, tout comme les écrits relevant du chiisme 'extrémiste' (*ghulū*),[2] confirment les analyses de notre collègue de Leyde. Les quatre notions qu'elle a relevées s'y retrouvent en effet dans un contexte lié à la génération spontanée, ou du moins à la génération 'sans père, ni mère' d'un être vivant à partir d'une matière préexistante. En outre, l'ensemble se profile sur l'arrière-plan d'une croyance à la métempsycose et à la réincarnation des âmes humaines dans des corps métamorphosés. Métempsycose et réincarnation forment ainsi une cinquième doctrine déterminant la conception de la génération spontanée en milieu musulman, que nous voudrions ajouter à la liste dressée par Remke Kruk.

---

[1] R. Kruk, 'A frothy bubble. Spontaneous generation in the medieval Islamic tradition,' *Journal of Semitic studies* 35 (1990), pp. 265–282.

[2] Sur le sens et l'évolution des termes *ghulū* et *ghulāt*, qui se réfèrent à l'exagération en matière religieuse, voir W. al-Qadi, 'The development of the term *ghulât* in Muslim literature with special reference to the Kaysâniyya,' *Abhandlungen der deutschen Akademie der Wiss. zu Göttingen, Philol.-hist. Klasse* 98 (1976), pp. 295–319; A. Straface, 'Il concetto di estremismo nell'eresiografia islamica,' *AION* 56 (1996), pp. 471–487.

D'inspiration foncièrement gnostique, les courants 'radicaux' et 'extrémistes'[3] au sein du chiisme conçoivent la transmigration des âmes comme la conséquence inéluctable de la catastrophe cosmique qui a produit notre monde: enfermées dans des corps matériels, les âmes déchues doivent se libérer de leurs attaches matérielles en se purifiant par la connaissance salvatrice. Cette purification s'effectue au cours d'innombrables cycles, pendant lesquels les âmes transmigrent d'un corps à l'autre. Si elles ont été vertueuses et pures, elles passent dans un autre corps humain, plus 'lumineux' que le précédent, jusqu'au moment où elles échappent au cycle des réincarnations successives et revêtent un corps de camphre (*jism kāfūrī*), un 'temple de lumière' (*haykal nūrānī*) élaboré par une complexe alchimie cosmique, loin des souillures de la génération sexuelle. Cette voie 'ascendante' est désignée par le terme de *nasūkhiyya*. En revanche, si les âmes se laissent dominer par leurs passions, elles devront revenir dans un corps plus ténébreux: c'est la voie descendante, appelée *masūkhiyya*, puisqu'elle prend la forme d'une métamorphose, d'abord dans un corps féminin,[4] puis dans des corps d'animaux de plus en plus abjects. De métamorphose en métamorphose (*maskh*), l'âme

[3] Toujours dans le sens de *ghulū* / *ghulāt*. A la suite de Heinz Halm, dont les travaux novateurs font autorité à juste titre, on établit généralement une distinction trop stricte entre les mouvements *ghulāt*, apparus en Irak dès le début du 8e siècle et dont les Nuṣayrīs actuels sont les héritiers directs, et les différentes branches de l'ismaélisme, auxquelles se rattache la doctrine druze; cf. H. Halm, *Die islamische Gnosis. Die extreme Schia und die 'Alawiten*. Bibliothek des Morgenlandes (Zürich [etc.], 1982), pp. 14–15. Tout en représentant sur le plan historique autant de traditions distinctes, tous ces courants d'inspiration chiite, loin d'être étanches les uns aux autres, ont puisé à un fonds commun, qu'Amir-Moezzi appelle avec raison 'le chiisme originel' (cf. M.-A. Amir-Moezzi, *Le Guide divin dans le Shī'isme originel* [Lagrasse, 1992]). De nombreux points de convergence existent en effet entre la doctrine des *ghulāt*, l'ismaélisme et certaines traditions appartenant au chiisme duodécimain, comme Halm l'a d'ailleurs lui-même relevé dans sa *Kosmologie und Heilslehre der frühen Ismā'īlīya. Eine Studie zur islamischen Gnosis*. Abhandlungen für die Kunde des Morgenlandes; 44/1 (Wiesbaden, 1978), pp. 142–168; voir en outre D. De Smet, 'Eléments chrétiens dans l'ismaélisme yéménite sous les derniers Fatimides. Le problème de la gnose ṭayyibite,' dans: M. Barrucand (éd.), *L'Egypte fatimide. Son art et son histoire. Actes du colloque organisé à Paris les 28, 29 et 30 mai 1998* (Paris, 1999), pp. 45–53. Le présent article offre une nouvelle illustration de cette interpénétration doctrinale au sein des différents courants du chiisme 'radical'.

[4] Certains mouvements *ghulāt*, comme le Nuṣayrisme, font parfois preuve d'une misogynie qui n'est pas sans rappeler Platon et la Gnose antique. Mais curieusement, ce mépris pour la femme s'avère absent de l'ismaélisme et de la doctrine druze, qui au contraire accordent à la femme et au féminin un statut privilégié, tant sur le plan social et religieux, que sur le plan cosmique; voir D. De Smet, 'Une femme musulmane ministre de Dieu sur terre? La réponse du *dā'ī* ismaélien al-Ḫaṭṭāb,' *Acta Orientalia Belgica* 15 (2001), pp. 155–164; Id., 'La valorisation du féminin dans l'ismaélisme ṭayyibite. Le cas de la reine yéménite al-Sayyida Arwā (1048–1138),' *Mélanges de l'Université Saint-Joseph* 58 (2005), pp. 107–122.

damnée échoit ainsi dans les corps immondes de scarabées, scorpions, cloportes, cafards et autres engeances répugnantes, générées spontanément à partir d'excréments et de matière en putréfaction.

Les deux extrémités de la *scala naturae*, le long de laquelle les âmes montent et descendent, sont donc peuplées par des êtres qui échappent à la génération ordinaire, mais sont produits de façon spontanée, par la matière en putréfaction ou par une subtile alchimie cosmique.

Nous décrirons successivement la voie descendante de la *masūkhiyya* et la voie ascendante de la *nasūkhiyya*.

'Masūkhiyya' *et génération spontanée des scarabées, scorpions et cloportes*

La métamorphose *(maskh)* ou 'transformation d'une forme en une autre plus laide'[5] est partie intégrante de la mythologie, des croyances et du folklore de nombreux peuples. Des hommes et des femmes sont transformés en animaux, voire en objets inanimés pour avoir encouru la colère et la jalousie des dieux suite à un acte d'orgueil, une transgression des commandements divins ou une faute morale.[6]

Aussi, le Coran menace les infidèles qui se sont laissé séduire par le Démon, d'être métamorphosés sur le champs (S. 36: 67, l'unique occurrence coranique du verbe *masakha*). Ayant transgressé la Loi divine, certains juifs idolâtres ont effectivement été changés en porcs et en singes, afin de servir d'exemple aux autres (S. 2: 65; 5: 60; 7: 166).[7] Plus explicites que le Coran, de nombreux *ḥadīth*s reconnaissent en certains animaux effrayants ou répugnants des êtres humains métamorphosés. Ainsi, le Prophète Muḥammad aurait soutenu que l'éléphant, la souris, le singe, le porc, l'araignée et l'anguille doivent leur existence à la

[5] Al-Jurjānī, *Kitāb at-Taʿrīfāt*, éd. I. al-Abyārī (Beyrouth, 1985), p. 272: *al-maskh: taḥwīl ṣūra ilā mā huwa aqbaḥ minhā.*

[6] Les *Métamorphoses* d'Ovide, l'*Ane d'Or* d'Apulée, l'épouse de Loth changée en statue de sel (*Genèse* 19: 26) en constituent des exemples célèbres.

[7] Voir U. Rubin, 'Apes, pigs, and the Islamic identity,' *IOS* 17 (1997), p. 90. Il est probable que ces versets laissent transparaître une tradition juive, puisque selon le Talmud (*Sanhedrin* 109a) une partie des ouvriers affectés à la construction de la Tour de Babel furent changés en singes; cf. Ch. Pellat, 'Maskh,' *Encyclopaedia of Islam* (New ed.), VI, pp. 725–727. I. Lichtenstadter, "'And become Ye accursed apes,'" *JSAI* 14 (1991), pp. 153–175, relève des cas analogues dans la mythologie de l'Egypte ancienne, mais il est hautement improbable que ceux-ci forment l'arrière-plan des versets coraniques sur la métamorphose.

transformation d'hommes pervers. Il aurait même distingué treize cas différents: l'éléphant serait la métamorphose d'un pédéraste, l'ours d'un homme efféminé, le porc d'un chrétien, le singe d'un juif, l'anguille d'un entremetteur, le lézard d'un voleur, la chauve-souris d'un escroc, le scorpion d'un menteur, le têtard d'un calomniateur, le lièvre d'une femme ayant négligé de se purifier après les menstrues, la constellation Suhayl (Canope) d'un collecteur d'impôts (!), Vénus d'une prostituée chrétienne et enfin l'araignée d'une femme vicieuse.[8]

Comme l'indique al-Jāḥiẓ, la tradition musulmane établit un lien direct entre la croyance en la métamorphose et certains tabous alimentaires, la consommation d'animaux 'métamorphosés' étant considérée comme répréhensible, si non illicite.[9] Ainsi, à en croire une tradition attribuée au Prophète, deux tribus juives auraient été changées l'une en lézards, l'autre en anguilles, raison pour laquelle leur chair devient *makrūh*.[10]

Bien que ces traditions reposent sur des bases coraniques, la notion de la métamorphose joue après tout un rôle plutôt marginal en islam sunnite, confinée principalement au folklore et à la religion populaire. Il en va tout autrement dans le chiisme, surtout dans ses courants 'radicaux' et 'extrémistes' où la métamorphose (*maskh*), liée à la métempsycose (*tanāsukh*), domine pour une large part l'eschatologie.

Dès le milieu du 8e siècle, Ibn Ḥarb, le chef de la secte chiite connue sous le nom de Kaysāniyya, professe une doctrine de la métempsycose qui s'articule autour d'une double voie, ascendante et descendante. L'hérésiographe de tendance muʿtazilite Nāshiʾ al-Akbar la résume comme suit:

> Cette secte prétend que la Résurrection (*qiyāma*) n'est autre que le passage de l'âme d'un corps à l'autre. Si les âmes se sont montré obéissantes, elles passent dans des corps purs et des formes harmonieuses et jouissent des délices éternels. Sans cesse, elles parcourent les différents rangs de la beauté, de la pureté et de la jouissance en fonction de leur mérite, jusqu'à devenir des anges et revêtir des corps purs de lumière (*abdān ṣāfiya nūriyya*). Si en

[8] R. Freitag, *Seelenwanderung in der islamischen Häresie*. Islamkundliche Untersuchungen; 110 (Berlin, 1985), p. 192. De nombreux exemples de métamorphoses sont rapportés par al-Jāḥiẓ, *Kitāb al-Ḥayawān*, éd. ʿA. Hārūn (Beyrouth, 1969), IV, pp. 68–74; cf. Ch. Pellat, *Arabische Geisteswelt. Ausgewählte und übersetzte Texte von al-Jāḥiẓ*. Bibliothek des Morgenlandes (Zürich [etc.], 1967), pp. 253–255.

[9] Al-Jāḥiẓ, *Kitāb al-Ḥayawān*, IV, pp. 74–77, au sujet de l'interdiction de manger du porc.

[10] M. Cook, 'Early Islamic dietary law,' *JSAI* 7 (1986), pp. 222–223, 252–253.

> revanche les âmes ont été rebelles, elles passent (*naqalat*) dans des corps immondes, des formes hideuses et des créatures répugnantes, comme les chiens, les singes, les porcs, les serpents et les scorpions. Aussi, ils disent que le Paradis et l'Enfer sont les corps. Ils interprètent allégoriquement le verset: 'La dernière demeure est vraiment la vie. S'ils savaient' (S. 29: 64), en prétendant que l'Au-delà où les gens arrivent après la mort n'est autre que le passage de l'âme d'un animal à l'autre, jusqu'à aboutir finalement soit dans des corps noirs et brûlés, soit dans des corps purs et lumineux (...). Ils prétendent que les fourmis, les scarabées (*khanāfis*) et les cloportes (*ji'lān*) qui se promènent dans leurs maisons, sont ceux que Dieu a fait périr à des époques antérieures, que Dieu a métamorphosés et dont il a fait transmigrer les âmes dans ces corps clairement visibles.[11]

Une doctrine très analogue est attribuée par l'hérésiographe imamite al-Nawbakhtī à la secte appelée Khurramdīniyya: '[Les âmes damnées] sont châtiées dans les corps répugnants et hideux des chiens, singes, porcs, serpents, scorpions, scarabées et cloportes. Elles prennent possession d'un corps après l'autre et y subissent leur châtiment, en toute éternité: voilà leur paradis et leur enfer. (...) Les corps ne sont que des réceptacles (*qawālib*) et des demeures (*masākin*), semblables à des vêtements que l'on porte: une fois usés, on les jette et on en met d'autres. Il en va de même des maisons que l'on construit: dès qu'on les abandonne pour en construire de nouvelles, elles tombent en ruines'.[12]

Le témoignage des hérésiographes est confirmé par les textes mêmes issus des mouvements *ghulāt*. Ainsi, l'*Umm al-Kitāb* décrit la déchéance progressive des âmes damnées à travers les rangs de la *scala naturae*: des différentes espèces d'oiseaux, de bêtes féroces et d'animaux domestiques, elles passent dans les insectes – fourmis, sauterelles, abeilles, mouches, moustiques, cafards – pour finalement sombrer dans la matière inanimée, comme les pierres et les métaux.[13] Dans un contexte analogue, le *Kitāb al-Haft wa 'l-aẓilla* lie la voie descendante des métamorphoses à une remarquable hiérarchisation du règne animal: les animaux dont la

[11] Nāshi' al-Akbar, *Kitāb Uṣūl al-niḥal*, éd. J. van Ess, *Frühe mu'tazilitische Häresiographie. Zwei Werke des Nāši' al-Akbar*. Beiruter Texte und Studien; 11 (Beyrouth, 1971), pp. 38–39; cf. Halm, *Islamische Gnosis*, pp. 71–72; Id., 'Das *Buch der Schatten*. Die Mufaḍḍal-Tradition der Ġulāt und die Ursprünge des Nuṣairiertums,' *Der Islam* 58 (1981), pp. 22–24; Freitag, *Seelenwanderung*, pp. 12, 15.

[12] Al-Nawbakhtī, *Firaq al-Shī'a*, éd. M.Ṣ. Baḥr al-'Ulūm (Najaf, 1936), pp. 36–37; cf. Halm, *Islamische Gnosis*, pp. 73–74.

[13] *Umm al-Kitāb*, éd. W. Ivanow, *Der Islam* 23 (1936), par. 403, p. 20. Sur l'*Umm al-Kitāb*, texte persan transmis par les Ismaéliens du Pamir mais qui remonte à un archétype irakien très ancien (8e s.), voir W. Ivanow, 'Notes sur l'Ummu'l-Kitab des Ismaëliens de l'Asie Centrale,' *REI* 6 (1932), pp. 419–481; Halm, *Islamische Gnosis*, pp. 113–117.

chair est licite et qui, par conséquent, peuvent être sacrifiés et abattus, sont suivis par les espèces impropres à la consommation, qui par conséquent ne peuvent être que tuées.[14] Cette dernière catégorie se divise, par ordre décroissant de noblesse, en animaux marins et terrestres, pour finir par les reptiles et insectes rempants: lézards, scorpions, serpents, scarabées et cafards.[15]

A en croire le *Kitāb al-Haft*, la punition ultime pour l'âme impie consiste à être métamorphosée en ver (*dūda*), car celui-ci se situe à l'échelon le plus bas du règne animal: 'le ver veille toujours, il ne dort jamais; il ne copule pas, il ne produit aucune descendance, il ne génère pas de petits et il ne pond pas d'œufs. Pour cette raison, le Coran dit: 'Puis nous l'avons renvoyé au plus bas des degrés' (S. 95: 5). Le ver n'a pas de progéniture, pas d'enfants: c'est la créature la plus imparfaite et la plus vile', située à l'extrémité de la 'chaîne de soixante-dix coudées' évoquée par le Coran (S. 69: 32).[16]

En d'autres termes, l'animal le plus méprisable auquel aboutissent les métamorphoses successives est produit par génération spontanée. La théorie aristotélicienne selon laquelle certains animaux surgissent de la terre par génération spontanée,[17] est doublée ici d'un jugement de valeur: ce mode de génération semble limité aux espèces les moins développées, les plus nuisibles et les plus répugnantes.

Les écrits druzes confirment une telle conception de la génération spontanée. Dans *al-Sīra al-mustaqīma*, la douxième des *Rasāʾil al-Ḥikma*,[18] Ḥamza b. ʿAlī rejette la thèse biblique et coranique selon laquelle Adam aurait été créé d'argile ou de terre, en invoquant comme argument: 'la poussière naturelle ne produit guère d'autres créatures que

---

[14] Nous retrouvons ici le rapport avec les tabous alimentaires; cf. Freitag, *Seelenwanderung*, p. 68; D. De Smet, 'Les interdictions alimentaires du calife fatimide al-Ḥākim: marques de folie ou annonce d'un règne messianique?,' dans: U. Vermeulen & D. De Smet (éds.), *Egypt and Syria in the Fatimid, Ayyubid and Mamluk eras*. Orientalia Lovaniensia analecta; 73 (Louvain, 1995), pp. 64–66.

[15] *Kitāb al-Haft wa 'l-aẓilla*, éd. M. Ghālib, *al-Haft al-sharīf* (Beyrouth, 1977), par. 24, pp. 64–68, par. 61, pp. 151–153; par. 65, pp. 180–184. Le *Kitāb al-Haft* est un texte *ghulāt* transmis par les Nuṣayrīs et les Ismaéliens de Syrie; voir Halm, *Islamische Gnosis*, pp. 240–242.

[16] *Kitāb al-Haft*, par. 61, p. 152.

[17] Voir les références dans Kruk, 'A frothy bubble,' pp. 268–270.

[18] Sur les écrits qui constituent le Canon druze, voir D. De Smet, 'Le Canon druze et ses arcanes,' dans D. De Smet, G. de Callataÿ [et al.] (éds.), *al-Kitāb. La sacralité du texte dans le monde de l'Islam*. Acta Orientalia Belgica. Subsidia; 3 (Bruxelles [etc.], 2004), pp. 403–424.

des vers, des serpents, des scorpions, des scarabées et autres animaux de ce genre. Un homme, par conséquent, ne pourrait provenir de la poussière'. Dès lors, ajoute l'auteur druze, il est absurde de prétendre, comme font les Ismaéliens,[19] qu'Adam n'a ni père, ni mère. Car la génération spontanée est un mode de production démoniaque, dont le prototype est la création d'Iblīs, comme l'indiquerait d'ailleurs son nom, dérivé de *ab laysa*: 'celui qui n'a pas de père'.[20] Dans sa polémique contre les Nuṣayrīs, Ḥamza b. ʿAlī réfute leur thèse que le châtiment des impies consisterait à être métamorphosé dans certains corps d'animaux, notamment ceux qui sont produits par génération spontanée: 'Il y a des cas encore plus remarquables, comme les scarabées noirs, les scorpions, les vers, les fourmis et autres animaux similaires: ils sont générés sans le sperme d'un mâle et sans la chaleur d'un utérus, mais doivent leur formation aux éléments et aux corps matériels'. Prétendre toutefois, comme le font les Nuṣayrīs, que les âmes des ennemis de ʿAlī et des adversaires retournent dans ces animaux répugnants, ou 'dans des chiens, des singes et des porcs, pour finalement entrer dans du fer et y être chauffées et frappées sous le marteau' ou 'soutenir que d'autres entrent dans des oiseaux et des hiboux et que d'autres encore reviennent dans le corps d'une femme ayant perdu ses enfants', s'avère contraire à la raison et à la justice divine: pour l'auteur druze, le transgresseur passe après sa mort dans un autre corps humain, afin qu'il soit pleinement conscient du châtiment qui lui est infligé.[21]

Au cours de son exposé de la doctrine d'Ibn Ḥarb,[22] l'hérésiographe Nāshiʾ al-Akbar mentionne, d'une manière à vrai dire assez confuse, que le processus des réincarnations successives se déroule en sept cycles, inaugurés par sept Adams. Au terme de chaque cycle, les âmes purifiées revêtent une forme angélique et montent au ciel qui correspond au cycle en question; les âmes damnées sont métamorphosées en des corps difformes et ténébreux et regagnent la terre propre à leur cycle. La voie

---

[19] Notamment Jaʿfar b. Manṣūr al-Yaman, *Sarāʾir wa asrār al-nuṭaqāʾ*, ed. M. Ghālib (Beyrouth, 1984), p. 27. Sur la génération spontanée d'Adam dans l'ismaélisme, voir *infra*.

[20] Ḥamza b. ʿAlī, *al-Sīra al-mustaqīma*, MS Paris, BN Fonds arabe 1408, f. 74b–76a; cf. D. De Smet, *Les Epîtres sacrées des Druzes. Rasāʾil al-ḥikma.* Vols. 1 et 2. Intr., éd. critique et trad. annotée des *traités attribués à Ḥamza b. ʿAlī et à Ismāʿīl at-Tamīmī.* Orientalia Lovaniensia analecta; 168 (Louvain, 2007) [à paraître].

[21] Ḥamza b. ʿAlī, *al-Risāla al-Dāmigha lil-fāsiq*, éd. R. Strothmann, 'Drusen-Antwort auf Nuṣairī-Angriff,' *Der Islam* 25 (1939), pp. 276–277; voir notre édition et traduction du traité dans De Smet, *Les Epîtres sacrées des Druzes* [à paraître].

[22] Voir *supra*.

ascendante parcourt ainsi sept cieux différents, la voie descendante sept terres.[23] Plus explicite est le *Kitāb al-Haft*, lorsqu'il soutient que les infidèles sont châtiés dans sept corps d'une impureté croissante, en lesquels ils transmigrent en fonction de leurs forfaits, au gré des sept cycles dont se compose l'histoire du monde.[24] Il y a ainsi sept formes ou degrés de *masūkhiyya*, conformément aux sept portes de l'Enfer évoquées dans le Coran (S. 15: 44).[25] Or, l'*Umm al-Kitāb* identifie ces sept grades de *masūkhiyya* aux sept climats de la terre, allant du climat le plus propice au climat le plus malsain, celui des Nègres (*Zanj*) qui abrite 'les corps noirs, compacts, grossiers et obscurs' des rebelles incorrigibles.[26]

Ainsi, les *ghulāt* développent leur conception de la métempsycose, qui aboutit à la métamorphose en des corps produits par génération spontanée, à base des deux notions qui, selon l'analyse pénétrante de Remke Kruk, forment précisément le cadre philosophique de la génération spontanée: la *scala naturae* et l'influence du climat sur la génération, conformément à la théorie des sept climats (*iqlīm*) du monde, empruntée par les géographes arabes à Ptolémée.[27] Nous retrouverons tous ces éléments, qui demeurent assez diffus vu le caractère lacunaire de nos sources sur les mouvements *ghulāt*, sous une forme plus systématique dans les écrits attribués aux Ikhwān al-Ṣafā', dont dépendent directement les Ismaéliens ṭayyibites.

Dans leur célèbre 'Encyclopédie', les 'Frères de la Pureté' (*Ikhwān al-Ṣafā'*) divisent le monde habité en sept climats, qui se succèdent selon un ordre hiérarchique. Les conditions climatiques et météorologiques propres à chaque climat déterminent la qualité morale et la couleur de la peau des hommes peuplant cette partie du monde, ainsi que la nature des animaux qui y vivent. Le premier climat, le degré le plus bas de la hiérarchie, habité par les Nègres (Zanj), les Ethiopiens, les Nubiens et les Berbères, est régi par Saturne. Planète maléfique par excellence, associée au noir, elle préside à la génération des animaux nuisibles: dragons,

---

[23] Nāshi' al-Akbar, *Uṣūl*, p. 39; cf. Halm, *Islamische Gnosis*, pp. 72–73.

[24] *Kitāb al-Haft*, par. 11, p. 38, par. 15, pp. 44–45; cf. Halm, *Islamische Gnosis*, pp. 261, 263–264.

[25] Freitag, *Seelenwanderung*, p. 66.

[26] *Umm al-Kitāb*, par. 175–177, pp. 70–71, par. 205, p. 64; cf. Halm, *Islamische Gnosis*, pp. 173, 182; D. De Smet, 'Les climats du monde et l'inégalité des races humaines. Une approche ismaélienne,' *Acta Orientalia Belgica* 16 (2002), pp. 77–78. Les traditions liées à la notion des sept climats et des sept lieux de châtiment que nous évoquerons dans la suite de notre article, reflètent le profond mépris de la civilisation arabo-musulmane pour les Noirs; voir à ce sujet De Smet, 'Les climats,' pp. 69–80.

[27] Kruk, 'A frothy bubble,' pp. 270–276; cf. De Smet, 'Climats,' pp. 73–74.

serpents, scorpions, vipères, renards, chats et souris, oiseaux nocturnes et autres engeances sournoises.[28] En remontant l'échelle des climats, la peau des hommes devient plus claire et leurs mœurs plus nobles, tandis qu'apparaissent des animaux de plus en plus parfaits et mieux organisés.[29] Les animaux répugnants du premier climat, générés sous l'influence de Saturne, comprennent ceux-là mêmes dont les Ikhwān disent ailleurs qu'ils sont produits par génération spontanée à partir d'une matière en putréfaction: il s'agit des serpents, dragons, crocodiles, sauterelles, vers et autres insectes rampants, comme les punaises, les scarabées et les cloportes.[30]

La *Risāla al-Jāmiʿa*[31] offre une élaboration significative sur le chapitre des *Rasāʾil* consacré aux sept climats. Au dire de son auteur anonyme, la terre entière, les villes, villages et îles au milieu des océans, ne sont que des prisons (*ḥubūs*), des caveaux souterrains (*maṭāmīr*), des cellules (*sujūn*) et des antres (*maḍāʾiq*) destinés à enfermer les âmes. De même, les corps des minéraux, des plantes et des animaux sont comme des liens (*quyūd*), des carcans (*aghlāl*) ou des ceps (*kubūl*), qui maintiennent l'âme liée à la matière: ce sont des *barzakhs*,[32] des lieux de châtiment, dans lesquels l'âme expie la faute qui a provoqué sa chute hors du monde intelligible.[33]

---

28 L'association de la maléfique Saturne avec la couleur noire et avec les animaux nuisibles et démoniaques, est un thème fort répandu dans l'astrologie antique; cf. A. Bouché-Leclercq, *L'astrologie grecque* (Paris, 1899), pp. 313–317.

29 Ikhwān al-Ṣafāʾ, *Rasāʾil* (Beyrouth, 1957), I, pp. 165, 170–179; cf. S.H. Nasr, *An introduction to Islamic cosmological doctrines* (Londres, 1978), pp. 87–89; A. Miquel, *La géographie humaine du monde musulman jusqu'au milieu du 11e siècle*, II, *Géographie arabe et représentation du monde: la terre et l'étranger*. Civilisations et Sociétés; 37 (Paris [etc.], 1975), pp. 60–70; De Smet, 'Climats,' p. 74.

30 Ikhwān, *Rasāʾil*, II, pp. 274–275; cf. Kruk, 'A frothy bubble,' p. 266.

31 Ce texte se présente à la fois comme une sorte de résumé des *Rasāʾil Ikhwān al-Ṣafāʾ* et comme un exposé de leur doctrine ésotérique; cf. I. Netton, *Muslim neoplatonists. An introduction to the thought of the Brethren of Purity*. Islamic Surveys; 19 (Edinburgh, 1991), p. 2. La *Risāla al-Jāmiʿa* me semble toutefois postérieure aux *Rasāʾil* et véhiculer une doctrine plus ouvertement ismaélienne.

32 Terme d'origine persane, *barzakh* apparaît trois fois dans le Coran (S. 23: 102; 25: 55 et 55: 20) dans le sens de 'barrière'. Compris par la tradition comme la barrière séparant le Paradis de l'Enfer ou ce bas monde de l'Au-delà, le *barzakh* est venu à désigner en eschatologie musulmane une sorte de purgatoire, un lieu provisoire de châtiment où l'impie réside entre le jour de sa mort et la Résurrection; cf. B. Carra de Vaux, 'Barzakh,' dans *Encyclopaedia of Islam* (New ed.), I (Leiden, 1960), pp. 1071–1072.

33 *Al-Risāla al-Jāmiʿa*, éd. M. Ghālib (Beyrouth, 1984), p. 102; cf. De Smet, 'Climats,' p. 75.

Outre qu'il exprime l'idée platonicienne du corps prison de l'âme, ce texte doit être compris à la lumière de la théorie des sept climats, dont il se veut l'exégèse: si toutes les formes corporelles sont des châtiments pour les âmes, il y a néanmoins une différence considérable entre les formes végétales, animales et humaines les plus grossières, les plus méprisables, et les corps les plus subtils, les plus proches de la perfection. En d'autres termes, les créatures vivant dans les climats les plus privilégiés de la terre représentent pour les âmes qui les incarnent un châtiment moins sévère qu'un séjour dans un corps maudit en un climat défavorable. Dès lors, bien que le texte ne le dise pas explicitement, il s'avère évident de mettre cette doctrine en rapport avec celle de la métempsycose, telle qu'elle est développée ailleurs dans la *Risāla al-Jāmiʿa*: les âmes transmigrent suivant une double voie ascendante et descendante, le long des degrés qui constituent la *scala naturae*. Les âmes insuffisamment purifiées pour échapper au cycle des réincarnations, restent liées à un corps matériel qui fait fonction de *barzakh*: lieu de châtiment et barrière qui les empêche de se libérer de leurs attaches corporelles.[34]

Les Ismaéliens post-fatimides de tendance ṭayyibite invoquent précisément ce passage de la *Risāla al-Jāmiʿa* pour soutenir leur conception du *maskh* ou *masūkhiyya*, à savoir la réincarnation des âmes impies, en guise de punition, dans des corps d'une impureté croissante.[35] Selon le

---

[34] Y. Marquet, *La philosophie des Iḫwān al-Ṣafāʾ*. Nouvelle édition augmentée. Textes et travaux de Chrysopœia; 5 (Paris [etc.], 1999), pp. 386, 390–391. Marquet remarque avec raison (p. 384) que la doctrine de la métempsycose est exprimée d'une manière beaucoup plus explicite dans la *Risāla al-Jāmiʿa* qu'elle ne l'est dans les *Rasāʾil*.

[35] Les Ismaéliens antérieurs, de tradition fatimide, rejettent le *tanāsukh* comme une doctrine hérétique propre aux *ghulāt* (ainsi, p. ex., Ḥamīd al-Dīn al-Kirmānī, *Kitāb Rāḥat al-ʿaql*, éd. M. Ghālib [Beyrouth, 1983], pp. 510–511). Néanmoins, Abū Yaʿqūb al-Sijistānī, tout en polémisant contre les *Ahl al-tanāsukh*, semble avoir professé une forme de métempsycose, tout comme al-Kirmānī d'ailleurs, dont l'eschatologie suppose la transmigration des âmes. L'ambiguïté est sans doute due au fait que ces auteurs prennent *tanāsukh* dans le sens de *maskh*, la réincarnation dans des corps non humains, animaux, végétaux ou minéraux. Sur cette question controversée, voir D. De Smet, *La quiétude de l'intellect. Néoplatonisme et gnose ismaélienne dans l'œuvre de Ḥamīd ad-Dīn al-Kirmānī (Xe/XIe s.)*. Orientalia Lovaniensia analecta; 67 (Louvain, 1995), p. 371; W. Madelung, 'Abū Yaʿqūb al-Sijistānī and metempsychosis,' dans D. Amin [et al.] (éds.), *Iranica varia. Papers in honor of Professor Ehsan Yarshater*. Acta Iranica, Série 3; 16 (Leiden, 1990), pp. 131–143; P. Walker, 'The doctrine of metempsychosis in Islam,' dans W.B. Hallaq & D.P. Little (éds.), *Islamic studies presented to Charles J. Adams* (Leiden, 1991), pp. 232–238. L'ismaélisme post-fatimide de tradition ṭayyibite, dont la doctrine a intégré non seulement la *Risāla al-Jāmiʿa* et les *Rasāʾil Ikhwān al-Ṣafāʾ*, mais tout un fonds de notions ultra-chiites émanant des mouvements *ghulāt*, professe ouvertement le *maskh*; cf. Freitag, *Seelenwanderung*, pp. 160–182. Sur la doctrine ṭayyibite, qui fut élaborée au Yémen à partir du 12e siècle, voir F. Daftary, *The Ismāʿīlīs. Their history and doctrines* (Cambridge, 1990), pp. 256–323.

Kitāb Kanz al-walad d'al-Ḥāmidī, la *Risāla al-Jāmiʿa*[36] nous apprend que la déchéance de l'âme damnée se fait à travers sept *barzakh*s, sept lieux de châtiment d'une impureté croissante. L'âme impie quitte 'la voie juste, humaine, sublime et arabe' (*al-ṣirāṭ al-sawī al-basharī al-sanī al-ʿarabī*), qui se situe au septième climat, pour s'incarner dans des enveloppes humaines de plus en plus viles. Descendant la hiérarchie des climats, elle échoit au premier climat dans le corps des Nègres, le *barzakh* le plus bas, à mi-chemin entre l'homme et la bête. Puis les *barzakh*s se succèdent dans le règne animal: au fil de l'échelle des climats, l'âme passe en des animaux de plus en plus misérables – chameaux, chèvres, bœufs, lions, loups, singes, porcs – pour aboutir dans les créatures répugnantes du premier climat, générées à partir d'excréments et de matières dégoûtantes: serpents, scorpions, cloportes; enfin, elle s'engouffre dans les excréments et la matière fétide. Toutes ces métamorphoses abjectes sont autant de *barzakh*s: l'âme y subit 'la punition inférieure' (*al-ʿadhāb al-adnā*, allusion à S. 32: 21) qui durera jusqu'au Jour de la Résurrection. Alors, le Résurrecteur jettera les âmes rebelles dans l'enfer de Sijjīn, lieu du châtiment suprême.[37]

A la suite d'al-Ḥāmidī, cette doctrine du *maskh*, dans laquelle nous retrouvons de nombreux éléments que nous avons relevés dans les écrits des *ghulāt*, sera reprise et élaborée par la plupart des auteurs ṭayyibites. A chaque fois, la déchéance de l'âme damnée aboutit aux créatures les plus humbles de la *scala naturae*, habitant le premier climat et dont les corps noircis sont générés sous l'influence maléfique de Saturne: les Nègres pour le règne humain, les scarabées, scorpions, cloportes et punaises pour le règne animal, ces derniers étant de surcroît produits par génération spontanée à partir d'une matière en putréfaction.[38]

---

[36] Outre la *Risāla al-Jāmiʿa*, al-Ḥāmidī se réfère à 'l'auteur du Traité sur les Animaux' (*ṣāḥib Risālat al-Ḥayawān*): cet ouvrage ismaélien anonyme, dans lequel sont cités, selon le témoignage d'al-Ḥāmidī, le Qāḍī al-Nuʿmān et al-Kirmānī, n'a pu être identifié; cf. Ibrāhīm b. al-Ḥusayn al-Ḥāmidī, *Kitāb Kanz al-walad*, éd. M. Ghālib. Bibliotheca Islamica; 24 (Wiesbaden, 1971), p. 300; Madelung, 'Abū Yaʿqūb,' p. 143 n. 32.

[37] Al-Ḥāmidī, *Kanz al-walad*, pp. 300–301, 309–310.

[38] Al-Ḥusayn b. ʿAlī b. Muḥammad Ibn al-Walīd, *Risālat al-Mabdaʾ wa 'l-Maʿād*, éd. et trad. H. Corbin, *Trilogie ismaélienne*. Bibliothèque Iranienne; 9 (Téhéran [etc.], 1961), par. 22, p. 109, par. 71–72, pp. 127–128; Idrīs ʿImād al-Dīn, *Zahr al-maʿānī*, éd. M. Ghālib (Beyrouth, 1991), pp. 327–328; R. Strothmann, *Gnosis-Texte der Ismailiten. Arabische Handschrift Ambrosiana H 75*. Abh. d. Akad. d. Wiss. in Göttingen. Philol.-hist. Kl. 3. Folge; 28 (Göttingen, 1943), pp. 11–13, 16–17, 53–54, 63–64, 126.

'Nasūkhiyya': *purification alchimique des corps et génération spontanée des corps camphrés*

Déjà Ibn Ḥarb, selon la présentation de sa doctrine par Nāshiʾ al-Akbar, avait doublé la voie descendante des métamorphoses d'une voie ascendante, parcourue par les âmes pieuses: au cours de leurs transmigrations – le terme employé en ce contexte est *naskh* ou *nasūkhiyya* – elles passent dans des corps d'une pureté et d'une subtilité croissantes jusqu'à devenir des anges et revêtir des corps lumineux (*abdān nūriyya*).[39] Cette vision forme le noyau de l'eschatologie gnostique du *Kitāb al-Haft*, ainsi que du Nuṣayrisme: sublimée par des purifications successives, l'âme monte à travers les sept cieux en revêtant des corps de plus en plus lumineux. A la fin de son périple, elle devient une forme angélique, un 'temple de lumière' (*haykal nūrānī*).[40]

Une fois de plus, ce sont les auteurs ismaéliens de tendance ṭayyibite qui offrent l'élaboration la plus complète de cette doctrine, en combinant les notions empruntées à la tradition des *ghulāt* avec la théorie de la *scala naturae* développée dans les *Rasāʾil Ikhwān al-Ṣafāʾ* et la *Risāla al-Jāmiʿa*.[41]

En effet, les Ṭayyibites ont introduit une remarquable dynamique et une dimension historique dans le système statique des Ikhwān al-Ṣafāʾ.[42] Ainsi, ils admettent que les espèces transmuent l'une dans l'autre, au cours des innombrables cycles qui rythment l'histoire de l'univers. Partant de la conviction que chaque être – minéral, végétal ou animal – est doté d'une âme qui aspire, chacune à son niveau, à se libérer du corps dans lequel elle est tenu prisonnière et à regagner le rang qu'elle occupait avant sa chute, les auteurs ṭayyibites ont élaboré une impressionnante sotériologie de la nature.

Par une complexe alchimie cosmique, l'influx (*mādda*) émanant du monde intelligible par l'intermédiaire des corps célestes, purifie graduel-

---

[39] Voir *supra*.

[40] Halm, *Islamische Gnosis*, p. 246; Strothmann, 'Seelenwanderung bei den Nuṣairī,' *Oriens* 12 (1959), pp. 92–93.

[41] Sur le thème de la *scala naturae* chez les Ikhwān al-Ṣafāʾ, voir D. De Smet, 'Al-Wāqwāq: un chaînon mythique dans le grand cercle de la nature,' *Acta Orientalia Belgica* 14 (2001), pp. 213–214; Id., 'The sacredness of nature in Shiʿi Ismaʿili Islam,' dans: A. Vanderjagt & K. van Berkel (éds.), *The book of nature in antiquity and the middle ages*. Groningen studies in cultural change; 16 (Louvain, 2005), pp. 87–89.

[42] Contrairement à une idée lancée au 19e siècle par Friedrich Dieterici et encore assez répandue de nos jours, qui reconnaît dans la *scala naturae* telle qu'elle est décrite

lement les âmes des minéraux les plus bas (qui sans doute servent de lieu de châtiment à des âmes ayant encouru un degré suprême de damnation), afin qu'elles atteignent, en une série de réincarnations successives, les minéraux les plus nobles – l'or et le rubis – pour passer insensiblement au règne végétal, puis animal. Au terme de cette longue chaîne de transmigrations, l'âme qui en cours de route est restée insensible aux attraits pervers du monde corporel, se purifie entièrement et devient un 'temple de lumière', revêtue d'un 'corps camphré' (*jism kāfūrī*).[43]

Dans la version ismaélienne, la *scala naturae* relie ainsi les êtres les plus vils, les substances noires et puantes comme le naphte et le goudron, au corps camphré de l'élu.[44] A mi-chemin du parcours, au degré le plus bas du règne animal, rampent les scarabées, scorpions, cloportes et punaises, surgis par génération spontanée de la matière végétale en décomposition. Or, au sommet de la hiérarchie, les 'corps camphrés' semblent eux aussi échapper à la génération sexuelle.

En effet, si les hommes sont fortement hiérarchisés par leurs traits physiques et leurs caractères moraux, déterminés par le climat qu'ils habitent, la 'crème' (*zubda*) du genre humain, l'élite en qui l'espèce humaine atteint sa finalité, possèdent un *jism kāfūrī*, un corps pur et blanc comme le camphre. A leur image, chaque croyant aspire à purifier son âme et à sublimer son corps en 'la forme camphrée la plus éminente' (*al-shabah al-afḍal al-kāfūrī*), afin de se libérer de ses attaches corporelles et d'échapper aux cycles des réincarnations.[45] Ces hommes d'élite aux corps camphrés ne sont autres que les Prophètes et les Imāms.

Déjà dans le *Kitāb al-Haft*, le sixième Imām Jaʿfar al-Ṣādiq enseigne à son disciple al-Mufaḍḍal que les Prophètes et les Imāms n'ont ni père, ni mère comme les simples mortels; qu'ils ne naissent point ni ne meurent,

par les Ikhwān un antécédent de la théorie darwinienne de l'évolution, leur système ignore toute forme d'évolution diachronique, mais repose entièrement sur une classification hiérarchique des êtres, marquée par une harmonie et une continuité totales entre les différents règnes; voir sur ce débat, De Smet 'Al-Wāqwāq,' p. 213; Id. 'Sacredness of nature,' pp. 85–87.

[43] Al-Ḥāmidī, *Kanz al-walad*, pp. 134–148; Idrīs 'Imād al-Dīn, *Zahr*, pp. 83–96, Strothmann, *Gnosis-Texte*, pp. 97–98, 105–107.

[44] Sur l'arrière-plan alchimique de cette théorie, voir D. De Smet, 'L'élaboration de l'élixir selon Ps.-Siğistānī. Alchimie et cosmogonie dans l'ismaélisme ṭayyibite,' dans: A. Fodor (éd.), *Proceedings of the 20th congress of the Union Européenne des Arabisants et Islamisants. Budapest, 10–17 September 2000*, II, *Islam, Popular Culture in Islam, Islamic Art and Architecture*. The Arabist. Budapest studies in Arabic; 26–27 (Budapest, 2003), pp. 30–32.

[45] ʿAlī b. Muḥammad Ibn al-Walīd, *al-Risāla al-Mufīda fī sharḥ mulghaz al-qaṣīda*, éd. H. Feki, dans: *Ḥawliyyāt al-Jāmiʿa al-Tūnisiyya* 17 (1979), pp. 172–173.

mais prennent possession d'un corps lumineux, exempt de toutes les souillures inhérentes aux corps matériels; au moment de la 'mort', ils quittent ce corps pour en revêtir un autre.[46]

Les auteurs ṭayyibites décrivent en détail la formation du corps camphré des Prophètes et Imāms. Trois jours après la mort apparente d'un des leurs, s'échappe de sa dépouille 'l'âme d'effluve' (*al-nafs al-rīḥiyya*), une âme inférieure liée au corps et distincte de l'âme 'supérieure', qui regagne directement le Temple de Lumière, le Plérôme des Intelligences ou archétypes, le Monde des Idées. Aspirée par la lune, cette âme d'effluve monte, par l'intermédiaire de Mercure et de Vénus, vers le Soleil et Jupiter, qui la purifient et la 'blanchissent'. Lorsqu'un nouvel Imām doit naître, des parcelles de cette âme d'effluve blanchie descendent, par l'entremise du Soleil, de Vénus et de Mercure, vers la Lune, qui les fait rayonner sur une eau absolument pure et sur des fruits d'une douceur incomparable. Ces fruits et cette eau servent de nourriture et de boisson aux parents de l'Imām, produisant ainsi en eux un germe spécial, qui engendre dans le sein de la mère – elle-même purifiée des souillures de la menstruation – un corps camphré (*jism kāfūrī*), qui est le corps de l'Imām, son *nāsūt*, sa figure humaine. En un délai variant entre quarante jours et quatre ans après la naissance apparente de ce corps camphré, le 'temple de lumière' qui est l'imamat, en prend possession: au *nāsūt* se joint alors le *lāhūt*, la 'divinité' de l'Imām.[47]

Certes, il convient de ne pas confondre la théorie de la métempsycose, qui forme l'objet du présent article, et la notion ultra-chiite, 'docétiste', du *ḥulūl*,[48] selon laquelle l'esprit divin du Prophète ou de l'Imām est 'infusé' dans un corps qui n'est qu'une 'enveloppe' (*ghilāf*), une 'chemise' (*qamīṣ*) ou un 'simulacre' (*shabah*). Il ne s'agit pas non plus d'une forme de génération spontanée au sens courant du terme, puisque le

[46] *Kitāb al-Haft*, par. 37–38, pp. 97–98.

[47] Al-Ḥāmidī, *Kanz al-walad*, pp. 191–194; Ibn al-Walīd, *Mabdaʾ*, par. 47–65, pp. 118–125 (éd.), 177–185 (trad.); Idrīs ʿImād al-Dīn, *Zahr al-maʿānī*, pp. 274–280; Strothmann, *Gnosis-Texte*, pp. 7–9, 55–56, 75; cf. H. Feki, *Les idées religieuses et philosophiques de l'ismaélisme fatimide (organisation & doctrine)*. Faculté des lettres et sciences humaines de Tunis, 6e Série: Philosophie-littérature; 13 (Tunis, 1978), pp. 203–205; De Smet, 'Eléments chrétiens,' pp. 48–49.

[48] Comme le souligne avec raison Walker, 'The doctrine of metempsychosis,' p. 220. Sur la notion de *ḥulūl*, voir A. Straface, '*Ḥulūl* and *tajassud*: Islamic accounts of the concept of incarnation,' dans: U. Vermeulen & J. M. F. Van Reeth (éds.), *Law, Christianity and modernism in Islamic society. Proceedings of the eighteenth congress of the Union Européenne des Arabisants et Islamisants*. Orientalia Lovaniensia analecta; 86 (Louvain, 1998), pp. 111–132.

corps camphré de l'Imām se développe dans le ventre de sa mère et naît d'une façon 'naturelle'.

Toutefois, le sein maternel peut être remplacé par un autre réceptacle. Cela eut lieu lors de la génération du premier homme Adam, qui inaugure la lignée des Prophètes et des Imāms, ainsi que de ses vingt-huit compagnons. Les auteurs ṭayyibites décrivent leur genèse, par génération spontanée, dans les sillons et alvéoles terrestres de l'île de Sarandīb (Ceylan), située à un endroit de la terre où le climat était parfaitement tempéré:

> Le Démiurge mit en mouvement la sphère; s'élevèrent les vapeurs issues de la partie la plus pure des minéraux, des végétaux et des animaux. Elles devinrent des nuages. Ces nuages se déversèrent sur la surface de la Terre en pluies dont l'eau était pure et bien tempérée. La terre se rida en sillons sans grande profondeur, et cette eau très pure se déposa dans leur cavité. Ensuite, cette eau s'éleva à son tour en formant une vapeur d'une qualité plus noble, plus raffinée et plus pure que la précédente. Puis elle se déversa en une pluie abondante qui joua le rôle de *semen virile*. Elle tomba dans ces alvéoles et dans ces sillons qui jouèrent le rôle de matrices; elle s'y mélangea avec l'eau qui s'y trouvait, laquelle fut l'analogue de l'humeur féminine. Il n'y eut plus alors qu'une seule chose. Ensuite, la chaleur de la Terre échauffa cette chose qui s'éleva pour fuir cette chaleur. Mais à l'extérieur des alvéoles terrestres elle rencontra la fraîcheur du *nasīm*, et elle redescendit pour fuir cette fraîcheur. Alors elle ne cessa tantôt de descendre, tantôt de monter, se contractant, se sublimant, se solidifiant, prenant forme en passant par tous les degrés que comportait sa nature constitutive, pendant neuf mois, soumise à la sage direction du Démiurge et à l'influx des énergies et des sphères qui agissaient sur elle, jusqu'à ce que le temps fût accompli pour elle. A ce moment elle ouvrit les yeux et ses sens; elle aspira le *nasīm*, et la vie organique, la vie sensible, se conjoignit à elle par l'intermédiaire de cette brise (traduction H. Corbin).[49]

Cette présentation ṭayyibite de la formation, par génération spontanée, du corps parfait d'Adam et de ses vingt-huit 'nobles' (*nujabā'*)[50] résulte une fois de plus de la combinaison ingénieuse de thèmes issus de la tradition des *ghulāt*[51] et des *Rasā'il Ikhwān al-Ṣafā'*. La scène primordiale

---

[49] Ibn al-Walīd, *Mabda'*, par. 28–31, pp. 111–112 (éd.), pp. 164–165 (trad.).

[50] L'évènement est également décrit par al-Ḥāmidī, *Kanz al-walad*, p. 154, et dans Strothmann, *Gnosis-Texte*, pp. 9–10; voir en outre Ibn al-Walīd, *Mabda'*, par. 38–39, p. 115 (éd.), p. 168 (trad.).

[51] Dans le *Kitāb al-Haft*, par. 41, pp. 103–114, les vingt-huit *nujabā'* partagent avec les Prophètes et les Imāms un même corps parfait, exempt des souillures matérielles et des contraintes biologiques, comme la faim et la soif. Les vingt-huit *nujabā'* apparaissent également dans l'*Umm al-Kitāb*, selon lequel ils créèrent vingt-huit îles, dont Sarandīb (*Umm al-Kitāb*, par. 176–177, p. 70).

décrite par Ibn al-Walīd rappelle immédiatement l'image évoquée par les Ikhwān de cette île idéalement située – en l'occurence Sarandīb – où Adam et Eve sont sortis de la terre par génération spontanée.[52] A son tour, cette histoire en rappelle une autre:[53] celle de l'île paradisiaque qui a vu naître par génération spontanée Ḥayy b. Yaqẓān, le héros du 'roman philosophique' d'Ibn Ṭufayl, si cher à Remke Kruk qu'elle nous en a donné une merveilleuse traduction néerlandaise.[54]

---

[52] Ikhwān, *Rasāʾil*, II, pp. 181–182.

[53] Kruk, 'A frothy bubble,' pp. 274–275; Ead., 'Neoplatonism and after. From Ibn Ṭufayl to Ibn an-Nafīs,' dans: A. Vanderjagt & D. Pätzold (éds.), *The Neoplatonic tradition. Jewish, Christian and Islamic themes*. Dialectica minora; 3 (Cologne, 1991), p. 77.

[54] Abu Bakr Muhammad ibn Tufayl, *Wat geen oog heeft gezien. De geschiedenis van Hayy ibn Yaqzan*. De oosterse bibliotheek; 23 (Amsterdam, 1985). Le passage en question s'y trouve, pp. 43–45.

THE RECEPTION OF AVICENNA'S *PHYSICS* IN THE LATIN MIDDLE AGES*

Jules Janssens

Of Avicenna's major work, *al-Shifāʾ*, *The Book of Healing*, the book dealing with the book *Physics* proper, called in Arabic, *al-Samāʿ al-ṭabīʿī*, has been translated into Latin, although not completely.[1] It became known by the title of the complete encyclopaedia, i.e., *Sufficientia*. There were two phases in the Latin translation. The first, which occurred most probably in the third quarter of the twelfth century, encompasses parts one and two, and the very beginning of part three (the 'prologus' [= ch. 1 in the Arabic editions] and approximately one third of ch. 1 [= ch. 2 of the Arabic editions]).[2] It stops abruptly in the middle of a sentence with the words 'per se notae'. In view of the similarities of vocabulary and translation technique with the Latin version of Avicenna's *De Anima*, one may place it in the Toledan milieu.[3] Gundissalinus, alone or in collaboration with someone else, may have been involved in this translation, but this cannot be affirmed with certainty for the moment. A systematic comparison of the style and vocabulary of the translations of both the *De Anima* and the *Physics* could clarify this issue, but clearly exceeds the limits of the present paper. Nor is there any obvious explanation for the sudden end of the translation. It could simply be the case that the translator, or his collaborator, died before completing the translation. But it might also be that James of Venice's translation of Aristotle's *Physics*, the so-called 'translatio vetus', became so influential (it is

* A French version of this paper has been presented at the SIHSPAI-conference, held at Florence, February 2006. I sincerely thank Peter Adamson for improving the English style.

[1] Besides the lithography of Tehran, 1886, there are two contemporary editions of the Arabic text: the one by S. Zayed (Cairo, 1983), and the other by J. Āl-Yāsīn (Beirut, 1996).

[2] The complete text of this first-phase partial Latin translation is available in a Renaissance-edition, printed Venice, 1508, repr. Louvain, 1961. The first and second parts are also available in a critical edition: S. Van Riet (ed.), *Avicenna Latinus. Liber primus naturalium. Tractatus primus. De causis et principiis naturalium* (Louvain-la-Neuve [etc.], 1992); S. Van Riet (†), J. Janssens & A. Allard (eds.), *Tractatus secundus. De motu et de consimilibus* (Bruxelles, 2006).

[3] Van Riet 1992, p. 53*.

present, together with many scholia, in the MS Avranches 221, dated approximately 1180 and in Hugues de Honau's *Liber de homoysion et homoeysion*, to be dated before 1180) that there was no longer any need for Avicenna's treatise.[4] This would, however, not explain the abrupt ending of the translation. Note moreover that this partial translation of the *Physics*, unlike that of the entire *de Anima*, seems not have had any influence before 1230, at least as far as I can see (I have moreover no knowledge of any author ever having mentioned such influence). It is striking that the vast majority of the 22 surviving manuscripts offering the translation date back to the second half of the thirteenth century when somewhat more systematic attention – although still in a limited way, as we shall see – was paid to Avicenna's text. One manuscript, i.e., Worcester, Chapter Library, Q. 81, can be dated with certainty somewhat earlier, i.e., between 1230 and 1240. Also the manuscript Paris, BN Lat. 16604, might belong to that period, but there is only strong evidence that it predates 1278.[5] The Parisian condemnations of 1210 and 1215 might have hindered the spread of the translation, and of its study. But there is no strong evidence, given the apparently complete absence of any use before these dates, and given also the fact that the copyist of the Dubrovnik manuscripts probably had access either to the autograph of the translator himself, or at least to a copy closely related to it.[6] In any case, Avicenna's treatise became more influential after 1250, although it would be used extensively for the first time only in Albert the Great's *Comments on the Physics*. This might explain why our partial translation was carried out between 1275 and 1280 in Toledo, at the request of the prelate Gonzalo Garcia Gudiel. Master John Gonsalvez of Burgos, in collaboration with someone named Salomon (given his name, probably a Jew), is explicitly designated as the author of this work, as

---

[4] F. Bossier and J. Brams, *Praefatio* to *Aristoteles Latinus. Physica. Translatio Vetus* (Leiden [etc.], 1990), p. xxi–xxvii.

[5] Van Riet 1992, p. 59*, affirms that both manuscripts belong to the first half of the thirteenth century, but H. Kisschlat, *Studien zur Verbreitung von Übersetzungen arabischer philosophischer Werke in Westeuropa 1150–1400* (Münster, 2000), p. 117, more prudently indicates 'second or third quarter of the thirteenth century' with respect to the Parisian MS BN 16604.

[6] The Dubrovnik-manuscript entails many more double translations than the other manuscripts. These double translations may result from a revision by the translator himself. See J. Janssens, 'L'Avicenne latin. Particularités d'une traduction,' in J. Janssens & D. De Smet (eds.), *Avicenna and his heritage. Acts of the international colloquium Leuven – Louvain-la-Neuve, September 8–September 11, 1999* (Leuven, 2002), pp. 113–129.

well as of four other Natural Books of the *Shifā'*.[7] However, he failed to complete the translation of the whole *Physics*. In fact, his translation actually stops with chapter 9 (chapter 10 of the Arabic editions) of part 3. Hence, he omits the last four chapters of part three (dealing with the infinity of motion and time, the essentially limited character of natural bodies, and the issue of directions) and the entire fourth part, which discusses properties related to the natural things, mainly motion (e.g., how it is one, and how multiple, etc.). This time, there is little doubt that the translator himself, or his team, decided to put an end to the translation. But the unique extant manuscript of this translation, Vat. Urb. Lat. 186, contains no justification for this omission. Had the new, revised translation by William of Moerbeke of Aristotle's *Physics* gained such influence that Avicenna's 'paraphrastic' commentary was considered to be no longer authoritative? But why then was the project ever started? A better hypothesis seems to be that the translator, being aware of the condemnations of 1270, and especially 1277, was reluctant to present the text of Avicenna's chapter 9 that defends the eternity of the world, even though it insists on God's essential priority over the world. Moreover, he may have judged that even with the partial translation the reader had access to all of Avicenna's essential physical insights: the status of natural science and the issue of fortune and chance (part one); motion, place and time (part two); atomism, and finiteness and infinity (part three). In any case, this translation seems not have had any serious influence, and it is significant that it has been preserved in only a single manuscript.

Although no explicit mention is made of Avicenna's *Physics*, it is likely that both Richard Rufus and Robert Grosseteste, in their commentaries on Aristotle's *Physics*, which both are to be dated between approximately 1230 and 1235, derived some ideas directly from it. In a discussion of the definition of motion in *In Physicam Aristotelis*, Rufus uses the idea of a double notion of act, linked with a double notion of potency, i.e., a 'potentia in formam' and a 'potentia in viam ad formam'.[8] Generally speaking, he seems in all this to follow Averroes' analysis. However, when he states 'sciendum tamen quod motus secundum quod est actus mobilis significatur par hoc "motus", secundum autem quod est actus motivi

---

[7] See Van Riet 1992, p. 54*; see also S. Van Riet, *Avicenna latinus. Liber tertius naturalium. De generatione et corruptione* (Louvain-la-Neuve [etc.], 1987), pp. 65*–68*. This part of the translation has been conserved in the single MS Vat. Urb. 186, f. 61a–83a.

[8] Richard Rufus of Cornwall, *In Physicam Aristotelis*, ed. R. Wood (Oxford [etc.], 2003), III, 1, especially pp. 141–145, §§ 8–15.

significatur per hoc nomen "motio" vel aliquid tale', one may wonder whether he was not inspired by Avicenna's description of motion as a double perfection, which was itself inspired by the Greek commentators Themistius and Philoponus, and, perhaps, Alexander of Aphrodisias.[9] Insofar as Rufus probably had no direct access to these authors or their Greek texts, one may suspect an influence from Avicenna, despite the inexact parallel. Admittedly Rufus, thanks to Averroes' commentary, was almost certainly familiar with the Greek commentators' basic conception of motion. But a possible Avicennian influence cannot be totally excluded. In this respect it is not without interest to note that Rufus' relational concept of place might ultimately stem from Avicenna.

But evidence that he used, and hence had read Avicenna's *Physics* is stronger in his exposé on the causes, i.e., in II, 4, p. 124, § 4. When examining the relation between efficient and final cause – the former being prior at the operational, the latter at the intentional level, he affirms: 'efficiens facit finem' (l. 273). In Avicenna's *Physics*, I, 9, p. 95, l. 3, we find: 'cum efficiens sit quod facit finem esse'. Hence, Rufus offers an almost literal quotation. Certainly, Avicenna's *Metaphysics*, VI, 5, directly inspires the rest of the paragraph. But there one does not find the present affirmation as such. Moreover, Avicenna, in his *Physics* (*ibid.*, l. 18–19), explicitly refers to his *Metaphysics* for a more detailed and profound account. Hence, Rufus' statement offers a strong indication that he directly used Avicenna's *Physics*.

Robert Grosseteste's *Notes on the Physics* are a posthumous compilation of his notes for a commentary. These notes can be dated somewhere between 1220 and 1240, but most if not all of the work would have been written during the 1220s, which means during his teaching activities at Oxford.[10] Of the three explicit references to Avicenna, two concern the *Canon* and one is derived from Averroes. However, when he deals with the motion of the outermost sphere, he stresses that its parts have a local motion in an improper sense.[11] Consequently, he concludes: 'solet autem motus localis parcium continuorum circulariter motarum vocari

[9] See A. Hasnawi, 'La definition du mouvement dans la *Physique* du *al-Shifāʾ*,' *Arabic sciences and philosophy* 11 (2001), pp. 224–226, esp. p. 225, n. 15 (indicating further literature).

[10] N. Lewis, 'Robert Grosseteste's *Notes on the Physics*,' in E.A. Mackie & J. Goering (eds.), *Editing Robert Grosseteste* (Toronto, 2004), pp. 103–134.

[11] Robert Grosseteste, *Commentarius in VIII. libros physicorum Aristotelis*, ed. R.C. Dales (Boulder, Colorado, 1963), liber IV, p. 83.

motus situalis quia illi est manifesta situs altero (...). Sed istud ubi non dicit locum sed situm certum et determinatum'. Grosseteste clearly prefers to qualify the motion of the outermost sphere as being according to 'situs' rather than to 'place'. As far as I can see, Avicenna is the only author to do so before him. In his *Physics*, II, 3, Avicenna insists that one has to admit the existence of motion in the category of 'situs', since otherwise one cannot explain the motion of the outermost sphere which is in no place. However, Grosseteste might have derived this affirmation from Averroes' Commentary on b. IV.[12] If this is the case, he clearly was not persuaded by Averroes' criticism against Avicenna's opinion. However, it is not clear whether Grosseteste, when writing the present notes, already had Averroes' translation at his disposal – though we know that he did for books 7 and 8.[13]

Around the middle of the thirteenth century, one finds traces of an explicit use of Avicenna's *Physics* in several works. The commentaries on the *Physics* of Adam of Bocfeld and Roger Bacon offer a good illustration.

In Adam's *Sententia super Physicam*, one finds the following assertion: '... vult Avicenna, quod natura principaliter non intendit speciem completam in qua salvetur esse divinum'.[14] Adam herewith offers a paraphrase of a passage of Avicenna's *Physics*.[15] A more literal quotation, covering lines 60–64 of the same passage, can be found in the anonymous *Notulae super libris Physicorum*, which clearly belong to the school of Adam and may be a little later than his commentary on the *Physics*: '... sicut dicit Avicenna, natura non intendit hoc individuum expressum signatum, quia, si intenderet, tunc destrueretur esse et ordo eius cum destrueretur hoc individuum et desineret esse. Item, natura non intendit formam communem et generalem per se et primo, quia, si intenderet, tunc perficeretur esse eius et ordo cum fieret huiusmodi forma, sicut si fieret corpus qualicumque modo vel animal qualicumque modo'.[16]

---

[12] See Averroes, *Opera omnia*, IV, *Commentarium (magnum) super libro* De physica auscultatione (Venice, 1562, repr. Frankfurt am Main, 1962), p. 144f.

[13] See Dales, p. xix.

[14] S. Donati, '*Physica* I, 1. L'interpretazione dei commentatori inglesi della *Translatio Vetus* e la loro recezione del commento di Averroè', *Medioevo* 21 (1995), p. 221, l. 127. I wonder whether one has to read 'individuum' instead of 'divinum', given the context as well as Avicenna's statement referred to? However, I admit that the reading 'divinum' is not necessarily mistaken.

[15] See Van Riet 1992, pp. 8, 53–59, 66.

[16] Donati, p. 230, l. 233–239.

However, references to Avicenna's *Physics* remain extremely rare in Adam's works, and in works related to his school.

A somewhat more systematic use, although still limited, is detectable in Roger Bacon's *Quaestiones alterae supra libros Physicorum*. One finds seven explicit references (a cursory reading did not reveal the existence of implicit ones), among which is that mentioned by Adam, but in different words: 'si natura intendit individuum signatum, destrueretur ordo; quare, cum non destruatur, necesse est quod prius intendit universale'.[17] This quotation is somewhat more literal than Adam's, but less than the one found in the *Notulae*. However, in all other cases Bacon's wording is paraphrastic, not to say extremely paraphrastic. They cover several parts of Avicenna's *Physics*. Three references concern Avicenna's *Physics* I, 1: besides the one already quoted, there are two more [p. 16, 5 and p. 17, 34] – both of them focusing on the issue of the priority of the universal or the particular in natural knowledge. The four other paraphrases refer to four other chapters of Avicenna's work: one [p. 132, l. 38] evokes the double character of nature as universal and particular and is based on Avicenna's *Physics*, I, 7; another [p. 119, l. 26] points to Themistius' doctrine on coincidence, *casus*, as indirectly alluded to by Avicenna in I, 13; still another [p. 298, l. 16] specifies the double character of rest in terms of natural or violent, deriving elements from Avicenna's *Physics*, II, 4; a fourth, and final, paraphrase [p. 278, l. 28] concentrates on time as accident and as measure, while it takes inspiration from the very same work, II, 13.[18] Note however that Bacon most of the time criticizes Avicenna's ideas, except when they are clearly in agreement with Aristotle's opinion, as is the case for the very first quotation. But Bacon seems to have read Avicenna's *Physics* very carefully. In his late work, *De multiplicatione specierum*, which is mainly of a biological nature, one finds an explicit reference to Avicenna's *Physics* I, 7 (pp. 68, 73–74) 'insuper ipsa mors singulorum secundum statum presentis vite necessaria est ex lege nature universalis, ut dicit Avicenna secundo (sic!) *Physicorum*'.[19]

---

[17] R. Bacon, *Questiones supra libros octo physicorum Aristotelis*, eds. F.M. Delorme & R. Steele (Oxford, 1935), pp. 14, 3–4.

[18] The numbers between brackets refer to the edition mentioned in the preceding note.

[19] See D.C. Lindberg, *Roger Bacon's philosophy of nature. A critical edition, with English translation, introduction, and notes, of* De multiplicatione specierum *and* De speculis comburentis (Oxford, 1983), p. 86, l. 157–159. The reference is not to part two, but to part one, chapter seven of Avicenna's *Physics*, see Van Riet 1992, p. 68, l. 73–74.

Among the scholastics, Albert the Great is undoubtedly the one who paid the most attention to Avicenna's *Physics*. In his early work *De IV Quoaequaevis*, dated 1246, he already shows a great familiarity with the contents of Avicenna's treatise. As shown by U. Jeck, Albert accepts with Avicenna that time is a continuum outside the soul.[20] But the most systematic use is found in his *Commentary on the Physics*, written most likely in 1251 or 1252.[21] One finds approximately thirty explicit references, and many more implicit ones. This certainly does not mean that Albert always, or even usually agrees with Avicenna's views. He sometimes accepts, sometimes rejects them. Despite a slight modification in the wording, Albert approvingly reproduces Avicenna's typology of theories of time, especially the models of comparison and aggregation. But on the other hand, he categorically denies that motion exists in the category of *situs*, as Avicenna claimed. Throughout one finds materials derived from almost all chapters of Avicenna's *Physics*, libri I and II. But even more important is the fact that Albert did not refer to only a few isolated phrases, but took into account Avicenna's innovative views in their totality. A full assessment of the significance of Avicenna's *Physics* for Albert's physical theory would indeed require a monograph, and hence falls outside the scope of the present investigation. Furthermore, it is obvious that Albert's commentary had a major impact on the later scholastic tradition. When looking at these authors, one must always carefully examine whether they based their information on Avicenna's text directly, or were dealing with Avicenna's views according to their wording and/or approval by Albert. When e.g., Ulrich of Strasbourg, in his *De summo Bono*, e.g., IV, 2, 23, affirms: 'liquet etiam ex omnibus praemissis, quod tempus species est quantitatis, et ex hoc arguit Avicenna, quod sit quid existens in rerum natura', whereby he explicitly refers to Avicenna, he offers no literal quotation of the latter's *Physics*.[22] So, one may wonder whether he derived this Avicennian view from Albert, although the present wording is as such not present in the latter's *Physics*.

---

[20] U.R. Jeck, *Aristoteles contra Augustinum. Zur Frage nach dem Verhältnis von Zeit und Seele bei den antiken Aristoteleskommentatoren, im arabischen Aristotelismus und im 13. Jahrhundert* (Amsterdam, 1994), pp. 219–232.

[21] The text is now available in a critical edition by P. Hossfeld (Aschendorf, 1988). Both the explicit and implicit references to Avicenna's *Physics* are given in the index of sources.

[22] I owe the quotation to Jeck, p. 290, n. 6. Ulrich vaguely refers to Avicenna's *Physics*, II, 11.

Besides Albert the Great, Thomas Aquinas also merits special mention. Already in his *In Boethium de Trinitate*, probably the direct product of his teaching activities in Paris in 1256, and hence of the young professor Thomas, one finds two explicit and two implicit references to Avicenna's *Physics*.[23] The two explicit references (the one stating that two natural bodies cannot be in the same place due to their very nature – TH. IV, 3, pp. 129, 176 = Av. Lat. II, 7 (near the end) and the other that the common principles can be said either according to predication or to causality – TH. V, 4, pp. 153, 113–9 = Av. Lat. I, 2, pp. 22, 81–23, 92) summarise Avicenna's wordings on these issues while respecting their essential outline. Moreover, Thomas fundamentally agrees with their contents. With respect to the two implicit ones (one affirming that the more common things are first known – TH I, 3, pp. 87, 142 = Av. Lat., 1, 11, pp. 12–14; another stressing that the science of astrology partly belongs to that of mathematics – TH V, 3, 144 = Av. Lat. I, 8, pp. 72, 40–1) it must be observed that Thomas remains faithful to Avicenna when evoking the first, but rather misuses the second when he concludes that mathematics does not abstract completely from matter and motion, since this is an idea that Avicenna clearly would have rejected. About ten years later, around 1268, Thomas wrote his extant commentary on Aristotle's *Physics*. Surprisingly, it contains only six explicit references to Avicenna's *Physics*.[24] Implicit references seem to be completely absent, or in any case exceedingly rare. But even when Thomas quotes Avicenna, it is not with approval. E.g., he objects to (I, 15, § 137–8) Avicenna's triple criticism of Aristotle's affirmation that matter naturally seeks and desires form (presenting a modified version of Av. Lat, *Physics*, I, 2, pp. 32, 69–34, 94) as being not really to the point, and simply rejects (IV, 7, § 475) Avicenna's acceptance of the existence of motion in the category of 'situs' (Av. Lat., *Physics*, II, 3). In none of the six cases where he offers an explicit reference to Avicenna does Thomas completely endorse the latter's point of view. One gets the impression that he does not consider Avicenna to be a great authority in the field of physics.

---

[23] In what follows, the references are to the critical edition.

[24] See C. Vansteenkiste, 'Avicenna-citaten bij S. Thomas,' *Tijdschrift voor filosofie* 15 (1953), pp. 477–478, nos. 184–189. The reader can also consult the English translation: St. Thomas Aquinas, *Commentary on Aristotle's* Physics, translated by R.J. Blackwell, R.J. Spath [et al.] (New Haven, 1963) (for the references, see the Index, p. 597 sub 'Avicenna').

Regarding the partial translation of part 3 of Avicenna's *Physics*, which as indicated earlier was made near the end of the thirteenth century and has survived in a single manuscript, I have looked thus far in vain for clear signs of influence. Certainly, Duns Scotus might have been influenced by it, when he elaborated one of his geometrical arguments in order to show that a quantum cannot be composed of discrete points.[25] Before Scotus, Peter Olivi, in his Commentary on the *Sentences* (to be dated in the 1270s), might already have used this partial translation of part three. In fact, he adheres to the theory that indivisibles do not exist at all, and he claims to derive this view from Avicenna's *Physics*.[26] The editor, B. Jansen, refers to II, 5 and II, 9, but there Avicenna gives at most hints for such an opinion. Therefore, I wonder whether III, 5 ('Quod spatium et motus <et tempus> a se invicem dependent et quod nihil horum habet primam partem') does not constitute the immediate source?[27] To adequately settle this delicate question, a very detailed examination of the passage, and of its possible sources in Avicenna, would be required, and this unfortunately once more exceeds the limits of the present investigation. However, if my impression is correct, it would signify that Olivi was in all probability the first author to have used the later part of the translation.

In sum, the scholastics, even those living in the second part of the thirteenth century, generally speaking paid little attention to Avicenna's original but 'unorthodox' views in the physical domain. They obviously preferred the Averroist-Aristotelian interpretation. Nevertheless some particular items, as e.g., his realist conception of time, or his idea that the elements keep their substantial form when being part of a mixture, received more particular attention. The former of these views was even adopted by so eminent a scholar as Albert the Great (and probably, through him, by his disciple Ulrich of Strasbourg), and also by the early fourteenth-century English scholar Walter Burley.[28] The latter view is

---

25 See R. Cross, *The Physics of Duns Scotus. The scientific context of a theological vision* (Oxford, 1998), pp. 121–123, and especially p. 123, n. 23. Scotus' argument occurs in his *Ordinatio*, 2.2.2.5, n. 320–326.

26 P. Olivi, *Quaestiones in secundum librum Sententiarum*, ed. B. Jansen (Quaracchi, 1922–1926), I, pp. 554 and 557. See also Cross, p. 133, where it is stressed that no thirteenth century author, with the exception of Olivi, accepted the theory that indivisibles do not exist at all (a theory called by Cross 'non-entitism').

27 See MS Vat. Urb., 186, f. 71b–72a (III, 5).

28 See C. Trifogli, *Oxford physics in the thirteenth century (ca. 1250–1270). Motion, infinity, place and time* (Leiden [etc.], 2000), pp. 231–237; see also A. Maier, *Metaphysische Hintergründe der spätscholastischen Naturphilosophie* (Rome, 1955), pp. 86–89.

mentioned as one among other theories by the early fourteenth-century scholars Meister Eckhart and Gregory of Rimini (the former taking no position; the latter clearly rejecting it).[29] But, all in all, the influence of Avicenna's *Physics* remained limited, especially compared to that of his *de Anima* or *Metaphysics*.

[29] See A. Maier, *An der Grenze von Scholastik und Naturwissenschaft* (Rome, 1952), pp. 93, 112.

A NEW LOOK AT THE BARBER'S ASTROLABE IN THE *ARABIAN NIGHTS*

Jan P. Hogendijk

Introduction

In one of the humorous stories of the *Arabian Nights*, a young man saw the daughter of the *qāḍī* of Baghdad, and immediately fell in love with her. Through the good services of an old woman, the young man was able to schedule a meeting with his beloved in the house of the *qāḍī*, during his absence for Friday noon prayers. Before going to the appointment, the young man called for a barber to shave him. The barber turned out to be extremely talkative, and the young man could not get rid of him. Due to the stupid behaviour of the barber, the young man and the girl were caught during their meeting, and the young man broke his leg during the escape.

The barber claimed to be, among other things, an expert in astrology, astronomy and mathematics, and he used his astrolabe in order to check whether the moment was suitable for shaving the young man. The passage on the barber and his astrolabe contains some astronomical data which were studied by Vernet in 1960 on the basis of the Calcutta and Būlāq (Cairo) editions of the *Arabian Nights*, which editions appeared in 1833 and 1835 respectively.[1] Because these editions refer to a 'conjunction' between Mars and Mercury, Vernet called his article *La conjunción del barbero de Bagdad.* A brief article by Maddison[2] is based on Vernet's analysis.

---

[1] See: J. Vernet Ginés, 'La conjunción del barbero de Bagdad,' in P. Gallais & Y.J. Riou (eds.), *Mélanges offerts à René Crozet* (Poitiers, 1966), pp. 1173–1175, repr. in J. Vernet Ginés, *Estudios sobre historia de la ciencia medieval* (Barcelona, 1956), pp. 301–303. Calcutta edition: W.H. Macnaghten (ed.), *Alif laila or book of the Thousand nights and one night. The original Arabic, from an Egyptian manuscript*, I (Calcutta, 1839). Būlāq edition: 'Abd al-Raḥmān al-Sharqāwī (ed.), *Alf layla wa-layla* (Būlāq, 1251/1835).

[2] F.R. Maddison, 'The barber's astrolabe,' *Interdisciplinary science reviews* 17 (1992), pp. 349–355.

In 1984, Muhsin Mahdi published a critical edition of the *Arabian Nights*, based on the oldest manuscripts.[3] Unlike the Calcutta and Būlāq editions, the manuscript texts do not refer to a 'conjunction'. Hence it is time for a new look at the barber's astrolabe. The following analysis of the passage on the astrolabe will confirm Vernet's main hypothesis – that the passage ultimately goes back to sensible astronomical and astrological data, probably taken from a horoscope. I will argue that the horoscope was drawn up in 763 AH/ 1361 AD rather than in 653 AH/ 1255 AD as suggested by Vernet. I would like to dedicate this brief article on the combination of science and literature to Remke, who has combined these two interests so fruitfully during her career.

The text

In his edition of the *Arabian Nights*, Mahdi distinguishes two manuscript families, which he calls the Syrian and the Egyptian branch. I first quote the passage on the barber and his astrolabe in Mahdi's edited text, which is based on the Syrian branch. The passage occurs in almost the same form in the Breslau 1826 edition of the *Arabian Nights*.[4] I keep the orthography of Mahdi's edition; below I note some differences with the Egyptian branch and variants in the individual manuscripts, based on Mahdi's critical apparatus. In my translation I have labeled the calendarical and astronomical statements [1], [2], etc. for future reference.

فادخل المزين يده الى حرمدانه فادا فيه اسطرلاب سبع صفايح مطعم بالفضه، فاخذه ومضى الى وسط الدار ورفع راسه الى شعاع الشمس ونظر فيه مليًا • تم آنه قال اعلم يا سيدى انه مضى من يومنا هدا وهو يوم الجمعه تامن عشر صفر سنه تلات وخمسين وستمايه للهجره وسبع الاف وتلتمايه وعشرين من تاريخ الاسكندر والطالع في يومنا هدا على ما اوجب في الحساب من المريخ تمان درجات وست دقايق اتفق رب الطالع عطارد وفي تلت الاسطرلاب والمريخ معه في الطالع وهو داخل معه فى تسديسه يدل على ان اخد الشعر جيد ويدل دلك ايضا على انك تريد الآتصال بنفس وهو مدموم الحال فيه مفسود •

[3] M. Mahdi, *The Thousand and one nights (Alf layla wa-layla) from the earliest known sources. Arabic text edited with introduction and notes*. 3 vols. (Leiden, 1984–1994).

[4] See Mahdi 1984–1994, I, p. 335. The story occurs in the 144th night. Breslau Edition: M. Habicht & H.L. Fleischer, *Alf layla wa-layla. Tausend und eine Nacht. Arabisch nach einer Handschrift aus Tunis*. 12 vols. (Breslau, 1826), see II, p. 227.

Translation

> The barber put his hand in his bag and produced an astrolabe with seven plates, inlaid with silver. He took it and went to the middle of the house (i.e., the courtyard). He raised its upper part to the rays of the sun and looked at it for a while. Then he said: "Sir, know that [1] there has passed from this day of ours, namely Friday, the eighteenth of Ṣafar of the year six hundred thirty-five of the Hijra, [2] and seven thousand three hundred twenty of the Alexander Era; [3] the ascendent on this day of ours is, according to computation, from Mars eight degrees and six minutes; [4] the lord of the ascendent happens to be Mercury; [5] and it is in one-third of the astrolabe; [6] and Mars is with it in the ascendent; [7] and it enters with it in its sextile (aspect). It indicates that it is (a) good (moment) for cutting hair, and that also indicates that you want to be in contact with someone, but this is inauspicious and the situation in this case is hopeless."

Before analysing this passage, I note the differences in the various manuscript versions, referring to Mahdi's critical apparatus.[5]

[1] In the three manuscripts of the Egyptian branch the Hijra dates are Friday the twelfth of Ṣafar 763, Friday the eighteenth of Ṣafar 763 and Friday the eighteenth of Ṣafar 773 respectively. The date Friday the tenth of Ṣafar 763 occurs in the Būlāq and Calcutta editions.[6] In Mahdi's later discussion of the whole passage,[7] the dates 766 and 776 are misprints for 763 and 773 respectively.

[2] The Alexander date is omitted in the manuscripts of the Egyptian branch and in the Būlāq edition. The version 7300 in one of the manuscripts of the Syrian branch is clearly a scribal error.

[3] The Būlāq edition has 'and its ascendent, as required by the science of computation, is Mars 7 degrees and 6 minutes'.

[4] Two manuscripts in the Egyptian branch have an interesting variant here: 'It happens that the lord of the ascendent of the right hair (*rabb ṭāliʿ al-shaʿr al-yamāniyya* [sic]) is Mercury', and 'it happens that the lord of the hairy (?) ascendent (*rabb al-ṭāliʿ al-shaʿra* [sic]) is Mercury'. The 'right hair', *al-shaʿr[a] al-yamāniyya*, is reminiscent of the Arabic name of the star Sirius *al-shiʿrā al-yamāniyya*.

---

[5] See Mahdi 1984–1994, II, p. 134.

[6] See al-Sharqāwī, p. 90, l. 4, and Macnaghten, I, p. 238, l. 17.

[7] M. Mahdi, *The Thousand and one nights* (Leiden, 1995), see p. 125 and p. 252 n. 135.

The Būlāq edition has, instead of [4]: 'it happens that Mercury is in conjunction with it' *(ittafaqa annahu qāranahu ʿuṭārid)*. The same variant occurs in the Calcutta edition and in Vernet's and Maddison's translations.

The astrological statements [5]–[7] are omitted in the Calcutta and Būlāq editions and also in the translations by Vernet and Maddison. Mahdi does not say anything in his critical apparatus about the statements [5]–[7], so I assume that they were in the manuscripts in both the Syrian and Egyptian branches. In [5], the Breslau edition reads (astrological) 'house' (*bayt*) instead of 'one third' (*thulth*).

Al-Bīrūnī on the Astrolabe

For reasons of space, I will not give a detailed discussion of the astrolabe. The reader can refer to numerous publications, which are easily available and well illustrated.[8] For our purpose it is sufficient to enumerate the parts of the astrolabe (see figs. 1, 2, 3).[9] (1) A metal circular disk, engraved on the back side (see fig. 1), and with a hole in the centre, in which a central axis can turn. A ring is attached to the disk so the disk can be held vertically. (2) The central axis. (3) A metal ruler (*alidade*) provided with two sights, and rotating about the central axis on the back side of the circular disk. The two endpoints of the ruler point to a circular scale on the outer edge of the disk. See fig. 1, but note that the two sights are perpendicular to the plane of the paper so they are only visible as two short straight lines. (4) A set of circular metal plates (*ṣafāʾiḥ*), of which both sides are engraved with a net of circles (including the horizon, altitude circles or *almuqantar*s, azimuthal circles) for a specific geographical latitude. See fig. 3. (5) The spider or web is an intricate metal

[8] Examples are: J. North, 'The Astrolabe,' *Scientific American* 230/1 (January, 1974), pp. 96–106, H. Michel, *Traité de l'astrolabe* (Paris, 1947), R. d'Hollander, *L'astrolabe. Histoire, théorie et pratique* (Paris, 1999), D.A. King, *In synchrony with the heavens. Studies in astronomical timekeeping and instrumentation in medieval Islamic civilization*, II, *Instruments of mass calculation* (Leiden, 2005), B. Stautz, *Untersuchungen von mathematisch-astronomischen Darstellungen auf mittelalterlichen Astrolabien islamischer und europäischer Herkunft* (Bassum, 1997).

[9] Figures 1, 2 and 3 have been taken from J. Frank, *Die Verwendung des Astrolabs nach al Chwârizmî*. Abhandlungen zur Geschichte der Naturwissenschaften und der Medizin; 3 (Erlangen, 1922), repr. in F. Sezgin (ed.), Arabische Instrumente in orientalistischen Studien, IV (Frankfurt, 1991).

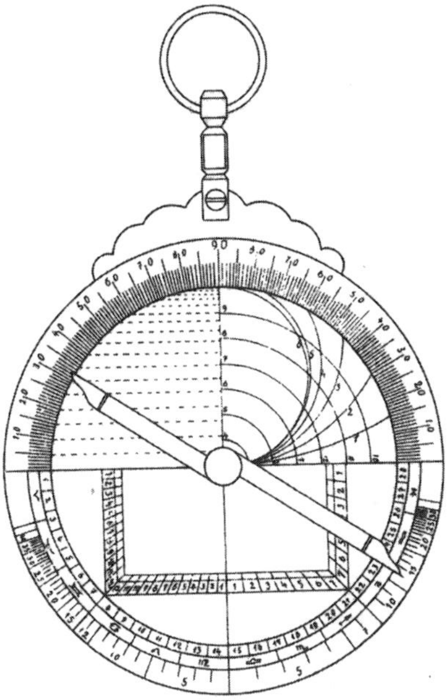

Fig. 1

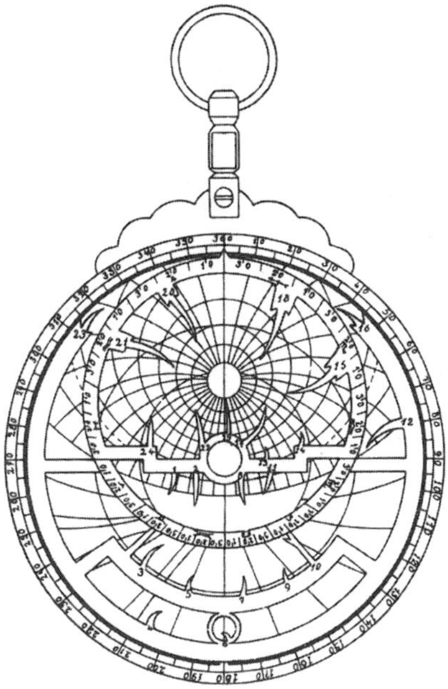

Fig. 2

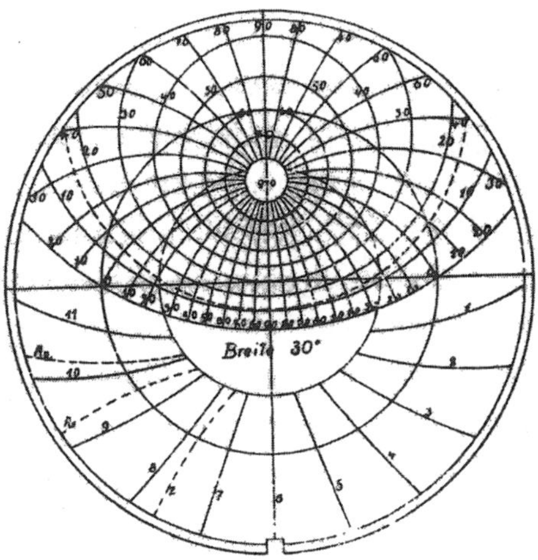

Fig. 3

network displaying around 30 star pointers and a ring which represents the ecliptic. The ring is divided into 12 signs, which are further subdivided into intervals of 5 or 1 degrees.

The plates should be attached to the front side of the disk such that the plate corresponding to the geographical latitude of the observer's locality is visible. The spider has a hole in the centre and can be attached to the central axis on top of the plate corresponding to the geographical latitude. All parts of the spider are connected, but there are many gaps in the metal, so that a large part of the engravings on the plate can be seen. The result is displayed in fig. 2. The spider is a map of the stars, and the plate is a map of the heavens with respect to the observer, on which the horizon, zenith, cardinal directions etc. can be seen. The rotation of the spider around the axis now simulates the rising, setting, culmination etc. of stars and other celestial bodies in the sky.

As a preliminary to my analysis, I will quote (in my own translation) from the *Introduction to Astrology* of al-Bīrūnī (972–1048) the descriptions of the operations with the astrolabe which the barber must have performed. The Arabic text can be found in the facsimile edition of Wright.[10] My own explanatory additions are in parentheses. The quotations will help us to interpret some of the terminology in the astrolabe passage in the *Arabian Nights*. I have omitted non-essential technicalities from my translations. The italicised titles of the chapters correspond to titles in boldface in the Arabic manuscript in the facsimile edition.

Al-Bīrūnī says

> *How the (solar) altitude can be found by means of the astrolabe.* Stand opposite the sun, suspend the astrolabe vertically from your right hand, and hold the quadrant of altitude opposite the solar disk; then the face of the astrolabe is towards you. Then turn the alidade (movable ruler on the back) up and down until the shadow of the sight which is adjacent to the sun falls on the sight adjacent to the earth, and the ray through the hole in the upper sight falls exactly on the hole of the lower sight. If that happens, leave the alidade in its position and do not move it. Look at the pointed end (of the ruler) which (turns) on the (scale for the) degrees of altitude (on the rim of the back side), (find) at which degree (mark) it is pointing.

[10] R.R. Wright, *The Book of instruction in the elements of the art of astrology by al-Bīrūnī* (London, 1934), repr. in F. Sezgin (ed.), Islamic mathematics and astronomy; 29 (Frankfurt, 1998). See p. 199 no. 330, p. 200 no. 332, p. 201 no. 333, p. 331 no. 527.

Then find the... (number of) degrees of altitude of the sun at that moment (corresponding to the mark). And know whether it (the sun) is Eastern of Western, that is, before noon it is Eastern, and after noon it is Western.

Determining the ascendent from the altitude of the sun. Turn the astrolabe towards your face and place the face of the plate whose latitude corresponds to the latitude of your locality,... (on the astrolabe) such that it is visible to the eye. Then find the *almuqantar* (circle on the plate) whose number is equal to the altitude of the sun which you have found; and if (the sun) is Eastern, then (look at only) the Eastern (part of the) almuqantar, and if it is Western, then (look at only) the Western (part of the) almuqantar, and mark it...

Then we determine from the yearly almanac (*daftar*) the position of the sun for that moment, and we find the (number of) degrees (*darajāt*) of it on the ecliptic on the spider..., and if we have found the degree of the sun, we put a mark on it. Then we (rotate the spider over the plate and) place it (the mark for the sun) on the almuqantar (on the plate) which we have marked. Then we look at the Eastern horizon, and we look which (degree) of the ecliptic coincides with it, and that is the sign of the ascendent and its degree.

How the elapsed time (al-māḍī) *of day can be determined.* If the ascendent with its (correct number of) degrees has been placed on the Eastern horizon (by rotating the spider if necessary), look at the pointer, and this is (at) the beginning of Capricorn, and see where it is among the degrees on the rim. Put a mark on its place. Then turn the spider contrary to the succession of the zodiacal signs (i.e., counterclockwise)..., until the degree of the sun, which you have marked on it, coincides with the Eastern horizon, and look which (degree) the pointer has reached (on the rim). Count from the first mark (on the rim) to it, and the result is the elapsed part from sunrise to the present moment, in time-degrees (*azmān*) of the equator. Then take for each 15 time-degrees (*azmān*) one hour, and for the remainder, for each time-degree (*zaman)* four minutes of an hour. The result is the elapsed (part) of the day in equal hours and fractions (of hours).

As far as the fourth subdivision (of astrological prediction, of horary questions) and its principles, namely the ascendents of the beginnings (of actions)...

The essence of the matter in it is that the cardinal points (ascendent, descendent, midheaven, Imum Coeli) should be in good position, that is, they should be away from maleficient (planets) and their lights (i.e., astrological rays), and they should be lighted by beneficient (planets), and especially the ascendent, and its lord, and the moon, and the governing (planet) of its house, and the (planet) governing the business. Place them in aspects with the ascendent, or else the chosen moment will turn out badly. This is a long and wide subject which cannot be exhausted here.

Astrological aspects and rays will be explained below.

Analysis of the Passage in the Arabian Nights

It is now time to analyse the passage in the *Arabian Nights* on the barber's use of the astrolabe. First I discuss a number of inconsistencies.

The sentence in [1] and [2] is incomplete, since it is not stated how much time of day has passed. Time of day, as measured by an astrolabe, is measured in time-degrees (*azmān*) or hours (*sāʿāt*) and minutes, as in al-Bīrūnī's description, never in 'degrees' (*darajāt*) and minutes. Thus the 8 degrees and 6 minutes in [3] cannot refer to the amount of time of the day which has passed. This 8 degrees 6 minutes must be the result of a computation by means of an astronomical handbook or almanac, since an astrolabe cannot possibly be read off with an accuracy of minutes of arc. The term 'ascendent for this day' in [3] does not make sense because the ascendent progresses through all signs of the ecliptic during a single day. It is likely that we have to interpret the information in [3] as relating to the sun rather than the ascendent, because the barber must have used the position of the sun to set the astrolabe, as we have seen in al-Bīrūnī's description. The statements [6] and [7] are inconsistent, because two celestial bodies cannot be in conjunction (according to [6]) and in sextile (according to [7]), that is to say, approximately 60 degrees apart.

These inconsistencies make one wonder whether the data in the passage are ultimately dependent on a genuine astronomical and astrological computation for a specific moment. We will now investigate whether this can be the case. We begin with the calendarical information in [1] and [2]. Because the variant '773' appears only once in the manuscripts studied by Mahdi, only the years '653' and '763' of the Hijra calendar are serious possibilities. The month is always Ṣafar, and we will now consider the weekday of the day numbers 10, 12 and 18, which occur in the manuscript (written out in words).

The new moons preceding Ṣafar 653 AH and 763 AH took place on March 10, 1255 AD around noon (Greenwich Mean Time) and November 28, 1361 AD in the late morning (GMT). The crescent was first sighted in the Middle East on the evenings of Thursday March 11, 1255 AD and Monday November 29, 1361 AD.[11] Days in the Hijra calendar begin at sunset. Consequently, 1 Ṣafar 653 AH probably corresponded

[11] I thank Dr Rob van Gent for checking the first visibility of the crescent on these dates by the program MoonCalc.

to March 11/12, 1255 AD and 1 Ṣafar 763 to November 29/30, 1361 AD. In these two cases, one obtains the same result if the beginnings of the months are computed according to the schematic Hijra calendar (not by actual observation of the crescent). Hence 18 Ṣafar 653 corresponded to Sunday March 28 / Monday March 29, 1255 AD and 18 Ṣafar 763 corresponded to Thursday December 16 / Friday December 17, 1361 AD. The dates 12 and 10 Ṣafar 653 and 12 Ṣafar 763 cannot possibly be fitted to the weekday 'Friday'. 10 Ṣafar 763 only corresponds to Thursday Dec. 9 / Friday Dec. 10, 1361 if we assume that the crescent was sighted one day later and the month Ṣafar 763 began on November 30 / December 1, 1361 AD. Thus we conclude that the date was probably 18 Ṣafar 763, with 10 Ṣafar 763 as a less likely alternative.

We will skip [2] for the moment, and turn to the astronomical data in [3] and [6]. The following table contains the positions of the Sun, Mercury and Mars for the relevant dates, 6 am GMT (corresponding to mid-morning in the Middle East). The positions have been computed from Tuckermann's tables[12] and rounded to degrees, since the Islamic astronomical handbooks used in computing planetary and solar positions were not more accurate. Ar., Sag. and Cap. are my abbreviations of the ecliptical signs Aries, Sagittarius and Capricornus.

Date AH	Ṣafar 653			Ṣafar 763		
	10	12	18	10	12	18
Date AD	March 1255			Dec. 1361		
	21	23	29	9	11	17
Sun	7 Ar.	9 Ar.	15 Ar.	26 Sag.	28 Sag.	4 Cap.
Mercury	6 Ar.	5 Ar.	1 Ar.	12 Sag.	15 Sag.	24 Sag.
Mars	5 Ar.	6 Ar.	12 Ar.	20 Sag.	22 Sag.	26 Sag.

The corresponding positions for 18 Ṣafar 773 (Saturday August 30, 1371 AD) 6 am (GMT) are: Sun 14 Virgo, Mercury 29 Virgo, Mars 23 Capricorn.

We now ask to what extent these results can be harmonized with the astronomical information in [3] and [6]. On 18 Ṣafar 763 and on 12 and 10 Ṣafar 653 (the date of Vernet's *conjunción del barbero*), Mars is 'with'

[12] B. Tuckermann, *Planetary, lunar and solar positions. A.D. 2 to A.D. 1649* (Philadelphia, 1964). More recent methods are a little more accurate than in the 1960s, but the difference can be ignored for our purpose, because we are only interested in positions rounded off to degrees.

Mercury, in agreement with [6]. If the date is 18 Ṣafar 763, the '8 degrees 6 minutes' in [3] can be interpreted as the distance between the sun and Mars. No such interpretation is possible on the other days, Thus, Friday 18 Ṣafar 763 emerges as the only possibility.

The fact that the same date agrees with the calendrical and with specific astronomical information is probably not accidental. Thus, the astrolabe passage in the story of the barber was in all probability based on a genuine astronomical computation for 18 Ṣafar 763. This interpretation is also plausible hom a paleographical point of view. The variant 'twelfth' *thānī ʿashara* (ثانى عشر) can easily be explained as a scribal error of 'eighteenth' *thāmin ʿashara* (ثامن عشر), and the reading 'tenth' *ʿāshir* can be explained by the omission of the word *thāmin*. In the Syrian branch, the date 763 may have been changed to 653 because the story of the young man and his beloved contains a reference to the caliph of Baghdad. The last caliph in Baghdad was murdered during the sack of the city by the Mongols in 656 AH / 1258 AD.

The dates suggest that the manuscripts in the Egyptian branch contain the more authentic version of the astronomical data. This impression is confirmed by another passage on the astrolabe, which we have not yet mentioned, and which is different in the manuscripts of the Syrian and Egyptian branches. Some time later, the barber tells the young man that the father of the young man had asked him to let blood. In the manuscripts of the Syrian branch, the barber says that on that occasion: 'I took the ascendent and found it to be a good ascendent' (*wa-akhadhtu al-ṭāliʿ fa-wajadtuhu ṭāliʿ jayyid*). In the manuscripts of the Egyptian branch, the whole story is slightly different, and the following, more detailed, passage occurs: 'I brought out the astrolabe and took for it (the bloodletting) the (solar) altitude, and I found the ascendent inauspicious' (*akhrajtu al-aṣṭurlāb wa-akhadhtu lahu al-irtifāʿ fa-wajadtu al-ṭāliʿ madhmūman*).[13] Here again, the Egyptian branch preserved a technical detail which was lost in the Syrian branch.

We will now discuss how the rest of the information in the earlier passage on the barber's astrolabe can be interpreted.

The Alexander Era, mentioned in [2], is the Seleucid era (SE), which began in 312 BC, so the date 7320 SE in the manuscripts of the Syrian branch is clearly impossible. The months Ṣafar 653 and 763 fall in the

[13] See Mahdi 1984–1994, I, p. 337, l. 12 (night 145), II, p. 134, l. 2 from bottom.

years 1566 SE and 1673 SE respectively. Another relevant chronology,[14] often used in astronomical works, is the Persian Yazdgerd Era (YE): Ṣafar 653 corresponds with 624 YE, 10–12 Ṣafar 763 with 730 YE, and 18 Ṣafar 763 with 731 YE. No easy explanation is possible. Perhaps the '7320' in the manuscripts in the Syrian branch was somehow inspired by the numbers '1673' or '731' which may have been found in an archetype of the surviving manuscripts in the Egyptian branch.

The astrological information in [4]–[7] is more fluid in nature, and many different interpretations are possible. Above we have mentioned that in two manuscripts in the Egyptian branch, Mercury is said to be 'the lord of the ascendent of the right hair', and 'the lord of the hairy(?) ascendent', where *al-shaʿra al-yamāniyya* could be explained as a pun on *Shiʿrā al-yamāniyya*, 'Sirius' (see above). Neither of these readings makes sense. Mercury is the 'lord' of Gemini and Virgo, and hence it cannot be 'the lord of the ascendent of Sirius', whose ascendent was in the end of Cancer or the beginning of Leo (depending on geographical latitude) in the 14th century AD. It is likely that 'right hair' *al-shaʿra al-yamāniyya* is somehow a deliberate corruption of the technical astrological term: 'right (astrological) ray' (*al-shuʿāʿ al-ayman*). Such a corruption is easy to explain in a story about a barber. In ancient and medieval astrology, each planet (including the sun) was believed to cast six visual rays to points on the ecliptic. These are the left sextile, quartile and trine rays, and the right trine, quartile and sextile rays. The distances of the points where these rays hit the ecliptic was on average 60, 90, 120 degrees for sextile, quartile and trine rays respectively. If such a ray is close to another planet or point (such as the ascendent), the other planet or point is said to be 'looked at' (Latin: adspectus); if the point is also a planet, the two bodies are said to form an aspect. In [7] the 'sextile aspect' is mentioned, so astrological rays were used in the horoscope on which the astrolabe passage must have been based. The two variants in the manuscripts of the Egyptian branch suggest that essential information was lost in the process of transmission.

According to Ptolemy and later authors, the 'home' of Mercury is in the signs of Gemini and Virgo, so one can interpret [4] to the effect that

[14] For these chronologies see F.K. Ginzel, *Handbuch der mathematischen und technischen Chronologie* (Leipzig, 1906), I; for tables see B. Spuler, *Wüstenfeld-Mahler'sche Vergleichungs-Tabellen zur muslimischen und iranischen Zeitrechnung* (Wiesbaden, 1961).

the ascendent is in either of these signs.[15] 'One-third of the astrolabe' in [5] does not make sense, but the sign Gemini is the 'third' sign of the ecliptic, so [5] may mean that the 'lord of the ascendent' is Mercury since the ascendent is Gemini. Many other interpretations are also possible. Thus I have not been able to derive the time of day of the horoscope which underlies the passage on the barber's astrolabe.

In conclusion, the interpretation shows that the 'authentic' version of the information entered the Egyptian branch, and appears in modified form in the manuscripts of the Syrian branch. In his analysis of the manuscripts of the *Arabian Nights*, Mahdi introduced a lost *dustūr*-version, which immediately precedes the Egyptian and the Syrian branches, Mahdi places the composition of this *dustūr*-version in the eighth century AH/fourteenth century AD.[16] This agrees well with the date of 1361 for the horoscope which underlies the passage on the barber's astrolabe.

[15] Planets as 'lords' of signs occur, for example, in Māshāllāh's *Astrological history*. See E.S. Kennedy & D. Pingree (eds.), *The astrological history of Māshā'allāh* (Cambridge, Mass., 1971), p. 91 and passim. Compare F.E. Robbins (ed.), *Ptolemy. Tetrabiblos* (Cambridge, Mass., 1980), I, Ch. 17, pp. 78–82, and Wright, p. 256.

[16] See Mahdi 1984–1994, I, pp. 29–30.

CONGRUENT NUMBERS IN THE TENTH AND IN THE TWENTIETH CENTURY[1]

Frans Oort

1. *Introduction*

In an anonymous Arabic manuscript[2] written before 972 we find a problem which is as follows in modern terms:

> *Find an integer number N and a fraction δ such that δ^2 – N and δ^2 + N are squares.*

We will call this problem the Congruent Number Problem. In the Arabic manuscript, 30 values for *N* are found[3] for which the problem has a positive answer. (We have adopted notation according to the text which will follow.) If *N* is a positive integer for which a solution δ exists, we will call *N* a *congruent number.*

The Congruent Number Problem was studied in Arabic mathematics in the tenth century, not only in the anonymous manuscript mentioned before, but also by Abū Jaʿfar Muḥammad b. al-Ḥusayn al-Khāzin.[4] It is not known whether the problem had an earlier origin (Indian?); in the

---

[1] Dedicated to Remke, for her sixty-fifth birthday.

[2] MS Paris, Bibliothèque Nationale, Fonds Arabe 2457, f. 82–86a, see F. Sezgin, *Geschichte des arabischen Schrifttums*, V (Leiden, 1974), pp. 304–305. A French translation of the Arabic text was published in pp. 211–227, 241–269 of F. Woepcke, 'Recherches sur plusieurs ouvrages de Léonard de Pise. III: Traduction d'un fragment anonyme sur la formations des triangles rectangles en nombres entiers, et d'un traité sur le même sujet par Aboû Dja'far Mohammed Ben Alhoçaïn,' *Atti dell'Accademia Pontificia dei Nuovi Lincei* 14 (1858–9), pp. 211–227, 241–269, 301–324, 343–356, reprinted in F. Woepcke, *Études sur les mathématiques Arabo-Islamiques. Nachdruck aus den Jahren 1842–1974*, hrsg. von F. Sezgin (Frankfurt, 1986), II, pp. 47–130.

[3] See L.E. Dickson, *History of the theory of numbers. Vol. II: Diophantine analysis.* Repr. (New York, 1952), p. 459, and Woepcke, pp. 256–257.

[4] See J. Sesiano, *Books IV to VII of Diophantus' Arithmetica* (New York [etc.], 1982), p. 83. The Arabic text is also in the manuscript Paris, Bibliothèque Nationale, Fonds Arabe 2457, in f. 204a–215a. The Arabic text was published by A. Anbouba, 'Un traité d'Abu Jaʿfar [al-Khāzin] sur les triangles rectangles numériques,' *Journal for the history of Arabic science* 3 (1979), pp. 134–156. A French translation appeared in Woepcke, pp. 301–324, 343–356.

new book by Kim Plofker[5] I find no traces of the problem. Sesiano says regarding the work by al-Khāzin and the work of Leonardo Fibonacci (see below): 'Both works undoubtedly stem from some common source, which may itself have been based on some Greek work.'[6]

Diophantus of Alexandria wrote his famous *Arithmetica* around 250 AD in thirteen 'books' or large chapters. Six of the thirteen books have survived in Greek, and four more in Arabic.[7] One of the main themes of the *Arithmetica* was to find solutions of equations in integer numbers or fractions. In the Greek books of the *Arithmetica* V.7, III.19 (compare also III.7 and II.19), we find the question to find δ and μ such that $\delta^2 \pm \mu$ are squares;[8] this was formulated as solving the two equations $s^2 + w = u^2$ and $s^2 - w = v^2$ in four unknowns. Diophantus also mentions the connection with rectangular triangles. So one could say that the Congruent Number Problem originally is due to Diophantus. However it seems that the author of the above-mentioned anonymous Arabic manuscript was the first to note that by choosing $N = \mu$, and making Pythagorean triples one could construct examples of congruent numbers. This point of view was also adopted by al-Khāzin.[9] Thus one could say that the Congruent Number Problem appeared for the first time in history in tenth-century Arabic mathematics. We note that work by Diophantus was already known in tenth century Arabic mathematics.[10]

Here are two examples to show you that the Congruent Number Problem is not easy. Take $N = 5$. One can see that for this value of N the problem can be solved as follows. Take $\delta = 41/12$. We see:

$$\delta^2 - 5 = \frac{1681}{144} - 5 = \frac{961}{144} = (\frac{31}{12})^2 \textit{ and } \delta^2 + 5 = \frac{1681}{144} + 5 = \frac{2401}{144} = (\frac{49}{12})^2.$$

Johannes Panormitanus di Palermo asked[11] Leonardo di Pisa (Fibonacci) around 1220 the Congruent Number Problem for $N = 5$. Fibonacci (1170–1250) solved the problem in the same way as we have seen above,

---

[5] K. Plofker, *Mathematics in India, 500 BCE–1800 CE*, to appear with Princeton University Press in 2007.

[6] Sesiano, p. 83.

[7] See Sesiano.

[8] For these propositions see, e.g., T.L. Heath, *Diophantus of Alexandria* (Cambridge 1910), pp. 205, 166–167, 158–159, 151. Compare Woepcke p. 252. Woepcke cites Diophantus' propositions in a different numbering as V.9, III.22, III.9, and II.20.

[9] Anbouba, p. 136.

[10] Anbouba, p. 136.

[11] Dickson, p. 460.

and took this as a starting point for his book *Liber Quadratorum* (1225). According to Picutti is is 'unlikely' that Fibonacci was aware of Arabic sources mentioning this problem,[12] but Sigler states 'that he must have used material from Arabic language sources...'[13]

Choose $N = 1$. The question whether this is a congruent number has been studied for at least seven centuries. Fibonacci said he had a proof that $N = 1$ is not a congruent number, but we doubt whether he really could show this. It took the genius of Fermat to settle this case. And we will see in which way this problem was a catalyst in mathematical research.

In the rest of this paper, we will use the term 'number' for integer numbers and fractions only. For example: 41/12 is a fraction (mathematicians call fractions 'rational numbers'). However, for example, $\sqrt{2}$ cannot be written in the form of a fraction m/n, where m and n are integers, so $\sqrt{2}$ is not a rational number; such cases will not be considered in this paper.

Rational numbers will be labeled as: $\alpha, \beta, \gamma, \delta, \ldots$ and integers as: $N, a, b, c, d, \ldots$. Note that 5/1 is a fraction which is also an integer.

We close this introduction with a suggestion to the reader. In this paper only elementary mathematical methods will be used. However, even so, the reader could be discouraged by the formulas below. Please try to understand the definitions in Section 2, the questions in Section 3, and then go to Section 11, where some conclusions are formulated. Sections 4–9 contain more mathematical material related to this question, and can be skipped on first reading.

2. *Definitions*

Definition I. A positive integer N is called a *congruent number* if there exists a rational number δ such that the three numbers

$$\delta^2 - N,\ \delta^2,\ \delta^2 + N$$

are squares of rational numbers.

---

[12] Conclusion D, p. 170 in E. Picutti. 'Sui numeri congruo-congruenti di Leonardo Pisano,' *Physis* 23 (1981), pp. 141–170.

[13] See the preface, p. xii, of Leonardo Pisano Fibonacci, *The book of squares. An annotated translation into modern English*, by L.E. Sigler (San Diego, 1987).

The name 'congruent number', which was introduced by Fibonacci, seems strange at first sight. It stems from the fact that these three squares form an arithmetic progression. Since the consecutive differences are equal, this common difference (Latin: *congruum*) N between the three squares is called a congruent number.[14]

From an algebraic point of view this definition seems mysterious. However, a geometric approach helps:

Definition II. A positive integer N is called a *congruent number* if there exists a rectangular triangle whose sides are rational numbers and whose area equals N.

Let us call the lengths of the sides α, β and γ (which are fractions). With the help of the theorem of Pythagoras we see that the definition amounts to:

$\alpha \cdot \beta/2 = N,$
$\alpha^2 + \beta^2 = \gamma^2;$
an example is : $\alpha = 9/6,\ \beta = 40/6,\ \gamma = 41/6,\ N = 5.$

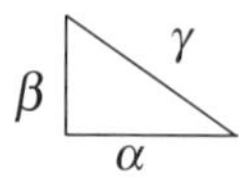

Let us show the well-known fact that these two definitions amount to the same.

Suppose $\delta^2 - N = \xi^2$ and $\delta^2 + N = \lambda^2$. Write $\gamma = 2\delta$, and $\alpha = \lambda + \xi$ and $\beta = \lambda - \xi$. Then

$$\alpha \cdot \beta = \lambda^2 - \xi^2 = 2N$$

and

$$\alpha^2 + \beta^2 = \lambda^2 + 2\lambda\xi + \xi^2 + \lambda^2 - 2\lambda\xi + \xi^2 = 2\lambda^2 + 2\xi^2 = 4\delta^2 = \gamma^2.$$

Hence Definition I implies Definition II.

Conversely if α, β, γ, N as in Definition II are given, define $\delta = \gamma/2$. Then

$$\delta^2 \pm N = \frac{1}{4}(\gamma^2 \pm 2\alpha\beta) = (\frac{1}{2}(\alpha \pm \beta))^2.$$

Hence δ and N satisfy Definition I. We see that the two definitions amount to the same.

We say that an integer M is '*square free*' if 1 is the largest square of an integer dividing M, or equivalently, if M is not divisible by p^2, for any

[14] Leonardo Pisano Fibonacci, pp. 53–54, especially p. 54, l. 13.

prime number p. A square free congruent number is called a *primitive congruent number.*

If N is a congruent number, and D a positive integer, then D^2N is a congruent number. For example the cases $N = 15$, and $D^2N = 60$, respectively $D^2N = 240$ are discussed by al-Khāzin.[15]

3. *Three Questions*

We start with the basic question: *Which are the congruent numbers?*

We will see that the wording is not precise enough: the answer to this question depends on the way in which it is phrased. Therefore we will formulate several, more precise questions. We will discuss how much is known at the moment. If you want a mental exercise: please read the questions below, and try to feel or to guess which of them is easy, and which seems difficult to you.

Question A. Can we make a list in which precisely all congruent number appear?

As we will see this question indeed can be answered affirmatively: it is not very difficult to make such a list.

Does this solve our problem? In order to decide whether a number is congruent, we could consult the list. However, this list is infinitely long. Suppose we would like to know whether $N = 1$ is a congruent number (a problem considered many times, and for the first time solved in the negative by Pierre de Fermat in the 17th century). If after a long consultation of the list you still do not see this number, what can you conclude? *We will see examples where N is a rather small integer, which is a congruent number, but where you have to go very far in the list in order to find that specific number.*

Question B. Does there exist an effective way to decide whether a given integer is a congruent number?

By this we mean: does there exist a 'formula' which gives for every integer N the amount of time (or, the number of necessary steps to perform) in order to decide whether N is a congruent number or not?

---

[15] Anbouba, p. 149.

As a preparation for Question C, we give a definition. The pair (δ, N) as in Definition I, or, equivalently $((\alpha, \beta, \gamma), N)$ as in Definition II, is called a 'presentation' of the congruent number N.

Question C. How many presentations does any given congruent number have?

Here again consultation of the list is deceiving. It seems that any congruent number appears only at most a few times. However, we will show (and this is easy) that every congruent number does have an *infinite number of presentations.*

In the next section we will introduce numbers m, n and D, used in order to construct a congruent number N. In the anonymous Arabic manuscript we find for given m and n the number $4ND^2$ (in the notation used here). Later on in the manuscript the author argues that we can divide by squares: indeed in this manuscript we find a construction of square free congruent numbers N.

About the methods developed in the tenth-century anonymous manuscript Woepcke remarks:[16] 'C'est en effet la meilleure méthode possible...les divers moyens particuliers qui permettent dans certains cas de reconnaître immédiatement si un nombre donné est ou n'est pas nombre congruent.' [This is indeed the best possible way...the various special ways which allow one to immediately recognize in certain cases whether a given number is a congruent number or not.] One can argue whether any method is the best possible (later on much better ones can be found, as is indeed the case here). My main criticism against 'reconnaître immédiatement...ou n'est...' is as follows: any finite part of the list does not allow us to decide whether $N = 1$ is a congruent number or not. For infinitely many integers it is still unknown whether they are congruent numbers.

It is not known whether the author of the above mentioned Arabic manuscript knew the *Arithmetica* by Diophantus.[17] Al-Khāzin knew the work of Diophantus, but perhaps not in the same form we have now at our disposal.[18] Later mathematicians however knew Diophantus' *Arithmetica* very well, as for example Fermat who wrote his 'Last Theorem'

[16] Woepcke, p. 252.
[17] Dickson, pp. 459–460 and Sesiano, especially pp. 9–10.
[18] Anbouba, pp. 152, 154.

in the margin of a page in his copy of a translation into Latin by Bachet; see also Section 8.

4. *Pythagorean Triples*

We say that a triple (a, b, c) of positive integers is a *Pythagorean triple* if $a^2 + b^2 = c^2$. Note that, by Pythagoras' theorem, this amounts to saying that there exists a rectangular triangle having sides of length a, b, c. We say that this triple is a *primitive* Pythagorean triple if moreover the greatest common divisor (gcd) of a and b equals one. It is easy to see that in a primitive Pythagorean triple either a or b is even. Sometimes we will interchange a and b (if necessary) in order to achieve that b is even.

We will see that Question A can be solved if we know all Pythagorean triples. And these have been known already for at least 23 centuries. In Proposition 28a of Book X of the *Elements* of Euclid (*c.* 300 BC), it is shown that a pair of integers m, n gives rise to a Pythagorean triple $(m^2 - n^2, 2m \cdot n, m^2 + n^2)$. This can be made more precise as follows:[19]

Theorem. For every primitive Pythagorean triple (a, b, c) *with a odd (and hence b even), there exist coprime positive integers m and n with* $m + n$ *odd such that*

$$a = m^2 - n^2, \quad b = 2m \cdot n, \quad c = m^2 + n^2.$$

This theorem constructs for us all Pythagorean triples.

Remark. In Section 5 of the anonymous Arabic manuscript we find a description of this result: the length c of the hypotenuse of a rectangular triangle is the sum of two squares, $c = m^2 + n^2$ (and then proceed as above). It seems clear that the above-mentioned theorem by Euclid was known to the author of that manuscript.

[19] G.H. Hardy & E.M. Wright, *An introduction to the theory of numbers* (Oxford, 1975), p. 190, Theorem 225 in 13.1.

Some examples:

$$a = m^2 - n^2, \quad b = 2mn, \quad c = m^2 + n^2.$$

n	m	a	b	c
1	2	3	4	5
1	4	15	8	17
2	3	5	12	13
1	6	35	12	37
2	5	21	20	29
3	4	7	24	25
1	8	63	16	65
2	7	45	28	53
4	5	9	40	41
1	10	99	20	101
2	9	77	36	85
3	8	55	48	73
4	7	33	56	65
5	6	11	60	61
1	12	143	24	145
2	11	117	44	125
3	10	91	60	109
4	9	65	72	97
5	8	39	80	89
6	7	13	84	85
etc.	etc.	etc.	etc.	etc.

If $\alpha^2 + \beta^2 = \gamma^2$ describes a triangle with area $\alpha\beta/2$, then for any $\rho > 0$ we have $(\rho\alpha)^2 + (\rho\beta)^2 = (\rho\gamma)^2$, describing a triangle with area $\rho^2\alpha\beta/2$. Thus, if N is a congruent number, then for any $D > 0$, the number D^2N is also a congruent number. Therefore we only consider integers as congruent numbers, and moreover we usually restrict ourselves to square free integers.

We can list 'all congruent numbers' by writing down 'all Pythagorean triples'. For such a triple (a, b, c) choose an integer D such that D^2 divides $ab/2$. Define $\alpha = a/D$, $\beta = b/D$, $\gamma = c/D$ and $N = \alpha\beta/2 = ab/(2D^2)$. Clearly N is a congruent number, as in Definition 2. By choosing D the largest integer such that D^2 divides $ab/2$ we obtain a *primitive congruent number.*

5. *Answer to Question A*

Hence we can answer the first question: *There does exist a list* (which is infinite) *which precisely contains all primitive congruent numbers.*

$$a = m^2 - n^2,\ b = 2mn,\ c = m^2 + n^2,\ ND^2 = (m^2 - n^2)mn.$$

n	m	a	b	c	D	N
1	2	3	4	5	1	6
1	4	15	8	17	2	15
2	3	5	12	13	1	30
1	6	35	12	37	1	210
2	5	21	20	29	1	210
3	4	7	24	25	2	21
1	8	63	16	65	6	14
2	7	45	28	53	3	70
4	5	9	40	41	6	5
1	10	99	20	101	3	110
2	9	77	36	85	3	154
3	8	55	48	73	2	330
4	7	33	56	65	2	231
5	6	11	60	61	1	330
1	12	143	24	145	2	429
2	11	117	44	125	3	286
3	10	91	60	109	1	2730
4	9	65	72	97	6	65
5	8	39	80	89	2	390
6	7	13	84	85	1	546
etc.	etc.	etc.	etc.	etc.	etc.	etc.

Here we start by writing numbers n and m in the two left columns such that

$$0 < n < m,\ \gcd(m, n) = 1,\ m + n \text{ is odd.}$$

Choose D^2, the largest square dividing $ab/2 = (m^2 - n^2) \cdot m \cdot n$. Write

$$\alpha = a/D,\ \beta = b/D,\ \gamma = c/D \quad \text{and} \quad N = \alpha\beta/2 = (m^2 - n^2) \cdot m \cdot n/D^2,$$

which is a primitive congruent number.

Note that the previous theorem translates the Congruent Number Problem for a given value of N into a new problem of finding positive integers $m > n$ and D such that

$$N \cdot D^2 = m \cdot n \cdot (m^2 - n^2).$$

6. *Question B: A Conjecture*

Surprisingly, Question B in general is still unsolved. In practice this means that in many cases we must find ad hoc methods in order to decide whether a given integer N is a congruent number. Advanced abstract techniques have been found, and in this way some cases have been solved. Some cases have been settled by the use of very fast computational programs.

In 1983 Tunnell formulated a conjecture which states precisely which integers are *expected* to be a congruent number. When you see the conjecture you will be surprised. *This is not something you would guess by consulting tables.* The mathematics behind the conjecture is deep and contains one of the most interesting and unsolved problems of the 20th century. Here is the conjecture.

Let N be a square free positive integer.

First suppose that N is *odd*. Define $L(N)$ as the number of triples (x, y, z) of integers such that $N = 2x^2 + y^2 + 32z^2$, and define $R(N)$ as half the number of triples (x, y, z) of integers such that $N = 2x^2 + y^2 + 8z^2$.

Next suppose that N is square free, positive and *even*. Define $L(N)$ as the number of triples (x, y, z) of integers such that $N/2 = 4x^2 + y^2 + 32z^2$, and $R(N)$ as half the number of triples (x, y, z) of integers such that $N/2 = 4x^2 + y^2 + 8z^2$.[20]

Remark. For N given, it is a finite and an easy task to compute $L(N)$ and $R(N)$.

Theorem (Coates and Wiles). *Let N be a square free positive integer. If N is a congruent number then $L(N) = R(N)$.*

---

[20] N. Koblitz, *Introduction to elliptic curves and modular forms* (New York etc., 1984), p. 221.

Conjecture (Tunnell). *Suppose N is a square free positive integer. If* $L(N) = R(N)$ *then* (?) *N is a congruent number.*

Application. Choose $N = 1$. We see that $L(N) = 2$ and $R(N) = 1$; in fact in both cases the only solutions are $x = 0$, $y = \pm 1$, $z = 0$. The theorem implies that $N = 1$ is *not a congruent number*. Note that theorem by Coates and Wiles does prove this fact, which was already proved (in an easier way) by Fermat more than three centuries ago.

Another application. choose $N = 157$. It is not difficult to prove that in this case $L(N) = 0 = R(N)$. The conjecture, if true, would imply that $N = 157$ indeed is a congruent number; a difficult computation can settle this case. Indeed, this was done by D. Zagier.[21]

A variant: Suppose that N is a square free positive integer which by division by 8 leaves either 5 or 6 or 7 as a remainder (e.g. $N = 13$, or $N = 157$, or $N = 263$, or $N = 10374$). Again, also in these cases it is not difficult to prove that $L(N) = 0 = R(N)$. The conjecture, if true, would imply that such a number N indeed is a congruent number. This covers an infinite number of cases, each of which can be settled directly, sometimes by a quite difficult computation.

Remark. Note that the criterion proposed by Tunnell indeed is effective; for a given N, it suffices to check (x, y, z) with $|x| < \sqrt{N/2}$, $|y| \leq \sqrt{N}$ and $|z| < \sqrt{N/8}$; only very few operations are necessary, and the number of these can be bounded explicitly in terms of N, as required in Question B.

P. Monsky proved indeed that a *prime number N* which by division by 8 leaves either 5 or 7 as a remainder, is indeed a congruent number.[22] This settles cases such as $N = 13$ and $N = 157$ by 'pure thought' (that is, abstract methods without difficult computations). According to a theorem by Dirichlet, there are infinitely many prime numbers in the arithmetic progression $5, 13, 21, \ldots, 5+8i, \ldots$. Thus the result of Monsky proves that there exist infinitely many primitive congruent numbers. The theorem by Dirichlet is proved by advanced analytic methods. Is there an elementary proof of the infinitude of primitive congruent numbers?

---

[21] Koblitz, p. 5.

[22] For this and several other cases see P. Monsky, 'Mock Heegner points and congruent numbers,' *Mathematische Zeitschrift* 204 (1990), pp. 45–67.

One of the outstanding conjectures of modern mathematics was made by Birch and Swinnerton-Dyer.[23] If this conjecture would be true, the Congruent Number Problem, as coined in Question B, would have a positive answer, because Tunnell's conjecture would be true. This situation is typical for modern mathematics: in trying to answer a question we try to find a much more general (and more difficult) problem, which explains the mathematical structure behind the original question. We have seen so often that this can lead to unexpected progress. The conjecture by Birch and Swinnerton-Dyer is one of the seven Millennium Prize problems of the Clay Mathematics Institute. That institute offers $1,000,000 for a solution of each of these.[24]

7. *Answer to Question C*

We start with an easy example, which will be a special case of formulas given later. We see that $3^2 + 4^2 = 5^2$; hence (3, 4, 5) is a Pythagorean triple. We see that $ab/2 = 3.4/2 = 6$ is a congruent number (taking $D = 1$).

Let us choose $A = 49$, $B = 1200$, $C = 1201$. As $49^2 = 2401$, indeed we see that

$$1201^2 = 1200^2 + 2.1200 + 1 = 1200^2 + 49^2.$$

Choosing $E = 70$ we have $AB/(2E^2) = 49 \times 1200/(7^2 \times 10^2 \times 2) = 6$. We have produced a *new presentation* of the congruent number 6.

Here is another example. We have seen that ($n = 4$, $m = 5$, $D = 2$) is a presentation of the congruent number 5. Moreover ($V = 720$, $U = 1681$, $E = 747348$) gives

$$720 \times 1681 \times (1681 - 720) \times (1681 + 720) = 5 \times 747348^2,$$

and we have *another presentation* of $N = 5$.

These are particular cases of the following more general formulas.

Suppose given $m > n$ as in the theorem characterizing Pythagorean triples, and D such that this gives a presentation of the congruent number $N = m \cdot n \cdot (m^2 - n^2)/D^2$, i.e.

$$m \cdot n \cdot (m^2 - n^2) = D^2 \cdot N, \quad ab = 2ND^2.$$

[23] See B. Birch, H. Swinnerton-Dyer, 'Notes on elliptic curves II,' *Journal für die reine und angewandte Mathematik* 218 (1965), pp. 79–108.

[24] See: http://www.claymath.org/millennium/ http://planetmath.org/encyclopedia/BirchAndSwinnertonDyerConjecture.html.

Choosing

$$U = c^2 = (m^2 + n^2)^2, \quad V = 2ab = 2(m^2 - n^2)2mn,$$

we have:

$$\begin{aligned} &U \cdot V \cdot (U - V) \cdot (U + V) = c^2 \cdot 2ab \cdot (a^2 + b^2 - 2ab) \cdot (a^2 + b^2 + 2ab) = \\ &= 2ab \cdot c^2 \cdot (a - b)^2 \cdot (a + b)^2 = \\ &= \{2 \cdot c \cdot D \cdot (a - b) \cdot (a + b)\}^2 \cdot N. \end{aligned}$$

Conclusion: if we start with a presentation (m, n, D), proving that N is a congruent number, choosing

$$U = c^2, \quad V = 2ab, \quad E = |\{2 \cdot c \cdot D \cdot (a - b) \cdot (a + b)\}|$$

gives a *new presentation* of N as a congruent number.

Question C is answered by the following.

Theorem. For every congruent number the number of presentations is infinite.

Note that the numbers U, V, E etc. involved grow rapidly; that explains why we do not see many presentations for the same congruent number in a finite list.

An afterthought: the formulas given above provide a simple proof of this theorem. But, how could we find these formulas? A full explanation would lead us too far. But let me tell you that a presentation of a congruent number can be seen as a point on a so called *plane elliptic curve*. If we draw the tangent line in that point, it can be shown that the tangent intersects the elliptic curve in one other intersection point (the point of tangency is counted as two intersection points). Computing the coordinates of this third intersection point and translating back into a presentation gives the formulas above. Again, easy geometry leads the way to complicated mathematical formulas.

The *formulas* finding the third intersection point of the tangent to an elliptic curve at a given point were known to Diophantus. A *geometric* interpretation seems to occur for the first time in history in the work of Newton.[25] Although the methods were fully within reach of, for example, Diophantus, the above-mentioned formulas and the answer to Question C only appear as late as in 20th-century mathematics.

[25] See A. Weil, *Number theory. An approach through history from Hammurapi to Legendre* (Basel etc., 1984), p. 108.

Remark. Sometimes there are more presentations which cannot be obtained from each other by the mechanism above. Such is the case for $(n, m) = (1,6)$, $(n, m) = (2,5)$ with $D = 1$ and $(n, m) = (7,8)$ with $D = 2$. These give three different presentations for the congruent number $N = 210$. Such phenomena can be explained by a sophisticated mathematical theory which is called 'the arithmetic on elliptic curves'.

We consider $N = 1$. For a long time it was an open problem whether $N = 1$ is a congruent number. For many centuries attempts were made to settle this case, and false proofs were offered.[26] After more than 7 centuries Pierre de Fermat proved the following theorem:[27]

$N = 1$ is not a congruent number.

Remark. Diophantus considered the two equations $s^2 - w = u^2$, $s^2 + w = v^2$ in four variables. Instead of working with numbers and considering integral solutions, as Diophantus did, let us consider this situation geometrically. We will use some technical terminology which may not be familiar to the reader, but we nevertheless hope that the reader will get the flavor of the argument. The two equations define a surface in four-dimensional space. Fixing one value for w, as we do in the Congruent Number Problem, we obtain two equations in three variables. Each of these two equations defines a surface in three-dimensional space. These surfaces is of a particular type, they are 'quadric surfaces'. Their intersection defines a curve in three-dimentional space. This curve has a certain property that it is 'non-singular'.[28] Modern geometry tells us that a non-singular intersection of two quadric surfaces is an elliptic curve. Hence it is no surprise that a solution of these two equations for a given $N = w$ amounts to a point on an elliptic curve. Writing down the formulas we see that *every presentation of a congruent number N is the same as a rational point with $y \neq 0$ on the elliptic curve given by $Y^2 = X^3 - N^2X$.*[29] Again, manipulation of formulas is necessary, but the geometric insight explains what happens.

---

[26] See Dickson, p. 462, and J.H. Coates, 'Congruent number problem,' *Quarterly journal of pure and applied mathematics* 1 (2005), pp. 14–27, especially p. 20.

[27] See A.W. Knapp, *Elliptic curves* (Princeton, 1992), Coroll. 4.20.

[28] It is easy to see that this curve is non-singular, even if we consider the projective compactification.

[29] See for example Koblitz, I.9, Proposition 19.

8. *Fermat's Last Theorem and* $N = 2$

Pierre de Fermat (1608–1665) proved that the numbers $N = 1$, $N = 2$ and $N = 3$ are not congruent numbers. In his proof for $N = 2$ he used the method which he had developed for proving a special case of what we call now 'Fermat's Last Theorem'; this theorem will be abbreviated as FLT. It might very well be that this connection was a source of inspiration for Fermat.

FLT says that the equation $x^n + y^n = z^n$ has no solutions in positive integers x, y, z if $n \geq 3$; this question/conjecture/theorem was a stimulating obsession for many mathematicians for more than 350 years. Fermat stated in the margin of a page of his copy of the Latin translation of Diophantus' *Arithmetica* by Bachet, that he had a proof for this 'theorem'. However, he did not tell us how he proved it, and thus he left his successors with a beautiful and difficult task. An enormous amount of work was done, special cases were settled, and new theories were developed in order to crack this nut. Andrew Wiles finally proved this theorem in 1995.[30] The case $n = 4$ is the only one of which we know that Fermat had a true proof.

Theorem (Fermat). *The equation $x^4 + y^4 = t^2$ has no solutions in positive integers.*[31]

This implies, by writing $z^2 = t$, *that* $x^4 + y^4 = z^4$ has no solutions in positive integers.

Corollary (Fermat). $N = 2$ *is not a congruent number.*

We suppose that $N = 2$ is a congruent number and we show, using the theorem, how we arrive at a contradiction, thus proving the corollary. Indeed suppose $\delta = c/d$ has the property $\delta^2 - 2 = (u/d)^2$ and $\delta^2 + 2 = (v/d)^2$. Write $x = uv$, $y = 2cd$ and $t = c^4 + 4d^4$. Because

$$u^2 = c^2 - 2d^2, \quad v^2 = c^2 + 2d^2$$

we obtain

$$x^4 + y^4 = (uv)^4 + (2cd)^4 = ((c^2 - 2d^2)(c^2 + 2d^2))^2 + 16c^4d^4 = (c^4 + 4d^4)^2 = t^2.$$

---

[30] See A. Wiles, 'Modular elliptic curves and Fermat's Last Theorem,' *Annals of mathematics* 141 (1995), pp. 443–551.

[31] For example see Knapp, III.d, or D. Shanks, *Solved and unsolved problems in number theory* (London 1978), p. 144, Th. 62.

This is a contradiction with the theorem above; hence $N = 2$ is not a congruent number.

Did the Congruent Number Problem inspire Fermat to formulate his FLT?

9. *More Examples*

Above we have seen some easy methods to find congruent numbers. We continue with some examples which are not so easy to find in that way.

$N = 13$. Choosing $m = 325$, and $n = 36$ we see

$$m \cdot n \cdot (m^2 - n^2) = 325 \cdot 36 \cdot 289 \cdot 361 = \\ = 13 \cdot 5^2 \cdot 6^2 \cdot 17^2 \cdot 19^2.$$

Conclusion: $N = 13$ is a congruent number.

We see that $\delta = 106921/19380$ has the property $\delta^2 - 13 = (80923/19380)^2$ and $\delta^2 + 13 = (127729/19380)^2$. Results not so easy to find....

$N = 23$. Choosing $m = 24336$, and $n = 17689$ we see $m = 156^2$, $n = 133^2$, $m - n = 6647 = 17^2 \times 23$, and $m + n = 42025 = 205^2$. This shows that 23 is a congruent number.

$N = 157$. This small integer indeed is a congruent number (as predicted by the conjecture by Tunnell, and as proved by D. Zagier by a computation and by Monsky by 'pure thought'). Trying to find $\delta = c/d$ such that $\delta^2 \pm 157$ are squares, the one with *smallest d* has many decimal digits:[32] choose $m = 443624018997429899709925$, and
$n = 166136231668185267540804$.

$N = 219$. This is a congruent number because
$48 \times 73 \times (73 + 48) \times (73 - 48) = 219 \times (4 \times 5 \times 11)^2$. Note that in the sequence

$$3, 11, 19, \ldots, i8 + 3, \ldots, 211 \text{ with } 0 \leq i \leq 26$$

all square free integers are non-congruent numbers. However $219 = 3 \times 73 = 27 \times 8 + 3$ is a congruent number, although 3 and 73 are not

[32] See Koblitz, p. 5.

congruent numbers. Moreover $N = 171 = 9 \times 19 = 21 \times 8 + 3$ is a congruent number.

Bastien proved that every prime number of the form $i8 + 3$ is not a congruent number.[33]

Note that $49 \times 48 \times 1 \times 97 = 28^2 \times 291$, showing that 291 is a congruent number, and the same for 299 as $36 \times 13 \times 23 \times 49 = 42^2 \times 299$.

We see the rather unpredictable way of behavior of integers with respect to being congruent or non-congruent.

$N = 263$. The choice $m = 24150469654071998864724443950150\,56$ and $n = 21965899725314208513405213564709\,69$ proves that this prime number is a congruent number (as predicted by the conjecture by Tunnell, and as proved as a special case of a theorem by Monsky). Again we see that a relatively small number only admits a very large presentation.[34]

$N = 10374$. This is the largest of the 30 primitive congruent numbers obtained in the Arabic manuscript discussed earlier. By choosing $n = 3$ and $m = 13$ we indeed obtain

$$m \cdot n \cdot (m + n) \cdot (m - n) = 13 \times 6 \times 19 \times 7 = 10374.$$

Here a relatively 'large' N admits a small presentation.

By means of the list containing all choices of m and n as above with $m + n \leq 19$, these 30 primitive congruent numbers (and several more) are found. Note that the number 10374 has a small presentation, and the smaller number 263 has a very large presentation; this is typical for the behavior of congruent numbers in the list produced in the answer to Question A.

If the remainder r of division by 8 of a square free integer N equals 5, 6 or 7, the conjecture mentioned in Section 6 predicts that N is not a congruent number. However for other values of r such an easy criterion does not hold:

---

[33] L. Bastien, 'Nombres congruents,' *Intermédiaire des mathématiques* 22 (1915), pp. 231–232.

[34] For a computation settling all cases $N < 1000$, hence also this case, the reader may consult: http://www.asahi-net.or.jp/KC2H-MSM/mathland/math10/matb2000.htm. or G. Kramarz, 'All congruent numbers less than 2000,' *Mathematische Annalen* 273 (1986), pp. 337–340.

$r = 0$ 8 is NC and 24 is C;
$r = 1$ 1 is NC and 41 is C;
$r = 2$ 2 is NC and 34 is C;
$r = 3$ 3 is NC and 219 is C;
$r = 4$ 4 is NC and 28 is C.

10. *Some References*

In the past ten centuries an enormous amount of literature has been published on the Congruent Number Problem. We have indicated a small selection of these publications in the notes.[35]

11. *Some Conclusions*

We give a survey of some statements and conclusions around the Congruent Number Problem.

a. In ancient civilizations Pythagorean triples played an important role. Already on the old Babylonian clay tablet Plimpton 322 (between BC

[35] In the second volume of Dickson, we find in Chapter 16 an ample survey of early attempts to solve the Congruent Number Problem. We find a survey of solutions, and several more modern references in Problem D27 in R.K. Guy, *Unsolved problems in number theory*. Third ed. (New York etc., 2004). In Sesiano we find a description of the connection between the *Arithmetica* of Diophantus and Arabic mathematics in the Middle Ages. In Koblitz we find a survey of modern methods, especially of the road to the conjecture by Tunnell formulated in J.B. Tunnell, 'A classical diophantine problem and modular forms,' *Inventiones mathematicae* 72 (1983), pp. 323–334.

For surveys and questions see e.g. R. Alter, 'The congruent number problem,' *American mathematical monthly* 87 (1980), pp. 43–45 and the paper of Coates cited above. See especially Knapp for a lucid explanation of several concepts needed for a modern treatment.

For more specialized modern treatments see N.M. Stephens, 'Congruence properties of congruent numbers,' *Bulletin of the London Mathematical Society* 7 (1975), pp. 182–184; J.H. Silverman, *The arithmetic of elliptic curves* (New York, 1986), and the papers of Kramarz and Monsky cited above.

For a treatment on an elementary level see A.H. Beiler, *Recreations in the theory of numbers. The Queen of Mathematics entertains* (New York, 1964).

Here are some links where information on congruent numbers can be found at this moment:
http://en.wikipedia.org/wiki/Congruent_number
http://mathworld.wolfram.com/CongruentNumber.html
http://www.asahi-net.or.jp/~KC2H-MSM/mathland/math10/matb2000.htm
http://www.math.harvard.edu/~elkies/compnt.html

1800 and BC 1650) such numbers appear.[36] Some people think that when building the pyramids right angles were constructed by using the Pythagorean triple (3, 4, 5). We are unaware of any historical or archeological evidence for this.

b. In the *Arithmetica* of Diophantus we find the question of solving the two equations $\delta^2 - \mu = \xi^2$ and $\delta^2 + \mu = \lambda^2$ in four unknowns, and the connection with rectangular triangles.

c. In an anonymous Arabic manuscript written before 972 we find a method of constructing congruent numbers; 30 primitive congruent numbers result from computations in that manuscript. Also al-Khāzin studied this problem in the same century. Clearly, the solution of quadratic equations and the construction of Pythagorean triples were known in ancient civilizations, and seem to have been handed down through Greek and Indian mathematics. But it might be that the more precise formulation of the Congruent Number Problem was given for the first time, and studied intensively, in tenth-century Arabic mathematics.

It is not clear to me whether the Congruent Number Problem has an Indian origin as some authors seem to think. In the standard literature[37] we do not find any evidence that the Congruent Number Problem has an Indian origin (although several related quadratic equations were studied in medieval India).

d. Mathematicians have tried to answer the question whether $N = 1$ is a congruent number. This problem remained unsolved for many centuries.

e. Leonardo di Pisa (Fibonacci) encountered the question of finding congruent numbers; it was the starting point of his 'liber quadratorum' (1225). It is not clear whether Fibonacci was aware of the earlier Arabic sources on this problem. Fibonacci claimed he had a proof that $N = 1$ is not a congruent number.

---

[36] See O. Neugebauer & A. Sachs, *Mathematical cuneiform texts* (New Haven, 1945) and D. Fowler & E. Robson, 'Square root approximations in old Babylonian mathematics. YBC 7289 in context,' *Historia Mathematica* 25 (1998), pp. 366–378.

[37] See Plofker and also B. Datta & A.N. Singh, *History of Hindu mathematics*. 2 vols. (Delhi, 1935–1938).

f. Pierre de Fermat (1601–1665) was the first to prove that $N = 1$ is not a congruent number. His proof that $N = 2$ is not a congruent number was connected with the fact that the equation $x^4 + y^4 = z^4$ has no solutions in positive integers.

This is the only know case of a proof by Fermat of 'Fermat's Last Theorem' (FLT, formulated around 1637): *$x^n + y^n = z^n$ has no solutions in positive integers x, y, z if $n \geq 3$.* This problem was studied for a long time, special cases were settled, and finally the theorem was proved in full generality in 1995 by Andrew Wiles. Did the Congruent Number Problem inspire Fermat to formulate his FLT?

g. As in the anonymous Arabic manuscript mentionend before we see that it is easy to produce an infinite list containing all primitive congruent numbers. Does this solve the Congruent Number Problem? Yes and No. 'Yes' in the sense that we have a list of all of them. 'No' in the sense that deciding whether a given number is congruent or not may take a very long time. We see that some 'small' integers appear 'very late' in the list. This behavior is not well understood as yet.

h. We do not know an effective method of deciding whether any given integer is a congruent number.

i. However, if a deep conjecture in mathematics would be true, namely the conjecture formulated by Birch and Swinnerton-Dyer in the 1960s, we would have an effective method, formulated by Tunnell, of deciding for every integer whether it is a congruent number or not. A modern, rather abstract theory gives access to this 'elementary' problem.

j. We have seen on several occasions that geometric methods give clear insight, provide formulas which seem difficult, and solve problems of an algebraic nature. This is a characteristic example of 'cross-fertilization' between different branches of mathematics.

k. The conjecture by Birch and Swinnerton-Dyer is still open. A particular case of that conjecture would imply:

Let N be a positive integer such that division by 8 leaves as a reminder either 5, or 6 or 7. Then N is a congruent number.

l. We see that the Congruent Number Problem in the history of mathematics was a stimulating catalyst; it induced mathematicians to formulate new methods, results and conjectures. However:

m. The Congruent Number Problem, studied in Arabic mathematics in the tenth century, remains unsolved up to now.

AL-FĀRĀBĪS ARISTOTELES GRUNDLAGEN SEINER ERKENNTNISLEHRE[1]

Hans Daiber

Fārābīs Regent des Musterstaats, der Prophet und Philosoph sein soll, teilt manche Züge mit Platos Vorstellung vom Philosophen, dem man den Staat anvertrauen kann, weil er 'Lust und Liebe' 'zu solchem Lerngegenstande' hat, der ihm 'den Schleier zu lüften vermag von jenem Sein, das ewig ist und keiner Veränderung unterworfen ist durch Entstehen und Vergehen'.[2] Dieser Wahrheitsliebende hat seinen Blick auf eine Welt gerichtet, 'worin eine ewige Ordnung und Unwandelbarkeit herrscht, worin die Wesen weder Unrecht tun noch voneinander leiden, und worin alles nach einer himmlischen Ordnung und Vernunftmäßigkeit geht'; er strebt darnach, diese Welt der 'ewigen Ordnung und Unwandelbarkeit' 'nachzuahmen und soviel als möglich davon in seinem Leben ein Abbild darzustellen'.[3] Die Ursache reiner Vernunfterkenntnis und Wahrheit bezeichnet Plato als das eigentliche höchste Gut, das über der Vernunfterkenntnis und Wahrheit stehe.[4]

Diese Aussagen sind Fārābī vielleicht in Form einer Paraphrase von Platons Republik bekannt geworden, haben aber im Kontext islamischer Theologie ein neues Gewand bekommen, das aristotelisch-peripatetisches und neuplatonisches Kolorit verrät. Dies möchte ich zu zeigen versuchen, indem ich Fārābīs Erkenntnislehre als eine von der aristotelischen Schule und vom Neuplatonismus angeregte modifizierende Weiterführung von Plato skizziere.

---

[1] Remke Kruk gewidmet mit guten Erinnerungen an unsere gemeinsame Zeit in Holland!

Diese Arbeit führt Überlegungen eines 1986 erschienenen Aufsatzes weiter: *The ruler as philosopher. A new interpretation of Fārābī's view.* Mededelingen der Koninklijke Nederlandse Akademie van Wetenschappen, afd. Letterkunde, Nieuwe reeks; dl. 49/4 (Amsterdam [etc.], 1986).

[2] *Rep.* 485A (Übers. W. Wiegand in: Platon, *Sämtliche Werke*, II (Heidelberg, o.J.)).

[3] *Rep.* 500C (Übers. W. Wiegand).

[4] Vgl. *Rep.* 508D.

In einer 1994 erschienenen Studie über *Al-Fārābī and Aristotelian Syllogistics* hat Joep Lameer versucht, Fārābīs Gegenüberstellung von Philosophie und Religion als Nachahmung der Philosophie auf die platonische Unterscheidung zwischen Wissen und Meinung zurückzuführen, die in einen aristotelischen Kontext eingefügt worden sei; Platos *Staatsmann* sei letzlich die Quelle für Fārābīs Auffassung von Religion gewesen, weil dort von mehr oder weniger guten Nachahmungen des wahren Staates bzw. des 'Wahren' die Rede sei (297C ff.; 300C).[5]

Doch dieser Vergleich ist bereits durch eine Aussage in Platos *Staatsmann* (301B) in Frage gestellt, wonach derjenige, der 'nach Gesetzen herrscht' und 'den Wissenden nachahmt' König genannt wird, wobei kein Unterschied sei zwischen dem 'mit Erkenntnis allein Herrschenden' und dem 'nach guter Meinung den Gesetzen gemäß allein Herrschenden' (*Staatsmann* 301B). Erst der Alleinherrscher, der vorgibt Wissender zu sein, sich aber nicht an Gesetz und Gewohnheit hält, wird zum Tyrannen.

Demnach ist bei Plato nicht *apriori* ein Gegensatz zwischen Wahrheit und Gesetz. Im Gegenteil, Plato berührt sich mit Fārābīs Auffassung von Religion bzw. Gesetz als philosophische Wahrheit, wobei Religion ebensowenig wie Platos Gesetz ein unvollkommenes Abbild von philosophischer Wahrheit sein muß.

Fārābī[6] hatte die Lehre entwickelt, daß die Religion 'die Philosophie nachahme' (*muḥākiyatun li ʾl-falsafa*). Hierbei umfassen beide, Religion und Philosopie, dieselben Objekte, beide vermitteln 'das Wissen um das erste Prinzip und die erste Ursache' der seienden Dinge; beide geben vor, was das Ziel des Menschen ist, nämlich die Glückseligkeit. Doch beide unterscheiden sich in der Art und Weise der Vermittlung des Wissens um das erste Prinzip. Philosophie stütze sich auf philosophische Beweise, indem sie mit Begriffen des Verstandes arbeite (*maʿqūlan mutaṣawwiran*)[7] – Religion jedoch suche zu überzeugen (*iqnāʿ*), indem sie die Prinzipien in Form von Bildern (*mithālāt*) vorstelle (*tukhayyiluhā*), die 'den körperartigen Prinzipien entnommen sind'; Religion 'ahme die Prinzipien mit dem ihnen Entsprechenden (*naẓāʾir*) nach. So 'ahme sie

---

[5] J. Lameer, *Al-Fārābī and Aristotelian syllogistics*. Islamic philosophy, theology and science; 20 (Leiden [etc.], 1994), S. 259ff.

[6] *Kitāb taḥṣīl al-saʿāda*, ed. Jaʿfar Āl Yāsīn (Beirut, 1981), S. 90,14f.; Übers. Muhsin Mahdi, *Alfarabi's philosopy of Plato and Aristotle*. Rev. ed. (Ithaca, New York, 1962), S. 44.

[7] *Kitāb taḥṣīl*, ed. Āl Yāsīn, S. 90,19.

die göttlichen Handlungen (*al-afʿāl al-ilāhiyya*) mit den Wirkungen der Prinzipien des (Muster)staates (*afʿāl al-mabādiʾ al-madaniyya*) nach und die Wirkungen der Kräfte und Prinzipien der Natur mit dem ihnen Entsprechenden, nämlich mit den vom Willen bestimmten Fähigkeiten, Fertigkeiten und handwerklichen Tätigkeiten'.[8] Fārābī fügt interessanterweise hinzu, daß 'Plato im *Timaeus* in dieser Weise verfahren sei'. Fārābī bezieht sich hier offensichtlich in lockerer Weise auf eine Stelle im *Timaeus*,[9] wonach die Welt ein Abbild von etwas ist, mit dem es verwandt bleibt und nachdem sie von ihrem Schöpfer geschaffen worden ist. Galens Kompendium von Platons *Timaeus* – das vielleicht von dem Ḥunainschüler ʿĪsā b. Yāḥyā b. Ibrāhīm übersetzt wurde – kürzt hier stark[10] und kann schwerlich Vorlage von Fārābī gewesen sein.

Für Fārābī ist der Musterstaat ein Abbild göttlichen Handelns – er spricht auch von der 'Musterreligion' (*al-milla al-fāḍila*),[11] die 'der Philosophie ähnlich sei' und deren 'Universalien' in der 'Philosophie' existierten. Hierbei lasse sich die theoretische Philosophie mit den 'Meinungen' (*ārāʾ*) der Religion parallelisieren und die praktische Philosophie mit den 'vortrefflichen Gesetzen' (*al-sharāʾiʿ al-fāḍila*).[12]

Abgesehen von der aristotelischen Unterscheidung zwischen theoretischer und praktischer Philosophie wird man hier an den bereits genannten platonischen Begriff der Nachahmung des 'Wissenden' durch den König erinnert, an die Nachahmung der 'ewigen Ordnung und Unwandelbarkeit', wovon der 'Wahrheitsliebende' in seinem Leben ein Abbild darzustellen trachtet. Hierbei ist, wie Plato im *Timaeus* an der von mir bereits genannten[13] und von Fārābī offensichtlich benutzten Stelle (29Af.) feststellt, dieses Abbild mit dem Abgebildeten 'verwandt': die 'Darlegung' des 'Bleibenden und Beständigen und im Lichte der Vernunft Erkennbaren selber' 'trägt das Gepräge des Bleibenden und Umumstößlichen an sich'. Dies schließt nicht aus, daß nachfolgende Abbilder – Plato rechnet dazu die Dichtung – lediglich 'Wahrscheinlichkeit' besitzen.

---

[8] *Kitāb taḥṣīl*, ed. Āl Yāsīn, S. 91,1ff.

[9] 29Bf.

[10] Vgl. Galen, *Compendium Timaeu Platonis*, ed. P. Kraus und R. Walzer. Plato arabus; 1 (Londinii, 1951), S. 5,4ff.

[11] Zur Austauschbarkeit beider Begriffe vgl. Ilai Alon, *Al-Fārābī's philosophical lexicon*. E.J.W. Gibb memorial trust (Warminster, Wilts., 2002), I, S. 454.

[12] *Kitāb al-milla*, ed. Muhsin Mahdi (Beirut, 1968), S. 46,22ff.

[13] S. oben Anm. 9.

Platos These der 'Verwandtschaft' von Original und Abbild und ihre Begründung durch die Sache, die Original und Abbild gemeinsam haben, erscheint ähnlich in Fārābīs *Kitāb al-alfāẓ al-mustaʿmala fī 'l-manṭiq* in einer Diskussion über die Hilfsmittel des Lernens.[14] Fārābī weist darauf hin, daß schwer vorstellbare Dinge durch leichter vorstellbare ersetzt werden können. Er stellt fest, daß 'das einer Sache Ähnliche gleichfalls deutlich sei'; 'das in unserer Seele vorhandene Vorstellungsbild (*khayāl*) einer Sache entspricht dem Vorstellungsbild dessen, was ihr ähnlich ist. Und zwei Dinge gleichen sich, indem sie eine einzige Sache gemeinsam haben, die in allen beiden gleichzeitig nachgeahmt wird'.[15] Es gibt trotz Unterschiedlichkeiten im Detail gemeinsame Grundstrukturen.

Diese strukturelle Ähnlichkeit begründet die Ersetzbarkeit des Originals durch ein Abbild. Sie ist Fārābī zufolge[16] jedoch ausgeschlossen, wenn überaus komplexe Gebilde' durch vereinfachende und verfälschende Vorstellungsbilder ersetzt werden, sodaß 'der Zuhörer und der Lernende von der gemeinsamen Sache überaus entfernt sind'. Solche falschen Substitutionen schreibt Fārābī[17] in Anlehnung an Aristoteles' *Metaphysik*[18] griechischen Philosophen zu, den Pythagoräern, Plato und Empedokles; es handle sich hier um 'Allegorien' und 'Rätsel', die im Philosophieunterricht vermieden werden sollten und allenfalls in der Rhetorik oder in der Politik erlaubt seien.

Hierbei hat die Rhetorik durchaus eine positive Funktion,[19] nämlich in Übereinstimmung mit Aristoteles[20] 'die Masse vieles von den theoretischen Dingen zu unterrichten; sie werde in politischen 'Ansprachen' benutzt,[21] wobei sie lediglich die Aufgabe habe, zu 'überreden' (*taqannuʿ*)

---

[14] Ed. Mahdi (Beirut, 1968), S. 88ff.; vgl. F.S. Haddad, *Alfarabi's theory of communication* (Beirut, 1989), S. 135 (= id., 'Early Arab theory of instruction,' *International journal of Middle East studies* 5 (1974), S. 240–259, hier S. 248); D.L. Black, *Logic and Aristotle's rhetoric and poetics in medieval Arabic philosophy*. Islamic philosophy and theology; 7 (Leiden [etc.], 1990), S. 67ff.

[15] *Kitāb al-alfāẓ*, ed. Mahdi, S. 88,14–16. Vgl. Black, *Logic and Aristotle's rhetoric*, S. 188ff.

[16] *Kitāb al-alfāẓ*, ed. Mahdi, S. 91,6ff.; vgl. L.V. Berman, *Oriens* 23/4 (1974), S. 512. Die Unterschiedlichkeit von Original und ähnlichem Abbild hat Fārābī nochmals thematisiert in *Kitāb al-siyāsa al-madaniyya al-mulaqqab bi-mabādiʾ al-mawjūdāt*, ed. F.M. Najjar (Beirut, 1964, 2.A., 1993), S. 85,3ff.; Übers. F. Dieterici, *Die Staatsleitung von Alfārābī*. Aus dem Nachlass hrsg. v. P. Brönnle (Leiden, 1904), S. 68f.

[17] *Kitāb al-alfāẓ*, ed. Mahdi, S. 91,10ff.

[18] 1001a10ff.; vgl. 1000a9–19.

[19] Vgl. im einzelnen Black, *Logic and Aristotle's rhetoric*, Kp. 4 u. 5.

[20] *Rhetorik*, 1357a1ff.

[21] Fārābī, *Kitāb al-ḫaṭāba*, ed. J. Langhade & M. Grignaschi, *Deux ouvrages inédits sur la réthorique* [sic] (Beyrouth, 1971), S. 57, bes. Z. 7–9. Vgl. Fārābī, *Falsafat Arisṭūṭālīs*,

und nicht für die 'Reflexion' (*rawiyya*) bzw. für Schlußfolgerungen herangezogen werde.[22] Wie in Aristoteles' Rhetorik[23] hat sie die Partikularien und nicht die Universalien bzw. das Notwendige im Auge;[24] sie vermittle in erster Linie 'Meinungen' (*ẓunūn*), nicht aber 'Gewißheit' (*al-yaqīn*).[25]

Für Fārābī ist die rhetorische Überredung der erste Schritt zum Akzeptieren des Inhaltes als wahr (*taṣdīq*). Wenn die 'sicheren Beweise' (*al-barāhīn al-yaqīniyya*) hinzukommen,[26] dann verdiene das aus der Verbindung von Rhetorik und 'sicheren Beweisen' erworbene 'Wissen um die seienden Dinge' (*ʿilm al-mawjūdāt*) den Namen 'Philosophie'.

Nun ist für Fārābī, wie bereits gesagt (s.o.), auch die Religion ein Weg, um zu überzeugen (*iqnāʿ*), indem sie die Prinzipien in Form von Bildern (*mithālāt*) vorstelle (*tukhayyiluhā*), die 'den körperartigen Prinzipien entnommen sind.' Das Wissen ist auf Bilder gerichtet, die etwas 'nachahmen'. Fārābī zufolge[27] wird es von den antiken Philosophen 'Religion' (*milla*) genannt.

Ist also die Religion eine Vorstufe der Philosophie, ebenso wie die Rhetorik der erste Schritt zum 'Wissen um die seienden Dinge' ist? Die Antwort hierauf fällt komplex aus. Zunächst müssen wir darauf hinweisen, daß Religion bei Fārābī nicht nur die 'Meinungen' umschreibt, sondern auch 'die Handlungen, die in der Form von Regeln auferlegt und festgelegt sind, die der erste Regent der Gesellschaft vorgeschrieben hat.'[28]

---

ed. Muhsin Mahdi (Beyrouth, 1961), S. 84f.; Übers. Mahdi, *Alfarabi's philosophy*, S. 92f. Vgl. J. Kraemer, 'Alfarabi's opinions of the Virtuous City and Maimonides' Foundations of the Law,' in J. Blau [et al.] (eds.), *Studia Orientalia memoriae D.H. Baneth dedicata* (Jerusalem, 1979), S. 119, Anm. 21; Ch. Butterworth, 'The rhetorician and his relationship to the community. Three accounts of Aristotle's Rhetoric,' in M.E. Marmura (ed.), *Islamic theology and philosophy. Studies in honor of George F. Hourani* (New York, 1984), S. 112ff.

22 *Kitāb al-Khaṭāba*, ed. Langhade & Grignaschi, S. 59,5f.; vgl. F.W. Zimmermann, 'Al-Fārābī und die philosophische Kritik an Galen von Alexander zu Averroes,' in A. Dietrich (ed.), *Akten des 7. Kongresses für Arabistik und Islamwisssenschaft*. Abhandlungen der Akademie der Wissenschaften in Göttingen, Philol.-hist. Kl., 3. Folge; nr. 98 (Göttingen, 1976), S. 402f.

23 1355b26.

24 *Kitāb al-khaṭāba*, ed. Langhade & Grignaschi, S. 33,11f.; 14.

25 Vgl. *Kitāb al-khaṭāba*, ed. Langhade & Grignaschi, S. 59,11f.

26 Vgl. Fārābī, *Kitāb taḥṣīl*, ed. Āl Yāsīn, S. 90,3f.; Übers. Mahdi, *Alfarabi's philosophy*, S. 44.

27 *Kitāb taḥṣīl*, ed. Āl Yāsīn, S. 90,10ff.; Übers. Mahdi, *Alfarabi's philosophy*, S. 44.

28 *Kitāb al-milla*, ed. Mahdi, S. 43,3f.; vgl. L.V. Berman, 'Maimonides, the disciple of Alfārābī,' *Israel Oriental studies* 4 (1974), S. 159f.

Nehmen wir hierzu den bereits aus Fārābī zitierten Passus, wonach der Musterstaat ein Abbild göttlichen Handelns sei. Fārābī sprach auch von der 'Musterreligion' (*al-milla al-fāḍila*),[29] die 'der Philosophie ähnlich sei' und deren 'Universalien' in der 'Philosophie' existierten.[30] Hierbei lasse sich die theoretische Philosophie mit den 'Meinungen' (*ārā'*) der Religion parallelisieren und die praktische Philosophie mit den 'vortrefflichen Gesetzen' (*al-sharā'i' al-fāḍila*).[31] Analog zum Parallelismus des praktischen Teils in Religion und Philosophie haben die 'theoretischen Meinungen' (*al-ārā' al-naẓariyya*) in der Religion ein Pendant in den Beweisen (*barāhīn*) der theoretischen Philosophie.[32] Wie bei den praktischen 'Handlungen' können die Regeln und Vorschriften, die 'theoretischen Meinungen' der Religion, bewiesen (*burhina*) und begründet werden (vgl. *a'ṭā 'l-asbāb*) durch die Philosophie, durch einen Vergleich mit ihren Universalien. Oder anders formuliert: Die Religion informiert die Masse über 'theoretische Meinungen' und praktische Handlungen, die sich in der Philosophie beweisen und begründen lassen und im Vergleich mit den Universalien sich als wahr erweisen.

Bemerkenswert ist hier die ethische Komponente in Fārābīs Begriff von Philosophie und Religion. Wir hatten bereits darauf hingewiesen, daß einerseits 'philosophische' Beweisführung und andererseits 'religiöse' Überzeugungskunst mit den Hilfsmitteln der Bildersprache von Rhetorik und Poetik zum Ziel haben, dem Individuum ethische Instruktionen zu geben, Instruktionen über den richtigen Weg zur 'höchsten' Glückseligkeit.

Religion ist Instruktion in der Gestalt von 'Vorschriften' und 'Gesetzen'; ihr Ausgangspunkt ist die Philosophie, die Fārābī zufolge der Religion vorausgeht[33] und die Aufgabe hat, eine für den Verstand akzeptable 'Begründung' (*burhān*) zu geben.

Diese Beobachtung gibt Anlaß zu folgenden entscheidenden Fragen: Entspringt diese philosophische 'Begründung' lediglich einem intellektuellen Bedürfnis des Gebildeten? Ist Religion daher einerseits zwar philosophisch beweisbar, andererseits aber nur für die ungebildete Masse, die Nichtphilosophen gedacht?

[29] Zur Austauschbarkeit beider Begriffe s. Anm. 11.
[30] *Kitāb al-milla*, ed. Mahdi, S. 46,22ff.
[31] *Kitāb al-milla*, ed. Mahdi, S. 47,5.
[32] *Kitāb al-milla*, ed. Mahdi, S. 47,6.
[33] Vgl. *Kitāb taḥṣīl*, ed. Āl Yāsīn, S. 91,13; Übers. Mahdi, *Alfarabi's philosophy*, S. 45; Fārābī, *Kitāb al-ḥurūf*, ed. Mahdi (Beyrouth, 1970), S. 131 und 154f.; Berman, 'Maimonides,' S. 156ff., 161f.

Wir werden zu zeigen versuchen, warum diese Fragen zu verneinen sind. Hierbei müssen wir vom aristotelischen Theorie-Praxis-Modell ausgehen, das Philosophie und Religion gemeinsam haben. Fārābī zufolge 'ahmt' Religion die Philosophie 'nach', die nach aristotelischem Vorbild als ein Miteinander von Theorie und Praxis, von 'wissenschaftlicher Erkenntnis und sittlicher Einsicht' verstanden wird.

In der Tradition von Aristoteles' *Nikomachischer Ethik* stehend und unter Heranziehung von Aristoteles' *De anima* sowie von Alexander von Aphrodisias' Kommentar hierzu hat Fārābī im 'Musterstaat' eine originelle Lehre entwickelt.[34] Er führt in der Nachfolge von Aristoteles' *Nikomachischer Ethik* die drei Seelenkräfte Sinneswahrnehmung, Verstand und Streben ein, die das ethische Handeln des Menschen sowie seine Erkenntnis des Richtigen steuern. Fārābī integriert hier den aristotelischen Begriff der *phronesis*, der praktischen 'Einsicht': Die Intelligibilien sind nicht nur ein Gegenstand wissenschaftlicher Erkenntnis, sondern vermitteln auch sittliche Einsicht, Erkenntnis des erstrebenswerten Guten und des zu meidenden Schlechten. Wissenschaftliche Erkenntnis und sittliche Einsicht hängen zusammen; hierbei kann man nicht allgemein tugendhaft sein, sondern nur, indem man das Gute praktiziert. Theorie und Praxis gehören zusammen, wobei einerseits die Vernunft, die philosophische Erkenntnis das tugendhafte Handeln bestimmt und andererseits die vernunftorientierte Einsicht in das erstrebenswerte Gute und das zu meidende Schlechte auf das Handeln des Menschen gerichtet ist; es orientiert sich an der Wirklichkeit des Musterstaates und ist nicht rein theoretisch.

In dieser an der Praxis orientierten Philosophie der Ethik[35] kann Fārābī sich auf eine aristotelische Lehre stützen, die in der Fārābīforschung bislang übersehen worden ist, nämlich auf die Interdependenz von Denken und Wahrnehmung: demnach ist das allgemein Gute, die Idee des Guten nicht denkbar ohne sinnliche Wahrnehmung. Daher haben Aristoteles und in seiner Nachfolge Fārābī die Einbildungskraft (*fantasía*) eingeführt; diese schickt als Vermittler zur Denkseele die

[34] Vgl. zu den Details H. Daiber, 'Prophetie und Ethik bei Fārābī (gest. 339/950),' in Chr. Wénin (ed.), *L'homme et son univers au Moyen-Age. Actes du septième congrès international de philosophie médiévale, 30 août–4 septembre 1982.* Philosophes médiévaux; 27 (Louvain-la-Neuve, 1986), S. 729–753.

[35] Vgl. Fārābī, 'Kitāb al-jamʿ bayn raʾyay al-ḥakīmayn Aflāṭūn al-ilāhī wa-Arisṭūṭālīs,' in F. Dieterici (ed.), *Alfārābī's philosopische Abhandlungen aus Londoner, Leidener und Berliner Handschriften* (Leiden, 1890), S. 20,22ff. und dazu Aristoteles, *Anal. post.* I 5; *De anima* III 1.424b.

'Sinneswahrnehmungen' (Aristoteles: *aisthemata*) bzw. die 'Vorstellungsbilder' (Aristoteles: *fantasmata*) des wahrgenommenen Objektes.

Diese 'Vorstellungsbilder' nennt Fārābī 'Nachahmungen' (*muḥākāt*) – eine terminologische Neuschöpfung des Fārābī. Alles sinnlich Wahrnehmbare, aber auch alle Intelligibilia (*al-maʿqūlāt*) werden von der Einbildungskraft nachgeahmt. Denn nicht das Wahrgenommene oder das Gedachte selbst gelangt in die Denkseele des Menschen, sondern lediglich eine Imitation, ein Bild. Die Seele denkt nur in Bildern.

Die hier zugrundeliegende Interdependenz von Denken und Wahrnehmung ist für Fārābī eine weitere Rechtfertigung für die aristotelische Kombination von wissenschaftlicher Erkenntnis (*episteme*) und sittlicher Einsicht (*phronesis*), von theoretischer und praktischer Vernunft. Dieses Miteinander von wissenschaftlicher Erkenntnis und sittlicher Einsicht kommt in Fārābīs Musterstaat (*al-madīna al-fāḍila*) und in seiner 'Musterreligion' (*al-milla al-fāḍila*) zum Ausdruck. Denn die Partikularien (*al-juzʾiyyāt*) der Religion entsprechen den Universalien (*kulliyyāt*) der Philosophie, die die Partikularien beweist.[36] Insofern erscheint die Musterreligion als Nachahmung der Philosophie, sie ist ihr 'ähnlich' (*shabīha*).[37]

Diese Ähnlichkeit beruht, wie wir gesehen haben und wie Fārābī in der von uns genannten Diskussion über die strukturelle Identität von Original und Abbild gezeigt hat, auf gemeinsamer Struktur. Damit entpuppt sich die Religion keineswegs als wertloses Abbild der Philosophie. Überdies kann sie allein die Bürger des Musterstaates überzeugen, das zu glauben und zum Erreichen der höchsten Glückseligkeit das zu tun, was sich von der Philosophie her beweisen, aber nicht von ihr herleiten läßt.

Dies bedeutet keineswegs, daß Philosophie eine Dienerin der Religion ist. Denn die Wirklichkeitsbezogenheit der Philosophie ist Fārābī zufolge nicht nur erkenntnistheoretisch beweisbar, nämlich mit der aristotelischen Lehre von der Interdependenz von Denken und Wahrnehmung; die Musterreligion ist auch ein praktisches, durch die Philosophie als gültig erwiesenes Beispiel für den Zusammenhang von wissenschaftlicher Erkenntnis und sittlicher Einsicht, insofern eine Nachahmung der Philosophie. Mit ihren Vorschriften und Regeln bestimmt sie die Praxisbezogenheit der Philosophie und hat das Ziel, die sittliche Einsicht der Philosophie zu verwirklichen.

---

[36] Vgl. *Kitāb al-milla*, ed. Mahdi, S. 47,12–17.

[37] Vgl. auch *Kitāb al-milla*, ed. Mahdi, S. 46,22.

Hier zeigt sich, daß der Zusammenhang von Theorie und Praxis in Philosophie und Religion auch Religion und Philosophie strukturell miteinander verbindet. Philosophie und Religion sind ebenso aufeinander bezogen, wie Denken und Wahrnehmung, Theorie und Praxis, wissenschaftliche Erkenntnis und sittliche Einsicht, religiöser Glaube ('Meinungen') und Handeln nach den Vorschriften der Gesetze. Der Nachweis dieser Zusammenhänge ist die originelle Leistung des Fārābī.

Die praktische Philosophie, die Aktualität der Religion im Musterstaat beinhaltet die Verwirklichung der Tugenden und die nachahmende Umsetzung der Idee bzw. des Begriffes Ethik in tugendhaftes Handeln. Religion erscheint als Instrument, das die Praxisbezogenheit der wahren Tugenden garantiert: wahre Tugenden existieren nicht als etwas Allgemeines, als Idee, sondern lediglich im ethischen Handeln.

Aus diesem Grund bedarf die Philosophie der Relgion als Instrument. In entsprechender Weise müssen wir eine Aussage des Fārābī in seinem *Kitāb al-ḥurūf*[38] interpretieren, wonach 'die Philosophie den Instrumenten in derselben Weise vorausgeht, wie der Benutzer der Instrumente den Instrumenten zeitlich vorausgeht.'

Diese Äußerung ist eine interessante Modifikation einer alexandrinischen Lehre,[39] die im 10. Jh. n. Chr. von christlichen Philosophen in Bagdad übernommen wurde:[40] hiernach ist Logik nicht ein Teil der Philosophie, sondern ihr Instrument, das den Menschen in die Lage versetzt, zwischen wahr und falsch in theoretischer Philosophie sowie zwischen gut und schlecht in praktischer Philosophie zu unterscheiden.

Fārābī übernahm diese Lehre in einer spezifischen Weise, indem er Logik durch Religion ersetzte: Er klassifiziert in seinem *Kitāb al-ḥurūf* Religion nicht als Teil der Philosophie, sondern als ihr Instrument; in dieser Eigenschaft setzt Religion in die Tat um, was allgemein in der Philosophie existiert, nämlich die philosophische Idee moralischer Einsicht, praktischer Klugheit, die zu höchster Glückseligkeit führt. Hiermit erscheint Religion nicht nur als ein Instrument der Philosophie; sie setzt die Philosophie auch in die Lage, zur praktischen Philosophie zu

---

[38] Ed. Mahdi, S. 132,7f.; Übers. Berman, 'Maimonides,' S. 172.

[39] Vgl. Elias' 'Kommentar zu Aristoteles' *Kategorien*,' in A. Busse, *Commentaria in Aristotelem Graeca*. XVIII/1 (Berolini, 1900, repr. 1961), S. 117,9ff.

[40] Vgl. N. Rescher, *Studies in Arabic philosophy* (Pittsburgh, 1966), S. 42; F. Zimmermann, *Al-Farabi's commentary and short treatise on Aristotle's De interpretatione* (London, 1981), S. 123; Chr. Hein, *Definition und Einteilung der Philosophie*. Europäische Hochschulschriften, R. 20; Bd. 177 (Frankfurt [etc.], 1985), S. 153ff.

werden. Philosophie ist auf die Religion angewiesen, ebenso wie Philosophie der Logik als Instrument bedarf.

Die Autonomie der Philosophie erscheint hier eingeschränkt, insofern als Philosophie und Religion aufeinander angewiesen sind – analog zur aristotelischen Lehre über Denken und Wahrnehmung, wissenschaftlicher Erkenntnis und praktischer Klugheit.

Diese Abhängigkeit der Philosophie von der Religion erscheint bei Fārābī untermauert durch seine Lehre von der prophetischen Eingebung als Quelle der Erkenntnis. Menschliche Erkenntnis hat ihre Grenzen und ist auf die Eingebungen des göttlichen Intellekt an den Regenten des Musterstaates, den Propheten, angewiesen.

Dies zeigt Fārābīs Lehre vom Traum. Sie knüpft an Alexander von Aphrodisias' Lehre des göttlichen aktiven Intellekts (*al-ʿaql al-faʿʿāl*) und an Aristoteles' Lehre von Traum und Weissagung an.[41] Er mag dabei Aristoteles' *Parva naturalia* konsultiert haben, die in arabischer Übersetzung erhalten ist.[42]

Für Fārābī sind die Träume das Resultat des Zusammenwirkens zwischen Wahrnehmung, nachahmender Einbildung und göttlichem 'aktiven Intellekt'. Wenn die nachahmende Einbildung auf die gegenwärtigen und zukünftigen Partikularien (*al-juzʾiyyāt*), auf die 'getrennten Intelligibilia' (*al-maʿqūlāt al-mufāriqa*) und auf 'alle erhabenen Dinge' gerichtet ist, kurzum auf dasjenige, was vom göttlichen Intellekt inspiriert ist, spricht man von Prophetie (*nubuwwa*). Diese tritt dann auf den Plan, wenn philosophische Erkenntnis dessen, was gut ist, nicht ausreicht und der Ergänzung durch die Eingebung (*waḥy*) des göttlichen aktiven Intellekts bedarf.[43]

Es ist daher kein Wunder, daß für Fārābī der Regent des Musterstaates nicht nur Philosoph, sondern auch Prophet sein muß. Da den Regenten die göttlichen Eingebungen des aktiven Intellekts in der Form von Bildern erreichen, die die Intelligibilia, die Universalien nachahmen, kann er sie auch in dieser Form an die Menschen, seine Untertanen weitergeben. Hierbei kann der Regent des Musterstaates sich der philosophischen Beweisführung bedienen. Er kann sich aber auch auf

---

[41] Vgl. Daiber, 'Prophetie,' S. 737ff.

[42] MS Rampur (Indien) 1752. Eine Edition wird von R. Hansberger, Oxford vorbereitet.

[43] Vgl. *Al-madīna al-fāḍila*, ed. F. Dieterici (Leiden, 1964), S. 51,14ff.; ed. (transl.) R. Walzer, *Al-Farabi on the Perfect State* (Oxford, 1985), S. 220,16ff.

prophetische 'Warnungen' beschränken.[44] In beiden Fällen muß er 'in der Lage sein', sein Wissen 'durch (seine) Äußerungen auf vortreffliche Weise vorzustellen'.[45] Er muß über rhetorisch-poetische Fähigkeiten verfügen. Er hat die Aufgabe, in philosophischen Beweisführungen den philosophisch Gebildeten zu überzeugen und die Masse, die nur ein 'bildhaftes Wissen' habe,[46] durch 'Warnungen' und 'Vorschriften' zu überzeugen. Als Philosoph kann er sich in seiner Instruktion des Bürgers verschiedener Hilfsmittel der Logik und Beweisführung bedienen, als Prophet bedient er sich der Metaphern der rhetorisch-poetischen Sprache.

Hierbei entpuppen sich die prophetischen Warnungen nicht als unvollkommenes, der Philosophie unterlegenes Wissen, weil es an die Anhänger der Religion und an solche gerichtet ist, die philosophische Wahrheit nur in vereinfachenden Bildern verstehen können. Eine solche vereinfachende bildhafte Vermittlung von Wissen mag es geben. Doch als Eingebung des göttlichen aktiven Intellekts an den Propheten ergänzt sie das philosophische Wissen. Dies geschieht in einer Weise, die der Ergänzung der theoretischen Erkenntnis durch die praktische Orientierung, die moralische Einsicht und praktische Klugheit. Die Religion und ihre Partikularien werden so zu einer 'Nachahmung', zu einem ergänzenden Bild der Philosophie, der Universalien, ohne mit dieser identisch zu sein. Sie ist ein Instrument der Philosophie und verhilft dieser, so zur praktischen Philosophie zu werden. Philosophie erscheint hier primär als praktische Ethik.

Die hier sichtbar werdende Grenze philosophischen Wissens bedeutet auch eine Einschränkung des Herrschers ausschließlich als Philosoph. Er muß auch ein Prophet sein – dies nicht nur im Hinblick auf die Anhänger der Religion, die Masse, sondern auch angesichts der Grenzen philosophischen Wissens.

Philosophisches Wissen muß durch den göttlichen intellectus agens, nämlich durch prophetische Eingebungen vermittelt und ergänzt werden. Das Wissen, das der Regent so erhält, kann nur in Form von nachahmenden Bildern an den Untertanen vermittelt werden. Diese

---

[44] Vgl. *al-mundhir* in *al-Madīna al-fāḍila*, ed. Dieterici, S. 59,1; ed. Walzer, S. 244,13 (vgl. S. 389).

[45] Vg. *al-Madīna al-fāḍila*, ed. Dieterici, S. 59,16ff.; ed. Walzer, S. 246,12ff. bes. ed. Dieterici, S. 59,5f.; ed. Walzer, S. 246,2f.

[46] Vgl. *al-Madīna al-fāḍila*, ed. Dieterici, S. 70,1ff.; ed. Walzer, S. 278,12ff.; Kraemer, 'Alfarabi's opinions,' S. 115f.

Nachahmungen ersetzen das Original, das nur in der Form von Abbildern wahrgenommen und weitervermittelt werden kann. Folglich erscheinen diese Abbilder orientiert an der Wirklichkeit – ebenso wie das philosophische Denken in seiner Interdependenz von Theorie und Praxis.

Dieses Wechselverhältnis zwischen Religion und Philosophie – Fārābī spricht von Religion als Nachahmung der Philosophie – erlaubt den philosophischen Beweis für religiöse Wahrheiten, die von der Eingebung an den Propheten gespeist werden und sinnfällig im Musterstaat theoretische Einsicht und praktische Philosophie symbolisieren. Der Regent im Musterstaat ist ein Philosoph und ein Prophet, der mit Hilfe von Gottes Eingebung – platonisch gesprochen[47] durch 'Angleichung' an Gott, d.h., indem er Gottes Vorschriften nacheifert[48] – den Staat regiert.[49]

Gleichzeitig hat Fārābī, wie wir schon gesehen haben, das Wissen des Regent-Philosophen und Regent-Propheten als Nachahmungen klassifiziert, als Abbilder göttlichen Handelns. Die bildhafte Gestaltung göttlichen Handelns entspricht Fārābī zufolge den von uns bereits genannten (s.o.) Wirkungen der Prinzipien des (Muster)staates (*afʿāl al-mabādiʾ al-madaniyya*).

Hier erscheint die Wirklichkeit des Menschen wie bei Plato und Aristoteles als politisches Wesen (ζῷον πολιτικόν) mit ethischen Verpflichtungen in der Gemeinschaft des Staates betont.[50] Philosophie ist nicht mehr ein Privileg der Spezialisten, der Elite, sondern kann durch den Regent-Philosophen an den Bürger vermittelt werden – nämlich in der Gestalt der Religion, der von ihr vorgeschriebenen Regeln und Gesetze, im Musterstaat. In der bildhaften Gestaltung der Philosophie erscheint Religion als einzig richtige Form des Philosophierens, die sich an der Praxis des Musterstaates orientiert, als Ethik.

Gleichzeitig ist das Original der bildhaften Gestaltung, der Religion, nämlich die Universalien der Philosophie, Gegenstand des Strebens

---

[47] Vgl. *Rep.* 613B.

[48] Vgl. Fārābī, 'Risāla fī mā yanbaġī an yuqaddam qabl taʿallum al-falsafa,' in F. Dieterici (ed.), *Alfārābī's philosopische Abhandlungen aus Londoner, Leidener und Berliner Handschriften* (Leiden, 1890), S. 53,13ff.; *Kitāb al-milla*, ed. Mahdi, S. 64–66; Übers. Berman, 'Maimonides on political leadership,' S. 122f.

[49] Vgl. *Kitāb al-milla*, ed. Mahdi, S. 64,16f.

[50] Vgl. M. Maróth, 'Griechische Theorie und orientalische Praxis in der Staatskunst von al-Fārābī,' *Acta antiqua* 26 (1978), S. 465–469; F.A. Sankari, 'Plato and Alfarabi. A comparison of some aspects of their political philosophies,' *Vivarium* 8 (1970), S. 1–9 (auch in *Moslem world* 60 (1970), S. 218–225 und *Studia islamica* 7 (1970), S. 9ff.).

nach philosophischer Erkenntnis. Diese philosophische Erkenntnis konstituiert sich nur in der bildhaften Gestaltung – wie Aristoteles in seiner in arabischer Übersetzung zugänglichen Schrift *Parva Naturalia*, in dem Buch über Gedächtnis und Erinnerung formulierte: 'ein Denken ohne die Bilder der Vorstellung ist nicht möglich' (449b30f.). So ist für Aristoteles 'ein auf dem Bild gemaltes Tier so gut wie ein Bild, ein und dasselbe ist dies beides, nur das Wesen ist beidemal nicht das gleiche, und man kann es betrachten als Tier und Bild' (450b20ff.).

Darüber hinaus zeigt Fārābī die Grenzen der Erkenntnis auf, die der Eingebung des göttlichen aktiven Intellekts an den Propheten bedarf. Dieser gibt als Regent sein philosophisches Wissen in der bildhaften Gestaltung der Religion weiter.

Fārābī zeigt sich hier beeinflußt von Plotin, Proclus, Aristoteles' *Metaphysik*, Themistius' Kommentar hierzu und vor allem von Alexander von Aphrodisias' Abhandlung über die 'Prinzipien des Universums'.[51] In deren Fußspuren geht er von einer hierarchischen Stufung zwischen dem göttlichen Einen, dem nachfolgenden Ersten Intellekt und den hieraus emanierenden neun Intellekten[52] aus. Entsprechend der in Fārābīs Quellen vorausgesetzten Stufung des Seins gelangen wir nicht zur letzten Wahrheit, sondern nur zum Ebenbild der Wahrheit.

Die Ursache dieser Wahrheit hatte Plato als das eigentliche höchste Gut bezeichnet, das über der Vernunfterkenntnis und Wahrheit stehe (s.o.). Fārābī operierte hier mit den neuplatonischen Emanationen, den Zwischenursachen, die die Präsenz des verursachenden göttlichen Intellekts in den verursachten Dingen garantierte. Sie erlauben aber keine Erkentnis der ersten Ursache, ihres 'Warum', sondern sind nur eine Indikation ihrer Existenz.[53]

So ist in einer für Ibn Sīnā vorbildlichen und weiter entwickelten Weise[54] Aristoteles' These einer Gleichheit von Ursache und Verursachtem (vgl.

---

[51] Vgl. M. Maróth, *Die Araber und die antike Wissenschaftstheorie*. Islamic philosophy, theology and science; 17 (Leiden [etc.], 1994), S. 203ff.; Der Traktat des Alexander ist hrsg. v. Ch. Genequand, *Alexander of Aphrodisias on the Cosmos*. Islamic philosophy, theology and science; 44 (Leiden [etc.], 2001).

[52] Vgl. Maróth, *Die Araber*, S. 199ff.

[53] Vgl. Maróth, *Die Araber*, S. 162ff.

[54] Vgl. Daiber, 'The limitations of knowledge according to Ibn Sīnā. Epistemological and theological aspects and the consequences,' in M. Lutz-Bachmann, A. Fidora [et al.] (eds.), *Erkenntnis und Wissenschaft. Probleme der Epistemologie in der Philosophie des Mittelalters* (Berlin, 2004), S. 25–34 (auch in Daiber, *The struggle for knowledge in Islam. Some historical aspects*. [Gefolgt von einer bosnischen Übersetzung]: *Borba za znanje u Islamu. Neki historijski aspekti* (Sarajevo, 2004), S. 87–104, bosnische Übers. S. 92–109).

Met. 1032a25) unter dem Eindruck der Tabula porphyriana eine Hierarchie von Ursachen und Verursachtem entwickelt worden, wonach die erste Ursache nur insofern alles Verursachte enthält, als dieses ihm ontologisch unterlegen ist und sich von ihm durch die Vielheit der Unterschiedlichkeiten unterscheidet. Fārābī sprach von den Partikularien der Religion, die als Abbild der Philosophie eine Indikation der philosophischen Wahrheit sind, ihrer Ursache.

Wenn Fārābī fernerhin, letztlich anknüpfend an die platonisch-neuplatonische[55] Unterscheidung zwischen Gottes Wesen und Wirken, von 'göttlichen Wirkungen' sprach, die sich in den 'Wirkungen der Prinzipien des Musterstaates', den *af ͑āl al-mabādi ͗ al-madaniyya* bildhaft gestalten (s.o.), so hat er Platos Modell des Stadtstaates und seiner hierarchischen Strukturen in ein aristotelisches Konzept von theoretischer und praktischer Philosophie integriert und unter dem Eindruck des Neuplatonismus die aristotelische Auffassung vom Denken in 'Bildern der Vorstellung' ergänzt mit der Vorstellung von der prophetischen Inspiration, die vom göttlichen Intellekt gespeist wird und vom Regenten, der Prophet und Philosoph ist, weitergegeben wird. Hierbei erscheint bei Fārābī Wissen in seiner hierarchischen Strukturierung primär an den Bedürfnissen der Gesellschaft orientiert.

[55] Vgl. Daiber, *Das theologisch-philosophische System des Mu ͑ammar Ibn ͑Abbād as-Sulamī (gest. 830 n. Chr.)* (Beirut, 1975), S. 166f.

MEDICINE & MAGIC

AL-RĀZĪ (D. 925) ON THE BENEFITS OF SEX
A CLINICIAN CAUGHT BETWEEN PHILOSOPHY AND MEDICINE[1]

Peter E. Pormann

In his book on spiritual medicine (*ṭibb rūḥānī*), the famous physician Abū Bakr Muḥammad b. Zakariyyā' al-Rāzī (*d.* 925) warns against the harmful effects of sexual intercourse, which, as he puts it, 'weakens the eyesight, wrecks and exhausts the body, speeds up aging, senility, and withering, damages the brain and the nerves, and renders the [bodily] strength weak and feeble, in addition to many other conditions which would take too long to mention.'[2] Like Ibn Sīnā's *Book of Curing* (*Kitāb al-Shifā'*), al-Rāzī's *Spiritual Medicine* is a philosophical work, even if the title has medical overtones. The strongly negative view which al-Rāzī expresses in this philosophical work is in stark contrast to the more nuanced picture which he paints in a treatise with the programmatic title *On Sexual Intercourse, its Harmful and Beneficial Effects, and Treatment* (*Kitāb al-Bāh wa-manāfi'ihī wa-maḍārrihī wa-mudawātihī*), belonging to the specialist medical literature 'On Sexual Intercourse (*fī 'l-bāh*)'.[3] Although he and other physicians writing on the topic list various

[1] The idea of broaching this subject came from conversations which I had with Emilie Savage-Smith, when we were finalising our *Medieval Islamic medicine.* Peter Adamson, Charles Burnett, and Simon Swain read earlier drafts of this article and made invaluable comments; moreover, Hans van de Velde and Arnoud Vrolijk kindly provided me with a photocopy of the Leiden manuscript discussed here. I would like to express my profound gratitude to all of them.

[2] Al-Rāzī, *Rasā'il falsafiyya*, ed. P. Kraus (Cairo, 1939; repr. Damascus, 2005), p. 75, l. 1–3; tr. A.J. Arberry, *The Spiritual Physick of Rhazes* (London, 1950), p. 81. The translation here is taken from P.E. Pormann and E. Savage-Smith, *Medieval Islamic medicine.* New Edinburgh Islamic surveys (Edinburgh, 2007), p. 62.

[3] Al-Rāzī's own treatise is preserved in a number of manuscripts, see M. Ullmann, *Die Medizin im Islam.* Handbuch der Orientalistik, I Abt.; Erg. VI/1 (Leiden, 1970), p. 194. It has been edited by H. 'Abd al-'Azīz and 'A. 'Abd al-Ḥamīd (eds.), *Al-Nisā'. Thalāth makhṭūṭāt nādira fī 'l-jins* (Cairo, 1999), pp. 149–176, from the photocopy of the MS Leiden, Universiteitsbibliotheek, Or. 1308, f. 1a–18b. In their preface, the editors only say that they 'have obtained a photocopy from the Manuscript Institute (*wa-qad ḥaṣalnā 'alā ṣūratin waraqiyyatin min ma'hadi 'l-makhṭūṭāti*)' (p. 19, ult.), without identifying their source as this Leiden manuscript.

diseases and debilities which can result from having sex, they also stress, in various degrees, its benefits. In this context, al-Rāzī is no exception: he has a whole chapter devoted to the topic, titled 'On the Benefits arising from Having Sex (*Fī 'l-manāfiʿ al-kāʾina fī 'stiʿmāl al-jimāʿ*)'. In the present contribution, al-Rāzī's views on the subject will be explored and contextualised in order to offer a solution to the apparent contradiction which exists between his medical and his philosophical writings.

Greek Antecedents

As is the case with most medical topics discussed in the medieval Arabic literature, Greek precursors of physicians in the medieval Islamic world set the agenda for later developments. Rufus of Ephesos (*fl. c.* 100) wrote a number of relevant treatises, one of which was a monograph *On Sexual Intercourse* (*Peri aphrodisiōn*).[4] In the latter Rufus states clearly: 'Sexual intercourse is a natural activity, and nothing natural is harmful.'[5] He lists a number of negative effects due to excessive sex, but then turns to its positive aspects. For instance, intercourse 'empties repletion, [...], and is useful against melancholy'.[6]

Galen (*d. c.* 216/17), who himself drew heavily on Rufus, appears to be the second important source of inspiration for Arabic authors. In various works, he dealt with the topic of sexual intercourse. In keeping with earlier medical and biological theory, Galen held that both men and women have semen (*sperma*).[7] A natural way to expel it is sexual intercourse. When this does not take place, the superfluous semen can putrefy and lead to certain diseases. A famous case in point is the

[4] M. Ullmann, 'Die arabische Überlieferung der Schriften des Rufus von Ephesos,' *Aufstieg und Niedergang der römischen Welt*, II 37.2 (Berlin [etc.], 1994), pp. 1293–1349, here pp. 1338–1339. *On Sexual Intercourse* is preserved in Greek and Arabic fragments; cf. Ch. Daremberg & Ch.É. Ruelle, *Œuvres de Rufus d'Éphèse* (Paris, 1879; repr. Amsterdam, 1963), fr. 60, pp. 318–323; and M. Ullmann 1970, p. 75; F. Sezgin, *Medizin-Pharmazie-Zoologie-Tierheilkunde bis ca 430 H.* Geschichte des arabischen Schrifttums; 3 (Leiden, 1970), p. 66.

[5] Ed. Daremberg & Ruelle pp. 318, l. 1–2; a quotation taken from Aëtius, Book III, ch. 6 (I, p. 265, l. 13–14 ed. Olivieri).

[6] As quoted in al-Rāzī, *al-Kitāb al-Ḥāwī*, 23 vols. (1st ed., Hyderabad, 1955–70), X, pp. 313, l. 6–8; see pp. 320, l. 1–10 (ed. Daremberg & Ruelle); I, pp. 266, l. 11–18 (ed. Olivieri).

[7] See L. Dean-Jones, *Women's bodies in classical Greek science* (Oxford, 1994), pp. 153–160. Ph. de Lacy (ed.), *Galen, On semen*. Corpus medicorum Graecorum; 3/1 (Berlin, 1992), pp. 48–51.

so-called 'suffocation of the womb (*husterikē pnix*)' occurring in women, especially young virgins and widows.[8] In a well-known passage from his *On the Affected Parts*, Galen ascribes the cause of this condition in women to the fact that either the menses or the sperm is not naturally expelled. He also makes a number of remarks about the deleterious effects of holding back semen in men.[9] Likewise, in the rest of his œuvre, Galen is acutely aware that not having sex can sometimes lead to illness, and that, conversely, intercourse can be an effective cure for certain disorders.

Al-Rāzī on the Benefits of Sex

Many of these ideas formulated by Rufus and Galen recur in al-Rāzī's treatise *On Sexual Intercourse* (*Kitāb fī 'l-bāh*). It contains the thirteen chapters, which address a whole range of topics. The first six deal with more general issues – Chapter One is entitled 'On the Harmful Effects Arising from Excessive Intercourse (*Fī dhikr al-maḍārr al-mutawallida ʿan al-isrāf fī 'l-bāh*)' – whilst the last seven are of a more practical nature and contain advice about different types of medication. Chapter Four is particularly relevant to our argument, and therefore deserves to be cited in full here:[10]

On the Benefits arising from Having Sex

> [I] Some people purport that sexual intercourse has no use at all under any circumstances. Others express a different opinion because of what is evident from sense perception. The excellent Hippocrates and Galen [also] testify to this.
> [II] For Galen said in the sixth book [*maqāla*] of his work known as *On the Affected Parts* (*Kitāb al-Aʿḍāʾ al-bāṭina*):[11] '[1] When young men who

---

[8] H. King, *Hippocrates' woman. Reading the female body in ancient Greece* (London, 1998), pp. 231–234; R. Flemming, *Medicine and the making of Roman women. Gender, nature, and authority from Celsus to Galen* (Oxford, 2000), pp. 333–340; A.E. Hanson, 'The Medical Writers' Women,' in D.M. Halperin [et al.] (eds.), *Before sexuality. The construction of erotic experience in the ancient Greek world* (Princeton, 1990), pp. 309–338.

[9] *De loc. aff.*, VIII, p. 417ff. (Kühn)/ p. 184ff. tr. R.E. Siegel, *Galen on the affected parts* (Basel [etc.], 1976).

[10] pp. 160, l. 11–161, l. 5 (ed. al-ʿAzīz & ʿAbd al-Ḥamīd)/ f. 8b6–9b2 (MS L); the numbers in square brackets are introduced to facilitate the discussion below.

[11] *De loc. aff.*, VIII, p 417, l. 16–418, l. 15 (Kühn)/ p. 184 tr. Siegel.

have a lot of sperm do not have intercourse, their heads are heavy, they become disturbed and sad, and they lose their appetite and joy. [2] I know people with a lot of sperm who deprived themselves of sex (*jimāʿ*) because of a kind of philosophy, asceticism or something else. Their bodies became cold, their movement difficult, they suffered from sadness without reason, displayed the symptoms of melancholy, lost their appetite, and had indigestion. [3] I saw a man abandon sex, although he had previously had constant sexual intercourse. He lost his appetite, so much so that even eating a small quantity was not agreeable to him. When he forced himself to eat only a little bit, he vomited immediately, and he had constant symptoms of melancholy. [4] After returning to his [previous] sexual habits, these symptoms ceased extremely quickly.'

[III] He also said:[12] 'Those who are used to having sex and then abandon it may suffer from the disease known as "priapism", i.e. constant tension of the penis accompanied by a severe pain; this sometimes occurs together with a spasm.

[IV] He said in his book *Epidemics* in the fifth book (*maqāla*) containing the commentary of the sixth:[13] 'Having a lot of sexual intercourse when the body is strong is useful against phlegmatic diseases.'

[V] In it, it is also said about it [?]: '[1] Sex is also useful for those in whose body there are humours which produce a vapour-like smoke [*bukhār dukhānī*]. [2] For it [sex] prevents these smokes from becoming congested in the body, which would generate acute and acrid fevers. One can observe that when the semen thickens, and a lot of it accumulates and becomes warm, then it increases the palpitation of the heart, trembling, tightening of the chest, craziness (*hawas*), and vertigo. [3] Moreover, the pain called 'suffocation of the womb (*ikhtināq al-arḥām*)' occurs in women only because of the loss of sexual intercourse (*fuqdān al-jimāʿ*), and no remedy is better than it [sex].'

[VI] Galen said in his book *The Small Art*:[14] 'Sex empties repletion, lightens the body, and imparts this [these benefits] to it [the body]. It dissolves vehement thinking, and appeases passionate anger. Therefore it [sex] is extremely useful against madness and melancholy, and a powerful remedy for diseases caused by phlegm. It [sex] makes certain people eat more, and improves their digestion.'

[VII] Elsewhere he says: 'It dissolves vehement thinking, brings one's opinions to rest and relaxation, and assuages the passion of lovers, even if they practise this [sex] with someone whom they do not desire.'

[VIII] In summary: [1] It is gratifying [*mumtiʿ*] that some kind of bodies do not benefit from rejecting it [sex] altogether, for nature arranges

---

[12] *De loc. aff.*, VIII, pp. 450, l. 6–9 (Kühn)/ p. 197 tr. Siegel.

[13] This is an extract from Galen's commentary to Hippocrates' *Epidemics; In Hipp. Epid. VI comment.* VI 23 (XVII pars 2, pp. 284, 2–4 Kühn). The Arabic title which the editors (pp. 161, l. 1) were unable to read is 'إبيذيميا' (MS L f. 9a1).

[14] This is Galen's *Technē iatrikē* (*Ars med.*), see discussion below.

> and organises things correctly, and does nothing in vain. [2] The benefits which one derives from it [sex] for preserving one's health are, however, achieved through moderation; [these benefits include,] for instance, emptying repletion and spasmodic weakness [*iʿyāʾ tamaddudī*], cooling bodies in which smoke-like vapours exist, and protecting the body from superfluities and hot vapours. [3] It is well known that only bodies which have a lot of blood, semen, and strong innate heat derive these benefits from it [sex]; the other [bodies] do not.

Most of this chapter consists of quotations, namely from Galen's *On the Affected Parts*, the *Small Art*, and his commentary on Hippocrates' *Epidemics*. Although he mentions Hippocrates at the beginning of the chapter as someone thinking that sex can have its benefits, he only cites the 'Father of Medicine' indirectly here. At the end, he sums up his own opinion in the light of these quotations. In al-Rāzī's case, quoting is hardly ever the modern scholarly technique by which one reproduces the exact words of one's source. For example, in his *Comprehensive Book* (*al-Kitāb al-Ḥāwī*, known in Latin as *Liber Continens*), al-Rāzī 'cites' Galen and Paul of Aegina in a variety of ways.[15] Sometimes, he is quite literal and hardly alters his source, whilst at other times his quotations are merely loose paraphrases in which even the order of the original idea has been altered. Jennifer Bryson has suggested that when al-Rāzī gives references to his sources, he may occasionally be quoting from a summary of the treatise in question rather than the original.[16] In the chapter just quoted, we can observe these different levels of 'citation'.

The longest and most faithful quotation is the first one [II]. Al-Rāzī names his source as Galen's *On the Affected Parts*, and even gives an exact reference to the number of the book (six). Although space does not permit to argue this point fully, comparison between this quotation and the Arabic translation attributed to Ḥunayn b. Isḥāq, which is extant in a number of manuscripts, shows that al-Rāzī quotes from this version

[15] See U. Weisser, 'Zitate aus De methodo medendi im *Ḥāwī*,' in G. Endress & R. Kruk (eds.), *The ancient tradition in Christian and Islamic Hellenism. Studies on the transmission of Greek philosophy and sciences, dedicated to H.J. Drossaart Lulofs on his ninetieth birthday*. CNWS Publications; 50 (Leiden, 1997), pp. 279–318; and P.E. Pormann, *The Oriental tradition of Paul of Aegina's* Pragmateia. Studies in ancient medicine; 29 (Leiden, 2004), pp. 60–92.

[16] J. Bryson, *The* Kitāb al-Ḥāwī *of Rāzī (ca. 900 AD). Book One of the Ḥāwī on brain, nerve, and mental disorders. Studies in the transmission of medical texts from Greek into Arabic into Latin*, Ph.D. diss., Yale University (New Haven, 2000), pp. 67–73, especially p. 69.

here.[17] The general context in the original is a discussion of the female womb, and notably its suffocation, to which we have already alluded above. Though overall faithful, al-Rāzī omits one sentence from the original at the end of [1], in which Galen says that 'Plato compares these [people suffering from not having sex] to trees which have more fruits than is equitable.'[18] The rest of the quotation is fairly literal, but there is one quite interesting point to be made. In the Greek original, it is 'out of shame (*hup' aischunēs*)' that the people in the first example [2] abstain from sex. This is rendered as *li-ḍarbin mina 'l-tabattuli* (lit.: 'because of a kind of living-in-chastity') in al-Rāzī's direct source, the translation attributed to Ḥunayn. Al-Rāzī, however, rephrases this expression and expands its meaning, saying *bi-ḍarbin mina 'l-falsafati wa 'l-taqashshufi wa-ghayrihī* (lit.: 'through a kind of philosophy, asceticism, and other things).' He thus elaborates on the term *tabattul* (chastity, retiring-from-the-world) by giving different reasons why one might seclude oneself such as philosophy or asceticism.

Paragraphs [III] and [IV] contain still fairly accurate, yet shorter quotations, although the definition of priapism in the second half of paragraph [III] is lacking in the original Greek. The next paragraph [V] constitutes a bit of a puzzle. It is introduced by the words *wa-qīla fīhi ayḍan ʿanhu* (lit.: 'It was also said about it/him from it/him'). In its present form this paragraph does not occur in any of Galen's works edited to date. There are, however, parallels in the Arabic medical literature, notably for the concept of vapour being expelled with the help of sexual intercourse.[19]

---

[17] I used London, Wellcome Library, MSS Arabic 14a and 501 to compare al-Rāzī's quotation with the translation. The relevant passage is found on f. 164a–b and 198b–199a respectively. See N. Serikoff, *Arabic medical manuscripts of the Wellcome library. A descriptive catalogue of the Ḥaddād collection (WMS 401–487)*. Sir Henry Wellcome Asian Series; 6 (Leiden, 2005), no. 501; and I. Garofalo, 'La traduzione araba del *De locis affectis* di Galeno,' *Studi classici e orientali* 45 (1995), pp. 13–63.

[18] *De loc. aff.*, VIII, pp. 418, 2–3 (Kühn)/ p. 184 tr. Siegel; this reference appears elsewhere in the literature on sexual hygiene, e.g. Qusṭā Ibn Lūqā, *Kitāb fīl-bāh wa-mā yuḥtāgu ilaihi min tadbīr al-badan fīstiʿmālihi = Das Buch über die Kohabitation und die für ihre Ausübung notwendigen körperlichen Voraussetzungen. 1. Abhandlung*, ed. [...] G. Haydar. Doctoral dissertation (Erlangen-Nürnberg, 1973), pp. 40, l. 3–5.

[19] E.g. Ibn Sīnā, *al-Qānūn fī 'l-ṭibb*, 3 vols. (Būlāq, 1877), II, pp. 535, l. 5–6; G. Bos, *Ibn al-Jazzār on sexual diseases and their treatment. A critical edition of Zād al-musāfir wa-qūt al-ḥāḍir*. Sir Henry Wellcome Asian series; 3 (London, 1997), ch. 11, p. 155 (txt.), p. 274 (tr.); cf. E. Montero Cartelle, *Constantini* Liber de Coitu. *El tratado de andrologia de Constantino el Africano* (Santiago de Compostela, 1983), ch. 9 and 10.

Nor does the quotation from the *Small Art* (*Ṣināʿa ṣaghīra*) [VI], as Galen's treatise *On the Art of Medicine* was known in Arabic, represent a faithful version of the Greek original. In the source, we read:[20]

> According to Epicurus, sex has no health benefit whatsoever. In truth, however, [it is useful] when practised in such long intervals, that one does not feel any feebleness, but rather has the impression to be lighter than one is, and breathe more freely. The right time to have [sex] is when the body is entirely in the middle of all external states, being neither too full nor too empty, neither too cold nor two warm, and neither excessively dry nor moist. If one departs from the mean, then the error should be small. For sex it is better to have a hotter rather than colder, a full rather than an empty, and a moist rather than a dry body.

There is therefore a great discrepancy here between what Galen actually said about sex in his *Art of Medicine*, and what al-Rāzī reports Galen to have said. On the one hand, the information given by Galen – that Epicurus thought that sex provides no health benefit, but that this is not so, provided that one has sex in moderation – is basically absent from al-Rāzī's quotation. Conversely, al-Rāzī talks about sex emptying and lightening the body; helping against melancholy, madness, and phlegmatic diseases; and improving appetite and digestion. Galen, however, did not mention any of these things in the *Art of Medicine*, although they do occur elsewhere in his œuvre.[21] A possible explanation for this disparity is that proposed by Bryson mentioned above: al-Rāzī may be quoting here from a summary of this Galenic treatise rather than the original itself. The author of the intermediate source would have included information which reflects Galenic theory developed in other treatises. The *Alexandrian Summaries* often contain additional material which goes beyond the treatise they 'summarise'.[22] It may therefore be the case here that al-Rāzī quotes the *Alexandrian Summary* of the *Small Art*.

---

[20] *Ars med.*, I. pp. 371, l. 14–372, l. 7 (Kühn).

[21] *De loc. aff.*, VIII, pp. 418–19 (Kühn)/ p. 184 tr. Siegel [as in n. 8]; *in Hipp. Epid. VI comment.*, XVII/2, p. 294 (Kühn).

[22] See I. Garofalo, 'Aspetti della trasmissione del sapere anatomico greco nel mondo islamico,' in C. Sarnelli Cerque [et al.] (eds.), *Atti del simposio internazionale: La civiltà islamica e le scienze* (Napoli 1995), pp. 63–70, here on p. 65; A.M. al-Dubayan, *Galen: 'Über die Anatomie der Nerven'. Originalschrift und alexandrinisches Kompendium in arabischer Überlieferung*. Islamkundliche Untersuchungen; 228 (Berlin 2000), pp. 51–62; and P.E. Pormann, 'The Alexandrian Summary (*Jawāmiʿ*) of Galen's *On the sects for beginners*. Commentary or Abridgment?' in: P. Adamson [et al.] (eds.), *Philosophy, science and exegesis in Greek, Arabic and Latin commentaries*. Bulletin of the Institute of Classical Studies. Supplement; 83, 2 vols. (London, 2004), II, pp. 11–33, here on pp. 16–17.

After a final vague quotation ([VII]) from 'elsewhere (*fī mawḍiʿin ākhara*)' in Galen's works, al-Rāzī himself summarises his ideas about the beneficial effects of sex. Given that the chapter mostly consists of quotations from Galen (and possibly summaries of Galen's works), it comes as no surprise that al-Rāzī generally agrees with the great Greek physician. Sex can help keep the body in balance, and is useful against certain diseases such as melancholy and madness. It is, however, only advised for a specific group of people who have a lot of blood, sperm, and innate heat. He even goes so far as to say that it is *mumtiʿ* – giving *mutʿa* (enjoyment, pleasure, delight) – that some bodies derive no benefit from abandoning sexual intercourse, that is to say, that sex is good for certain people. He quotes the well-known phrase that nature does nothing in vain, an idea originally formulated by Aristotle which had gained wide-spread acceptance in Late Antiquity and the medieval Islamic world.[23]

Al-Rāzī also repeats his conclusions in a somewhat modified and shortened form in his *Book for al-Manṣūr* (*al-Kitāb al-Manṣūrī*), the fourth treatise (*maqāla*) of which contains a chapter entitled 'On the Beneficial and Harmful Effects of Sexual Intercourse and How to Engage in It (*Fī manāfiʿ al-jimāʿ wa-maḍārrihī wa-jihati ʾstiʿmālihī*)'.[24] We can therefore be certain that these conclusions express al-Rāzī's medical view on the subject. It would appear, however, that this medical opinion on the subject can hardly be reconciled with the stance against sex which he takes in the chapter 'On Sexual Intercourse' in his *Spiritual Medicine*. For in the latter, al-Rāzī does not talk at all about possible health benefits to be derived from sex. On the contrary, he insists that any man of intellect (*ʿāqil*) should abstain at all cost from intercourse. He especially rejects the analogy between sex and food or drink: without the latter one cannot survive for the obvious reasons, but in the case of the former, the situation is different. He says:[25]

> Moreover, this desire [*lidhdha*, i.e. for sex] is the one which deserves most to be rejected, for it is not necessary for staying alive as food and drink are. Abandoning it [sex] does not cause any visible or palpable pain [*alam*] as in the case of hunger or thirst.

---

[23] J.G. Lennox, 'Nature does nothing in vain,' in id., *Aristotle's philosophy of biology* (Cambridge, 2001), pp. 205–223.

[24] Ḥ. al-Bakrī al-Ṣadīqī (ed.), *al-Manṣūrī fī ʾl-ṭibb* (Kuwait, 1987), pp. 220–221.

[25] Al-Rāzī, Rāzī, *Ras. falsaf.*, p. 76, l. 11–13.

I shall shortly suggest a solution for explaining this apparent contradiction between al-Rāzī's position in his philosophical and his medical works respectively. First, however, it is useful to look briefly at previous Arabic writers on sexual intercourse in order to contextualise al-Rāzī's views.

Al-Rāzī's Predecessors: al-Kindī and Qusṭā b. Lūqā

The earliest treatise preserved today with the title *On Sexual Intercourse* (*fī 'l-bāh*) is that by the celebrated Arabic polymath al-Kindī.[26] This treatise, however, is relatively short and is not really concerned with the question whether or not one should have sex. Rather, al-Kindī simply explains the physiological causes for impotence and lack of sperm production, and provides a number of remedies to cure these problems. Within al-Kindī's pharmacological writings, *On Sexual Intercourse* occupies a position between the more theoretical *On Degrees*, a work on the properties of different simple drugs, and the mostly practical *Formulary*, a recipe book. As Peter Adamson put it:[27]

> [...] The interrelated *On Degrees, On Coitus* and *Formulary* show that al-Kindi was capable of approaching a single topic like pharmacology from a theoretical perspective, a practical perspective, or a perspective halfway in between.

We shall return to the problem of different approaches to a single topic below; for now, let us look at the second, and more important precursor to al-Rāzī.

The Christian author of Greek origin Qusṭā b. Lūqā wrote two treatises on the subject, one *Book on Sexual Intercourse* (*Kitāb fī 'l-Bāh*) in twenty two chapters (*bābs*) and a longer *Book on Sexual Intercourse and the Regimen of the Body Which One Needs in order to Engage in Sex* (*Kitāb al-Bāh wa-mā yuḥtāju ilayhi min tadbīr al-badan fī 'sti'malihī*), consisting of two treatises (*maqālas*), the first comprising seven chapters, and the second consisting on a long collection of recipes.[28] Both

---

[26] It has been edited by G. Celentano, *Due scritti medici di Al-Kindī* (Naples, 1979), pp. 21–36. For al-Kindī in general, see P. Adamson, *Great medieval thinkers. Al-Kindī* (New York [etc.], 2006).

[27] Adamson, *Great medieval thinkers: Al-Kindī*, p. 166.

[28] Both have been edited and translated into German in a number of M.D. theses

works contain chapters which are particularly relevant to us here. The last two chapters (21–22) in the shorter *Kitāb fī ʾl-bāh* by Qusṭā are entitled 'Which are the diseases that reportedly benefit from sexual intercourse (*Ayyu ʾl-amrāḍi ʾllatī yuqālu annahū yantafiʿu fīhā ʾl-bāhu*)' and 'Which illnesses occur because of abstention from sexual intercourse (*Ayyu ʾl-āfāti taḥduthu ʿani ʾl-imtināʿi mina ʾl-bāhi*)'.[29] In chapter 21, Qusṭā lists hemiplegia (*fālij*), epilepsy (*ṣarʿ*), facial paresis (*laqwa*), and melancholy (*waswās sawdāwī*) as diseases against which sex is useful. In chapter 22, Qusṭā says that women suffer more than men from not having sex because they are afflicted by the suffocation of the womb (*ikhtināq al-raḥim*) and apnoea (*buṭlān al-nafas*) because of an excess of semen.

The last (seventh) chapter of the first treatise in Qusṭā's longer work on sexual intercourse is entitled 'about the benefits resulting from having sexual intercourse'. The Christian physician makes a number of interesting remarks, repeating, for instance, the story of Diogenes of Sinope masturbating for medicinal reasons and thus avoiding sex with a prostitute, which is originally narrated by Galen.[30] In general, Qusṭā is profoundly influenced by the Greek medical ideas outlined at the beginning of this contribution. A particularly interesting example is the following:

> I read a treatise by a man belonging to the ancient physicians [*mutaṭabbibūn qudamāʾ*], in which he urges virgin slave girls [*jāriya*s] to marry. He describes the ailments [*āfāt*] which occur to many of them when they abstain from sex, just as Galen and Rufus describe them. Rufus even has

produced during the 1970s, which appear to have been a fertile time for this sort of topic. They are: N.A. Barhoum, *Das Buch über die Geschlechtlichkeit* (Kitāb fī ʾl-Bāh) *von Qusṭā Ibn Lūqā*. Edition und Übertragung des arabischen Textes nach der Handschrift Nr 242 der Universitätsbibliothek Istanbul. Doctoral Dissertation (Erlangen-Nürnberg: Seminar für die Geschichte der Medizin, 1975); G. Haydar, *Kitāb fīl-bāh wa-mā yuḥtāgu ilaihi min tadbīr al-badan fīstiʿmālihi des Quṣṭā Ibn Lūqā = Das Buch über die Kohabitation und die für ihre Ausübung notwendigen körperlichen Voraussetzungen. 1. Abhandlung*. Edition, Übertragung und Bearbeitung des arabischen Textes auf der Grundlage der Handschrift der Universitätsbibliothek Istanbul Nr. 243. Doctoral Dissertation (Erlangen-Nürnberg: Seminar für die Geschichte der Medizin, 1973); and F. Abdo, *Kitāb fīl-bāh wa mā yuḥtāğ ilaihi min tadbīr al-badan fī stiʿmālihi des Qusṭā Ibn Lūgā = Das Buch über die Kohabitation und die für ihre Ausübung notwendigen körperlichen Voraussetzungen. 2. Abhandlung*. Edition, Übertragung und Bearbeitung des arabischen Textes auf der Grundlage der Handschrift der Universitätsbibliothek Istanbul Nr. 242. Doctoral Dissertation (Munich: Technical University, Fachbereich Medizin, 1978).

29 Ed. Barhoum, pp. 25, l. 7–26, l. 4.

30 Ed. Haydar, p. 46, l. 7-ult.; Galen, *De loc. aff.*, VIII, p. 419, l. 6-ult. (Kühn)/ pp. 184–185 tr. Siegel.

> a separate treatise, the scope [*rasm*] of which is virgins. He wants them to marry, and outlines the regimen which they ought to follow if they are prevented from marrying [...]

Qusṭā b. Lūqā therefore seems to have had a particular concern for the sexual health of women. He wrote in an age where slave girls (*jāriya*s) were extremely common, especially in the upper echelons of society, and in the harems of the caliphs.[31]

The literary genre of treatises on sexual hygiene enjoyed great popularity also in later times. Authors such as Abū Naṣr al-Samawʾal b. Yaḥyā b. ʿAbbās al-Maghribī al-Isrāʾīlī (d. 1180)[32] and al-Tīfāshī (d. 1253)[33] wrote monographs on the topic, as did Ibn al-Jazzār (d. 980), who also devoted a chapter in his medical encyclopedia *Provisions for the Traveller* (*Zād al-musāfir*).[34] His monograph was probably the basis of Constantine the African's *On Coitus* (*De coitu*), which circulated widely in the medieval Latin world.[35]

---

[31] J. Bray, 'Men, women, and slaves in Abbasid society,' in L. Brubaker & J.M.H. Smith (eds.), *Gender in the early medieval world: East and West, 300–900* (Cambridge, 2004), pp. 121–146, see pp. 136–137; and N.M. El Cheikh, 'Gender and Politics in the Harem of al-Muqtadir,' ibid., pp. 147–164.

[32] T. Haddad, '*Kitāb nuzhat al-aṣḥāb fī muʿāšarat al-aḥbāb fī ʿilm al-bāh = Das Buch der Unterhaltung der Freunde über den vertrauten Umgang der Liebenden mit der Wissenschaft von der Sexualität des Abū Naṣr as-Samauʾal Ibn Yaḥyā Ibn ʿAbbās al-Maġribī al-Isrāʾīlī. Teil eins, Abschnitt sechs bis acht.* Edition, Übertragung und Bearbeitung des Textes auf der Grundlage der Handschriften Gotha 2045 und Berlin 6381. Doctoral Dissertation (Erlangen-Nürnberg: Seminar für die Geschichte der Medizin, 1976); F. Mansour, '*Kitāb nuzhat al-asḥāb fī muʿāšarat al-aḥbāb fī ʿilm al-bāh = Das Buch der Unterhaltung der Freunde über den vertrauten Umgang der Liebenden mit der Wissenschaft von der Sexualität des Abū Naṣr as-Samauʾal Ibn Yaḥyā Ibn ʿAbbās al-Maġribī al-Isrāʾīlī. 2. T., 1. bis 5. Abschnitt.* Edition, Übertragung u. Bearbeitung d. Textes auf d. Grundlage d. Handschrift Istanbul (Şehid Ali Paşa) 2145 unter Hinzuziehung d. Handschriften Berlin 6381 u. Gotha 2045. Doctoral Dissertation (Erlangen-Nürnberg: Seminar für die Geschichte der Medizin, 1975).

[33] H.M. El-Haw, *Risāla fīmā yaḥtāǧ ilaihi ʾr-riǧāl wan-nisāʾ fī stiʿmāl al-bāh mimmā yaḍurr wayanfaʿ des At-Tīfāšī = Abhandlung darüber, was Männer und Frauen zur Ausübung des Geschlechtsverkehrs brauchen, was dabei nützt und schadet.* Edition des arabischen Textes auf der Grundlage der Handschrift Al-Azhar (64 Majāmiʿ). Doctoral Dissertation (Erlangen-Nürnberg: Seminar für die Geschichte der Medizin, 1970). This text was incorporated in the author's more ample *The return of the old man to his youth as regards his sexual strength* (*Rujūʿ al-shaykh ilā ṣibāh fī quwwat al-bāh*), recently (2001) printed in Damascus in a series entitled 'Sexual literature among the Arabs (*Adab al-jins ʿinda ʾl-ʿArab*)'.

[34] The latter has been edited and translated by G. Bos.

[35] See Montero Cartelle, pp. 19–22; see also, D. Jacquart, Cl. Thomasset, *Sexualité et savoir médical au Moyen Âge* (Paris, 1985).

Conclusion

Al-Rāzī took a dim view about authors who had written treatises on sexual intercourse before him. In the introduction to this own work, al-Rāzī says that although works on the topic abound, those which he has seen are 'either weak and deficient, or confused and unclear'.[36] He does not refer to his predecessors by name, but he might well have thought of al-Kindī, the author of a very short treatise, and Qusṭā b. Lūqā, who had written a relatively long monograph. Al-Rāzī states that his own aim is to produce an easily understandable volume, not for theoreticians and philosophers, but physicians and the common crowd. As he himself puts it: 'My aim and objective is [to provide] things useful to people who practise and work, not for those engaged in research and theory (*gharaḍī wa-qaṣdī ilā mā yantafiʿu bihī ahlu 'l-ʿilāji wa 'l-ʿamali, lā ahlu 'l-baḥthi wa 'l-naẓari*)'.[37] These stated aims and objectives are interesting in their own right. We have seen that al-Rāzī draws largely on Greek medical theory in the chapter cited above. Yet, as he has outlined in his introduction, his target audience is practitioners, not theoreticians. His treatise therefore appears to belong to the genre of medical writings in which Greek theory finds practical applications.[38]

Genre, however, may also hold the key to explaining why al-Rāzī expresses such different opinions about sexual intercourse in his *Spiritual Medicine* and his medical writings. We have seen that al-Kindī could pitch the amount of explanation about drug functions according to the purpose of his work. Al-Rāzī takes things one step further: he adjusts his views about sexual intercourse to the potential expectations of his audience. In the literature about practical ethics to which his *Spiritual Medicine* belongs, it has become commonplace by Late Antiquity to enjoin that one should avoid all passions (*pathē*), of which love, lust, and sexual desire are prime examples. In this context, sexual intercourse acquires entirely negative connotations, except for the preservation

[36] P. 149, l. 7–8 (ed. al-ʿAzīz/ʿAbd al-Ḥamīd)/ f. 1b, l. 3–4 (MS L), reading *munaththaratan* instead of *mubtathiran*.

[37] Quotation here on pp. 149, l. 5f. (ed. al-ʿAzīz/ʿAbd al-Ḥamīd)/ f. 1b, l. 9–10 (MS L).

[38] For a discussion of the medical theory versus practice, see P. Horden & E. Savage-Smith (eds.), *The year 1000. Medical practice at the end of the first millennium*. Social history of medicine; 13/2 (Oxford, 2000); and P.E. Pormann, 'Theory and practice in the early hospitals in Baghdad: al-Kaškarī on rabies and melancholy,' *Zeitschrift für Geschichte der arabisch-islamischen Wissenschaften* 15 (2002–3), pp. 197–248.

of the species. The medical views, however, are much more nuanced. Already Hippocrates, as we have seen, recognised that sex could be beneficial under certain circumstances. Rufus categorically stated that, as all natural things, intercourse is not, in and of itself, harmful. His views were espoused by the great Galen, who set the agenda for so much later medical theory and practice.

When al-Rāzī adopts different positions in his medical works and his *Spiritual Medicine*, he adjusts what he says to the expectations of his audience. It is hard to guess where his own personal preferences in this matter lie. In the conclusion to his chapter on the benefits of sex cited above, he remarks that it is 'gratifying (*mumti'*)' that some people should have sex. Moreover, the fact that al-Rāzī changed the possible reasons for harm derived from not having sex from *tabattul* (chastity) to *falsafa* (philosophy) and *taqashshuf* (asceticism) seems highly significant. It is as if al-Rāzī criticised philosophy as something best avoided occasionally, thus perhaps providing a caveat for his own philosophical works. Lenn E. Goodman has describe al-Rāzī's ethical outlook as Epicurean, but Epicurean with a twist: in his philosophical works, he adopts an ascetic hedonism, which, at first, appears to be a contradiction in terms. Goodman explained the ascetic tendencies in al-Rāzī's otherwise hedonistic philosophy by suggesting that the latter may not have dared to express his own views entirely frankly, given the general intellectual climate.[39] Al-Rāzī apparently had less compunction about being taken to task for his opinions in his medical works.

[39] L.E. Goodman, 'The Epicurean ethic of Muḥammad ibn Zakariyâ' al-Râzî,' *Studia Islamica* 34 (1971), pp. 5–26.

'PRIDE AND PREJUDICE, PRAISE AND BLAME'

'ABD AL-LAṬĪF AL-BAGHDĀDĪ'S VIEWS ON BAD AND GOOD MEDICAL PRACTITIONERS

N. Peter Joosse

The First Statement[1]

'He was as lean as a rake, short of stature, of a repulsive appearance, and had furrowed and sunken cheeks. When (Tāj al-Dīn) Zayd b. al-Ḥasan (Abū 'l-Yumn) al-Kindī saw him, he compared him to something that had just come out of a frying pan, and (therefore) called him the shrivelled one. (From then on) the nicknames fell down from heaven [= they spread like wildfire] and became public knowledge to everyone. After that, he was only known by his nickname. He claimed to write books containing original materials, but merely occupied himself with compiling (other) books. He either summarized them or made unnecessary additions to them. His writings are inadequate and radiate emotional coldness. When he met a person who was specialized in a particular kind of knowledge, he avoided discussing that branch of knowledge with him and changed the subject. He was always uncertain about anything he claimed or proclaimed. I used to meet him on a regular basis and knew him well. So, I was able to observe him from nearby and put him to the test with regard to the matters in which he claimed to be a specialist, but in which he actually groped in the dark just as a blind person who pretended to be quick-sighted (...). He resided in Aleppo at the end of his life. There he tried to make a livelihood in medicine, but he did not have (sufficient) knowledge of it. In one of the months of the year 628 [AH] it came to his mind to travel to Iraq before making the pilgrimage to Mecca, but he fell ill in Baghdad. He started to apply his own medical treatment, but passed away – just as Allah had intended – in one of the months of the year 629 [AH]. His books were sold in

[1] Round brackets denote an addition to the text; square brackets denote an explanation of the text; squiggly brackets plus a question mark denote a doubtful reading.

Aleppo. Some of them fell to my share, but they were of extreme inferiority and far removed from the level of perfection. May Allah protect us from the temptation of becoming pretentious.'[2]

Commentary

The above statement, though not written in a very elegant style, was taken from the *Inbāh al-ruwāt ʿalā anbāh al-nuḥāt* ('The information of the narrators on renowned grammarians'), a work by the prominent thirteenth-century chronicler Jamāl al-Dīn Ibn al-Qifṭī (b. 568/1172 – d. 646/1248). Ibn al-Qifṭī was born in Upper Egypt. He followed his father, the judge al-Ashraf Yūsuf, to Jerusalem. From there he went to Damascus and Aleppo. In Aleppo he became the personal secretary of the amir Ṣalāḥiyya, Fāris al-Dīn Maymūn al-Qaṣrī. After the amir's death, he was appointed as the private treasurer of the prince al-Malik al-Ẓāhir Ghāzī b. Yūsuf and later he became the head of the office of finances (*dīwān al-māl*) of the entire Ayyūbid principality. After the prince's death he was invested with the office of vizier, a function which he held until the end of his life.[3] The person who is insulted by al-Qifṭī and whose pride must have been severely wounded by the latter's hurtful remarks, is no one else than the well-known Muslim intellectual ʿAbd al-Laṭīf b. Yūsuf al-Baghdādī (1162–1231), a fellow townsman, a contemporary and apparently an old acquaintance of al-Qifṭī, if we are inclined to believe the above statement. Al-Qifṭī's personal and rather violent attack on al-Baghdādī is quite remarkable. A reason for it is not given here, so for now we can only guess at the motive. Was it (professional) jealousy or did al-Qifṭī perhaps harbour some grudge against al-Baghdādī?[4] In any case, al-Qifṭī's criticism of al-Baghdādī is certainly not limited to a personal attack on the latter's character, but must also

[2] Jamāl al-Dīn Ibn al-Qifṭī, *Inbāh ar-ruwāt ʿalā anbāh al-nuḥāt*, ed. Muḥammad Abū 'l-Faḍl Ibrāhīm, 4 vols. (al-Qāhira, 1369–1374/1950–1955; repr. Beirut, 2004), II, pp. 193–196. The translation is my own. A very short paraphrase can be found in P. Ghalioungui, 'The legend of Abdullatif al-Baghdady's spirit,' *Annales islamologiques* 13 (1977), pp. 257–267, p. 258.

[3] A.-M. Eddé, *La principauté ayyoubide d'Alep (597/1183–658/1260)*. Freiburger Islamstudien; 21 (Stuttgart, 1999), pp. 312–315.

[4] ʿAbd al-Laṭīf al-Baghdādī was often attacked by his contemporaries, but he states that 'he will not be deterred thereby from saying his mind openly,' cf. S.M. Stern, 'A collection of treatises by ʿAbd Al-Laṭīf al-Baghdādī,' *Islamic studies* 1 (1962), pp. 53–70, p. 66. ʿAbd al-Laṭīf at the same time felt that he did not want to take a very hard line

be considered a serious attack on his scholarly work and educational methods. Al-Qifṭī severely blamed al-Baghdādī for his dull and mind-numbing teaching, his one-track mind and his use of incorrect, that is, ungrammatical language. This is accounted in a brief anecdote in which is stated that al-Baghdādī made an attempt to teach the Arabic language to the two sons of the army scribe Ismāʿīl b. Ḥajjāj al-Muqaddasī: he appeared to have taught them in a very dry and uninspiring manner which almost bored the poor youngsters to death.[5] Still one also needs to read al-Qifṭī's statement with caution, because al-Qifṭī himself is widely criticized by other parties: Ibn al-ʿImād felt that al-Qifṭī exaggerated the negative description of al-Baghdādī's person and works and that he treated him unjustly by spreading incorrect information about him.[6] Ibn Maktūm emphasized al-Qifṭī's strong prejudice against al-Baghdādī and mentioned that he purposefully maligned his contemporaries, that he diminished their standing in the community, and that he made the people believe that he was intimately acquainted with important scholars and so was granted the privilege of entering the higher echelons of certain sections of society. Ibn Maktūm denied that this was the case and even declared that it was far from the truth, to which he added: 'May Allah forgive him [Ibn al-Qifṭī] (for his lies).'[7] It appears that apparently Sultan Ṣalāḥ al-Dīn himself was not that enchanted by al-Qifṭī. There is evidence of a poem, written by the ruler for his personal use, which is not flattering to al-Qifṭī at all. In it the ruler describes the vizier as an evil scribe who was resolved to bring harm and misfortune upon the people (*kātib sūʾ... qad ajmaʿa al-nās ʿalā naḥsihi*).[8]

---

against his contemporaries, even though quite a few of them were quacks and charlatans who had to be exposed. He often believed that he was too judgmental and for this he had to make his apologies to Allah, cf. F. Allemann, *ʿAbdallaṭīf al-Baġdādī. Ris. fī Mudjādalat al-ḥakīmain al-kīmiyāʾī wan-naẓarī ("Das Streitgespräch zwischen dem Alchemisten und dem theoretischen Philosophen"). Eine textkritische Bearbeitung der Handschrift: Bursa, Hüseyin Çelebi 823, fol. 100–123 mit Übersetzung und Kommentar*, [diss.] (Bern, 1988), p. 121; § 97.

[5] Cf. Jamāl al-Dīn Ibn al-Qifṭī, *Inbāh*, p. 194.

[6] Ibn al-ʿImād al-Ḥanbalī, *Shadharāt al-dhahab fī akhbār man dhahab*, 8 vols. (al-Qāhira, 1350/1958), V, p. 132.

[7] Cf. P. Ghalioungui & S. Abdou, *Maqālatān fī 'l-ḥawāss wa-masāʾil ṭabīʿiyya. Risāla li l-Iskandar fī 'l-faṣl. Risāla fī 'l-maraḍ al-musammā diyābīṭis, taʾlīf ʿAbd al-Laṭīf Al-Baghdādī*. The Arab Heritage series, Wizārat al-Iʿlām; 18 (Kuwait, 1392/1972), p. 164; likewise Jamāl al-Dīn Ibn al-Qifṭī, *Inbāh*, p. 196.

[8] Ghalioungui & Abdou, p. 165.

In the above statement it has also been mentioned that one Tāj al-Dīn Zayd b. al-Ḥasan Abū 'l-Yumn al-Kindī (d. 613/1217), a Hanafitic grammarian and prominent reciter of the Koran, compared al-Baghdādī to something that had just come out of a frying pan, and that he therefore had called him the shrivelled one. This remark of al-Kindī should be considered a sign of sour grapes since al-Kindī is mentioned in Ibn Abī Uṣaybiʿa's biography of al-Baghdādī under the name al-Kindī al-Baghdādī al-Naḥwī: al-Baghdādī had met al-Kindī before when he was still a student and in search of inspirational mentors. He found al-Kindī to be an intelligent and witty shaykh, but 'he was very self-satisfied and troublesome to his associates.' Al-Baghdādī neglected to attend his classes and further declared that 'the annoyance I caused him by my neglect of him was greater than the annoyance that he caused to others.' Surprisingly enough all of al-Qifṭī's views about al-Baghdādī were taken over into the analysis of present day scholars: the medical doctors Paul Ghalioungui and Said Abdou who, at the end of their edition of al-Baghdādī's *Treatise on Diabetes* (Kuwait, 1972), present us with an outrageously obtuse study of al-Baghdādī's character, which unfortunately bears more resemblance to a character assassination. Besides telling us that he was an ugly and mean little man with a speech impediment, that is to say, a heavy stutter, they accuse him of wanderlust and call him arrogant, conceited, haughty and loquacious.[9] He had a sharp tongue, was a mudslinger and must clearly have been hyper-active. He's supposed to have had a morbid predilection for exaggeration and a tendency towards aggression and enmity. He took pleasure in other people's hardship and enjoyed their pain. They state that he was afflicted by a particular type of sadism, which made him overstate all the suffering he witnessed. Ghalioungui and Abdou finish off by saying that al-Baghdādī showed a propensity to introversion, which obviously made it very hard for him to open up and unbend and that he even might have suffered from severe depression.[10]

[9] It is rather strange to discover that the modern Arabic authors Ghalioungui and Abdou consider wanderlust (*kathrat at-tajwāl*) a medical condition. After all, the Arabs introduced the famous saying *Uṭlubū al-ʿilm wa-law bi 'l-Ṣīn* ('Seek knowledge, even if it is in China'). This saying can be found in al-Ghazālī's work *Iḥyāʾ ʿulūm al-dīn* in the chapter called *Faḍīlat al-taʿallum*. Apparently it has been derived from Abū Bakr al-Bayhaqī's large work on the Prophetic tradition.

[10] Ghalioungui & Abdou, pp. 176–179.

The Middle Statement

'In the years 613 and 614 [AH] a group of noblemen and notable citizens passed away. I witnessed (the death of) a few of them, but information concerning (the death of) the others reached me through reports. Among those who passed on in the year [6]13 was the governor of the city [Aleppo], [the prince] al-Malik al-Ẓāhir Ghāzī b. Yūsuf. He had become feverish on Saturday the 26th of (the month) *Jumādā* I [= 10th September] of the year mentioned above [= 613 AH/1216 AD]. He was plump and overweight and had already surpassed the age of forty. Thoughts and worries about the battle array of his army, (ways of) forcing (his) enemies to surrender [= battle strategies] and the costs of war had taken root in his mind, which caused him to abandon the drinking of fluids and to decrease the intake of food, whereas (at the same time) his insomnia increased and the thoughts kept spinning around in his head. Therefore, a physician who had embraced Islam suggested bleeding him, to which all the (other) doctors agreed unanimously. But this one damned devil prevented them [the other doctors] from bleeding him by saying to them in secret that bleeding him would cause the particular doctor to enjoy the favours and good graces (of the prince) and that he [the former] then would become their chief. Thus they avoided bleeding him. Accordingly, on the sixth day of his illness (this damned devil) gave him [the prince] his (own) approved potion [= his laxative] to drink, (which was administered to the prince) {before the illness [or: fever] reached its breaking point?}.[11] It made him walk [run] to the latrine twenty-four times, but during the twenty-fourth time, his strength broke down, his body cooled off and his movements stopped for approximately thirty hours. They (continued) nursing him, but this damned devil sat there laughing and enforced his judgment on all those who were next in line to examine him [the prince]. A man, who was dedicated and experienced, stood up from their midst to examine (the prince's) strength.[12] They watched him closely (examining the prince)

---

[11] Optional renderings of this sentence may read: 'before the crisis' or 'before it was well-cooked.'

[12] This proves that ʿAbd al-Laṭīf also spoke highly of some of his direct colleagues. He had a strong predilection for the ancient Greek doctors Hippocrates, Dioscorides, Galen and Rufus of Ephesus, but he also praised Arabic physicians: Amīn al-Dawla Hibat Allāh Ibn al-Tilmīdh and his son, and most of all Aḥmad b. Abī ʾl-Ashʿath (d. *c.* 360/970), a physician who was saved from oblivion through the work of Remke Kruk. ʿAbd al-Laṭīf also cherished two Arabic medical handbooks. He considered the *Kunnāsh* by Yūḥannā

and said (to him): "Do you want to quicken the dead?," to which he replied: "There is still hope that he may live, the more so because (the reason for) the breakdown of his powers is (due to) the treatment and not to the malignancy of the disease". Once the prince had managed to open his mouth with great difficulty, they gave him a little chicken broth to drink. Thereupon his pulse came back and he regained his strength and opened his eyes and spoke, after which he asked for water and food. He then rose up (from the bed) all by himself. The physicians all returned (to visit the patient), but this damned devil saw that his arrow had not reached its target and thereupon he administered (to the prince) one enema after another [that is, via his anus],[13] and in between every two enemas he (also) gave him multiple (soap-like) suppositories. Yet, the doctors, who were (also) attending him [the prince], held him back and prevented him from (undertaking further actions), which made his face turn gloomy before their very eyes. He abused them, called them idiots and said: "My intention was rather to cleanse his body". With (remarks like) this and with similar examples he tried to persuade the servants and the kinfolk (of the prince) to tie him up and surrender him to fate.[14] Then, the retentive power declined and poured out from him. Thereupon, they [the doctors] began to administer astringents in an excessive manner and believed the drugs to work by themselves without (the intervention) of a power within the body which directs the drugs and is able to influence them.[15] But I tell you that the drugs do not work by themselves, but rather are tools for the bodily powers. Whenever that which is governing the tools [the drugs] becomes corrupted, then the tool [the drugs] itself (also) becomes void, shall cease to function, and shall therefore not be of any use (anymore). But the tool [the drugs] did not cease to flow [= kept on being administered] until he [the prince] – may Allah have mercy upon him – passed away on the twenty-fourth day of his illness. Each one of the physicians who had stayed with him gathered for a meeting and spoke out frankly that he [this damned evil] killed (the prince), and

---

Ibn Sarābiyūn the best of its kind; the *Kitāb al-mi'a fī 'l-ṣināʿa al-ṭibbiyya* by Abū Sahl al-Masīḥī came in second place.

[13] Enemas were sometimes also administered via the urethra; cf. M. Algera, *Mens en medicijn. Geschiedenis van het geneesmiddel* (Amsterdam, 2000), pp. 120–123.

[14] That is, the vicious doctor wanted to prevent the prince from rising up on his own accord!

[15] That is, the doctors first started administering purgatives and when these did not have the desired effect, they resorted to giving the patient the opposite, namely astringents!

(thereupon) every one of them gave his own description of a specific aspect of the fatal treatment which he [this damned devil] pursued. One of the subtle tricks (which this damned devil employed) in (the course of) the disease, is that he made the one in charge of the servants believe that the majority of the doctors was displeased with him (and hostile towards him). Thereupon, the doctors (who were allowed to stay at the prince's bedside) were limited to two (persons only). He suggested himself and an elderly, venerable gentleman who was intelligent, friendly and kind, and well-known for his good medical treatment.[16] If he [this damned devil] would have been in the position to dismiss him [the elderly gentleman], then he would (certainly) have done it. However, he stayed alone with the prince and the elderly doctor over which he had control would remain with him, so that he could order him to do whatever he pleased until the prince regained consciousness and demanded another person from their midst to keep him company. This person was a man of mature age, well-known for his sound treatment and his correct views.[17] He (then) took over the medical attendance, but afterwards this damned devil was again given precedence to (the other doctor), and thus surrendered the prince (to the fate of death). Of the many things that I have seen from this damned devil, was that he wished to preserve health by means of the opposite and remove disease by applying something similar. In case of phlegmatic [= mucous] diseases, he gave (the patients) a fluid to drink which was made from the seeds of the cucumber, barley water and rose syrup. In case of choleric, (melancholic), [= respectively black and yellow bilious] diseases and sanguine [= blood] diseases and to those affected and weakened by the disease phthisis [or: consumption] he prescribed {pure and precious?} hot purgatives and warm(ing) aperients. Some wise persons believed that he did all these things intentionally because he wanted to destroy human lives and took pleasure in the distortion of forms and shapes.'[18]

---

[16] Cf. n. 12 *supra*.

[17] Cf. n. 12 *supra*.

[18] ʿAbd al-Laṭīf al-Baghdādī, *Kitāb al-naṣīḥatayn min ʿAbd al-Laṭīf b. Yūsuf ilā ʾl-nās kāffatan* (MS Bursa, Turkey, Hüseyin Çelebi 823, f. 62a–100b) [medical section: f. 62a–78a; philosophical section: f. 78b–100b], medical section f. 69a, line 11–70a, line 15. The translation is my own, although I would like to express my sincere thanks to Dr. Peter E. Pormann (Warburg Institute, London) and Dr. Wim Raven (Universität Marburg) for their very helpful suggestions. Thanks is also due to my erudite *kātib*, Rob 'Abū Iskandar' Hilders M.D. (Schellinkhout) for his most thorough medical analysis of the middle statement. The middle statement has been translated here for the first time. A paraphrase can be found in Stern, p. 61.

Commentary

In the second of the three statements presented here, this time from 'The Book of two advices to the public at large' (*Kitāb al-naṣīḥatayn [...] ilā 'l-nās kāffatan*), written by ʿAbd al-Laṭīf al-Baghdādī himself, we have a detailed eyewitness account of the last illness of prince al-Malik al-Ẓāhir Ghāzī b. Yūsuf of Aleppo. The fragment under consideration highlights the treatment of his illness by the court-physicians and their relation to one another. The prince al-Malik al-Ẓāhir was the middle son of Sultan Ṣalāḥ al-Dīn, the brother of al-Afḍal ʿAlī and al-ʿAzīz ʿUthmān, the husband of Ḍayfa Khātūn bint al-ʿĀdil, and the father of his at that time minor son and would-be successor Muḥammad al-ʿAzīz. A small number of theories exist about the prince's death: certain authors attribute it to an intestinal disorder, others were of the opinion that it was caused by a sudden high fever. Ibn Naẓīf al-Ḥamawī in his *Ta'rīkh al-Manṣūrī* states that the prince after having returned from a hunting party, ate salted meat with lentils and had too much wine to drink with it, which made him fall ill.[19] A modern author, Manfred Ullmann, accepted the hypothesis that the prince died as the result of diabetes. Ullmann based his hypothesis on a passage from ʿAbd al-Laṭīf al-Baghdādī's treatise *(Fī) 'l-maraḍ al-musammā diyābīṭā*. This treatise is a *consilium*, a confidential opinion, by al-Baghdādī written only for colleagues who were personally connected to the case. In it al-Baghdādī defines his attitude about a doctor, 'a man from the *Maghrib*, an old man in years, but a child in knowledge and insight,' and refutes the latter's views. Ullmann believes this doctor to be the same person as 'the damned devil' [*al-malʿūn*] described in the statement above. Since the question at issue was discussed 'at the court of the Sultan' and since al-Baghdādī lived in Aleppo from 1216 until 1220, Ullmann concluded from this that the tractate must have been composed in the year 1216, and that the patient in question is prince al-Malik al-Ẓāhir.[20] Ullmann might be correct in stating that al-Baghdādī addressed the same doctor in both of his treatises. However, his argument is difficult to prove: on the one hand not all the physicians at the court of al-Malik al-Ẓāhir are known to us; on the other hand the term *Maghribī* [North African] is often used in a general and derogatory manner to depict fortune-

[19] Eddé, pp. 84–85.

[20] M. Ullmann, *Islamic medicine*. Islamic surveys; 11 (Edinburgh, 1978), pp. 79–80.

tellers, quacks, and charlatans.[21] However, Ullmann's further argument that prince al-Malik al-Ẓāhir died of the disease diabetes seems incorrect to me. The picture which emerges from the text is that the prince had abandoned the drinking of fluids and had decreased the intake of food. In the case of seriously unregulated diabetes, the prince would have in fact taken recourse to excessive drinking and this would have resulted in excessive urinating. Otherwise, the clinical picture is not totally clear, but the author has described it exceptionally well by especially indicating the relevant medical phenomena without digressing too much towards trivial matters. Two treatments are weighed against another here: bleeding [= bloodletting] the patient or purging him. One of the doctors did not agree to the treatment of bleeding proposed by one of the other doctors, but insisted on administering laxatives and persuaded the other doctors by pointing out that if the other doctor's treatment was followed, he would gain the favour of the prince. Regardless of the exact nature of the disease, by purging him very thoroughly, the patient, who already had drunk very little or nothing at all, became dehydrated. Technically this of course means that there was a serious loss of water and salt. The prince had all the signs of dehydration: when he tried to get up and walk to the latrine, he collapsed; he had a weak pulse and shivered and felt cold. The administering of water and salt, chicken broth, was in this specific case the only relevant *i.e.* effective treatment. But, by giving him laxatives once again, the entire process started anew and resulted in the prince's untimely death. Against the background of the Galenic doctrine, purging and bleeding were without a doubt legitimate choices. If we disregard al-Baghdādī's insinuations for a moment, we may draw the conclusion that the events described above proceeded in a completely logical and understandable manner and, therefore, the doctor who has been portrayed as diabolical, may have faithfully 'cleansed' and killed his patient at the same time. If we value al-Baghdādī's insinuations, 'the damned devil' is charged with the worst crime a medical doctor can be accused of: deliberately causing the death of a patient, which is murder. I am inclined to attach a certain importance to al-Baghdādī's words because of the seriousness of the accusation, and unlike Stern and others I find it conceivable that there is some form of truth in it. However, a tendency which can be perceived when studying al-Baghdādī's works is

[21] For the causes for the migration of *maghāriba* to Syria in the 7th/13th century, see: L. Pouzet, 'Maghrébins à Damas au VIIe–XIIIe siècle,' *BEO* 28 (1975), pp. 167–199.

that a significant part of his oeuvre seems to consist of three elements: truth, half-truth and *topos*. To distinguish between these three elements is an arduous task which can only be performed after a detailed and thorough analysis of al-Baghdādī's entire oeuvre. The above statement is for instance also largely in conformity with a report on the death of Sultan Ṣalāḥ al-Dīn by Sibṭ b. al-Jawzī. Sibṭ b. al-Jawzī was the only historian to accuse the high-ranked physician Raḍī al-Dīn al-Raḥbī of being the cause of the death of the Sultan, saying that when the Sultan became ill, 'all the physicians agreed not to bleed him, but al-Raḥbī disagreed with them and bled him and thus was the cause of his death.' This incident is also mentioned by al-Baghdādī, but he does not mention the name of the erring physician, but merely states that the Sultan was bled by someone who had no knowledge of bleeding, that the procedure weakened the Sultan's strength and thus brought on his death. A different account of the Sultan's death is offered by the Sultan's military judge and biographer Bahā' al-Dīn Ibn Shaddād, who states that the physicians all agreed to bleed the Sultan and that when they did so, his illness became worse. He does not mention al-Raḥbī.[22]

The Third Statement

'I believe that the one who allotted these shameful deeds and despicable practices among them, was a *Maghribī* shaykh who had been converted to Islam, but returned to the Jewish religion outwardly – though they (the Jews themselves) firmly believed that he was neglectful with regard to their religion, thought little of observing their devotional duties, was careless in (his) obedience to (the regulations of) the *sharīʿa* [= the revealed or, canonical, law of Islam], and could not be believed at all (even) if he swore on the Torah and the Ten Commandments. He used to be a poor man who was travelling through the (various) countries in the service of merchants, and who (only) took a serious interest in the art of medicine at a (relatively) late age. He had a strong desire for worldly belongings and possessed the strength, tenacity and avidity of a dog, which made him annihilate human lives and attack human beings

[22] S. Jadon, 'The physicians of Syria during the reign of Ṣalāḥ Al-Dīn 570–589 AH 1174–1193 AD,' *Journal of the history of medicine and allied sciences* 25 (1970), pp. 323–340, especially p. 333.

in the manner in which a lion attacks his prey. People agreed on the fact that he does not fear the Creator or (any) created being, and that he attached no importance whatsoever to the massacre of a thousand souls in one hour. They said that he takes much pleasure in such things.'[23]

Commentary and Conclusion

The third and last statement, again from ʿAbd al-Laṭīf al-Baghdādī's 'Book of two advices,' aims at describing the 'damned devil.' Although al-Baghdādī does not mention his name, there cannot be any mistake that the devil in disguise is the highly esteemed Jewish doctor Abū 'l-Ḥajjāj Yūsuf b. Yaḥyā b. Isḥāq al-Sabtī al-Maghribī, the favourite pupil of the famous Jewish philosopher/physician Moses Maimonides and the intimate friend of the chronicler/vizier Ibn al-Qifṭī.[24]

Abū 'l-Ḥajjāj is better known by his usual name Ibn Shamʿūn, or his Hebrew name Rabbi Joseph ben Judah. Ibn al-Qifṭī also dedicated an entry to his good friend in his biographical dictionary the *Taʾrīkh al-ḥukamāʾ* of which the outlines match al-Baghdādī's statement. Joseph ben Judah and Ibn al-Qifṭī apparently met each other at social gatherings [*majālis*] at the house of the amir Fāris al-Dīn Maymūn al-Qaṣrī in Aleppo. They were not just good friends, but real 'buddies' as the following anecdotes may show: In the first anecdote, Joseph tells Ibn al-Qifṭī in confidence that after having had two girls, he would now like to have a boy. Thereupon, Ibn al-Qifṭī gives him 'a recipe,' a method which Joseph has to employ during sexual intercourse. In the following years indeed three sons were born to him. In the second anecdote, the two men promise each other that the one who passes away first should inform the other about the hereafter. Joseph is the first to die and after his death he appears to Ibn al-Qifṭī in a dream. He is completely dressed in white and speaks the enigmatic words: 'the universal has joined the universal and the particular stayed behind in the particular.' This

[23] ʿAbd al-Laṭīf al-Baghdādī, *Kitāb al-naṣīḥatayn*, f. 69a, lines 1–10. The translation is my own. A paraphrase can be found in Stern, pp. 60–61.

[24] S. Munk, 'Notice sur Joseph ben-Iehouda ou Aboul'hadjādj Yousouf ben-Ya'hya al-Sabti al-Maghrebi, disciple de Maïmonide,' *Journal asiatique*, Série 3, 14 (1842), pp. 5–70; Jamāl al-Dīn Ibn al-Qifṭī, *Taʾrīkh al-ḥukamāʾ*, ed. J. Lippert (Leipzig, 1903), pp. 392–394; Ibn Abī Uṣaybiʿa, *K. ʿUyūn al-anbāʾ fī ṭabaqāt al-aṭibbāʾ*,' ed. Nizār Riḍā (Beirut, 1965), p. 696; Barhebraeus, *Taʾrīkh mukhtaṣar al-duwal*, ed. A. Ṣalḥānī (Beirut, 1890), p. 242.

indicates to Ibn al-Qifṭī that the soul has returned to the universal, while the body has rested in the grave.[25]

By attacking al-Baghdādī in such a vicious manner (see the *First Statement*), Ibn al-Qifṭī wanted to take revenge for al-Baghdādī's accusation that Joseph intentionally killed his patient, the prince al-Malik al-Ẓāhir. Ibn al-Qifṭī of course intended to protect his best friend's honour, but what was al-Baghdādī's motive for blaming Joseph? Was it frustration, professional jealousy or hate against a Jew, who had once apostatized and lived as a Muslim, but practiced the Jewish faith in secret? Stern and Eddé seem to hold the view that it was an act of racism or xenophobia. I rather believe that there is some truth in the old saying that there is no smoke without fire and consequently I would like to raise some slight reservations about the flawlessness of Joseph's character. In a letter addressed to Joseph, Maimonides had to restrain his student and warned him against insulting Samuel ben Eli, the principal of the rabbinic academy in Baghdad, a man who, despite all his limitations, was much older, held a dignified post, and was respected by society.[26] In another letter, Maimonides responds to Joseph's dark complaint that he had been the subject of false tales spread by his enemies [*sic*!]. Thereupon, his master advised him to learn to control his temper and to forgive his foes.[27] Joseph was apparently not 'very forgiving' for on another occasion his master had to reprimand him again. The most interesting, but unfortunately somewhat murky statement comes from the *Taḥkemoni* by the poet and translator Judah al-Ḥarizi, who visited Aleppo in the beginning of the year 1217 not so long after the death of prince al-Malik al-Ẓāhir. He found Joseph at the peak of his glory, but the physician was in a state of distress because one of his disciples(?) had shown ingratitude towards him in a rather vicious and malicious way.[28] The testimonies of Moses Maimonides and Judah al-Ḥarizi provide us with new food for thought and indicate that the final word about this minor episode in Arabic medical history has not yet been spoken. ʿAbd al-Laṭīf al-Baghdādī may not have been the most perfect of all men,

---

[25] Eddé, p. 476.

[26] H.A. Davidson, *Moses Maimonides. The man and his works* (Oxford, 2005), p. 520; cf. Moses ben Maimon, *Epistulae* (Jerusalem, 1946), pp. 61–62.

[27] Davidson, pp. 547–548; cf. Moses ben Maimon, pp. 49–50.

[28] Munk, p. 54; cf. Judah al-Ḥarizi, *Taḥkemoni* (Tel Aviv, 1952); Davidson, p. 265, p. 427. Judah al-Ḥarizi also mentioned another Jewish physician who was present at the court of al-Malik al-Ẓāhir Ghāzī: a man called Eleazar, cf. Eddé, p. 467.

but it is rather unfair to accuse him of intentionally ruining someone's reputation. ʿAbd al-Laṭīf fought against the dangerous increase of utter simplicity and gullibility among the masses, and was most of all concerned about breaking peoples' barriers down in order to ensure a fruitful discussion and to focus on the real problems in society.[29] He felt that society was in danger of being caught off guard and thereby not focusing on the fundamental issues threatening it. He must have believed that if society remained asleep, it would be lost forever.

[29] ʿAbd al-Laṭīf's way of debating was often counter-productive, but this did not concern him too much as long as there remained a fruitful discussion.

ʾĀXIR IṬṬIBB INNĀṚ

TEXTE ZUR VOLKSMEDIZIN AUS ILBAŠANDI IN DER OASE DAKHLA

Manfred Woidich

Fragen zur Volksmedizin gehören zum festen Bestandteil eines dialektologischen und volkskundlichen Fragebogens zu den ägyptischen Dialekten. Themen wie bestimmte Heilpflanzen oder allgemein gebräuchliche Heilverfahren bieten gerade in den ländlichen Gegenden Ägyptens ein weites Feld für unverfängliche Gespräche, kennt doch jeder Bauer oder jede Bauersfrau ein Hausmittel, eine *waṣfa baladi* zu all den vielen Krankheiten, die garantiert hilft: *isʾal mugaṛṛab wala tisʾal ṭabīb*, wie das Sprichwort sagt. Daher bieten dialektologische Arbeiten, vor allem Textsammlungen zu Ägypten nicht selten nützliche Informationen zu diesem Thema und zu anderen anthropologischen Gegenständen – s. etwa die beiden Textbände zum ägyptischen Dialektatlas[1] –, die leider noch nicht zusammenfassend ausgewertet wurden.[2]

Aus dem Material, das ich seit 1997 in dem Dorf ilBašandi [ɪlbɪʃɪn'de:j] in der ägyptischen Oase Dakhla habe sammeln können, möchte ich hier

[1] P. Behnstedt & M. Woidich, *Die ägyptisch-arabischen Dialekte*, Bd. 3, Texte I: *Nildelta* (Wiesbaden, 1987); Bd. 3, Texte II: *Niltaldialekte*. III: *Oasendialekte* (Wiesbaden, 1988). Speziell Volksmedizinisches findet sich darin in den Texten Nr. 94, 108, 110.19, 138, 139. F.E.E. Elder, *Egyptian colloquial Arabic reader* (London, 1927), S. 62–67. J.S. Willmore, *The spoken Arabic of Egypt* (London, 1919), S. 350 Text 11 (Sonnenstich). Zahlreiche Informationen findet man natürlich auch in den Standardwerken zur ägyptischen Volkskunde wie W.S. Blackman, *The fellāhīn of Upper Egypt* (New Impression London, 1968), S. 211; H.A. Winkler, *Ägyptische Volkskunde* (Stuttgart, 1936); ders. *Bauern zwischen Wasser und Wüste* (Stuttgart, 1934).

[2] Die Volksmedizin an sich ist nicht der Forschungsgegenstand des Dialektologen, denn diesem geht es in erster Linie darum, die Gewährsleute mit einem vertrauten Thema zum Sprechen zu bringen, um Texte in möglichst originaler Sprache aufzunehmen. Im Gegensatz dazu untersucht die Ethnomedizin diesen Themenkomplex vom medizinischen, pharmazeutischen und botanischen sowie sozialwissenschaftlichen Blickwinkel aus, s. dazu die http://www.agem-ethnomedizin.de und die EthnoMedical DataBase http://emdb.lettere.unige.it/emdb/ethnomed/2_0ethnomed.shtml. Als beispielhafte Studien zur ägyptischen Ethnomedizin seien hier S.A. Morsy, *Gender, sickness, & healing in rural Egypt* (Boulder, 1993) und M.C. Inhorn, *Quest for conception. Gender, infertility, and Egyptian medical traditions* (Philadelphia, 1994) angeführt.

in der Festschrift, die unserer Kollegin Remke Kruk gewidmet ist, eine Auswahl von Texten im lokalen Dialekt beitragen, die solche volksmedizinische Themen zum Inhalt haben. Ich hoffe, daß dieses Thema, dem sie auch in ihren Arbeiten großes Interesse gewidmet hat, sie ansprechen wird, zumal sie selbst ja auch in letzter Zeit in Dakhla forschend tätig geworden ist und das Dorf ilBašandi aus eigener Anschauung kennt.

Volksmedizin läßt sich unter verschiedenen Aspekten betrachten und ist oft verbunden mit magischen Praktiken. Nicht selten haben sich hier vorislamische Elemente erhalten, wie etwa in ilBašandi bei dem Fruchtbarkeitsritual *ilmulāgā́*, das mit dem Aufgang des Morgen- oder Abendsterns *niğmit ilmaġrēb wu niğmit ilfağir* ausgeführt werden muß. Andererseits gibt es auch Praktiken, die an Zeitpunkte gebunden sind, die durch die islamische Zeiteinteilung festgelegt sind, wie das Ausräuchern des Hauses gegen böse Geister,[3] das während der *tafkīrt iğğumʿa* geschieht, ein vor dem *ʾaḏān* zum freitäglichen Mittagsgebet vom Minarett gesungener Lobpreis des Propheten.[4] Die Kindbettdämonin *tabīʿa*, andernorts *garīna* genannt,[5] ist ebenso hierher zu zählen. Um den Rahmen eines Festschriftbeitrags nicht zu sprengen, beschränke ich mich hier auf Texte zum medizinischen Nutzen einiger Pflanzen und zu den Heilverfahren *kayy, xarṭ* und *tadʿīk*.

Die Informanten, die hier nicht mit vollem Namen angegeben werden, sind alle in ilBašandi aufgewachsen, sind aber im benachbar-

---

[3] Aus einem hier nicht wiedergegebenen Text aus ilBašandi: *illi yigdēr yibixxēr yōm iğğumʿa, da wagt ittafkīra* „wer kann, der räuchert [das Haus] am Freitag aus, das ist [zur] Zeit der *tafkīra*".

[4] Zur *tafkīra* s. M. Woidich, 'Neue Volkslieder aus ilBašandi/Dakhla,' *Estudios de dialectología norteafricana y andalusí* 8 (2006) (Festschrift Peter Behnstedt). Hierbei dürfte es sich um den *salām* handeln, den Lane erwähnt, E.W. Lane, *An account of the manners and customs of the modern Egyptians. The definitive 1860 edition.* Intr. by J. Thompson (Cairo [etc.], 2003), S. 83. Heute wird die *tafkīrt iğğumʿa* in ilBašandi nicht mehr praktiziert, da nicht dem orthodoxen Islam entsprechend, s. M. Woidich, 'Giṭʿa's from il-Bašandi / Dakhla-Oasis,' in C. Holes (ed.), *Proceedings of an international conference on Middle Eastern popular culture, Magdalen College, Oxford, 17–21 September 2000* (Oxford, 2001), S. 199–206 zur Unterdrückung volksislamischer Bräuche in den Oasen. Das Freitagsgebet als ein für magische Praktiken besonders geeigneter Zeitpunkt wird auch an anderen Orten erwähnt, etwa in il-Ašmunēn in Mittelägypten, wo sich die Frauen zur Ausführung eines Fruchtbarkeitsrituals zu eben dieser Zeit an eine bestimmte Stelle im Ruinenfeld begeben (*iddakaṛ wi nnitāya*), s. P. Behnstedt & M. Woidich, *Die ägyptisch-arabischen Dialekte.* Bd. 4: *Glossar Arabisch-Deutsch* (Wiesbaden, 1994), S. 139b. In Kharga, wo ein der *mulāgā́* entsprechendes Ritual stattfindet, wird dieses ebenfalls zur Zeit des Freitaggebets ausgeführt.

[5] M. Meyerhof, 'Beiträge zum Volksheilglauben der heutigen Ägypter,' *Der Islam* 7 (1917), S. 319.

ten größeren Ort Balāṭ geboren und im Kindesalter an ihren jetzigen Wohnort gekommen. Ihre alten, nunmehr dem Verfall anheimgegebenen Häuser sind noch im historischen Zentrum von Balāṭ zu finden, auch haben die Familien ihre Grabstellen noch auf dem dortigen südlichen Friedhof beim Šēx ilQaḍēy. Dialektologisch geben die Texte das Arabisch wieder, das im Osten der Oase Dakhla geprochen wird. Dieses hebt sich deutlich vom Zentrum und vom Westen ab und ist bereits in früheren Publikationen kurz beschrieben,[6] so daß sich eine nochmalige Behandlung der dialektologischen Besonderheiten erübrigt. Die Transkription der Texte folgt der in diesen Publikationen geübten Praxis.[7] Die Informanten wurden teils nach bestimmten Pflanzen gefragt, deren medizinale Wirksamkeit bekannt ist, teils gaben sie auch selbst bei der systematischen Aufnahme der Flora an,[8] welche Anwendungen die einzelnen Pflanzen haben. Nach den beiden Heilverfahren *kayy* und *šilg* wurde entweder speziell gefragt oder sie kamen bei Fragen zu Krankheiten von Mensch oder Tier zur Sprache. Zum Vergleich werden hier auch gelegentlich Angaben aus dem Dorf Barīs im Süden der Oase Kharga herangezogen. Aus Platzgründen können wir hier nicht alle diesbezüglichen Texte publizieren und müssen uns auf einige illustrative Beispiele beschränken.

Über die folgenden Heilmittel wird hier berichtet:[9]

---

[6] Zum Dialekt von Ost-Dakhla s. M. Woidich, 'Aus den Erinnerungen eines Hundertjährigen. Ein Text im Dialekt von Balāṭ / Dakhla, Ägypten,' *Estudios de dialectología norteafricana y andalusí* 3 (1998), S. 7–33 und M. Woidich, 'The Arabic dialect of ilBašandi in Dakhla-Oasis (Egypt),' in M. Mifsud (ed.), *Proceedings of the 3rd international conference of L'Association internationale pour la dialectologie Arabe held at the University of Malta* (Malta, 2000), S. 113–118. Ferner Woidich 2006 mit weiterer Literatur.

[7] Sekundäre Längung von Vokalen unter dem Akzent wird mit einem [·] angegeben wie in *ṣa·məġ*, Spirantisierung von /b/ als /ḇ/ [ß], der Aufsprengungsvokal am Wortende als ᵊ.

[8] Zu den Namen der Pflanzen s. P. Behnstedt & M. Woidich, *Die ägyptisch-arabischen Dialekte*. Bd. 5: *Glossar Deutsch-Arabisch* (Wiesbaden, 1999), S. 219–224.

[9] Lateinische Namen nach L. Boulos, *Flora of Egypt*, vol. 4 (Cairo, 2005).

Arabisch[10]	Deutsch [wissenschaftlicher Name]	Anwendung
ʿušāṛ ~ iʿšāṛ	Sodomsapfel [Calotropis procera], milchiger Saft	Hämorrhoiden; wenn bei Zahnschmerz angewendet, führt er Zahnausfall[11] herbei
ḥanḍāl	Koloquinte [Citrullus colocynthis]	Rheuma, Durchfall[12]
šuxxēra	Kohl-Gänsedistel [Sonchus oleraceus]	Wundversorgung[13]
sakrān	Bilsenkraut [Hyoscyamus muticus L.]	beruhigt Zahnschmerz[14]
qarāṭ	Akazienfrüchte [Acacia nilotica L.]	Durchfall[15]
xabāṭ	Akazienblätter	Durchfall (Ziege)
šaʿbūṛa, šaʿabīr	Akazienästchen	Durchfall (Ziege)
baṣāl	Zwiebel [Allium cepa L.]	allgemein gesund; Wundversorgung, beruhigt Wunden; beseitigt das Gefühl, kratzen zu müssen; Beruhigung der Augen[16]
baṣal iswād	schwarzer Zwiebelsamen	Zahnweh
tōm	Knoblauch [Allium sativum L.]	in Öl gebräunt bei Hufverletzungen des Esels, Wundversorgung[17]
ṣaˑməġ	Akazienharz, Gummi arabicum	Wundversorgung, Pflaster[18]

[10] In lokaler Aussprache notiert mit der für Dakhla typischen Betonung der Ultima.

[11] S. auch L. Boulos, *Medicinal plants of North Africa* (Algonac, 1983), S. 27 für ägyptische Beduinen. Daneben zahlreiche andere Anwendungen, gefährlich für Haut und Augen.

[12] Boulos 1983, S. 73 meldet zahlreiche Anwendungen, unter anderem gegen Schlangenbisse und Skorpionstiche; zur Ethnogynäkologie s. Inhorn, S. 174. Aus den Samen kann ein Öl (*gaṭarān*) gewonnen werden, das beim Gerben, bei Behandlung von Wunden und als Abführmittel bei Tieren Verwendung findet [Barīs]. Zur Ölgewinnung s. Z.E. Yaniv [et al.], 'Colocynth. Potential arid land oilseed from an ancient cucurbit,' in J. Janick (ed.), *Perspectives on new crops and new uses* (Alexandria, VA, 1999), S. 257–261.

[13] S. http://www.liberherbarum.com/Pn4574.HTM, dort nicht zur Schließung von kleinen Wunden, sondern für Umschläge, bei Fieber, zur Aktivierung der Menstruation.

[14] Boulos 1983, S. 167. S. auch http://www.liberherbarum.com/Pn0300.HTM für die Verwendbarkeit gegen Zahnschmerz.

[15] Von *qrḍ, vgl. oberäg. *garaḍ*. Nach Boulos 1983, S. 116, wirksam bei Diabetes; weiteres zum volksmedizinischen Gebrauch s. http://www.hort.purdue.edu/newcrop/duke_energy/Acacia_nilotica.html. Von alters her zum Gerben sowie zum Färben von Textil benutzt.

[16] Boulos 1983, S. 23. Zum Gebrauch s. Elder, S. 62,8; Blackman, S. 261; Inhorn, S. 172f.

[17] Boulos 1983, S. 25.

[18] S. auch Blackman, S. 211.

Table (*cont.*)

Arabisch	Deutsch [wissenschaftlicher Name]	Anwendung
šišlān	dorniges Gebüsch, mit nierenförmigen violett-braunen Samenkapseln [Prosopis farcta].	Diabetes[19]
quṭṭāb	der Beschreibung nach ein Stäubling (?) [eine Lycoperdon-Art] oder Kartoffel-Bovist (?) [Scleroderma citrinum]	Wundversorgung[20]
xaṛwāʿ	Rizinus [Ricinus communis L.]	Eiterbeulen
šīḥ	Wermut [Artemisia genus]	Wundberuhigung, beseitigt das Gefühl kratzen zu müssen[21]
ṭīn	Lehm	Wundversorgung
ḥallāba	Steinstaub, Steinmehl	für die Haut bei Kindern, s. Text 4 Fn.45
zetḥār	Leinsamenöl	zur Stärkung des Esels
fardit ḥamām	Täubchen	Schlangenbiß
laban bizza	Muttermilch	Kinderkrankheiten

Bei den Heilverfahren, die alle auch in Oberägypten gängig sind, ist in erster Linie die Kauterisierung (*kayy*) anzuführen,[22] die bei Mensch wie Tier durchgeführt wird. Es handelt sich um das Brennen mit einem glühenden, breitköpfigen Nagel (*ṭubbāʿa, ṭabābīʿ*), das beim Menschen vor allem den Kopfschmerz vertreiben soll. Der Schmerz vertreibt hier den Schmerz.[23] Es gibt bestimmte Stellen am Körper für bestimmte Anwendungen – gegen Kopfweh wird man z.B. im Nacken kauterisiert – und es hinterläßt die entsprechenden Narben.[24] Bei Tieren kommt ein gröberes

[19] Prosopis farcta wird auch unter den in Israel zur Behandlung von Diabetes gebräuchlichen Pflanzen aufgeführt, Z. Yaniv [et al.], 'Plants used for the treatment of diabetes in Israel,' *Journal of Ethnopharmacology* 19/2 (March–April 1987), S. 145–51.

[20] Als blutstillend beschrieben im Liber herbarum, http://www.liberherbarum.com/Pn0151.HTM.

[21] Zum Gebrauch in der Ethnogynäkologie s. Inhorn, S. 172–173, 189.

[22] Meyerhof, S. 343; ohne Namensnennung kurz beschrieben in Blackman, S. 211.

[23] J. Walker, *Folk-medicine in modern Egypt* (London, 1934), S. 107. Zum Gebrauch in der Ethnogynäkologie s. Inhorn, S. 182 und zur Verbreitung S. 388 Fn.7 mit weiterer Literatur.

[24] Einen mit *kayy* behandelten ʿAbbādi beschreibt und ein Foto von ihm mit den entsprechenden Narben zeigt H. Barnard, 'Geneeskunst geïnspireerd door armoede,' *Nederlands tijdschrift voor geneeskunde* 144/20 (2000), S. 949–951.

Instrument zur Anwendung, das Brenneisen (*miḥwāṛ*), auch hier gibt es für jedes Krankheitssymptom eine dafür vorgesehene Stelle am Körper, etwa auf den Bauch beim Nabel, wenn der Esel auf eine bestimmte Art hinkt, s. Text 2.14–2.15. Dementsprechend empfiehlt das Sprichwort: *in ḍalāʿ iğḥēšak ikwīh* „wenn dein Esel hinkt, kauterisier ihn!".

Anders, aber sicher nicht weniger schmerzhaft ist der *xarṭ* (= *šalg*, *šlūg* in Oberägypten),[25] der beim Esel, aber auch bei Kühen und Kamelen angewendet wird. Hier wird ein etwa 10 cm langes Strück Schnur aus Palmfasern (*līf*) durch die Haut z.B. an der Schulter gezogen. Die Enden werden verknotet oder zusammengebunden, damit die Schnur an ihrem Platz gehalten wird. Die Wunde eitert, und ab und zu zieht man die Schnur hin und her, damit der Eiter abfließen kann. Auch dies soll helfen, wenn der Esel z.B. Lasten nicht tragen will, weil ihn etwas drückt. Aus al-Mūšiyya in West-Dakhla wird berichtet (1980), daß *kayy* und *šiluwwaʾ* durch einen Spezialisten, *ġazēʾ* genannt, ausgeführt werden, der zu Festeszeiten aus al-Qaṣr kommt und mit Trommelschlag auf sich und seine Dienste aufmerksam macht. Er bekommt einen *ġoʾmər*[26] Weizen bei der Getreideernte.

Zur Kinderheilkunde gehört das Einreiben und Massieren mit Muttermilch,[27] ersatzweise mit Butter (*tadʿīk*, in Barīs *mirrīs*), was gegen durch Zerrungen und Verstauchungen hervorgerufenes Unwohlsein der Kinder – das Kind ist dann *mamzūg*, in Barīs *mamlūx, magṭūʿ* – helfen soll. Die Massage wird von einer älteren, erfahrenen Frau ausgeführt, die im Dorf bekannt ist für ihre heilenden Hände.[28] Danach wird das Kind fest eingewickelt und warm gehalten.

Während den oben angegebenen planzlichen Heilmitteln sicher eine gewisse Wirksamheit zuzusprechen ist, s. die Angaben dazu in der ethnobotanischen Literatur, kann man sich bei anderen Mittel und

---

[25] Meyerhof, S. 343 führt die gleiche Prozedur als *xuzām* „Haarseil" an; ohne Namensnennung kurz beschrieben in Blackman, S. 210; zum Gebrauch in der Ethnogynäkologie s. Inhorn, S. 182, dort (nordwestliches Delta) *fatla* „Schnur" genannt. *šalg* oder *šilg* ist in Oberägypten allgemein Schnur oder Strick aus Palmfasern.

[26] D.h. eine Garbe so dick, wie ein Mann mit beiden Armen umfangen kann. Der *ġazēʾ* gehört also zu den *xayyāna*, d.h. zu den Handwerkern wie Schmied, Zimmermann, Friseur und Flickschuster, die einmal im Jahr bei der Ernte durch Naturallohn (*xīna*) für ihre Bemühungen entschädigt werden.

[27] Letztlich dürfte es sich dabei doch um eines der vielen Gegenmittel für den bösen Blick und Verhexung handeln, wie auch der Gebrauch der Muttermilch in der Ethnogynäkologie nahelegt, s. Inhorn, S. 136 und passim.

[28] Vgl. für Barīs in Kharga Text 139 in Peter Behnstedt u. Manfred Woidich 1988, S. 424–431. In Kharga-Stadt muß dies eine Frau sein, die Zwillinge geboren hat.

Verfahren schwer vorstellen, daß mit ihrer Anwendung einiger Nutzen, es sei denn ein magischer und damit placebo-Effekt, verbunden sein könnte. Erwähnt sei hier das Auflegen eines frisch geschlachteten Täubchens bei Schlangenbissen, wovon auch für andere arabische Länder berichtet wird.[29] Eigenartig ist auch das Ausräuchern der Würmer (*sūs*), die nach landläufiger Meinung Karies verursachen und so Zahnschmerzen hervorrufen, daher heute auch im MSA *ḍirs musawwas* „kariöser Zahn".

Natürlich stellt sich auch die Frage, in wie weit die hier beschriebenen Heilmittel und Heilverfahren heutzutage noch Anwendung finden, in einer Zeit in der die neuere Medizin in die entlegensten Gebiete vorgedrungen ist. In den ägyptischen Oasen ist wohl in jedem größeren Dorf eine *waḥda ṣiḥḥiyya* zu finden ist, während zentrale Orte wie die Stadt Mūṭ in Dakhla über ein Krankenhaus verfügen und niedergelassene Ärzte dort ihren Praxen betreiben. Die Gewährsleute beeilen sich auch meistens zu vermelden, daß was sie hier berichten, für frühere Zeiten gelte – *kān zamān* –, und man heute auf die moderne medizinische Versorgung zurückgreife. Das mag zum Teil stimmen – s. Text 3.48, ist doch festzustellen, daß jüngere Leute sehr viel weniger über die Pflanzen und ihre heilende Wirkung wissen als die Generationen davor. Andererseits wird auch durchaus von persönlichen Erlebnissen bei der Anwendung volkzmedizinischer Rezepte berichtet, s. etwa Text 4.04 und 4.09. Nach meiner, zugegebenermaßen eher impressionistischen Einschätzung wird die erste Hilfe in einfacheren Fällen oft doch bei den traditionellen Mitteln gesucht, deren Ingredientien für jeden verfügbar in der freien Natur zu finden sind, s. Text 3.27. Wenn diese nichts fruchten, wendet man sich an den Arzt, denn dieser kostet schließlich Geld. Dies gilt sicher für die Tiermedizin und wer mit offenen Augen durch die Dörfer geht, dem können die Spuren von *kayy* und *xarṭ* nicht entgehen. H. Barnard's kurzer Erfahrungsbericht aus der östlichen Wüste mit dem Titel „Geneeskunst geïnspireerd door armoede" (Durch Armut inspirierte Heilkunde)[30] trifft, wenn auch in geringerem Maße, auch auf die Oasen zu.

---

[29] S. auch Peter Behnstedt & Manfred Woidich 1988, S. 205f. und 386.

[30] S. Fn. 24.

1. *Sodomsapfel, Koloquinte und Akaziensamen*

Muḥammad[31]

1.01. di lʿušāṛ dēy, di tinfáʿ, fi lli ʿandu lbawaṣīr. ma-fīš dakātra walla lkalām dī. yigṭaʿ ilwarága gdī, tixurr labān katīra miš ziyy iššixxēra kdā baṣīṭa, dīk labān katīra.[32]

1.01. Der Sodomsapfel, der ist nützlich, bei Leuten mit Hämorrhoiden. Es gibt keine Doktoren oder sowas. Er reißt so Blätter ab, da kommt viel Milch heraus, nicht so ein wenig wie bei der Kohl-Gänsedistel, diese [gibt] viel Milch.

1.02. yiǧībha fi qzāza gdī, wu yídhunbahā̊ maṭraḥ ilbawaṣīr, yiṭihhirhī̊ ʿād. wu kān ma-yiʿmilšᵊ ʿamaliyyāt walla šiyy, ma-kānš fī ʿamaliyyāt ziyyᵊ dilwa·kt. dilwā·kt ʿama yiʿmélu ʿamaliyyāt.

1.02. Er tut sie in eine Flasche und reibt damit die Stelle ein, wo die Hämorrhiden sind, er desinfiziert sie so. Und man pflegte auch keine Operationen zu machen oder sonst was. Es gab keine Operationen so wie jetzt. Jetzt machen sie Operationen.

1.03. ilḥanḍāl, da li lli ʿandu rrumatizm. illi ʿandu rrumatizm, yiǧīb ilḥanḍála, ilḥanḍála, wu yiftaḥḥī̊ nuṣṣēn. yiḥuṭṭᵊ kull nuṣṣ fi kaʿb, min taḥt. yisḥib irruṭūba. yisḥib irruṭūba lli fi ǧismu kullih.

1.03. Die Koloquinte, die ist für den, der das Rheuma hat. Wer das Rheuma hat, nimmt eine Koloquinte, eine Koloquinte, und teilt sie in zwei Hälften. Er tut jede Hälfte an eine Ferse, unten. Das zieht die Feuchtigkeit[33] heraus. Das zieht sämtliche Feuchtigkeit in seinem Körper heraus.

1.04. wu li lli ʿandu ʿād disinterya, illi ʿandu disinterya, yiǧīb ilqarāṭ. ʿārif ilqarā̊ṭ? btā̊ḥt iṣṣanṭa dēy. ilqarā̊ṭ wu yduqqu, wu yiġlīh wu yišrábu. tiʿmél maʿāh ilʾimsāk. timnáʿ iddisinterya di xāliṣ.

1.04. Und wer nun Durchfall hat, wer Durchfall hat, nimmt Akazienfrüchte. Du kennst die Akazienfrüchte? Die von der Akazie. Die Akazienfrüchte und er zerstößt sie, und kocht sie und trinkt sie. Das verursacht Verstopfung bei ihm. Es stoppt den Durchfall völlig.

1.05. iṣṣanṭ da minnu fāyda katīra, ʿama yuḥbukō̊w, ilʿidān bitāʿithī̊, ʿama nḥuṭṭu… niḥbukbahā̊ fi lbēt. wi ʿamma niʿmil minnu lbibān, wi lgabḍa btaḥt ilmanǧā̊l, wu nnuṣāb ibtāʿ ilmisḥāya, wu nqīdbō̊w, fi lbēt.

1.05. Die Akazie ist sehr nützlich, man macht Dächer, [mit] ihren Stämmen, wir legen… wir machen damit das Hausdach. Und wir machen daraus die Türen, und den Griff der Sichel, und den Stiel der Hacke, und machen damit Feuer, im Haus.

[31] Ein etwa 35jähriger Lehrer und Landwirt.

[32] Eine Reihe von Nomina, die gewöhnlich mask. sind, gelten in ilBašandi als feminin. Dazu gehören u.a. *labān* „Milch", *ḥalg* „Kehle", *mošṭ* „Kamm", *ṭarīg* „Weg", *ǧorn* „Dreschgut", *ṣēf* „Sommer", *ṭabāg* „Teller".

[33] Krankheiten werden vielfach der Feuchtigkeit *ruṭūba* zugeschrieben, s. auch Inhorn, S. 168.

1.06. nuṭbuxbṓw. fawāyid katīra. w ilxabā́ṭ, ʿārif ilxabā́ṭ? ilwarága btāḥtu dīkha lxaḍrā́ʾ dēh, yaklūh ilġanā́m, wi lxawarīf wi lkalām dī. wu lqarā́ṭ ziyyᵊ ma gùltlā́k. illi ʿandu disinterya yiğīblu šwayya yiduqqəhṓm baʿdᵊ ma ʿād yibasṓw, yiduqqəhṓm lamma yinʿimu kidī, yiġlihōm wu yišṛabhṓm.

1.06. Wir kochen damit. Viel Nutzen. Und die Akazienblätter, kennst du die Akazienblätter? Ihre Blätter, diese grünen, die fressen die Ziegen, und die Schafe und so weiter. Und die Akazienfrüchte, wie ich dir gesagt habe. Wer Durchfall hat, holt sich ein wenig [davon], zerstößt sie, nachdem sie natürlich getrocknet sind, er zerstößt sie, bis sie ganz fein geworden sind, er kocht sie und trinkt sie.

1.07. wi lxabā́ṭ barḍu. ḥaṭab xabā́ṭ barḍu, ilxabā́ṭ yinfáʿ ḥatta law ġanáma hāṛṛa, ġanáma law harrit, barḍu yiğīblaha lbtāʿa dēy. šaʿabūr šōk šaʿabūr, ilxaṭā́b takṓlu ʿād timnáʿ … habuṣṣᵊ ʿa lṃuyya w ağīlā́k.

1.07. Und auch die Akazienblätter. Auch das Holz der Blätter, die Akazienblätter sogar sind von Nutzen, wenn eine Ziege Durchfall hat, wenn eine Ziege Durchfall bekommen hat, holt er ihr auch das Dings da. Die kleinen Ästchen, die Dornen von kleinen Ästchen, die Blätter fressen sie auch, sie verhindern …. ich schau schnell nach dem Wasser und komme gleich wieder zu dir!

2. *Kohl-Gänsedistel, Gummi arabicum, Kauterisierung,* xarṭ *und Zwiebel*

Muḥammad

2.01. iššixxēra zamān dīy, nigṭaʿhā́ gdī́, wu llabán yiṭláʿ minhī, nudhun bihā́ šgūg. itmuwwḗt iššgūg, illabān bitāʿ iššixxēra. wu zamān illi kān yiğğirḗḥ, illi kān yiğğirḗḥ, ʾīdu tiğğirḗḥ kidī́ yibga fihā́ ğa·rḥ, niʿmélu.[34]

2.01. Die Kohl-Gänsedistel, die rissen wir früher so auseinander, und da kam dann die Milch heraus, damit rieben wir Schrunden ein. Sie heilt die Schrunden, die Milch der Kohl-Gänsedistel. Und früher, wenn sich da einer eine Wunde zuzog, seine Hand so verwundet wurde, daß da eine Wunde war, machten wir das.

2.02. niğīb danšit ṭīna, niğīblu danšit ṭīn balīz, wu nḥuṭṭu ʿalēh kidī́. wu bi ṭṭīn ma-yğībšᵊ damm tānḗy. wu baʿdēn yiṣe·ḥḥ. w illi kān yitkisḗr nnaggīlu ğōz ṣa·məġ. nnaggīlu ğōz ṣa·məġ min iṣṣanṭa,

2.02. Wir nahmen ein wenig Lehm, wir holten dafür ein wenig feinen Boden, und taten diese so darauf. Mit dem Lehm darauf, blutet sie nicht mehr. Dann heilt sie. Wer sich was gebrochen hatte, für den sammelten wir ein wenig Gummi

[34] Aus einem anderen Text: *zamān kānu ygūlu ʿalehā labàn ḥimāra* „früher nannte man sie ‚Eselsmilch' ".

wu nbúllihom fi lṃuyya, wu nǧīblu waṛáqa wu nudhunhī̃ ʿala gdī̃, wu nḥuṭṭəhālō̃ ʿa lkasᵊr wu nsību.

arabicum. Wir sammelten für ihn ein wenig Gummi arabicum von der Akazie, und feuchteten es an in Wasser und holten ein Stück Papier und schmierten es so darauf, taten es ihm auf den Bruch und ließen ihn ruhen.

2.03. yugʿód kām yōm kidā̃ ʿād yugʿō̃d qaddᵊ ma yugʿōd. wu baʿdēn yiṣeˑḥḥ. ilkasᵊr yiṣeˑḥḥ. miš ǧeˑbs ziyy dilwaˑkt, ma-kanšᵊ fī ǧeˑbs. kān da nǧīb ǧōz sáˑməġ, mi ṣṣanṭa, wu nbulluhō̃m fi danšit ṃuyya, fī ḥāǧa gdī̃, fi šqēfa walla btāʿa dē̃y.

2.03. Das bleibt so ein paar Tage, es bleibt so lange, wie es eben bleibt. Dann heilt es. Keinen Gips so wie heute, es gab keinen Gips. Wir pflegten ein wenig Gummi arabicum zu nehmen, von der Akazie, feuchteten es an in ein wenig Wasser, in irgendwas, in einem Scherben oder sonst was.

2.04. wu awwal ma ylīnō̃w, inǧīb waṛáqa, wu nbúllaha min iṣṣaˑməġ dī, ṃuyyat iṣṣaˑməġ dēy. wu nǧē̃ ʿa l… ilkasər, maṭraḥ ma maksūṛa n kānit ʾīd walla riǧil, wu nidʿakha bi lṃuyya lʾawwā̃l, wu nḥuṭṭ ʿalēha ṣṣamġa wu xalāṣ ʿala gdī̃.

2.04. Sobald es weich geworden war, nahmen wir ein Stück Papier, und befeuchteten es mit diesem Gummi arabicum, dem Wasser davon. Wir taten das auf die Bruchstelle, wo es gebrochen war, sei es Hand oder Fuß, massierten es zuerst mit dem Wasser ein, taten den Gummi arabicum darauf und fertig war es.

2.05. tùgʿodlahā̃ qaddᵊ ma tugʿō̃d, túgʿod yomēn talāta ʾasbūʿ šahər, lamma yṣeḥḥ yišīlhī ʿād. wu yibga ṣaˑḥḥ, xalāṣ, ma-kānš fī dakātra walla btāʿ ziyy dilwaˑkt. lamma nrūḥ ilbēt ʿād nubga nugūl tānē̃y.

2.05. Es blieb solange, wie es eben brauchte, es blieb zwei, drei Tage, eine Woche, einen Monat, wenn es geheilt war, nahm er es ab. Dann war es geheilt, fertig, damals gab es keine Doktoren noch sonst was wie heute. Wenn wir nachhause gehen, erzählen wir weiter.

2.06. fīh kaˑyy. nikwē̃y. lamma ykūn wāḥid ḍahru wāǧiʿ gāʿid ḍahru wāǧiʿ miš gādir yitʿidē̃l ḍahru wāǧiʿ igdī̃ walla ḥāǧa, ənǧīblu ṭṭubbāʿa, ʿārif iṭṭubbāʿa? ã, iṭṭubbāʿa dē, wu nsixxínha fi nnāṛ malīḥ lamma tḥimmē̃r. lamma tḥimmir malīḥ.

2.06. Es gibt da die Kauterisierung. Wir kauterisieren. Wenn einem der Rücken weh tut, er sich nicht mehr aufrichten kann, sein Rücken tut so weh oder sonst was, dem bringen wir den Brennagel, kennst du den Brennagel? Ja, diesen Brennagel, den erhitzen wir gut im Feuer, bis er glüht. Bis er schön glüht.

2.07. wu yúrgud ʿala baṭnu, wu yǧīlu wāḥid illi ʿamyikwē̃y, yirūḥ ḥāṭithālō̃ maṭraḥ ilwaǧā̃ʿ. maṭraḥ ma yuwǧáʿ. tiḥuṭṭahalō̃w. ṭubbāʿa, itnēn, talāta, wu yugʿō̃d ʿād. ma-yištiġēlš ʿād yāxō̃d ṛāḥa.

2.07. Er legt sich auf den Bauch, dann kommt einer, der kauterisiert, und drückt es ihm auf die schmerzende Stelle. Wo es weh tut. Er tut es ihm darauf. Ein Brennen, zwei, drei, und er bleibt [im Haus]. Er arbeitet nicht und ruht sich aus.

2.08. yimši gdā bass, da ma-yištiġēlšᵊ xāliṣ ʿašān lamma ʾē:, iṭṭaḇaḇīʿ itṣe·ḥḥ ʿād. iṭṭababīʿ dē ʿād… iṭṭubbaʿāt dēy, iṭṭubbaʿāt dī ʿād ʾē, itfiqfēq. in kān ḍahru wāǧiʿ, tfiqfēq. wu tisḥēb ilṃuyya lli fīh.

2.08. Er läuft nur so rum, der arbeitet gar nicht, bis was? Die Brennstellen dann heilen. Diese Brennstellen dann… diese Brennstellen, diese Brennstellen tun was? Sie platzen auf. Wenn ihm der Rücken weh tut, platzen sie auf. Und ziehen das Wasser heraus, das darin ist.

2.09. tisḥéb ilṃuyya lli fīh tlāgihī ʿama tne·zz igdā ṃuyya gdī. ziyy ilǧe·sər kidī, tiṭílleʿ ṃuyya barṛā. nhāṛ ma yilgahōm ʿama ygirriṣōw, yiġlō gdī, yiǧīblihom baṣála, baṣála yābsa, wi yigsimha nuṣṣēn, yiduqqha gdī daqqa fi lhōn.

2.09. Sie ziehen das Wasser heraus, das darin ist, du merkst, daß sie so nässen. So wie dieser Damm, sie ziehen Wasser heraus. Wenn er merkt, daß sie ihn jucken, sie kribbeln so, da nimmt er eine Zwiebel, eine trockene Zwiebel, halbiert sie, und zerstößt sie im Mörser.

2.10. yiduqqha daqqa malīḥa, walla ʿala ḥāǧa gdī. yiduqqəhī, wu yniʿʿimhā yikirrifhī yikirrif iṭṭababīʿ dēy. iṭṭababīʿ illi ʿala ḍahru dēy. yuḥuṭṭu ʿalēha lbaṣála gdā ʿašān ma-tiġlīš.

2.10. Er zerstößt sie gut, oder auf so etwas. Er zerstößt sie, und macht sie fein, und er beseitigt das Jucken,[35] er beruhigt [damit] diese Brennstellen. Die Brennstellen auf seinem Rücken. Sie tun die Zwiebel so darauf, damit sie nicht juckt.

2.11. yimkin šammit rīḥa, ʿiṭir, walla ṣabūnit rīḥa walla ḥāǧa gdī. tugʿōd tigirrēṣ kidā ʿašān yúgʿod yukruš fīhī? lā! yiʿmíllaha lbaṣála dē ʿašān ma-yukrošš fīhī.

2.11. Vielleicht haben sie einen Geruch angenommen, Parfüm, oder parfümierte Seife, oder so etwas.[36] Da bilden sie dann so Beulen, daß er dann dauernd daran herumkratzt? Nein! Er gebraucht dafür diese Zwiebel, damit er nicht daran herumkratzt.

2.12. iǧǧa·ḥš in kān magṭūʿ yiǧīblu lmiḥwār. ʿārif ilmiḥwār? šuftu? ilmiḥwār silk áhu gdī. ḥadīda rfayyʿa gdī ziyyᵊ ṣbāʿak dī, wu yiḇga tanīhī gdī. tiḇga ʿāmla gdī. māši? gidī. wu nimsiku mi nnáḥya dēy, māši, yiḇga ʿamal kīf, txayyal ilmawḍūʿ dī?

2.12. Der Esel, wenn er überanstrengt ist, nimmt man das Brenneisen. Kennst du das Brenneisen? Hast du es gesehen? Das Brenneisen ist ein Draht, so wie das. Ein dünnes Eisen so wie dein Finger, und der ist so gebogen. Der sieht dann so aus. O.K.? So. Und wir halten ihn an dieser Seite, dann macht er wie? Kannst du dir diese Sache vorstellen?

[35] Das Jucken wird im Volksglauben durch fremden und starken Geruch hervorgerufen, vgl. Kairenisch *kārif* „den Geruch von etwas angenommen habend". S. auch Text 7.

[36] Starke Gerüche behindern die Genesung, s. Meyerhof, S. 329.

2.13. buṣṣ, áha ʿāmla gdī. bassᵊ maftūḥa ʿala wāsiʿ ziyy ma hī gdī. māši? yiḥuṭṭu fi nnār, niqīd innār, wu nḥuṭṭᵊ fīh dī, ilmiḥwār dī, lamma yḥimmēr barḍu ziyy iṭṭubbāʿa.

2.13. Schau, so sieht er aus. Nur ist er ganz weit offen, so wie das hier. In Ordnung? Sie tun ihn ins Feuer, wir zünden ein Feuer an und legen es da hinein, dieses Brenneisen, bis es auch glüht wie der Brennagel.

2.14. wu nğību n kān magṭūʿ, ʿama ykōḥḥ kḥkḥkḥ... gdī, yiʾē, yisíxxin da malīḥ tamām gidī wu lḥitta lmaʿwūğa dē, ziyy ma hī maʿmūla gdī, wu yğībha wu yḥuṭṭəhalō ʿala baṭnu gdī. ta·ḥt fi baṭnu gdī.

2.14. Wir holen es dann, wenn er [der Esel] überanstrengt ist, hustet er *kḥkḥkḥ*... so, er tut was? Er macht das richtig schön heiß, und dieses gewundene Stück da, so wie es gemacht ist, und dann nimmt er es und tut es ihm so auf den Bauch. Unten an seinem Bauch.

2.15. wāḥid, itnēn, ā:, wu yğīblu wāḥid tanē bi lʿa·ks. wāḥid kidī bi ṭṭūl wu wāḥid bi lʿaks ziyyᵊ dī. fi baṭnu min taḥt, yibga ʿāmil ziyy iṣṣalība kidī, māši? ā:, wu yugʿod kidī. wu n kān dēlu maksūr min warā, barḍu yḥuṭṭlu ʿalē.

2.15. Einmal, zweimal, ja, und dann macht er ihm noch einen entgegengesetzt. Einen so der Länge nach und einen entgegengesetzt wie dieser. An seinem Bauch unten, das sieht dann aus wie ein Kreuzschlüssel, o.k.? Ja, und das bleibt so. Und wenn er hinten den Schwanz gebrochen hat, auch dann setzt er ihm ein [Brenneisen].

2.16. wu n kān min guddām, ʿama-ykoḥḥ ṣidru ḍiyyig min guddām, ʿama-yitxabbāl fi lmašy, yixróṭu. yixróṭu kīf? ʿārif ilmisilla? šuftihī lmisilla? ilmisilla dēy, barḍu nsixxinha fi nnār, wu yğībhī, wu yimsik iğğilda btāḥtu guddām, guddām fi ṣidru huna, wu yrūḥ ʾē·, yib̲ga fātil še·lg, šilgᵊ rfiyyiʿ gdī.

2.16. Und wenn es vorne ist, er hustet, ihm ist die Brust vorne eng, er verheddert sich beim Laufen in den eigenen Beinen, dann wendet er den *xa·rṭ* an. Wie wendet er den an? Kennst du die Sacknadel? Hast du die Sacknadel gesehen? Diese Sacknadel, die erhitzen wir im Feuer, und man nimmt sie und packt seine Haut vorne, vorne hier an der Brust, und dann was? Dann ist da ein Schnürchen aus Palmfaser, so eine dünne Palmfaserschnur.[37]

2.17. wu yğību yḥuṭṭu fi lmisilla, wu yimsik iğğilda gdī, yirūḥ maʿáddi fīha gdī. wu yṭilliʿ mi nnaḥya ttaniyya. yib̲ga maʿgūd. yiʿgidu min hōna wu yiʿgidu min hōni, yib̲ga dnēša ṣġiyyra gdī tiwga ẓāhra min hōna w hōni.

2.17. Die fädelt man in die Sacknadel ein, und man packt die Haut so, und führt sie [die Sacknadel] so hindurch. Und zieht sie an der anderen Seite wieder heraus. Dann wird er geknüpft. Er macht hier einen Knoten hinein und hier auch, dann ist noch so ein kleines Stück davon hier und hier zu sehen.

---

[37] S. auch Text 3.01–3.02.

2.18. wu nnaḥya ttaniyya hī barḍu. yiʿmillu tnēn talāta, wu yimsiku wu ysību gdī. yugʿod kām yōm kidā wu ybiṭṭēl kaḥḥ. wu mumkin yizgī zetḥār.

2.18. Und auch die andere Seite. Man macht ihm zwei oder drei, und er behandelt ihn und man läßt ihn so. Er bleibt so ein paar Tage und hört auf zu husten. Man kann ihm auch Leinsamenöl einflößen.

2.19. yiǧīblu šwiyyit zetḥār, wu yrùbbuhúm fi kūz, wu maʿāhum daḥyitēn, wu ṣo·bḥ, iṣṣobḥᵊ badrēy, bassᵊ ykūn fi ššti ziyy ilyomēn dōla, yikūn fi Ṭūba walla fi Mšīr, yikūn bard. wu yizgihumlō ʿa ṣṣobḥᵊ ʿa rrīg.

2.19. Er nimmt etwas Leinsamenöl, und verrührt es in einer Blechdose, zusammen mit zwei Eiern, und am Morgen, frühmorgens, freilich sollte es Winter sein, so wie jetzt, im [Monat] Ṭūba oder im iMšīr, es soll kalt sein. Und er flößt es ihm frühmorgens auf nüchternen Magen ein.

2.20. wu yriyyíḥu. ma-yšuddūš xāliṣ bi llukāf walla magāṭif walla ḥāǧa xāliṣ. yiriyyiḥu, yugʿúdlu ʾasbūʿ kidī yuwga yṣe·ḥḥ, ʿād yuwga ziyy ilḥuṣān. ʿād yuwga malīḥ ziyy ilḥuṣān. yirkábu ʿād wu yḥimmēl. yisəráḥbō wu yruwwēḥbō ʿād, ziyyᵊ ma kān lawwāl.

2.20. Und er läßt ihn ausruhen. Er sattelt ihn nicht mit dem Sattel oder mit den Körben, gar nichts. Er läßt ihn ausruhen, er bleibt so eine Woche und dann wird er wieder gesund, da wird er wieder bärenstark. Da geht es ihm wieder gut wie einem Pferd. Er reitet ihn wieder und belädt ihn. Er geht mit ihm aufs Feld und wieder nach Hause, wie es vorher war.

3. šilg, kayy, *Sodomsapfel, Koloquinte, Zwiebel, Knoblauch, Zwiebelsamen,*[38] *Gummi arabicum*

Ṭāha[39]

3.01. Ṭ: iššilg[40] *da nistixdámu fi lǧaḥš ʾaw ilḥimāṛ ʾaw ilbagáṛa yaʿni iǧǧaḥš masalan law ʿandu šwiyyit ṃuyya ʿala ṣidru inrūḥ ḍarbīnlu šilgēn, iššilgēn dōl biynizzílu lṃuyya illi ʿala ṣidru, yaʿnī byibʾa zayy iltihāb ʾaw zayyᵊ ḥāga kida tibʾa ʿala ṣidru.*

3.01. Ṭ: Die ‚Schnur' gebrauchen wir beim Esel oder der Kuh, d.h., beim Esel z.B. wenn er etwas Wasser auf der Brust hat, dann machen wir ihm ein paar ‚Schnüre', diese ‚Schnüre' holen das Wasser heraus, das er auf der Brust hat, d.h., das ist wie eine Entzündung oder so, wie etwas halt, das auf seiner Brust ist.

[38] Der Sprecher hat längere Zeit in der Armee verbracht hat und seine Sprache ist daher stark vom Standardägyptischen geprägt.

[39] Ein etwa 40 jähriger Koch und Landwirt.

[40] *šilg* (oder *šalg*) bedeutet in Oberägypten sowohl das Heilverfahren wie den Palmfaserstrick. In ilBašandi heißt das Verfahren *xarṭ*, während *šalg* nur den gewöhnlichen Palmfaserstrick bezeichnet.

3.02. lamma byāxud iššilgēn dōl, diyyat bitiʿmil ʿamaliyyit tanḍīf. bitṛawwaʾu muddit talát-arbaḥ t-iyyām kida w-titġassil. aġsilu bi ṣṣabōna kida w-itmašši iššilg bi lṃuyya ṣṣuxna yaʿnī li ḥadd masalan arbaḥ t-iyyām isbūʿ, baʿdə kida trūḥ ʾāṭiʿ iššilg, tʾuṣṣu b-ḥālu. W: trūḥ gāṭiʿ?

3.02. Wenn er die Schnüre bekommt, bewirken sie eine Säuberung. Sie erleichtern ihn so binnen drei, vier Tagen und müssen gewaschen werden. Ich wasche ihn mit Seife und bewege die Schnur mit heißem Wasser hin und her, etwa vier Tage oder eine Woche lang, danach schneidest du die Schnur durch, schneidest ihn ganz durch. W: Du schneidest durch?

3.03. Ṭ: mmm, wi trūḥ ʾē? sāybu. hūwa baʾa yiṣaḥḥ ʿala kida. da ʿamaliyyit iššilg.
3.04. innama ḥkāyt iṭṭubbāʿa, iṭṭubbāʿa di tiwga yaʿnī waḥda masalan ṛāsha wāǧʿa, riǧlēha maksūra, drāʿha maksūr,
3.05. fa ṭabʿan bitiʿmil ʿamaliyyit ʾē? taʾsīr fi lkūʿ ʾaw fi lʿaḍm illi hīya tibʾa fi lkasr wi yuwga ʿalēha ṃayya min ǧuwwā. fa lāzim tāxud ṭubbāʿa lihī, iṭṭubbāʿa di lāzim tiʿiyyēn.

3.03. Ṭ: Mmm, und dann was, dann läßt du ihn. Er wird auf diese Weise gesund. Das ist das Verfahren mit der Schnur.
3.04. Die Sache mit dem Brennagel aber, der Brennagel, da ist z.B. eine, die hat Kopfweh, ihr Beine sind gebrochen, ihr Arm ist gebrochen, 3.05. da macht sie natürlich das Verfahren von was? Eine Narbe auf den Ellenbogen oder auf den Knochen, die gebrochen sind und worin innen Wasser ist. Sie muß dafür einen Brennagel nehmen, der Brennagel muß genau sitzen.

3.06. wi baʿdᵊ kida truḥ ʾē? miṣiffiyya lṃuyya lli ʿalēhī. baʿdᵊ ma tṣaffit ilṃuyya illi ʿalēhī trūḥ ʿāmla ʿamaliyyit ʾē? bitṣiḥḥ ʿād. 3.07. bassᵊ biyibʾa makān iṭṭubbāʿa yibʾa līh taʾsīr, lāzim tiwga gāʿda bāyna, miʿillimā. yaʿnī sawāʾ ʾan kānit fi rāshí ʾaw fi riǧlihí ʾaw ʿala ʾīdhí ʾaw ʿala ḍahrihí masalan.

3.06. Was tut der dann? Der zieht das Wasser heraus, das darin ist. Nachdem das Wasser, das darin ist, herausgezogen ist, tut sie was? Sie wird dann gesund. 3.07. Nur hinterläßt der Brennnagel eine Narbe, er muß sichtbar bleiben, er hinterläßt einen Abdruck. Gleich, ob das nun auf ihrem Kopf oder an ihrem Fuß oder an ihrer Hand oder auf ihrem Rücken ist.

3.08. mumkin tikūn wugéʿit ʿala ḍahrihí, tilgāha ḍahrāha tkasár. lāzim tåxud laha ṭubbāʿtēn ʾaw talāta ʿala ḍahrāha min waṛā. 3.09. bi ʿamaliyyit iṭṭabab̠īʿ wi ṭṭabab̠īʿ di yaʿnī aḥla yaʿnī ṭibb fi ṭṭubbāʿa bizzāt yaʿnī. 3.10. ilǧaḥs ṭabʿan da lū miḥwāṛ, ilmiḥwāṛ ġēr iṭṭubbāʿa. ilmiḥwāṛ da masalan yikūn masalan iǧǧaḥs magṭūʿ yaʿnī ḥammaltu ḥiml tagīl, maṛṛa ʿala maṛṛa ʿala maṛṛa, fa biyrūḥ magṭūʿ.

3.08. Sie kann auf ihren Rücken gefallen sein, dann ist ihr Rücken gebrochen [geprellt?]. Sie muß zwei- oder dreimal gebrannt werden hinten auf ihren Rücken. 3.09. Mit den Brennägeln und die Brennägel sind nämlich die beste Medizin, ganz besonders die Brennägel. 3.10. Für den Esel gibt es das Brenneisen, das Brenneisen ist anders als der Brennnagel. Das Brenneisen ist dafür, wenn z.B. der Esel überanstrengt (*magṭūʿ*) ist, d.h., du hast ihn schwer beladen, ein ums andere Mal, da wird er dann überanstrengt.

3.11. ʿašān ṭabʿan yixallīh yirǧaʿ zayyᵊ mā kān… ʾaw yirūḥ kawīh bi lmiḥwāṛ. yiḍrub miḥwārēn talāta kida ʿala baṭnu min taḥt, yiṣiḥḥ ʿala ṭūl. da bi nnisba li lmiḥwāṛ. 3.12. ilḥāǧa ttánya lli hūwa masalan ʿamaliyyit ʾē? yaʿnī barḍak yikūn masalan drāʿu maksūr ʾaw draḥḥa maksūr yaʿnī bi nnisba li ssitt ʾaw rrāǧil.

3.11. Damit er ihn [den Esel] natürlich wieder so werden läßt, wie er war… oder er brennt ihn mit dem Brenneisen. Er drückt ihm zwei oder drei Mal das Brenneisen auf den Bauch unten, da wird er gleich gesund. Soweit das Brenneisen. 3.12. Die andere Sache ist z.B. die Arbeit mit was? D.h. ebenso für jemand, der den Arm gebrochen hat, oder eine, die den Arm gebrochen hat, also für Frau und Mann.

3.13. wugiʿ masalan w hūwa māši ʿala wirku ʾaw ʿala ḍahru ʾaw xabaṭ iṣbāʿu, da nigibluziblᵊ bi ṣṣaməġ, fi hāǧa smáha iṣṣaməġ. 3.14. da biyitball fi lṃuyya wi baʿdᵊ kida biyitdawwib kuwayyis w yithaṭṭ fi ḥāǧa wi ʾayyᵊ kasr fi ṣbāʿu ʾaw fi ḍahru ʾaw fi riǧlu ʾaw fi ʾayyᵊ ḥāǧa.

3.13. Beim Gehen ist er z. B. auf den Oberschenkel gefallen, oder auf den Rücken oder er hat sich den Finger angeschlagen, für den nehmen wir Mist mit Gummi arabicum. Es gibt etwas, das heißt Gummi arabicum. 3.14. Den macht man im Wasser naß und danach löst er sich gut auf und wird dann in etwas getan und jeder Bruch in seinem Finger oder Rücken oder Fuß oder sonst irgendwas.

3.15. da biyitʿālig yigīb masalan waraʾa wi yrūḥ ʿāyiṣha ḥittit ṣamġa wi yrūḥ lāziʾ ʿala ṣbāʿu. sawāʾ ʾan ʾayyᵊ ḥitta yaʿnī ʿalēha kasər yaʿnī tkūn masalan ma-tiḥtāġš iṭṭubbāʿa. 3.16. innama iza law tiḥtāǧ iṭṭubbāʿa lāzim ṭabʿan da baʿdᵊ lazg iṣṣaməġ, law ma ǧābitš lāzʾit b-ṣṣaməġ wi bitrūḥ ǧiyya ʾē? fi ṭṭubbāʿa. da bi nnisba li lʿilāǧ illi hūwa ʾē? ṭabīʿi ittāni.

3.15. Der wird behandelt, er nimmt z.B. ein Papier und beschmiert es mit etwas Gummi arabicum und klebt es auf seinen Finger. Oder auf irgendeine Stelle, wo ein Bruch ist, die aber doch keinen Brennagel braucht. 3.16. Wenn sie aber den Brennagel benötigt, dann muß das nach dem Kleben mit Gummi arabicum geschehen., wenn das Kleben nichts bringt, dann macht sie was? Sie geht über zum Brennagel. Soweit die Behandlung, die was ist? Auch natürlich.

3.17.W: fi lġēṭ fī ḥagāt tanya, fī ḥāga ismáha lʿušāṛ? 3.18. Ṭ: ilʿušāṛ di…W: ilʿušāṛ ʾaw iliʿšāṛ? T: lā, iliʿšāṛ di byiṭlāʿ min arḍu, ilwaṛaʾ bitāʿu kabīr, waṛágu ḳabīr, wi yiʿmil ḥāga ziyy ilburtuʾāna kida ʾaw ziyy ilbaṭṭīxa, bassᵊ di tibʿa ʿala ṣṣuġiyyar, ṣuġiyyra šwiyya kida, da ʾismu liʿšāṛ.

3.17. W: Auf dem Feld gibt es andere Sachen, gibt es etwas mit dem Namen Sodomsapfel? 3.18. Ṭ: Der Sodomsapfel …. W: *ilʿušāṛ* oder *ilʿišāṛ*? Ṭ: Nein, *iliʿšāṛ*, der kommt von alleine aus dem Boden, sein Blatt ist groß, sein Laub ist groß, er bringt etwas wie eine Orange hervor, oder wie eine Melone, nur dann in klein, so ein wenig klein, das heißt Sodomsapfel.

3.19. law wāḥid masalan ḍirṣu wāǧiʿ ʾaw waḥda ḍirsha wāǧiʿ, yirūḥ kāsr ilwáṛaga btāḥt iliʿšāṛ, wi yrūḥ ʿāyiṣ iḍḍiṛs ʿašān biyibʾa illi missuwwēs min ǧuwwá yrūḥ lāsʿu illában btāʿ iliʿšāṛ di, da mumkin tihaddīh šwayya. da bi nnisba ilwaǧʿa illiḍrās. 3.20. ʾaw šwiyyit baṣal, baṣal iswid, wi yitḥaṭṭ ṭabʿan ʿala sakkīna tkūn suxna fi nnāṛ wi yithaṭṭ ʿalēha […] ilqulla, wi yrūḥ ḥāṭiṭ fīha būzu, ʾaw yrūḥ fātiḥ būzu.

3.19. Wenn einem z.B. der Backenzahn wehtut, oder einer der Backenzahn wehtut, dann bricht er das Blatt des Sodomsapfel ab und schmiert den Backzahn ein, damit den, der von innen kariös ist, die Milch des Sodomsapfels versengt, das kann ihn etwas beruhigen.[41] Soweit die Zahnschmerzen. 3.20. Oder etwas Zwiebel, schwarzer Zwiebelsamen, der wird auf ein Messer getan, das im Feuer erhitzt worden ist und das wird darauf getan […] der Wasserkrug, und tut seinen Mund hinein oder macht seinen Mund auf.[42]

3.21. yiftaḥ būzu ʿala lḥamuw btāʿ ssakkīna wi yrūḥ nātir ilbtāʿ illi fi ḍirṣu, iza kān sūs, ʾaw iza kān dūd, ʾaw ayyᵊ ḥāǧa. 3.22. wi yrūḥ nātir kida ʿa lfilfil illi fi ṣiḥən, wi yrūḥ nātir kida ʿa ssakkīna. da bi nnisba li lʿilāǧ wi illuḍrūs wi ssinān. 3.23. W: kuwayyis. fī ḥanḍāl? 3.24. Ṭ: ilḥánḍāl da biyibʾa masalan da binḥuṭṭu yaʿnī wāḥid ʿandu ruṭūba ʾaw rumatizm ṭabʿan ma-fīš yaʿnī ʿilāǧ iḥna ma-ʿandināš ʿilāǧ yaʿnī.

3.21. Er öffnet seinen Mund der Hitze des Messers und die treibt heraus, was in seinem Backzahn sitzt, ob es nun Maden oder Würmer sind, oder sonstwas. 3.22. Sie wirft das hinaus auf den Pfeffer in der Schüssel und wirft es so auf das Messer. Soweit die Behandlung und die Backenzähne und die Zähne. 3.23. W: Gut. Gibt es Koloquinten? 3.24. Ṭ: Die Koloquinten, die gebraucht man z.B., wenn das einer ist, der die Feuchtigkeit oder das Rheuma hat, da gibt es natürlich kein Heilmittel, wir haben keine Behandlung.

3.25. iḥna binitʿāliǧ bi lḥanḍāl wu rrūḥ ingassim ilḥanḍála nuṣṣēn, yrūḥ ǧāyib fīha kaʿbᵊ riglē wu yrūḥ ḥāṭṭu fi lḥanḍála. ilḥanḍála bitmōṣṣ ilmarāṛa btāʿit ilḥanḍála, bitmoᵊṣṣ irruṭūba ʾaw irrutizm fi ǧǧism. 3.26. sawā ʾan kān fi ddrāʿ ʾaw fi lriǧl, fi kaʿb rrigl ʾaw fi… arrukba yaʿnī. da bi nnisba li rrumatizm, da lli lḥanḍāl.

3.25. Wir behandeln uns mit der Koloquinte, wir halbieren die Koloquinte, er tut da seine Ferse hinein, er steckt seine Ferse in die Koloquinte. Die Koloquinte saugt heraus, d.h. die Bitterkeit der Koloquinte saugt die Feuchtigkeit[43] oder das Rheuma im Körper heraus. 3.26. Gleichgültig ob es nun im Fuß, in der Ferse oder in… dem Knie ist. Soweit das Rheuma, das ist die Koloquinte.

---

[41] Ähnlich in Barīs/Kharga, führt zur Zerstörung und Beseitigung des Zahns.
[42] S. auch Text 8.
[43] S. oben Text 1.03.

3.27. fa yaʿnī ʾayyᵊ ḥāğa binʿāliğ bihā ṭabʿan bi lʿilāğ ilbaladī, ma-ʿandināš iḥna... ḥāğa smáha duktūr wala mustašfa wala ḥāğa. kull ʿilāğna b-kida ʿala ṭūl. da bi nnisba li lḥanḍāl. 3.28. W: ilbaṣāl, ilbaṣál yinfaʿ fi ʾē? 3.29. Ṭ: lā, ilbáṣal da yaʿnī iḥna bnizráʿu liʾinn iḥna bnākul maʿā yaʿnī bnākul bīh ʿala ṭūl.

3.27. Wir behandeln also alles mögliche damit, natürlich mit den Hausmitteln, bei uns gibt es keinen... Doktor oder Krankenhaus oder sonstwas. All unsere Behandlungsmethoden sind direkt so. Soweit die Koloquinte. 3.28. W: Die Zwiebeln, wozu nützen die Zwiebeln? 3.29. Ṭ: Nein, die Zwiebeln die bauen wir ja an, weil wir sie immer beim Essen haben, d.h. wir sie immerzu essen.

3.30. W: bi nnisba barḍu li ṭṭibb luh fayda miš kida? bi nnisba li ğğarḥ walla ḥāga. 3.31. Ṭ: yaʿnī mumkin law wāḥid ʿēnu wārma ʾaw masalan tkūn yaʿnī yʾullak ʿēnak kārfa, šāmma ṣinna. 3.32. W: šāmma ʾē? sinna. šāmma sinna. bi ssīn? Ṭ: bi ṣṣād. W: ṣinna. Ṭ: ṣinna.

3.30. W: Aber sie hat auch Nutzen für die Medizin, nicht wahr? Bei Wunden oder sowas. 3.31. Ṭ: D.h., möglicherweise, wenn einem das Auge geschwollen ist oder z.B., wie man sagt, dein Auge hat einen Geruch angenommen, es hat einen Gestank gerochen. 3.32. W: Was hat es gerochen? *sinna* ? *šamma sinna*? Mit *sīn*? Ṭ: Mit *ṣād*. W: *ṣinna*. Ṭ: *ṣinna*.

3.33. ykūn masalan it... yaʿnī kuntᵊ māši kida fi lhawa wi ʿēnak yaʿnī fīh ğaww kida yaʿnī masalan fīh buxāṛ ʾaw ḥāğa w šammit ḥāğa bitrūḥ wārma, wi trūḥ inta ğāyib wáṛagit baṣāl, wu trūḥ ḥāmilha fi lḥitta dēy. 3.34. bitxalli ʿēnak turūʾ yaʿnī lbaṣal da nistaxdamū. lākin iḥna ʿādatan binizráʿu liʾinn iḥna bnākul bīh ʿala ṭūl. yaʿnī istixdāmu ʿandina bnākul bīh ilʿēš wu rruzz, ʾaklína fi lbaṣal ʿala ṭūl.

3.33. Er ist z.B.... d.h., du gingst so an der Luft und dein Auge war da so im Wind, d.h., es war da ein Dampf oder sowas und es bekam davon etwas ab, da schwillt es an, da nimmt du dann ein Zwiebelblatt und tust es auf diese Stelle. 3.34. Das läßt dein Auge sich beruhigen, wir gebrauchen also diese Zwiebeln. Aber säen tun wir sie gewöhnlich, weil wir sie immerzu essen. D.h., ihr Gebrauch bei uns besteht darin, daß wir Brot und Reis dazu essen, wir essen immerzu Zwiebel.

3.35. liʾinn lbaṣal mufīd ʾaṣlan yiṣaḥḥ ʾaw baʿdēn bi... biyšīl ilḥagāt lli hiyya fi lgism. kuwayyis li ṣṣiḥḥa yaʿni kuwayyis li ṭṭabīx wu kuwayyis yaʿni. yaʿni da muhimmᵊ lāzim nizráʿu. 3.36. W: wi ttōm? Ṭ: ittōm barḍu kuwayyis yaʿnī. ittōm da binistixdámu fi lʾakl, wu fi... lamma yikūn masalan ʿandi ğaḥš masalan wu riğlu wāğʿā, biyidḷáʿ ʿalēhī, inta ʿārif yidḷāʿ? yaʿnī masalan byibʾa māšé wu biyidḷáʿ, biyuʿkuz ʿala riğlu.

3.35. Denn die Zwiebel ist von Haus aus nützlich und gesund oder dann.... sie holt die Sachen heraus, die im Körper sind. Gut für die Gesundheit also, gut zum Kochen und überhaupt gut. Sie ist wichtig und wir müssen sie anbauen. 3.36. W: Und der Knoblauch? Ṭ: Der Knoblauch ist auch gut. Den Knoblauch gebrauchen wir beim Essen und beim... wenn ich beispielsweise einen Esel habe und dem tut das Bein weh, er hinkt darauf, weißt du, was das ist, *yidḷāʿ*, er geht also z.B. und hinkt, er stützt sich auf sein [anderes] Bein.

3.37. biyimši kida w biyʿikkēz wi ṭabʿan […] ilḥāfir bitāʿu biyibga ḥafyān. wi nigūm niʿmil ʿamaliyyit…ʾē?	3.37. Er geht so und stützt sich und natürlich…sein Huf ist bloß. Da machen wir die….was?
3.38. nrūḥ nġassil riǧlu wu rrūḥ ǧāybīnlu wu binsixxinlu šwiyyit zēt wi šwiyyit tūm, wu yrūḥ ylissiʿhālu ʿala lḥāfir bitāʿ riǧlu min taḥt. da iza kānit ḥafyāna yṣaḥḥ ʿala ṭūl. da yaʿnī bi nnisba li ʿilāǧ ittūm.	3.38. Wir waschen sein Bein und nehmen ein wenig Öl und erhitzen es mit etwas Knoblauch, und schmieren ihm das leicht unten auf den Huf von seinem Bein. Wenn dieses unbeschlagen ist, wird es sogleich gesund. Soweit das Heilen mit Knoblauch.
4. kayy, *Knoblauch, Gummi arabicum Kartoffel-Bovist, Kohl-Gänsedistel*	*Ḥamīda*[44]
4.01. iṭṭubbāʿa wi lmiḥwāṛ, ilmiḥwāṛ da li lbahāyim. da yikwūbu la mʾaxza li rrukūba aw li lbagáṛa ʿama tuḇga maksūṛa fi riǧilhī, fi ʾayyᵊ ḥitta, yikwūhā bōh. wi ṭṭubbāʿa yikwubhā ʿa… il…ilʾinsān. bniʾādam. mumkin ʿala ṛāsu, mumkin ʿala baṭnu, mumkin ʿala ḍahru ʿala riǧlu ʿala ʾayyᵊ ḥitta maksūra dīhī.	4.01. Der Brennagel und das Brenneisen. Das Brenneisen ist für die Tiere. Damit brennt man – Verzeihung! – die Esel, oder die Kuh, sie hat einen Bruch an ihrem Bein, an sonst einem Teil, dann brennt man sie damit. Mit dem Brennagel kauterisiert man den Menschen…den Menschen. Es kann sein auf den Kopf, auf den Bauch, auf den Rücken, auf das Bein, auf jeden Teil, bei dem ein Bruch [Prellung?] vorliegt.
4.02. ʿala dṛāʿu, ʾayyᵊ ḥitta. ʿašān ṣudāʿ ʿašān kasᵊr miš lāzim yaʿni ʿašān kasᵊr. itǧīb šwiyyit xatī tiqīd kōm nāṛ, wi tǧīb ilmiḥwāṛ wi tḥuṭṭu fī. law ṭallaʿtu min innāṛ tšūfu ʾáḥmāṛ, mumkin tḥuṭṭu huna, yiʿillim makān.	4.02. Auf den Arm, jeden Teil. Wegen Kopfweh, wegen einem Bruch, es muß nicht unbedingt ein Bruch sein. Du holst ein paar Kuhfladen und zündest ein großes Feuer an, bringst das Brenneisen und legst es hinein. Wenn du es aus dem Feuer holst, siehst du es ganz rot, das kannst du hier hintun, es hinterläßt einen Abdruck.
4.03. maṭraḥ ma tḥuṭṭu, yiʿillim makān. ʾādi kayy innāṛ. tiwga ǧuḥr kida nǧīb min taḥt mi lli…min izzibil yaʿni lli min ilbahāyim wi tḥùṭṭuhōm wi	4.03. Wo du es hintust, hinterläßt es einen Abdruck. Das ist das Kauterisieren mit Feuer. Das ist so eine Grube, wir holen von unter den…, vom Mist, d.h. den von den Tieren, und du legst sie [die

[44] Eine etwa 45 Jahre alte Bauersfrau.

ṭṭabāg lamma ʿama-tḥimmēr šuftuhōm hnāk niʿmilbaha gdī.

Brenneisen] und den Teller dahinein, bis sie glühen, du hast sie dort gesehen, so tun wir es damit.

4.04. lammā tḥimmēr. w ilkān bahīm hayḥuṭṭha ʿalē w ilkān biniʾādim hayikwī. ilbiniʾādim yiǧību min waṛāh ʿašān ma-yšufhāš. law šafhī, haygūm miš hayitḥammāl dī. laʾinn ana ʾawwal šuftahā gult laʾ ana ma-gdirš aḥuṭṭəhī. ṣawwaṭt wu gult, miš kwayyis, dī ḥamṛā ḥamṛā ana xift minhī.

4.04. Bis sie glühen. Wenn es ein Tier ist, tun sie es darauf und wenn es ein Mensch ist, kauterisieren sie ihn. Beim Menschen tut man das von hinten, damit er es nicht sieht. Wenn er es sieht, steht er auf und hält das nicht aus. Weil ich, sowie ich das sah, sagte: nein, das kann ich nicht tun, ich schrie: das ist nicht gut, die ist ja ganz rotglühend, ich hatte Angst davor.

4.05. ma-gdirtiš aḥuṭṭəhī. lakān yiǧīlāk min waṛāʾ mumkin tugʿód, tibuṣṣᵊ kida w tiwga mumkin yiḥuṭṭūhī. ilbilām da li ssaʿy. bilām yaʿni mumkin innᵊ wala mʾaxza ǧaḥš irkūba bitʿuḍḍ, yigullak ḥuṭṭlahā blām.

4.05. Ich konnte es nicht tun. Aber wenn er von hinten kommt, kannst du sitzen bleiben, du schaust so, und dann ist es möglich, daß man sie anbringt. Der *bilām* ist für die Tiere. *bilām*, d.h. es kann z.B. sein, daß ein Reitesel beißt, man sagt dann, mach ihm einen *bilām*.

4.06. ḥadīda. miš yḥuṭṭihā fi lǧaḥš wu yrūḥ xārim ilbilām ʿašān da yiʿuḍḍ ilbiniʾādim? yirūḥ māsik tāll ilbilām ma-yiʿrifš yʿuḍḍ minhī. ḥadīda tiwga malwiyya. ʾādi lbilām. ilbilām wu lliǧām. inǧīblaha yimkin Ḥusba w yikwihī.

4.06. Ein Stück Eisen. Legt man es nicht dem Esel an und durchbohrt die Oberlippe, weil der den Menschen beißt? Man packt und zieht die Oberlippe heraus, er [der Esel] kann daher nicht beißen. Ein gebogenes Stück Eisen. Das ist der *bilām*. Der *bilām* und der Zügel. Wir holen dafür vielleicht Ḥusba[45] und der kauterisiert sie.

4.07. yikwihā w mumkin tṣeʾḥḥ. bi ttōm, miš bi lbaṣāl. yiduqqōh wu yḥuṭṭu ʿalēh ʿašān mumkin tiʿiyyēn, ʿašān mumkin law fī ṣadīd walla ʾayyᵊ ḥāǧa yinzēl. ittōm da law ḥaṭṭētu ʿala ʾayyᵊ ḥitta ḥatta miš makwiyya mumkin hū yxallīha tšūfhī tgūl di makwiyya.

4.07. Er kauterisiert sie und vielleicht wird sie gesund. Mit Knoblauch, nicht mit Zwiebel. Man stößt ihn fein und tut ihn darauf, damit sie eitern kann, damit, wenn es Eiter gibt oder sowas, das herauskommen kann. Der Knoblauch, wenn man den auf eine Stelle tut, selbst wenn die nicht kauterisiert ist, kann er sie so aussehen lassen, daß du denkst, sie ist kauterisiert.

[45] Koseform *fuʿla* für Mḥassāb, ein entfernter Verwandter der Sprecherin, der sich auf *kayy* und *xarṭ* versteht. S. auch unten Fn. 55, allgemein zu Koseformen s. M. Woidich, 'Kindersprache in ilBašandi. Ein Text aus der Oase Dakhla,' in Th. Bauer [et al.] (eds.), *Alltagsleben und materielle Kultur in der arabischen Sprache und Literatur. Festschrift für Heinz Grotzfeld zum 70. Geburtstag* (Wiesbaden, 2005), S. 419 Fn. 38.

4.08. yixallīha mumkin tiʿmil ʾayyᵊ ṣadīd yinizzílu. fawāydu hū fi ṭṭabīx, fi lkasər, yaʿni mumkin wāḥid riǧlu wāǧʿa, ma-lgīš ḥāǧa, mumkin yiduqq išwiyya, wu yrūḥ ḥāṭiṭhōm w aṛābiṭhōm ʿalehī. hatʿabbēy tiwǧāʿ. hatiwǧāʿ lwaḥdihā w baʿdēn hatnizzil iṣṣadīd.	4.08. Er macht und bewirkt, daß irgendein Eiter herauskommt. Er ist nützlich beim Kochen, bei einem Bruch, d.h. einem kann der Fuß weh tun, er findet nichts, da kann er ein wenig zerstoßen, und tut es darauf und bindet es darauf. Das füllt sich [mit Eiter] und tut weh. Das tut von allein weh und dann läßt es den Eiter herauskommen.
4.09. iṣṣaməġ da nilzígbō. inbullu, induqqu fi lhōn ilʾawwāl. wu baʿdᵊ ma nduqqu nubullu. bi šwiyyit ṃuyya, wu rrūḥ lazgīn bōh. mumkin wāḥid maksūr min huna w huna ʾayyᵊ ḥitta. iṣṣaməġ fayditu katīra ǧiddan. ʾayyᵊ ḥitta, mumkin lamma yiǧē ʿa lkasər biyfīd xāliṣ.	4.09. Das Akazienharz (gummi arabicum), mit dem machen wir Pflaster. Wir machen es naß, zuerst zerstoßen wir es im Mörser. Nachdem wir es zerstoßen haben, machen wir es naß. Mit ein wenig Wasser, und machen ein Pflaster damit. Jemand der hier oder hier, irgendwo etwas gebrochen hat. Das Harz hat viele gute Wirkungen. An jedweder Stelle, wenn es auf den Bruch[stelle] kommt, ist es äußerst nützlich.
4.10. ilḥaǧar ilḥallāb. ḥaǧār, ḥaǧār marmi fi lʾarḍ, da ṣawwāna dēy, da ḥaǧār ṣawwān. lākan dīk ḥallāb. ḥallāba kullihī, innās ʿama tilʿab fīhī. yiʿmilūha zawwāya. yilʿabu bō. tubga nuxrumhī, bi mismāṛ, wu niʿmilhā gdī. zawwāya. bi lḥaǧār. ilḥallāb lamma tiṭḥánu tib̲ga ṛamla.	4.10. Der *ḥallāb*-Stein. Ein Stein ist das, ein Stein auf die Erde geworfen, das ist ein Stück Feuerstein [sie zeigt auf einen im Hof liegenden Stein]. Aber der [den ich meine] ist Kalkstein. Alles Kalkstein, die Leute spielen damit. Sie machen einen Schwirrer[46] daraus. Sie spielen damit. Wir durchbohren ihn, mit einem Nagel, und machen so. Ein Schwirrstein. Aus Stein. Wenn man den *ḥallāb* mahlt, wird er zu Sand.
4.11. mumkin fīh fi lġēṭ li ǧǧild yaʿni law wāḥid maǧrūḥ fīh quṭṭāba. quṭṭāba kida ṣafra tǧībhī tiftaḥḥī, wu trušš minha ʾayyᵊ ǧarḥ.	4.11. Es gibt [etwas] im Feld, d.h. für die Haut, wenn einer eine Wunde hat, da gibt es einen Kartoffel-Bovist. Einen Bovist, den gelben, den holst du, machst ihn auf und bestreust damit jede beliebige Wunde.[47]

[46] Ein Kinderspielzeug, das an einen Strick gebunden und herumgewirbelt wird.

[47] Gleiches wird aus Barīs berichtet, dort heißt der Pilz *gaṭṭāb* und wird ebenfalls zur Wundversorgung gebraucht, s. Behnstedt & Woidich 1994, S. 383.

4.12. di hī lli kwayysa, miš ilḥallāba. šiklaha tuwga kída, bissᵊ ma-tiwgāš ilyōm tiwgā fi ṣṣēf. tiṭlaʿ maṭraḥ iddṛā', maṭraḥ illūbya, tuwga wāsʿa kida. wu ṭālʿa fi lʾarḍ mumkin tišùddihī. tiglaʿhī, di ʿād ʾayyᵊ fiǧǧa ʾayyᵊ ǧarḥ, turuššᵊ minhā wu ǧǧarḥ yuqṭōb wu yṣeʾḥḥ.

4.12. Der ist gut, nicht die *ḥallāba*.[48] Er [der Bovist] sieht so aus, aber es gibt ihn jetzt nicht, den gibt es im Sommer. Er kommt an Stellen mit Hirse hervor, an Stellen mit Weißen Bohnen, er bildet einen großen Fleck. Und er wächst aus dem Boden, du kannst ihn herausziehen. Du ziehst ihn heraus, und dann bestreust du jede Schrunde, jede Wunde damit, und die Wunde schließt sich und heilt.

4.13. iššixxēra, fayditha mumkin takulha lbahāyim, kwayysa. zamān kāt takulha nnās, laʾinnaha kwayysa mṣiḥḥa. lakan liyyām dēy ma-ʿādš ḥadd yaʿnē miš gawēy. di fi lġēṭ, di lxišna di, fī šixxēra xišna wi šixxēra nāʿma.

4.13. Die Kohl-Gänsedistel, ihr Nutzen ist, daß das Vieh sie fressen kann, sie ist gut. Früher haben die Menschen sie gegessen, da sie gut und gesund ist. Aber heutzutage niemand mehr, d.h. nicht viel. Die ist auf dem Feld, das ist die rauhe, es gibt eine rauhe Kohl-Gänsedistel und ein feine.

4.14. áha wāsʿa kida fi lġēṭ. yaʿni mumkin law wāḥid ariǧlu wāǧiʿa yaʿni ǧarḥᵊ kabīr kida yiḥuṭṭu ʿala riǧlu. mumkin yixallīha tlīn, lában iššixxēra dī.

4.14. Sie bildet einen großen Fleck auf dem Acker. D.h. wenn einem sein Fuß weh tut, d.h. eine große Wunde, kann er sie auf den Fuß tun. Sie kann ihn weich werden lassen, die Milch der Kohl-Gänsedistel.

5. *Koloquinte*, šišlān

Fuʾād[49] *und Muḥammad*

5.01. Mu: yāklu lwarāg, yiqirrimō lwarāg. Fu: tiʿrif ilḥilba? ilḥilba lxaḍrā'. tittakāl. ilḥanḍāl da mufīd li lʿayā'. yikūn wāḥid ʿandu ttaʿbān, irrṭūba yṭilliʿhī. Mu: yiʿmilha kīf y Abu Ḥmād?

5.01. Sie fressen die Blätter, sie stuzen die Blätter. Fu: Kennst du den Bockshornklee? Den grünen Bockshornklee? Den kann man essen. Die Koloquinte nützt bei Krankheit. Wenn einer die Müdigkeit hat, die Feuchtigkeit, sie treibt sie aus. Mu: Wie tut er das, Abu Aḥmad?

[48] *ḥallāba* ist der Steinstaub, den die Frauen aus den Quadersteinen antiker Gemäuer kratzen, um ihre Kinder damit einzureiben. Dies soll für eine gesunde und kräftige Haut sorgen. Spuren dieser Tätigkeit sind als tiefe Einkerbungen in diesen Steinen überall zu sehen, so etwa bei der Grabkuppel des Šēx ilBašandi.

[49] Ein etwa 60 Jahre alter Bauer.

5.02. Fu: in kān ʿandak irṭūba fi riǧlak ʾaw fi ʾayyᵊ ḥāǧa, itǧīb ilḥanḍāl dī, wu tiftáḥḥa fi ǧa·nb, wu tḥuṭṭ la muʾaxza riǧlak fihī trūḥ sāḥba rruṭūba lli ʿandak. Mu: itbiyyēt? Fu: ą̄, tḥuṭṭaha… tḥuṭṭᵊ riǧlak fīha šwiyya trūḥ sāḥba rruṭūba.

5.02. Fu: Wenn du Rheuma hast im Fuß oder sonstwo, nimmst diese Koloquinte, öffnest sie an der Seite und steckst, mit Verlaub, deinen Fuß hinein, da zieht sie die Feuchtigkeit bei dir sofort heraus, Mu: Über Nacht? Fu: Jawohl, du steckst sie … du streckst deinen Fuß ein wenig da hinein, dann zieht sie die Feuchtigkeit gleich heraus.

5.03. wu ššišlān! bálaḥ iššišlān. yiʿmilū ʿilāǧ barḍu… biss miš ʿārif yiʿmilūh kīf. Mu: illi ʿandu ssukkaṛ yiʿmílu? yišṛábu gāl yiġlī w yišṛábu. – hū ḥālig qrāʿ kida lē? – Mu: innhārda tʾaxxār, walla ma-sara·ḥš? Fu: lā nnahārda sarāḥ bidrē hū, ʿaǧā? ilḤaǧǧᵊ Saʿd sarāḥ bidrēy.

5.03. Und der *šišlān*! Die *šišlān*-Früchte, die macht man auch zu Medizin… aber ich weiß nicht, wie man das tut. Mu: Macht das, wer zuckerkrank ist? Der trinkt es, sagt man, er kocht es und trinkt es. – Warum ist er so kahl geschoren?[50] – Mu: Heute hat er sich verspätet oder ist er gar nicht aufs Feld gegangen? Fu: Nein, heute ist er früh aufs Feld, wieso? Der *Ḥaǧǧᵊ* Saʿd ist früh aufs Feld gegangen.

6. *Rizinus* — *Muḥammad*

6.01. ilxaṛwāʿ, ʿārif ilwaṛága btāḥtu? lamma ykūn fī dimmil fi ǧismak wu ʿāwiz ʿašān yistawēy, ʿašān yiftāḥ li lqēḥ illi fīha dī. trūḥ itǧīblu ʾē? wàragit xaṛwā̃ʿ. wu tduggha kdī, biss miš gawēy yaʿnēy kidī.

6.01. Der Rizinus, kennst du sein Blatt? Wenn du auf dem Körper eine Beule hast und willst, daß sie reift, so daß sie sich öffnet für den Eiter, der darin ist. Was nimmst du dann? Ein Rizinusblatt. Du zerstößt es so, aber nicht allzu sehr, d.h. so.

6.02. ʿašān yiṭillēʿ izzēt illi fīh. wu trūḥ ḥāṭiṭhī ʿalēh, wu tbiyyithā ʿalē li ṣṣo·bḥ. tlagīh ǧammaʿ illi f īdak kullu walla fi maṭraḥ ma ykūn. walla fi ǧismak kullu. ilqēḥ dī.

6.02. Damit es das Öl darin freigibt. Das legst du dann darauf, und läßt es über Nacht bis zum Morgen darauf. Da merkst du, daß es alles, was in deiner Hand ist oder wo denn auch, gesammelt hat. Oder in deinem ganzen Körper. Diesen Eiter.

[50] Jemand geht vorbei und veranlaßt Fuʾād zu dieser Bemerkung.

6.03. wu mumkin ma-nfēḥš tibga tǧīblu ʿalēh…fi lli ʿamma yiʿmillu ʿaǧīna b sukkāṛ, yibiyyithī ʿalēh, ʿaǧīnit me·lḥ, ʿašān tisḥib kull illi fi ǧǧisim da wu yǧimméʿu fi lḥitta dēy, ʿašān lamma tiftaḥu yiṣaffēy ʿād.

6.03. Es kann sein, daß es nichts nützt, dann nimmst du dafür…es gibt welche, die machen dafür einen Brei aus Zucker, lassen ihn über Nacht darauf, [oder] einen Brei aus Salz, damit er alles, was da im Körper ist, an dieser Stelle sammelt, damit, wenn du ihn öffnest, er dann trockengelegt wird.

6.04. yiwga ǧismak ma-fīš ḥāǧa. ilxaṛwāʿ ʿamma yiʿmélu minnu zzēt. ma-fīhš dilwāk fi ṣṣaydaliyyāt fī šaṛba smaha šaṛbit zēt ilxaṛwāʿ. yāxódu lwāḥid illi ykūn ʿandu ʾimsāk walla ykūn ʿandu dūd, yāxōd iššaṛba dēy.

6.04. Dann ist nichts mehr in deinem Körper. Vom Rizinus macht man das Öl. Das geschieht jetzt nicht mehr, in den Apotheken gibt es jetzt ein Abführmittel mit Namen Rizinusöl. Das nimmt derjenige, der Verstopfung hat oder der Würmer hat, der nimmt dieses Abführmittel.

6.05. wu fīh me·lḥ, fī šaṛba smaha šaṛbit ilme·lḥ, ʿāriffī? ʾáhi di lwāḥid yaxudha ʿa rrīg tiġsēl baṭnu ġasīl ziyy ilbīra walla ʾē? miš ʿārif ilbīra? miš ʿariffī? ʿamma yiṣnaʿuhā miš min iššaʿīr? ṭayb, di miš ʿamma tniḍḍéf ilbaṭin? bissᵊ dī law itxammárit tġiyyib ilʿaqil.

6.05. Und es gibt Salz, es gibt ein Abführmittel, das heißt Salztrunk, kennst du es? Schau, das nimmt man auf nüchternen Magen, das wäscht seinen Bauch wie das Bier oder wie? Kennst du nicht das Bier? Macht man das nicht aus Gerste? Gut, säubert das nicht den Bauch? Doch wenn das gärt, bringt es um den Verstand.

7. *Wermut* — *Muḥammad*

7.01. W: min faḍlak tigulli ʿala ššīḥ da byiʿmilu ʾēh bīh? M: iššīḥ, da dilwakt ʿama yḥuṭṭu, illi yiwga ʿamal ʿamaliyya, yiḥuṭṭu ʿala ǧǧaṛḥ, kīs kida ʿama yǧību ḥitta ṣġiyyra f-kīs, wu yḥuṭṭu ʿala ǧǧaṛḥ min baṛṛa xāliṣ, ʿa ṛṛibāṭ bitāʿ iǧǧaṛḥ.

7.01. W: Bitte, erzähle mir vom Wermut, was macht man damit? Mu: Der Wermut, den tut jetzt jemand, der eine Operation gemacht hat, auf die Wunde. So einen Beutel, man nimmt ein kleines Stück in einem Beutel und man tut ihn ganz außen auf die Wunde, auf den Verband der Wunde.

7.02. ilʿamaliyya, ʿašān law ḥaddᵊ ǧā zāṛu w ḥāṭiṭ rīḥa, ʿiṭir yaʿnēy, iǧǧaṛḥ ma-yšummiš wu ma-yuhrušš ᵊ fī. lakan lamma ḥaṭṭ

7.02. Die Operation, damit wenn jemand ihn besuchen kommt und sich etwa parfümiert hat, Essenz heißt das, die Wunde diese nicht ‘riecht’[51] und er nicht

[51] D.h., nicht zu jucken beginnt, s. oben Text 2.12, über die Zwiebel.

iššīḥ dī, mumkin iǧǧarˀ yšumm ilʿiṭir wu yxalli ṣāḥb ilʿamaliyya yugʿud yuhruš fīh kidī. law haraš fīh mumkin yiftaḥ ilʿamaliyya tanēy.......

daran kratzt. Wenn er aber den Wermut darauf gelegt hat, [...][52] die Wunde kann die Essenz ‘riechen’ und den Patienten dauernd die Wunde kratzen lassen. Wenn er daran herumkratzt, öffnet er die Operation aufs Neue....

8. *Bilsenkraut,*[53] *Zwiebelsamen* — *Muḥammad*

8.01. zamān lamma kān wāḥid ḍirsu yiwga msuwwēs, kān hayǧīblu ḥabbitēn sakrān, wu yiġlihōm wu yḥuṭṭ... yuqṭur fīh da ʿašān yiteˑxx wu yiskōt. wu kān fī ṭarīga taniyya. iyǧību btāʿ ... aràgabit qúlla, wu yǧību ˀē, sakkīna, wu yǧību siḥin fi šweyyt uṃṃuyya.

8.01. Früher, wenn da einer einen faulen Zahn hatte, nahm er dafür ein wenig Bilsenkraut, kochte das und tat es... und läßt es darauf tropfen, damit er sich beruhigt und Ruhe gibt. Und es gab noch eine andere Methode. Man nahm... den Hals einer Wasserflasche, und man nahm was? Ein Messer, und man nahm eine Schüssel mit etwas Wasser darin.[54]

8.02. yiḥuṭṭu ssakkīna ʿa nnār, lamma tsxān. yirūḥu ḥaṭṭīnha ʿa ṣṣiḥin igdī lli fīh ilṃoyya, wu yǧību fōg minha ràgabit ilqóˑlla. w illi ḍirsu msuwwēs dī ylaggi būzu gdā ʿala ràgabt ilqóˑlla. yaʿni būzu gdā ʿala ràgabt ilqóˑlla. yirūḥ misiqqēṭ issūsa lli fīh, tisqoˑṭ fi lṃoyya.

8.02. Man legte das Messer ins Feuer, bis es heiß wurde. Dann legten sie es so auf den Teller, worin das Wasser war, und taten darüber den Flaschenhals. Wer einen kariösen Zahn hatte, tat seinen Mund direkt über den Flaschenhals. D.h., sein Mund war direkt über dem Flaschenhals. So läßt er den Wurm im Zahn herausfallen, er fällt direkt ins Wasser.

8.03. wu yiḥga ǧaybīn kamān btāʿ... insītu ˀanī, ḥabbitēn baṣāl iswād. zarrīʿit baṣāl sudīˀ. yiḥúṭṭaha ʿala ssakkīna w hī lissa suxna. yiṭláʿ idduxxān dī. idduxxān dī, yilaggi būzu lli ḍirsu wāǧiʿ yilaggi būzu ʿalēh, yirūḥ imsiqqēṭ issūsa min būzu. tisqōṭ ʿala l... fi lṃoyya dēy. wu yiskōt.

8.03. Und sie nehmen dazu auch noch... das habe ich vergessen, etwas „Schwarze Zwiebel“. Schwarzen Samen der Zwiebel. Den tut er auf das Messer, solang es noch heiß ist. Da steigt dann Rauch auf. Dieser Rauch, auf den richtet der, dem der Zahn wehtut, seinen Mund. Der wirft ihm den „Wurm“ aus dem Mund. Der fällt auf den... in dieses Wasser. Und gibt Ruhe.

---

52 Hier fehlt die logische Fortsetzung [dann passsiert das nicht].

53 S. auch Text 3.19–3.22.

54 S. Text 3.20–3.22.

9. kayy

Ḥamīda

9.01. ʿamalt ağīb […] mismāṛ ḥaddādi, kabīr. ṛāsu tiwga wāsʿa gdī. wu tqīd šwiyyit ğilla. min taḥt ilbagāṛ. tiwga biss nāšfa. wu twulliḥ fihōm wu tḥuṭṭ iṭṭibbāʿa. tḥuṭṭ itnēn, tšīl waḥda, tikwēy, yigullak ʾē:, ḥāṭiṭ ē:, ḥāṭṭa arbáʿa ʿala ṛasēy.

9.01. Ich mache und bringe […] einen Schmiedenagel, einen großen. Der hat so einen breiten Kopf. Und du machst ein Feuer mit Kuhmist. Von unter den Kühen. Der muß freilich trocken sein. Du zündest ihn an und tust den Brennagel hinein. Du tust zwei hinein, nimmst einen heraus und kauterisierst, man sagt was? Was tust du? Ich habe vier auf meinem Kopf lassen machen.

9.02. yigullak ʾāxr iṭṭibb innāṛ. iyğībha w baʿdᵊ ma tiwga ḥámṛa ḥámṛa gdī, yimsikak wāḥid. ana kawātli ṃṃāk Ḥubba. maṛṛa kamān ilmarḥūm Ğābir wu maṛṛa kawātli Maḥbūba, waḥda se:tt…

9.02. Man sagt: Die höchste Medizin ist das Feuer. Er bringt es und nachdem es glühend geworden ist, hält dich einer. Mich hat deine Mutter Ḥubba[55] kauterisiert. Einmal auch der verblichene Ğābir und einmal kauteriserte mich Maḥbūba, eine Frau…

9.03. ḥaṭṭātli waḥda hōna, wu ʾādi waḥda hōna, ḥāṭṭa‿áṛbaʿa ʿala ṛasēy, w áṛbaʿa ʿala baṭnēy w áṛbaʿa ʿala ḍahrēy, yigullak ʾāxr iṭṭibb innāṛ. biss ē hī law hatiqšaʿhōm miš mumkin tḥuṭṭhī. liʾannaha tiwga ḥamṛa ṭālʿa min innāṛ ḥamṛa txuwwēf.

9.03. Sie machte mir einen hier, und auch hier einen, sie machte mir vier auf den Kopf, und vier auf den Bauch, und vier auf den Rücken, man sagt: Die höchste Medizin ist das Feuer. Freilich ist es so, daß, wenn du ihn [den Brennagel] siehst, sie ihn nicht plazieren kann. Weil er glühendrot ist, er kommt glühend aus dem Feuer und macht Angst.

9.04. miš hatigdēr tḥuṭṭaha ʿala ṛāsak. lamma… waḥda timsikak tikfīk ʿala wiššak kidā ma-txallīkš tšufhī tugʿod waḥda guddāmak wu tġaṭṭi wiššak wu timséku. wu wāḥid waṛāk, illi hayikwī hū lli yiwga waṛāk, lāzim tikwīk ʿala ṛāsak, ʿala maṭraḥ ma ntᵊ ʿāwiz tikwēy.

9.04. Da kann sie ihn nicht auf deinen Kopf plazieren. Wenn… eine dich festhält, bedeckt sie dein Gesicht, sie läßt dich nicht sie sehen, eine sitzt vor dir und bedeckt dein Gesicht und hält es fest. Und einer hinter dir, derjenige, der dich kauterisiert, ist hinter dir, sie muß dich am Kopf kauterisieren, dort, wo du wo du es willst.

[55] *Ḥubba* ist eine Koseform zu *Ḥabība* oder *Maḥbūba*, s. oben Fn. 45.

9.05. liʾinn law šuftu, miš hatigdēr tikwēy. W: tiǧirrēs. Ḥ: tṣirrēx, itǧirrēs ē:! ṭabʿan di ana awwal ma šufthī gult lā di ḥamṛā miš haḥuṭṭahā ʿala ṛāsēy. lā laʾ di ḥamṛāʾ! gālitli lā d-āxr iṭṭibb innāṛ.

9.05. Denn, wenn du ihn siehst, kann sie nicht kauterisieren. W: Da kriegst du Angst! Ḥ: Schreien tust, was heißt hier Angst kriegen! Natürlich habe ich gesagt, als ich sie zum ersten Mal sah, nein, die glüht ja, die tu ich nicht auf meinen Kopf. Nein, nein, die glüht! Sie sagte zu mir: Die höchste Medizin ist das Feuer.

9.06. di baʿdᵊ ma tiḥtēr[56] *xāliṣ lāzim hatikwēy. minṭarr*[57] *tikwēy. W: tamām, wi nafáʿit ilḥikāya? Ḥ: šwiyya, barḍaha šwiyya. ʾaṣlᵊ ʾana taʿíbni dilwāk ḍaġṭēy biyiwṭāʾ. W: tāʿbik ē? Ḥ: iḍḍaġṭᵊ btāʿi biyíwṭā, lamma biyiwṭā byiʿmilli ṣḍāʿ. min kutr iššuġil ya dúktur barḍu taʿāb.*

9.06. Das ist dann, wenn du nicht mehr ein noch aus weißt, dann mußt du kauterisiert werden. Du bist gezwungen zu kauterisieren. W: Gut, und hat die Sache auch genützt? Ḥ: Ein wenig, auch nur ein wenig. Denn jetzt leide ich unter meinem Blutdruck, er sinkt. W: Worunter leidest du? Ḥ: Mein Blutdruck sinkt, wenn er sinkt, macht er mit Kopfweh. Wegen der vielen Arbeit, Doktor, das macht müde.

10. tadʿīk

Ḥamīda

10.01. Ḥ: illabān? da lāzim... da sibir zamān. kān kullu gdī! lakān liyyām dī laʾ xalāṣ. W: miš kān zamān ḥāga smaha tamrīs? yimarrisō lʿayyil bi bizz... bi laban min bizz ilʾumm? Ḥ: da tadʿīk. W: ismu ʾē? Ḥ: tadʿīk miš tamrīs.... [sie lacht] *W: fi lXārǧa ʿamma ygūlu tamrīs, ʿama ymirrisōw. Ḥ: ydiʿʿikōw! W: da ʾē da? Ḥ: tadʿīk? yigūlu ʾē? ilʿiyyil mamzūg. ilʿiyyil mamzūg, ḍaḥḥaktnēy ya duktur, ilʿiyyil mamzūg.*

10.01. Ḥ: Die Milch? Das muß... das ist schon lange der Brauch. Da war alles so. Aber heutzutage, nein! Das ist abgelaufen. W: Gab es nicht früher was, das hieß *tamrīs*?[58] Man massierte das Kind mit Muttermilch. Ḥ: Das ist das Massieren. W: Wie heißt das? Ḥ: *tadʿīk* nicht *tamrīs*! W: In Kharga sagt man *tamrīs*. *ʿama ymirrisōw*. Ḥ: *yidiʿʿikōw*! W: Was ist das? Ḥ: Das Massieren? Was sagen sie da? Das Kind hat sich was gezerrt. Das Kind hat sich was gezerrt, du hast mich lachen lassen, Doktor, das Kind hat sich eine Zerrung zugezogen.

[56] *iḥtēr* < **iḥtār* „er geriet in Verwirrung" ist in den ersten Stamm übergegangen, wie die Konjugation zeigt: *hī ḥtérit* „sie geriet in Verwirrung", *hōm ḥitrōw* „sie gerieten in Verwirrung". Ebenso *rtēḥ* „er ruhte sich aus", *hī rtéḥit, hōm ritḥōw*.

[57] *minṭarr* = *muḍṭarr*.

[58] Ein Irrtum meinerseits, es ist *mirrīs* in Barīs.

10.02. buṣṣ ana gʿōd igdī, aywa gdī,[59] *aḥuṭṭu ʿala benāt riǧleyya, wu ʾēh, tiwga waḥda wālda. waḥda tiḥliblēy labān, ʿala … ʿala ṣidr ilʿeyyil w akšéfu. w adiʿʿéku ʾaywa gdih. baʿd ma adʿáku aḥliblu adʿak benāt iktāfu min warạʾ, w adʿāk ṣidru min guddām, w aǧīblu ašārib ṣūf w arḅuṭu ʿala ṣidru.*

10.02. Schau, ich sitze so da, ja, so! Ich lege es zwischen meine Beine, und was? Da ist eine, die niedergekommen ist. Eine, die mir Milch auf die Brust des Kindes melkt und ich decke es auf. Und ich massiere es. Ja, so! Nachdem ich es massiert habe, melke ich ihm, ich massiere hinten zwischen seinen Schultern, und vorne seine Brust, ich nehme einen Wollschal und binde ihn um seine Brust.

10.03. ma-fīš waḥda wālda, lāzim laban beˑzza, ma-fīš laban bizza lāzim zibda. izzibda di tiwga zoˑbda yaʿnēy nǧībha min iṛṛobʿa, wu barḍu nḥuṭṭaha lwaḥda tidʿāk ʿala ʾidehā kdī, wu tdiʿʿik ilʿeyyil. wu baʿdᵊ ma tdiʿʿéku hī tiʿmēl b-idēha gdī. timsik kullᵊ ʾīd wu riǧil xulfᵊ xlāf, wu tidʿāk wu tiʿmil igdī ʿašān tigūl ʿašān aḍummu.

10.03. Wenn keine da ist, die gerade niedergekommen ist, es muß Muttermich sein, gibt es keine Muttermilch, dann Butter. Die Butter, das ist die *zōbda*,[60] d.h. wir holen sie aus dem Butterkrug, und geben sie einer, die massiert mit ihren Händen, und sie massiert das Kind. Nachdem sie es einmassiert hat, macht sie mit ihren Händen so: sie nimmt jede Hand und jeden Fuß schräg und massiert, und sie macht das, damit ich ihn fest zusammenfüge, sagt sie.

10.04. wu nḥuṭṭ ʾē? nǧīb ašārib ṣūf, ʾaw laffit ṣūf, wu waḥda turbuṭ ʿala ṣidru huna gdī, lakān liyyām dēy laʾ! ʿād waráqit ǧurnān, wu tḥuṭṭlu ʾayyᵊ ḥāǧa šarīṭ xafīf, miš ziyyᵊ zamān, lákan zamān kunna … biss ittadʿīk gāʿid ildilwāk, hū da lli gāʿid mašēy. min zamān yaʿnēy.

10.04. Wir tun was? Wir nehmen einen Wollschal oder ein wollenes Wickeltuch, und eine bindet es hier so um seine Brust, aber heutzutage, nein! Einfach Zeitungspapier, und tut ihm irgendwas herum, ein leichtes Band, nicht wie früher, aber früher da haben wir … das Massieren ist bis heute üblich, das wird praktiziert, seit früheren Zeiten schon.

10.05. yigullak ʾē da ṭāyiḥ. yaʿni masalan ǧēt ašīlu, šaddētu gawi gdíh, ʿala kitfēy, da ṭāḥ minnēy. biyriǧǧēʿ wu biyishēl. da ṭāḥ, da ṭāyiḥ. kān ʾādi lʿilāǧ, ilʿiyyil ʿandna yishēl wu yriǧǧēʿ, ʿala ṭūl yiddūh ilwaḥda, waḥda kabīra, lāzim tiʿrif itdiʿʿéku.

10.05. Man sagt was? Das ist runtergefallen. ich wollte ihn z.B. tragen, ich habe es gezerrt, an seiner Schulter, das ist mir runtergefallen. Es übergibt sich und hat Durchfall. Das ist gefallen, runtergefallen. Das war dann die Behandlung, das Kind bei uns hat Durchfall und übergibt sich, sofort bringt man es zu einer, einer alten Frau, sie muß wissen, wie man es massiert.

---

[59] Sie tut, als ob sie ein Kind im Schoß liegen hätte.

[60] Sie verbessert *zibda* zum lokalen *zōbda*.

10.06. áha tidi"éku wu tluffu wu tšīlu wu tdi"éku talat tiyyām. tidi"éku bi llēl, wu ṣṣubḥᵊ tigullāk ʿašān ilʿamūd ibtāʿu yiwgā ... ʾitšūf uṃṃāl... ṭāyiḥ walla laʾ. ilʿamūd ibtāʿu wāgif walla miš ṭāyiḥ. il kān ilʿamūd bitāʿu miš wāgif, itgullak lā da ṭāyiḥ, di"ikūh. laban bizza, ya zibda, wu ndi"éku. talat tiyyām, ʾādi ttadʿīk ya duktur.

10.06. Da massiert sie es dann und wickelt es ein und massiert es drei Tage lang. Sie massiert es am Abend und am Morgen, sie sagt, damit sein Rückgrat... wird, sie sieht natürlich... ist es runtergefallen oder nicht. Ist sein Rückgrat gerade oder nicht gerade. Wenn sein Rückgrat nicht gerade ist, sagt sie, nein das ist runtergefallen, massiert ihn, Muttermilch, oder Butter, und wir massieren es, drei Tage lang, das ist das Massieren, Doktor.

THE *ARGAN* TREE OF SOUTH MOROCCO
AN ETHNOGRAPHIC NOTE

Harry Stroomer[1]

My contribution to the *Festschrift* in honour of my colleague Remke Kruk is of ethnographic nature and divided into two sections. In the first I will present a short and general description of the argan tree of South Morocco and its product: argan oil. In the second section I will present translations of ethnographic texts in Tashelhiyt Berber[2] about this tree, valuable testimonies from Tashelhiyt Berber informants, full of interesting details.

The argan *tree*

The argan tree is of great economic importance for the inhabitants of South Morocco, because they make argan oil out of its fruits. It is known in Latin as *Argania spinosa (L.)* or as *Argania sideroxilon* Roem.&Schult., and belongs to the family of the *Sapotaceae*. The word *argan* is Tashelhiyt Berber, it contains the root consonants *RG*, that can be found in a number of lexical items, all linked to the argan oil production.[3] This tree is indigenous

[1] I thank Mr John Cooper (Norwich, United Kingdom) for his kindness to correct the English text.

[2] The number of berberophones in North Africa is approximately 25–30 million. In Morocco, Berber is spoken in the Rif (*Tarifit*), in the Middle Atlas (*Tamaziyt*) and in the High Atlas, the Sous plains and the Anti-Atlas (*Tasusiyt, Tašlḥiyt*), estimated 45% of the total population of 30 million people. In Algeria, Berber languages are spoken in Kabylia, in the Aurès mountains and in the Mzab areas, a total of 25% of the population, also of 30 million people. Tuareg Berber is found in the south of Algeria, in Mali, Niger and Burkina Faso. Smaller groups of berberophones live in Tunisia, Libya and Egypt. Berber immigrant communities of various origins are settled in France, Belgium, The Netherlands, Germany and Israel. Of all Berber languages Tashelhiyt Berber is the one with the highest (estimated) number of speakers: some 8 to 9 million.

[3] Examples are: *rg* 'to break the argan nut'; *amrrag* 'someone who breaks the argan nut with the help of a stone; the stone with which one breaks the argan nut'; *awrag* 'the action of breaking the argan nut; the stone with which one breaks the argan nut'; *irg*, plural *irgn* 'pieces of the broken shell of the argan nut'; *tamrragt*, plural *timrragin* 'woman who breaks the argan nut with the help of a stone', *targant* 'a small argan tree'.

in South Morocco[4] (see Louis Emberger's map of its distribution, fig. 1, p. 181)[5] and considered a botanic relic.[6]

Inhabitants of South Morocco appreciate the argan tree and argan oil to such an extent that they see it as an emblem of their regional heritage. Argan trees and argan oil are to South Morocco what the maple and maple syrup are to Canada. The tree's special qualities and its importance for the local population were reasons for the UNESCO to include the argan tree, in 1999, in its World Heritage List.

One usually sees the argan trees scattered quite haphazardly in the South Moroccan landscape. They are not planted in gardens in the way people do with olives. In many cases they stand isolated rather than in groups. They can grow at altitudes up to 1700 metres. The roots of the argan tree are known to be capable of reaching water at great depths.

The argan tree usually has a short, twisted stem. Its crown can be quite big, five to six metres high and may even display a certain grandeur. Its prickly branches grow thickly together. The bark has the texture of the skin of a snake. When pruned, mutilated or even cut back to the minimum, the argan comes back vigorously with an abundance of shoots.

The argan tree blossoms in late spring, showing white, soft green or soft yellow flowers with five petals. The dark green shiny, long-oval leaves, standing on very short stalks and placed alternately on the two sides of a branch, persist throughout the year. Challot has observed that the argan tree may lose its leaves in times of great drought and he explains it as an appropriate reaction of the tree to prevent over-evaporation.[7]

The fruit of the argan tree looks like a dark green olive in the shape of a Brazil nut or in the shape of a date. It stands on such a small stalk as if it grows from the branch itself. The outside layer of the fruit is pulpy like the olive. When this layer is removed one sees what I will call here the argan 'nut' with its hard shell. Inside this 'nut' one finds often one, sometimes two almond-like argan kernels that contain the highly appreciated argan oil.

---

[4] J.-P. Challot, 'L'arganier,' *Revue du bois* (juin-juillet 1949), pp. 1–6. According to Challot, the argan has been succesfully introduced in the South of France in the vicinity of the cities of Bandol and Toulon. In recent years attempts have been made to introduce the argan in the Moroccan Rif. In Chouihia, in the Berkane region, two thousand argan trees have been planted in 2003.

[5] 'L'aire de l'arganier,' reproduced in Challot, p. 4.

[6] G. Camps, 'Arganier,' in *Encyclopédie berbère*, VI (Aix-en-Provence, 1989), pp. 879–880 (sub A267). See p. 879: 'un élément d'une flore relique, témoin d'une ancienne extension de la végétation tropicale au sud-ouest du Maghreb.'

[7] Challot, p. 1: 'Toutefois, lorsque l'arbre se trouve avoir à faire face à des conditions

Some weeks before the harvest of ripe argan fruits, the argan tree area is declared an *agdal*,[8] which means that it is closed to people and animals, in order to avoid damage to the developing argan fruits[9] and to prevent theft. In former times it was one of the tasks of the village council (*ljmaʿt*) to decide when to *agdal*-ize the argan grounds and when to annul this measure.

Children, men and women are involved in the harvest of the argan fruits. It is mainly the fallen argan fruits that are carefully collected. The actual production of argan oil is a job for women only and is known to be very labour-intensive. They first have to peel off the dried pulpy outside layer of the fruit. Then they have to break the shell of the 'nut', which they usually do with a stone, in order to reach the argan kernels. During this breaking operation women work together in groups.[10] And finally they have to squeeze out, with their hands, the argan oil from the ground argan kernel paste. The labor-intensive character of the argan oil production is demonstrated by data given by Challot. According to him one needs at least three kilograms of argan kernels to produce one litre of argan oil. For three kilograms of argan kernels one has to collect one hundred kilograms of argan fruits.[11]

The Laoust text, given in section 1 (below), states that also argan nuts from argan fruits digested by goats and subsequently excreted, are collected by people. These argan nuts are treated separately in the argan production process, but are also highly appreciated because of their special flavour.

Argan oil, with its soft and nutty taste, is used first of all in the kitchen as a cooking and salad oil. Moreover it plays an important role in the traditional health care of the region, in particular in the treatment of various

de sécheresse accentuée, il a un réflexe de défense contre l'évaporation et se défeuille entièrement.'

[8] The word *agdal* is frequently found in Moroccan toponomy, the best-known case being Rabat Agdal, a former extramural agricultural area, now a modern residential quarter of the city.

[9] The leaves of the argan tree are sought-after by goats. Everywhere in Morocco one can buy postcards showing goats grazing argan leaves while standing amazingly high on this tree.

[10] Argan tree owners who don't have enough women at their disposal to help them, may ask for help from neighbouring women (*tiwizi* 'the social institution of helping out one's neighbour') or they may hire them.

[11] Usually the argan fruits are dried before taking out the 'nut'; the dried argan fruits weigh half of their weight when harvested.

dermatological diseases.[12] As it is rich in polyunsaturated fatty acids, argan oil is considered good for the human heart and vascular system.

Some of the traditional uses of argan oil no longer exist. Argan oil used to be one of the ingredients in traditional soap making and it was used as fuel in oil lamps.

None of the by-products of argan oil production are wasted: the pulp debris (*alig*), left after removing the outside layer of the argan fruit, is excellent fodder for cattle; the broken nutshells (*irgn*) are used as fuel in the kitchen, and the debris cakes, left after pressing the argan oil out of the argan kernels, is a high protein cattle concentrate (*tazgmmut*).[13]

Another well-known product from the Moroccan South that cannot be made without argan oil is *amlu*, which looks like fluid peanut butter. It is the name for both the paste of ground roasted argan kernels, *and* the mixture of this paste with honey and argan oil.[14] For Tashelhiyt Berbers the latter is a traditional treat for travellers and convalescing sick people. It is eaten by simply dipping one's bread into it.

Most of the argan oil is consumed locally, relatively small quantities are sold to outsiders. Sellers of argan oil can be seen at the sides of the national roads in the High Atlas, e.g. alongside the road from Marrakech to Agadir. They ask high prices for their argan oil, often using two correct sales arguments: the labour-intensivity of the product, and the mentioned positive medical effects.

People in South Morocco report that local religious scholars disagree whether argan oil should be submitted to the same kind of tax (*zakât*) as olive oil. Many of them declare it exempt from any tax because its production is so labour-intensive. According to a translated *tertib* document published by G. Salmon in 1904, specifying tax rates of various agricultural products,[15] the owner of an argan tree had to pay two and half *douro* (an old coinage) per tree.

Translated Tashelhiyt Berber texts on the argan *tree and its oil*

The first translated text (see section 1) is taken from one of the publications of Emile Laoust, one of the best French specialists in the anthropology of

[12] An article about the traditional role of argan oil for medical purposes as described in the Tashelhiyt Berber manuscripts written in Arabic characters is in preparation.

[13] Pulp (*alig*) and cakes (*tazgmmut*) are sometimes offered for sale at local markets.

[14] One comes close to the taste of this *amlu* mixture, if one mixes peanut butter of excellent quality with pure honey.

[15] G. Salmon, 'Le tertib,' *Archives marocaines* 2 (1904), pp. 154–158, see p. 155.

the Berber populations of North Africa. In his *Cours de berbère marocain* (1921) he gives an interesting Tashelhiyt Berber text about the argan tree.[16] Though his transcription of Tashelhiyt Berber is far from flawless,[17] the text can easily be understood. His text is from the Ihahan, a Tashelhiyt Berber-speaking tribe south of Essaouira/Mogador (see the map) and treats the following subjects: the distribution of the argan tree; the development of the argan fruit and nomenclature for the various stages; measures to regulate the harvest; the harvest itself; pressing the argan oil, the product and the by-products; helping one's neighbour with the production of argan oil.

The second translated text (see section 2) is taken from Arsène Roux, *La vie berbère par les textes* (1955). It is a part of Roux's text 13 called *Lidam* ('Edible fat').[18] Roux (1893–1971) was a great specialist in the Berber language of the Middle Atlas and in Tashelhiyt Berber. His deep knowledge of Middle Atlas Berber and Tashelhiyt Berber is evident from the personal library and papers that he left behind after his death. Unfortunately he did not publish much. *La vie berbère*, an important collection of ethnographic descriptive texts in an excellent phonetic transcription, was his only book. He did not publish the translation volume that he promised in the *Avertissement* to this book.[19] The text is from the Ashtuken, a Tashelhiyt Berber-speaking tribe south east of Agadir. One may note some dialect differences in the argan terminology when one compares the texts below with the text of Laoust.

The third translated text (see section 3) is a short text, also from Arsène Roux, *La vie berbère par les textes* (1955).[20]

The fourth translation (see section 4) is of a small text from the Fonds Roux, box 30, text 1.1.,[21] a page in the handwriting of Roux. This unedited Tashelhiyt Berber text is presented here also. It is in the Ashtuken dialect and given by Roux's main informant, the well-known traditional scholar Si Brahim al-Kunki in a notebook of 1948–1949. This small text, describing succinctly the blessing of the argan, may serve as a conclusion.

---

[16] E. Laoust, *Cours de berbère marocain. Dialectes du Sous, du Haut et de l'Anti-Atlas*. 1ère éd. (Paris, 1921), pp. 264–266.

[17] Laoust is, for instance, inconsistent in the writing of long (double) consonants, which are quite important in Berber languages in general.

[18] A. Roux, *La vie berbère par les textes. Parlers du Sud-Ouest marocain (Tachelhit). Première partie: la vie matérielle. I. Textes* (Paris, 1955), p. 34.

[19] Cf. Roux, p. 5.

[20] Roux, p. 62.

[21] H. Stroomer & M. Peyron (avec la collaboration de C. Brenier-Estrine), *Catalogue des archives berbères du 'fonds Arsène Roux'*. Berber studies; 6 (Köln, 2003), p. 34.

1. *Laoust 1921, pp. 264–266 (Ihahan dialect)*

There are argan trees in the Ihahan region, in the Chiadma[22] region as well as in the Sous valley. Argan is not planted, it simply grows in the area where it occurs. When it grows, it is a small argan shoot (*tag^{w}ntaft*).[23] When it has started to produce, it is a small argan tree (*targant*).[24] Its leaves are like those of the pomegranate tree, its wood is like the wood of the apricot tree. It has thorns. It is a big tree, it grows well in good, well-manured soil. Where the ground is not (suitable), it remains small. It is found on mountain slopes, where they can be big or small, they are not of the same size everywhere.

In October[25] there are shoots on the argan, in November it makes flowers. Their flowers are like those of the olive tree, they are white. When it has thrown off its flowers, it makes buds (*aḥbub*). We then say: 'The argan shows its buds' (*ibiyyn wargan aḥbub nns*). Approximately one month after 'showing its buds', it produces young green argan fruits (*ag^{w}mmu*). When it has finished producing young fruits, it produces the big fresh argan fruits (*ablẓiẓ*). The *ablẓiẓ* is like the apricot. When it has become ripe, it is like (the shape of) a date and then we call it *afiyyaš*.

When it has produced *afiyyaš*, the fruits fall on the ground under the tree. Then there is a public announcement in the village. They say: 'The argan fruits have fallen down, access to the argan tree area is now prohibited (*agdal*), let no animals, no cows, no sheep, no camels go there, access is forbidden. If we find the animals of anyone in the argan area, then a big animal of his herd must be slaughtered (as a fine).'

The custom concerning *agdal* is as follows: they surround (the argan tree area) with cairns (*tikrkurin*), so that people know that access is forbidden.

The day they annul the prohibition (lit.: open the *agdal*), there is again a public announcement saying: 'The restriction is now lifted' (lit.: the *agdal* is now open).

---

[22] The *Chiadma* is an arabophone tribe north of Essaouira/Mogador (see map).

[23] Note that *tag^{w}ntaft* and the masculine equivalent *ag^{w}ntaf* also indicate a woman and a man from the well-known Tashelhiyt Berber-speaking Goundafa tribe of the High Atlas. The Goundafa, Glaoua and Mtougga tribes dominated South Moroccan politics of the pre-protectorate and early protectorate period (1850–1930).

[24] I.e. the word *argan* with the feminine circumfix *t- -t* to denote the diminutive.

[25] Farmers in Morocco make use of the Muslim calendar and the Julian calendar. The latter one is used in relation to agricultural life and festivals.

Then women, children and men go there and collect ripe argan fruits. They put in baskets what they have collected and bring it to a good place (where) they pour the argan fruits out and make a heap. When they have collected all the fruits, they place a fence around the heap to keep the animals at bay. Anyone who does not own argan trees, collects what has been left on the trees or under them.

People don't bother to harvest argan fruits from trees standing on steep cliffs. Animals climb to them and collect them. When these animals take a rest in the shade at midday, they digest these fruits in their stomach. People collect the 'nuts' digested by these animals, but they keep them apart, because there is good argan oil in them, surpassing that of argan fruits collected in the normal fashion. There is a nice strong flavour to them and that is why they keep them apart.

People take pack animals to the heap of argan fruits in the argan area. They fill the panniers with them and take them, on the back of their pack animals, to their homes, where they store them. When they have finished taking the argan fruits home, they remove the pulpy part of the fruit with a stone. They put the pulp (*alig*) on one side and the argan 'nuts' (*uẓlim*) on the other. The pulp is eaten by cattle and camels. They pound the argan nuts with a stone. When they have broken it, they take the 'argan kernel' (*tiẓnin*) out of the argan nut. They throw the broken nutshells (*irgn*) in the fire and they roast the 'argan kernels' on an iron plate above a fire. When they have roasted them, they pour the argan kernels onto a flat basket (*tasgg^{w}it*) and bring them to the argan hand mill, where they grind them. The ground argan kernels are like porridge, we call it the paste (*amlu*) of argan kernels because there is also the paste of almonds (*amlu n tẓnin n lluz*). The *amlu* is poured onto a plate for pressing. When they press with their hands, they add some water. At first it looks like sheep droppings, when it is mixed together is looks like dough.

Then it is pressed. They take the argan kernel pulp cake (*tazgmmut*) and shape it into loaves. They press and store (the argan oil). They take the argan oil and pour it into a gourd. They give the argan kernel pulp cakes (as a concentrate) to camels and bulls in the ploughing season. Camels usually eat *tazgmmut* in winter, because it is good (for them).

Cooperation (*tiwizi*) during the harvest of the argan fruits and the argan oil production. When people (are allowed to) go and collect (fallen argan fruits), the person who has a lot of ripe argan fruits (*afiyyaš*) may ask for help in collecting them. When the moment to collect them has come, there is a host of women and children around. The owner of the argan tree area rejoices. When they go for a siesta, he offers them a copious meal. When the women have finished, he notifies them saying: 'Please, (I would like to have the help of some) women (*timrragin*) who break argan "nuts".' If this is the case, there are many women (who are willing to do this). When they go to his house, there is joy and ululating. People say (to one another): 'What is going on today at the house of So-and-so?' 'Well, he has women around who (are helping him) break argan nuts.' Some of them do the breaking, others divide the argan kernels from the broken nutshells (*irgn*), again others roast the argan kernels on an iron plate. When they have roasted them, they pour them into a basket. One other among the women grinds them on an argan kernel hand mill. They spread the paste out on a plate for pressing (*taqṣṛiyt n iẓmi*) until the *amlu* paste comes out. One woman grinds, another presses. When she has pressed the argan out, she makes *lbsis*[26] from the newly made argan for the women who had been breaking the argan nut open. This is the custom towards the women who break the argan nut open.

2. *Roux 1955, p. 34 (Ashtukn dialect)*

As soon as the argan tree is in bloom, we say that it makes *ajdur* 'argan flower'. When its small fruit (lit.: child) is visible, we call it *ayray* 'newly developed fruit'. If it is big, but still white we call it *ẓrgmmu.*[27] When it has turned yellow and has become ripe, we call it *bilẓiẓ*. When shaken down from a tree, and dried, we call it *tifiyyišt*.

When people have collected the dry argan fruits that have fallen down, they remove the outside pulp layer, it is (separated between) pulp (*alig*) and nuts (*aqqayn*); they collect the nuts and give the pulp to the animals: the oxen, the camels and the sheep. Mules, donkeys and horses don't eat argan pulp, because it is too bitter for them. They take these argan nuts, break them open like (they do) with almonds and nuts, resulting in broken argan nutshells (*irgn*) and argan kernels (*tiẓnin*). They throw the nutshells in the fire, they roast the kernels on an iron plate like they do with grain, grind them on a handmill like they do with almonds and nuts. It has become a paste (*amlu*) similar to the dough of which the doughnut maker

[26] *Lebsis* is a paste made of flour made of roasted grains and argan oil, mixed with a variety of sweet or spicy ingredients. It is served on ceremonial occasions.

[27] Lexical items denoting the subsequent stages in the development of the argan fruit are numerous. Roux, p. 35 gives *admam*, plural *idmamn* 'fruit qui commence à se former' as a stage in between *ayray* and *ẓrgmmu*.

makes doughnuts. Women press that out. How do they do that? They add some water and knead it with their hands for some time and then add some water again. They mix it again until the argan oil and the kernel pulp separate. This pulp is like a compact mixed dough (*tummit*). They feed the young animals with it, that they want to fatten, it is better than fodder from barley or maize. The argan is handled with clean tools, people use it as edible oil in their meals like olive oil or butter. They bake meat with it and women oil their braids with it and people burn oil-lamps with it.

Argan is very useful: its wood can be used as firewood and as roofbeams in buildings. Young shoots and leaves are fodder for animals: camels and sheep, but not for other animals. Its fruit is money, the same for the one who sells it. He can separate fruit pulp from argan nuts, he can give the pulp to the beasts, the broken nutshells are fuel for the fire, the argan kernels can be ground, *amlu* can be pressed out of it. The cattle cake (*tazgmmut*) is for the animals. People either consume argan or sell it. From it argan soap can be made that has no equal.

People remove also the shells of almonds and nuts. One may sell it, another one may not want that, he (rather) eats it or grinds it so that it is *amlu*, (a paste that) he, or any guest, eats with bread. When they want to grind it, they roast it first, like roasted grain (*tirufin*). They grind it in a hand-mill, they pour argan (oil) into the 'eye' of the hand-mill in order to grind. If they don't do that, it will not grind, it will get stuck in the hand-mill.

3. *Roux 1955, p. 62 (Lakhsas dialect)*

Ripe argan fruits (*tafiyyušt*): When the time of the ripe argan fruits (*tafiyyušt*) has come, people publicly announce the closure of the area concerned, until all the argan fruits have fallen down and dry on the ground. If they were not to close the area in this way, people would take away all the fallen argan fruits, whether it is theirs or not. At that time (i.e. when access to the argan area is prohibited), the person who is caught, must pay a fine (*linṣaf*).

When the ripe argan fruits are dry, they announce that they open (the area for the harvest). Each person goes to his own argan trees and they collect the ripe argan fruits, store them (at home) in rooms, remove the outer pulp layer, one quantity after the other, they break the argan nut (*aqqayn*) open, roast the argan kernels (*tiẓnin*), grind them and press the argan paste (*amlu*) and separate the argan oil from the (squeezed out) argan kernel cake (*tazgmmut*). They (themselves) consume the argan oil and give the argan kernel cake to their animals.

4. *Fonds Roux 30.1.1. (Ashtukn dialect)*

Argan: Ix tssmγi tagant argan, ar grrun mddn tifyyišt nns; tifyyišt ann tga kullu lbaraka: ar ttuẓḍint tiẓnin nns, gint zzit n wargan; ar šttan laksab alig nns d tzgmmut nns, trγ lʿfiyt s irgn nns.

Argan: When the area outside the village has produced argan, people go and collect the ripe argan fruits (*tifyyišt*). These fruits are a true blessing (*lbaraka*). Women grind the kernels (inside) them and make argan oil from it. Cattle eat the pulp debris (i.e. the outside pulpy layer of the argan fruit) and the squeezed out argan kernel cake and they make a fire with the broken argan nut shells.

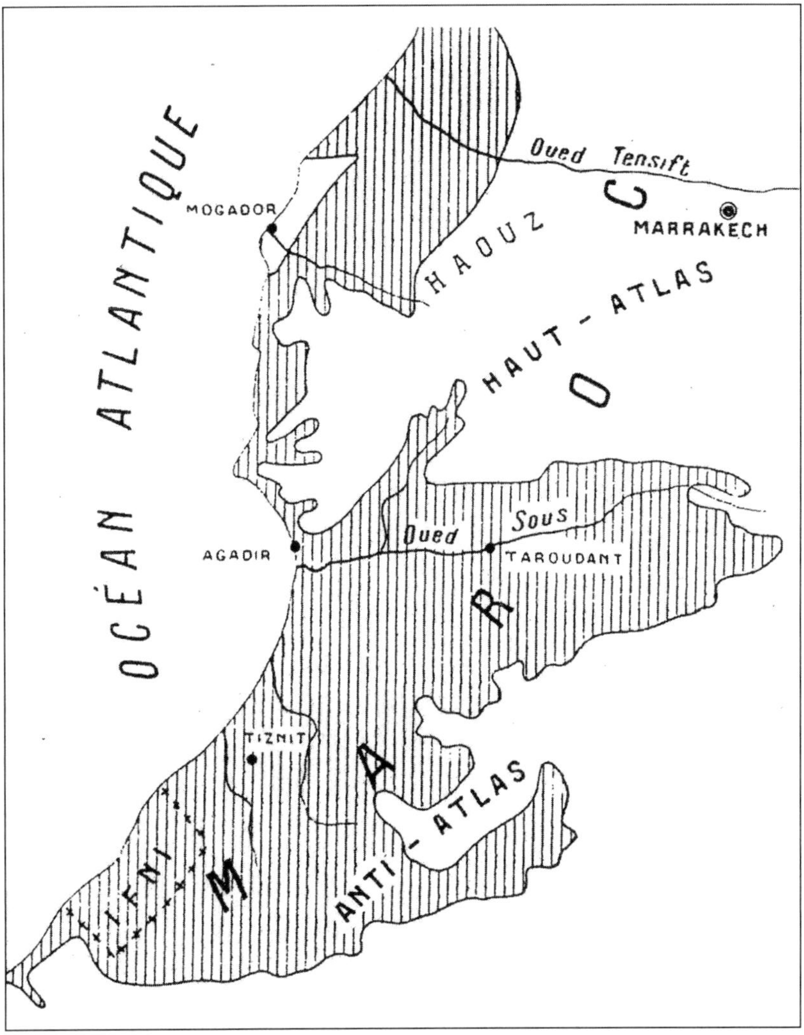

Fig. 1. L. Emberger, 'L'aire de l'arganier,' in J.-P. Challot, 'L'arganier,' *Revue du bois* (juin–juillet 1949), p. 4.

GAZING AT THE SUN

REMARKS ON THE EGYPTIAN MAGICIAN AL-BŪNĪ AND HIS WORK

Jan Just Witkam*

The Corpus Būnianum

Whoever leafs through the editions of the works of the Egyptian magician Aḥmad b. ʿAlī b. Yūsuf al-Būnī (d. 622/1225) is bound to be soon confused. In his search for some structure in the author's argument, the reader will instead find numerous repetitions, and not seldom will he discover that he is reading what proves to be a second or third handling of issues that have already been the subject of earlier discussions, even in the very same work by al-Būnī which he has at hand. The more one reads in al-Būnī's published works, the more the impression grows that we here have the result of a continuous process of permutation of constituent elements and ideas. This may lead us to the assumption that some, if not all, of al-Būnī's works were not written by himself. Instead, we may consider several of the titles that go under al-Būnī's name as part of a Corpus Būnianum, as the product of the work of several generations of practicing magicians, who arranged al-Būnī's work and thought and brought it out, probably while mixing these with elements of their own works. This does not merely imply that there is a pseudepigraphic Būnian literature, but also that some works by al-Būnī, or ascribed to him, may in fact constitute a composition of fragments of very diverse origin.[1] This should not shock the trained philologist, who usually thinks in terms of a fixated text with certain authorship and a reconstructable

* *Interpres Legati Warneriani* and Professor of 'Paleography and Codicology of the Islamic World' in Leiden University. Address: Faculty of Arts. P.O. Box 9515. 2300 RA Leiden, The Netherlands. E-mail: <j.j.witkam@let.leidenuniv.nl>.

[1] Mohamed M. El-Gawhary, *Die Gottesnamen im magischen Gebrauch in den al-Buni zugeschriebenen Werken* (Bonn, 1968), p. 17. El-Gawhary dates the composition of the *Shams al-maʿārif* to at least a century after al-Būnī's death. This is not contradicted by the datings of the Leiden manuscripts of the *Shams al-maʿārif* and the *Tartīb al-daʿawāt* (see the 'Bibliographical note' below for all references to manuscripts and printed works).

stemma of manuscripts, since in the case of popular texts, both magical and other, the author is unimportant and has often disappeared behind 'his' text. Users and readers of such popular classics treat the work as their own books, in the double sense of the word, and feel free to alter these.[2]

Western scholarship has not always been kind to al-Būnī. Ullmann characterizes his intellectual achievement, while contrasting al-Būnī's ingenuity in analyzing a finely structured cosmos to his practical approach of the unseen world, as a sign of 'colossal credulity', and al-Būnī's method of using the numerical and other properties of words and letters as 'stupid, formalistic arithmetic.'[3] When reading such qualifications, one wonders whether Ullmann by this remark had actually wished to propose that al-Būnī, if only he had worked in a less credulous and less formalistic arithmetical way, would really have provided his readers with meaningful answers to the enigma of the universe.

Ullmann is however correct in stating that al-Būnī's works do not constitute a source for our knowledge of the older Arabic literature on magic, and he proposes that they rather represent a cross section of practices and beliefs which were current before, during and after al-Būnī's lifetime, during which period the corpus, which now counts some forty[4] titles, has come into being. Al-Būnī is the figurehead of this corpus, nothing more. Personally, I think that if the assumption of a popular origin of the varieties of magic discussed can be documented, it would make the Būnian corpus even more interesting.

Ibn Khaldūn, who often takes a relaxed view on those fields of science which he does not personally endorse, quotes al-Būnī on his own view of the attainability of letter magic, the most conspicuous element in his works, as: 'One should not think that one can get at the secret of the

---

[2] See e.g. El-Gawhary, *Die Gottesnamen*, p. 14ff. for a full discussion on the genesis of the Būnian corpus.

[3] Manfred Ullmann, *Die Natur- und Geheimwissenschaften im Islam* (Leiden [etc.], 1972), p. 391.

[4] Brockelmann, *GAL*, G I, p. 497, in combination with S I, pp. 910–911, mentions 39 titles, but he has not even attempted to relate these to one another. His list still reflects the order in which he worked through the catalogues that constitute his primary sources for the bibliography of al-Būnī's works. Progress in manuscript cataloguing will reveal more titles. However, some of Brockelmann's 39 different titles will, upon closer examination, prove to be identical with others which go by other titles. The Corpus Būnianum is as yet far from being clearly defined.

be understood that the additions are the publisher's own. This example of the first, or of an early, edition became the norm, and many of the later editions, if not all, contain this extensified and over-illustrated version of the text. As a result, the *Shams al-maʿārif al-kubrā* consists of many parts that seem to have simply been patched together without much consideration for compository requirements and their mutual relevance. Hence the confusion which I noted at the beginning of this article.

Al-Būnī's Tartīb al-Daʿawāt

If the *Shams al-maʿārif*, whether *kubrā* or not, has become a popular and wide-spread text, because of its practical use and its matter-of-fact and non-intellectual approach of the mysterious world of the unseen, the opposite is true for a text by al-Būnī which so far has escaped the attention of scholars. It is a work entitled *Tartīb al-daʿawāt fi takhṣīṣ al-awqāt ʿalā ikhtilāf al-irādāt*, which may perhaps be translated as 'The Order of Invocative Prayers. On the Determination of the Moments, according to the Different Wishes'. It is a rare text, and I am not aware of any other copy than the manuscript in the Leiden University Library.[11] Although it was duly mentioned in all of this library's catalogues,[12] it somehow escaped Brockelmann's attention, and, as a result, was not mentioned among his thirty-nine titles, nor does it seem to have attracted any further scholarly attention. Already the indication of the name of the author, immediately after the *basmala*, makes it clear that it was written after his death, probably by an admirer or a pupil.[13] Its content is a mixture of the well-known Būnian themes on the magical use of letters, divine names and sentences from the Qur'an, but astrological and other

[11] MS Or. 1233. It has been in Leiden since the mid-18th century, first in the private collection of J.J. Schultens (1716–1778). It was auctioned off in 1780, and in 1781 it was registered by the University Library, together with other manuscripts of the Schultens collection. It may already have been in Europe before it came into Schultens' possession.

[12] Lastly in P. Voorhoeve, *Handlist of Arabic manuscripts in the Library of the University of Leiden and other collections in The Netherlands*. 1st ed. (Leiden, 1957) and 2nd ed. (The Hague, 1980), p. 374.

[13] Beginning on p. 1: قال الشيخ الأمام العارف العالم الربانى مزّى المريدين شيخ السالكين قدوة العارفين ابو العباس أحمد بن علي بن يوسف البونى القرشى قدس الله روحه ونور ضريحه .

considerations are given here as well. In the overall level of its intellectual approach, the work differs from the *Shams al-maʿārif*. Another difference consists in the fact that the application of amulets and recipes is made subservient to the appropriate moment, and such moments are determined by astrological considerations. Also concepts such as *ṭabīʿa*, 'nature', and *imtizāj*, 'mixing', of the natural elements of the letters are introduced and discussed. The four elements (fire, air, water and earth) are, among other things, brought into connection with the values of the letters of the Arabic alphabet, and consequently with the magical value of the words and sentences which they constitute. One may find such considerations and concepts in the *Shams al-maʿārif* as well, but in the *Tartīb al-daʿawāt* they seem to receive more attention, and, it would seem, on a higher intellectual level than is the case in al-Būnī's other works. This in fact may be the reason why the text is so rare. Another reason for its rarity may be the fact that the Leiden manuscript of the *Tartīb al-daʿawāt* stands out by a large number of expertly drawn and sophisticatedly constructed magical squares and figures. These, too, may have hindered the manufacture of more manuscripts and the spread of this text.

The general structure of the content of the *Tartīb al-daʿawāt* is familiar enough. After an introduction on the subject of the divine names and the letters of the names as their constituent elements, al-Būnī (if he is the author) treats the divine names according to a categorization of his own (pp. 14–100). In the following, which is the main part of the text, the author sets out to treat Qur'anic sentences (*mufradāt al-Qurʾān*) and the magical operations (*aʿmāl*) which he performs in connection with these. The Qur'anic sentences seem to be given more or less in their order of occurrence in the Qur'an, a habit which is highly recommended among readers of the Qur'an.[14] This order is not strictly adhered to, however. The first quotation is from *Sūrat al-Baqara* (p. 101), and the last one

[14] See for this recommended habit e.g., ʿAlī Muḥammad al-Ḍabbāʿ, *Kitāb fatḥ al-karīm al-mannān fī ādāb ḥamalat al-Qurʾān* (Cairo, 1353/1934), p. 10: 'It is a recommended custom that he [that is the *Qāriʾ* who recites fragments of the Holy Book, JJW] recites in the order of the Qur'an text, because that order is there for a wise reason (*ḥikma*).' This short text by a 20th-century author on handling the divine word is available in many editions. It is often published together with *al-Tibyān fī ādāb ḥamalat al-Qurʾān*, a similar text, by al-Nawawī. One of the editions which I have used is the one published in Cairo (Muṣṭafā al-Bābī al-Ḥalabī) in 1353/1934. An English translation of the *Fatḥ al-karīm al-mannān* by myself is forthcoming in the Yūsuf Ibish Memorial volume, to be published by the al-Furqān Foundation in London.

from *Sūrat al-Nās* (p. 332), and generally speaking the author adheres to the order of the *sūra*s and *āyāt*, but a random check reveals a few irregularities. One may view this part of the *Tartīb al-daʿawāt* as a selective exegesis of the Qur'an and the work of the author as a systematic attempt to make magic out of God's Word.

One feature may be discussed here in more detail. At the end of the work, the author treats the *Qalam al-ṭabīʿī*, the 'natural secret alphabet' (pp. 335–337), and he gives a table of this secret alphabet, which he connects with the four elements. The division according to the elements is as follows. Fire (*Nār*) has *alif, bāʾ, tāʾ, thāʾ, jīm, ḥāʾ, khāʾ*. Air (*Hawāʾ*) has *dāl, dhāl, rāʾ, zāy, ṭāʾ, ẓāʾ, kāf*. Water (*Māʾ*) has *lām, mīm, nūn, ṣād, ḍād, ʿayn, ghayn*. Earth (*Turāb*) has *fāʾ, qāf, sīn, shīn, hāʾ, wāw, yāʾ*. From the magical literature of the period in which the Corpus Būnianum has originated, at least two more examples of this particular 'natural script' are known from the survey of secret and esoteric scripts which was compiled by the *ḥurūfī* mystic al-Bisṭāmī (d. 858/1454) in his work *Kitāb mabāhij al-aʿlām fī manāhij al-aqlām*.[15] On f. 25a–b of the Leiden manuscript of this text, *al-Qalam al-ṭabīʿī* is explained as the *Qalam al-ḥikma*, the 'wisdom script', and its invention is attributed to a certain Ardīmūs al-Ḥakīm, whom I have not identified. Al-Bisṭāmī in his survey quotes yet another 'natural script', which was purportedly taken from the work *Kayfiyyat al-ittifāq fī tarkīb al-awfāq*, a compilation made by al-Shaykh Yaʿīsh b. Ibrāhīm al-Umawī al-Andalusī.[16] Although the explanations of the nature and use of these 'natural scripts' are entirely different between al-Būnī, al-Bisṭāmī and al-Shaykh Yaʿīsh, the signs of the 'natural script' as given by al-Būnī and al-Shaykh Yaʿīsh are remarkably similar in shape, whereas the script proposed by al-Bisṭāmī is entirely different from that of the other two authors. Whether or not the interest in such secret alphabets in different works dating from the 9/15th century would indicate that there is a multiple use from different sources of similar or related content, needs further investigation. Apart from more detailed information on the history of esoteric alphabets, it also may shed some

[15] He is ʿAbd al-Raḥmān b. Muḥammad b. ʿAlī b. Aḥmad al-Ḥanafī al-Bisṭāmī, see *GAL* G II, p. 232. His work is quoted here after the Leiden MS Or. 14.121. See Jan Just Witkam, *Catalogue of Arabic manuscripts in the Library of the University of Leiden and other collections in The Netherlands* (Leiden, 1983–1989), pp. 210–218, esp. p. 214. See also his work *al-Fawāʾiḥ al-miskiyya fī ʾl-fawātiḥ al-makkiyya* (Leiden MS Or. 947).

[16] Cursory mention of this 9/15th-century author is made in *GAL* S II, p. 379. The work *Kayfiyyat al-ittifāq* is mentioned in *GAL* S N II, ad p. 155.

light on the methods and chronology of the composition of the works which are part of the Corpus Būnianum.

Al-Būnī's spiritual genealogies

Ullmann seems to have expressed his unfavourable judgment of al-Būnī's working method exclusively on the basis of the printed editions of *Shams al-maʿārif al-kubrā*, since he mentions one particular section in the book which is absent in all manuscripts which I have seen. It is a number of *sanad*'s, lists of authorities whom the author considers to be his predecessors or teachers, a sort of spiritual or educational pedigree, which is far from a rare feature in Islamic scholarly literature.[17] At the end of part 4 of the *Shams al-maʿārif al-kubrā* is an epilogue which by content and form is different from the rest of the work. It is titled 'Epilogue in which is mentioned the chain of our *shaykhs* [...]'.[18] The tone of this section is much more matter-of-fact, sometimes even personal, which stands in contrast to the descriptions of the world of magic phenomena in the main part of the *Shams al-maʿārif*, in which the author remains rather impersonal towards his readers. One wonders when and why these *sanad*'s were added to the text, and what their origin might be. If they were not integral part of the *Shams al-maʿārif* from the very beginning (and the manuscript tradition does not seem to warrant this), they may have been part of the author's *Fahrasa*, the educational (auto)biography, a genre which has become particularly popular in the Maghrib. The present *sanad*'s may have been added to the printed text of the *Shams al-maʿārif*, and probably at a quite late date. This addition as well may have been the initiative of an early publisher of the printed text. However, its origin may also lie in one particular manuscript, or diploma text, which at some stage was added to the *Shams al-maʿārif*.

---

[17] See Jan Just Witkam, 'The human element between text and reader. The *ijāza* in Arabic manuscripts,' in Yasin Dutton (ed.), *The codicology of Islamic manuscripts* (London, 1995), pp. 123–136, for an overview of the characteristics of such lists.

[18] *Shams al-maʿārif al-kubrā*, in the al-Ḥalabī edition, pp. 530–535; the Murād edition, pp. 506–510; the Calcutta (or Bombay?) edition, pt. 4, pp. 135–139 (see below under 'Printed sources used or quoted') (خاتمة في ذكر سند مشايخنا رحمهم الله تعالى وقدس أرواحهم آمين). This epilogue was – partly, but quite literally – also added at the end of *Manbaʿ uṣūl al-ḥikma* (pp. 323–325), but its occurrence there is evidently a recent addition, whereas in the *Shams al-maʿārif al-kubrā* it makes at first the impression of being an integral part of the text.

The editions of the *Shams al-maʿārif* which I have seen all contain these pedigrees at the end, whereas they are absent in all manuscripts which I have seen. Interestingly enough, apart from a few occasional remarks in the course of al-Būnī's works, it seems to be the only longer text which contains a few details of his life.[19] An obvious reason for the inclusion of the *sanad* material must have been an attempt to authenticate al-Būnī's authorship of the *Shams al-maʿārif*.

The epilogue contains the author's spiritual pedigrees for four different sorts of knowledge (*Shahāda, ʿIlm al-Bāṭin, Ḥurūf, Awfāq*). A survey of the *Sanad* material follows here. After having stated that the Prophet's cousin and son-in-law, ʿAlī b. Abī Ṭālib, first received the *Kalimat al-Shahāda* (the words of the confession of faith) from the Prophet Muḥammad, he continues with the spiritual pedigrees, which are here fully reproduced.[20]

Pedigree A (*for the* Kalimat al-Shahāda)

1. Aḥmad b. ʿAlī al-Būnī (d. 622/1625), who took from:
2. *al-Imām al-ʿĀlim* Abū ʿAbd Allāh Muḥammad b. Maḥmūd b. Yaʿqūb al-Fakūnī[21] al-Tūnisī al-Mālikī, who took from:

---

[19] See Rosenthal's remark to his translation of Ibn Khaldūn's *al-Muqaddima*, III, note 807, concerning al-Būnī's lifetime. The date of copying of the Berlin MS Mf. 80 (Ahlwardt 4126), a work going by the rather non-descript title *Kitāb maʿānī asrār al-ḥurūf*, identified by Ahlwardt, III, pp. 510–511 as *Laṭāʾif al-ishārāt fī asrār al-falak wa ʾl-ḥurūf al-maʿnawiyyāt*), and apparently identical with work entitled *Laṭāʾif al-ishārāt fī asrār al-ḥurūf al-ʿulwiyyāt* (Ullmann, p. 391) is 669/1270, which would thereby be the oldest known historical indication of the existence of a work of the Būnian corpus. For the commonly accepted year of al-Būnī's demise (622/1225) there seems to be no other authority than the *Kashf al-ẓunūn* of Ḥājjī Khalīfa. I am not aware of any research done on the manuscripts of the *Shams al-maʿārif*, or any other text by al-Būnī for that matter. Many printed editions mention the fact that 'old and correct manuscripts' were used, but such remarks should not be given any credit, as they are made by the publishers for commercial reasons, and do not serve any philological purpose. Numerous popular books with doubtful antecedents claim on their title-page to be based on unique, old or valuable manuscripts. The title-page of the Leiden MS Or. 336 has some additional information on al-Būnī's parentage: his father is referred to as *al-Shaykh al-Ajall al-Muqriʾ* Abū ʾl-Ḥasan ʿAlī, the most striking element of which is the occupational indication, *al-Muqriʾ*, 'the Qur'an reader'. It suggests that al-Būnī may have been intimately familiar with the text of the Qur'an from an early age onwards.

[20] The structuring of this material, by designating the different pedigrees with letter A-K, by numbering the names within each pedigree, and by indicating the cross-references, is mine. From the collation between the three printed texts, it is clear that the Murād edition belongs to one branch, and the combined Ḥalabī and Calcutta/Bombay editions to another.

[21] Reading of Murād. Ḥalabī and Calcutta/Bombay have here: al-Kūfī.

3. *al-Shaykh* Māḍī ʾl-ʿAzāʾim, who took from:
4. *al-Shaykh al-Quṭb* Abū ʿAbd Allāh Muḥammad b. Abī ʾl-Ḥasan ʿAlī b. Ḥirzhum,[22] who took from:
5. *al-Shaykh al-Ṭawīq*[23] *wa-Maʿdin al-Taḥqīq* Abū Muḥammad Ṣāliḥ b. ʿAqbān al-Wākilī al-Mālikī, who took from:
6. *Ḥujjat al-Zamān wal-Wāḥid fī ʾl-ʿIrfān* Abū Madyan Shuʿayb b. Ḥasan al-Andalusī al-Ishbīlī (E-9, H-3),[24] who took from:
7. Abū Shuʿayb Ayyūb b. Saʿīd al-Ṣanhājī (H5), who took from:
8. *Shaykh al-ʿĀrifīn Quṭb al-Ghawth al-Fard al-Jāmiʿ* Abū Yaʿzā[25] al-Maʿarrī, who took from:
9. Abū Muḥammad b. Manṣūr, who took from:
10. Abū Muḥammad ʿAbd al-Jalīl b. Maḥlān, who took from:
11. Abū ʾl-Faḍl ʿAbd Allāh b. Abī Bishr (H-7), who took from:
12. Mūsā al-Kāẓimī,[26] who took from:
13. Abū Jaʿfar al-Ṣādiq (B-1?),[27] who took from his father:
14. Muḥammad al-Bāqir, who took from his father:
15. Zayn al-ʿĀbidīn,[28] who took from his father:
16. al-Ḥusayn, who took from his father:
17. ʿAlī b. Abī Ṭālib, who took from:
18. Muḥammad b. ʿAbd Allāh (B-3, H-14, I-13), the Prophet.

Pedigree B (ʿIlm al-Bāṭin)

1. *al-Imām* Jaʿfar al-Ṣādiq (A-13?), who took from:
2. Qāsim b. Muḥammad b. Abī Bakr al-Ṣiddīq, who took from:
3. Muḥammad b. ʿAbd Allāh (A-18, H-14, I-13), the Prophet.

---

[22] Reading of Murād. Ḥalabī and Calcutta/Bombay have here: Ḥarām.

[23] Reading of Murād. Ḥalabī and Calcutta/Bombay have here: *Shaykh al-Ṭarīq*.

[24] His hagiography in ʿAbd al-Wahhāb al-Shaʿrānī, *Lawāqiḥ al-anwār fī ṭabaqāt al-akhyār* (Cairo, 1373/1954), I, pp. 154–156 (No. 275).

[25] Reading of Murād. Ḥalabī and Calcutta/Bombay have here: Yaʿr. Under his name Abū Yaʿzā al-Maghribī his hagiography can be found in al-Shaʿrānī, *Lawāqiḥ*, I, pp. 136–137 (No. 257).

[26] The hagiography of Mūsā al-Kāẓim in al-Shaʿrānī, *Lawāqiḥ*, I, p. 38 (No. 58).

[27] The hagiographies of Abū Jaʿfar Muḥammad al-Bāqir and Jaʿfar al-Ṣādiq in al-Shaʿrānī, *Lawāqiḥ*, I, pp. 32–33 (Nos. 38, 39).

[28] His hagiography in al-Shaʿrānī, *Lawāqiḥ*, I, pp. 31–32 (No. 37).

Pedigree C (ʿIlm al-Ḥurūf):

1. *al-Shaykh al-Imām* Abū Ḥasan al-Baṣrī, who took from:
2. Ḥabīb al-ʿAjamī (H-11), who took from:
3. *al-Shaykh* Dāwūd al-Jabalī,[29] who took from:
4. *al-Shaykh* Maʿrūf al-Karkhī,[30] who took from:
5. *al-Shaykh* Sarī al-Dīn al-Saqaṭī (H-9),[31] who took from:
6. *Shaykh al-Waqt wa ʾl-Ṭarīqa Maʿdin al-Sulūk wa ʾl-Ḥaqīqa al-Shaykh* al-Junayd al-Baghdādī,[32] who took from:
7. *al-Shaykh* Ḥammād al-Dīnawarī, who took from:
8. *al-Shaykh* Aḥmad al-Aswad, who took from:
9. *al-Shaykh* Muḥammad al-Ghazālī (I-2), who took from:
10. *al-Shaykh* Abū ʾl-Najīb al-Suhrawardī,[33] who instructed (*laqqana*):
11. *al-Shaykh al-ʿĀrif al-Fāḍil* Aṣīl al-Dīn al-Shīrāzī, who instructed:
12. *al-Shaykh* ʿAbd Allāh al-Bāyānī, who instructed:
13. *al-Shaykh* Qāsim al-Sarjānī, who instructed:
14. *al-Shaykh* al-Sarjānī,[34] who instructed:
15. *al-Shaykh al-Imām al-ʿĀrif al-Ṣamadānī wa ʾl-Hammām al-Nūrānī* Jalāl al-Dīn ʿAbd Allāh al-Bisṭāmī, who instructed:
16. *Shams Wuṣlatī wa-Badr Qalbī Ṭawd al-Ḥaqāʾiq al-Shāmikh wa-Jabal al-Maʿārif al-Rāsikh Shams al-ʿĀrifīn wa-Sirr Allāh fī ʾl-Arḍīn* Abū ʿAbd Allāh Shams al-Dīn al-Iṣfahānī.

Pedigree D (ʿIlm al-Awfāq)

1. Also taken from *al-Shaykh al-Imām al-ʿĀrif bi ʾllāh* Abū ʿAbd Allāh Muḥammad b. ʿAlī.

Pedigree E (ʿIlm al-Awfāq), *taken from*

1. *al-Shaykh al-Imām al-ʿAllāma* Sirāj al-Dīn al-Ḥanafī, who took from:

---

29 Reading of Murād. Ḥalabī and Calcutta/Bombay have here: al-Jīlī.
30 His hagiography in al-Shaʿrānī, *Lawāqiḥ*, I, p. 72 (No. 142).
31 His hagiography in al-Shaʿrānī, *Lawāqiḥ*, I, pp. 74–75 (No. 144).
32 His hagiography in al-Shaʿrānī, *Lawāqiḥ*, I, pp. 84–86 (No. 164).
33 His hagiography in al-Shaʿrānī, *Lawāqiḥ*, I, p. 140 (No. 261).
34 Reading of Murād. Ḥalabī has for Nos. 13 and 14: *al-Shaykh* Qāsim al-Sarjānī, and *al-Shaykh* al-Sīrḥānī, whereas Calcutta/Bombay have here: *al-Shaykh* Qāsim al-Sīrḥānī, and *al-Shaykh* al-Sīrḥā.

2. *al-Shaykh* Shihāb al-Dīn al-Maqdisī,[35] who took from:
3. *al-Shaykh* Shams al-Dīn al-Fārisī, who took from:
4. *al-Shaykh* Shihāb al-Dīn al-Hamadānī (F-3), who took from:
5. *al-Shaykh* Quṭb al-Dīn al-Ḍiyāʾī (F-4), who took from:
6. *al-Shaykh* Muḥyī ʾl-Dīn Ibn al-ʿArabī (F-5), who took from:
7. *al-Shaykh* Abū ʾl-ʿAbbās Aḥmad b. al-Tūrīzī, who took from:
8. *al-Shaykh* Abū ʿAbd Allāh al-Qurashī (H-2, K-1),[36] who took from:
9. *al-Shaykh* Abū Madyan al-Andalusī (A-6, H-3).

Pedigree F (ʿIlm al-Awfāq), *taken from*

1. *al-Shaykh* Muḥammad ʿIzz al-Dīn b. Jamāʿa al-Shāfiʿī, who took from:
2. *al-Shaykh* Muḥammad b. Sīrīn (H-12),[37] who took from:
3. *al-Shaykh* Shihāb al-Dīn al-Hamadānī (E-4), who took also from:
4. *al-Shaykh* Quṭb al-Dīn al-Ḍiyāʾī (E-5), who took from:
5. *al-Shaykh* Muḥyī ʾl-Dīn Ibn al-ʿArabī (E-6).

Pedigree G (ʿIlm al-Ḥurūf waʾl-Wafq), *taken from*

1. *al-Shaykh al-Imām al-ʿĀlim al-ʿAllāma al-Faqīh al-Thiqa* Musāʿid b. Sāwī b. Masʿūd b. ʿAbd Allāh b. Raḥma al-Ḥawārī al-Ḥimyarī al-Qurashī, who took from:
2. *al-Shaykh* Shihāb al-Dīn Aḥmad al-Shādhilī, who took from:
3. *al-Shaykh* Tāj al-Dīn b. ʿAṭāʾ al-Mālikī al-Shādhilī, who took from:
4. *al-Shaykh* Abū ʾl-ʿAbbās Aḥmad b. ʿUmar al-Anṣārī al-Mursī.[38]

Pedigree H (ʿIlm al-Ḥurūf waʾl-Wafq), *taken from*

1. *al-Shaykh al-Imām al-ʿAllāma* Abū ʾl-ʿAbbās Aḥmad b. Maymūn al-Qasṭalānī,[39] who took from:
2. *al-Shaykh* Abū ʿAbd Allāh Muḥammad b. Aḥmad al-Qurashī (E-8, K-1), who took from:

---

[35] Reading of Murād and Ḥalabī, whereas Calcutta/Bombay have here: *Khalīfat* al-Maqdisī.
[36] His hagiography in al-Shaʿrānī, *Lawāqiḥ*, I, pp. 159–160 (No. 281).
[37] His hagiography in al-Shaʿrānī, *Lawāqiḥ*, I, p. 36 (No. 49).
[38] His hagiography in al-Shaʿrānī, *Lawāqiḥ*, II, pp. 12–20 (No. 310).
[39] Evident typesetting error in Murād: al-Saṭlānī.

3. *al-Shaykh al-Imām al-ʿAllāma Ustādh al-ʿAṣr wa-Awḥad al-Dahr* Abū Madyan Shuʿayb b. Ḥasan al-Anṣārī al-Andalusī *Raʾs al-Sabʿat Abdāl wa-Rāḥat al-Arbaʿat Awtād* (A-6, E-9), who took from:
4. *al-Shaykh al-Ustādh al-Kabīr* Dāwud b. Maymūn al-Harmīrī[40] (apparently a blind man, with an anecdote on Abū Madyan visiting him), who took from:
5. *al-Shaykh al-Imām Quṭb al-Ghawth* Abū Ayyūb b. Abī Saʿīd al-Ṣanhājī al-Azammūrī (A7),[41] who took from:
6. *al-Shaykh al-Walī al-Kabīr* Abū Muḥammad b. Nūr, who took from:
7. *al-Imām al-ʿĀlim* Abū ʾl-Faḍl ʿAbd Allāh b. Bishr (A-11), who took from his father:
8. Abū Bishr al-Ḥasan al-Jūjarī, who took from:
9. Sarī al-Dīn al-Saqaṭī (C-5), who took from:
10. Dāwud al-Ṭāʾī, who took from:
11. *al-Shaykh* Ḥabīb al-ʿAjamī (C-2), who took from:
12. *al-Shaykh* Abū Bakr Muḥammad b. Sīrīn (F-2), who took from:
13. Anas b. Mālik (I-12), who took from:
14. The Prophet Muḥammad (A-18, B-3, I-13).

Pedigree I (ʿIlm al-Ḥurūf waʾl-Wafq)

1. al-Būnī met with:
2. (with many epithets) *al-Shaykh* Abū ʾl-Ḥasan Muḥammad b. Muḥammad al-Ghazālī (C-9), who instructed (*laqqana*):
3. *al-Sirr al-Makhzūn wal-Durr al-Maknūn wa ʾl-Sirāʾ al-Qarīb Aḍʿaf ʿIbād Allāh wa-Aḥqar Khalq Allāh al-Mutamassik bi-Dhayl Karam Allāh* Aḥmad b. Yūsuf al-Qurashī (with whom al-Būnī himself is meant, but evidently not in his own words), saying that he had seen:
4. *al-Shaykh al-Imām* ʿAlī b. Sīnā, who took from:
5. *al-Shaykh* Muḥammad al-Dawrakī,[42] together with whom he had sat [...], and who had seen:
6. *al-Shaykh* Muḥammad al-Jazarī,[43] who had seen:

---

[40] Reading of Murād. Ḥalabī and Calcutta/Bombay have here: al-Harīrī, or al-Hurayrī.

[41] Reading of Murād. Ḥalabī and Calcutta/Bombay have here: al-Armūzī.

[42] Reading of Murād. Ḥalabī and Calcutta/Bombay have here: al-Dawūkī.

[43] Reading of Murād. Ḥalabī has al-Jarazī and Calcutta/Bombay has: al-Jararī.

7. *al-Ṣadr al-Kabīr al-Shaykh* ʿIzz al-Dīn Muḥammad ʿAbd Allāh Muḥammad b. Mūsā b. Sulaymān al-Anṣārī, who had seen:
8. *al-Ṣadr al-Ajall al-Shaykh al-Imām* Abū ʾl-Ḥasan ʿAlī b. Aḥmad b. ʿAbd al-Wāḥid al-Qudsī, who had seen:
9. Muḥammad b. ʿAbd Allāh b. Ibrāhīm b. Mūsā, who had seen:
10. Muslim b. Ibrāhīm b. ʿAbd Allāh al-Makkī, who had seen:
11. Ḥumayd al-Ṭawīl, who had seen:
12. Anas b. Mālik (H-13), the companion of the Prophet Muhammad, who had been accepted as the scribe of:
13. The Prophet Muḥammad (A-18, B-3, H-14; follows anecdote).

Pedigree K (ʿIlm al-Ḥurūf waʾl-Wafq), *taken from*

[other *Shaykh*s of al-Būnī]

1. *Shaykhunā* Abū ʿAbd Allāh al-Qurashī (E-8, H-2), *min Aʿyān Mashāʾikh al-Gharb wa-Miṣr*, who took from more than six hundred Shaykhs (follows anecdote).
2. *Shaykhunā al-Imām al-ʿĀrif bi ʾllāh al-ʿAllāma* Abū ʾl-Ḥasan ʿAlī b. Muḥammad b. Ibrāhīm b. Muḥammad al-Ḥarrānī (follow anecdotes), who lived in Ḥamā, where he died in 538 AH.

After having mentioned these two Shaykhs the author gives a short epilogue, which concludes the *Shams al-maʿārif al-kubrā*. I will desist, at least for the moment, from a detailed analysis of all authorities mentioned in these pedigrees. A few general remarks are in place, however. There are several Maghāriba and Andalusians in the different pedigrees, and this shows that there is indeed a Western origin in the Corpus Būnianum. Yet there are Oriental pedigrees as well. From pedigree A, especially from the mention of several of the earlier Shīʿa imāms (given as Nos. 12–16, though for some with curious variants in their names), it might be surmised that the spiritual forebears of al-Būnī could also be found in the Shīʿa tradition. But the Shīʿa connection is clearly not the only one, and al-Būnī should not, on the basis of pedigree A only, be identified as a Shīʿite.[44] Pedigree B is an evident attempt to sunnify

[44] As is done by El-Gawhary, p. 14. The fact that al-Būnī's *Shams al-maʿārif* is mentioned in the *Dharīʿa* is probably based on the occurrence of the names of the early imāms in pedigree A, and cannot, therefore, be adduced (as El-Gawhary seems to do) as additional proof for a Shīʿite affiliation of al-Būnī. The names of the early Shīʿa imāms are also mentioned in the orthodox canon, and they are also mentioned in mainstream orthodox hagiographies such as al-Shaʿrānī's *Lawāqiḥ al-anwār*.

the Shīʿa connection. Other pedigrees, such as pedigree H, show that the spiritual forebears of al-Būnī can also be found in the orthodox Ṣūfī *milieu*, with perfectly Sunni credentials, also for the early period. The *ensemble* of the pedigrees may be interpreted as an attempt to make al-Būnī into an author for all people, both Sunna and Shīʿa, both in the East and in the West of the Islamic world.

Bibliographical note

1. *Manuscripts of works by al-Būnī*

Shams al-maʿārif wa-laṭāʾif al-ʿawārif

Berlin We. 1210 (Ahlwardt 4125). Undated, but the date of copying estimated by Ahlwardt as c. 1100/1688, copied by al-Ḥājj ʿAlī al-Shāmī.

Leiden Or. 336. Dated the end of Sunday 9 Jumādā II 857/1453, copied by Muḥammad b. ʿUmar b. Muḥammad b. Muḥammad b. al-Buqrāṭ al-Ḥamawī al-Shāfiʿī, in Cairo.

Leiden Or. 666 (1), on f. 1a–233b. Dated 853/1449–1450, copied by al-Ḥājj b. al-Ḥājj Sāsān al-Jīlānī.

Leiden Or. 736. Dated Saturday 20 Ṣafar 981/1573, copied by Yaḥyā b. Kamāl al-Dīn Muḥammad b. *al-marḥūm* ʿAbd al-Ḥaqq al-Sibāṭī al-Shāfiʿī al-Ashʿarī al-Muqriʾ.

Leiden Or. 1055. An extract (*khulāṣa*), undated but possibly a 15th-century MS (in view of the paper).

Leiden Or. 8371 (1) ff. 1a–89b, slightly lacunous. Dated Wednesday 14 Ṣafar 1057/1647, copied by al-Ḥājj Muṣṭafā b. al-Ḥājj ʿAlī, commonly known as Ibn Anjīr (or Abkhīr).

Tartīb al-daʿawāt fī takhṣīṣ al-awqāt ʿalā ikhtilāf al-irādāt

Leiden Or. 1233. Dated 19 Rajab 812/1409, (partially, according to the colophon) copied by Aḥmad b. ʿAlī al-Anṣārī.

2. *Selected manuscripts containing texts with a possible connection to work by al-Būnī*

Leiden Or. 947. *al-Fawāʾiḥ al-miskiyya fī ʾl-Fawātiḥ al-makkiyya* by ʿAbd al-Raḥmān b. Muḥammad al-Bisṭāmī (d. 858/1454), *GAL* G II, 231. Undated, but an owner's note by Muṣṭafa b. al-Ḥājj Yūsuf al-Fāṣilī, with seal print with date 1034 AH. Ewald Wagner, describing MS Berlin Or. oct. 3931, says about this text: 'Auf 100 Kapitel geplante, aber nur bis zum 30. Kapitel vollendete Enzyklopädie aus der Sicht mystischer and magischer Weltanschauung.'[45] According to Wagner the work is.mainly based on al-Būnī's *Shams al-maʿārif* and Ibn al-ʿArabī's *al-Futūḥāt al-makkiyya*. This remark, however, shows a certain lack of familiarity with al-Būnī's works.

Leiden Or. 7283. This is a composite volume copied in the middle of the 19th century, most probably in Palembang, Sumatra, Indonesia. It contains the Malay translations

[45] Ewald Wagner, *Arabische Handschriften*. Teil I, unter Mitarbarbeit von F.-J. Dahlmanns [...] beschrieben von Ewald Wagner. Verzeichnis der orientalischen Handschriften in Deutschland; Bd. 17 B 1 (Wiesbaden, 1976), p. 79, No. 92.

of two substantial magical texts in Arabic, together with some shorter notes. The second text in the volume is characterized by the cataloguer, Teuku Iskandar,[46] as an al-Būnī-like text. The name of al-Būnī does not seem to be mentioned in this text, however, and the link with the *Shams al-maʿārif* remains unsubstantiated by the text itself. The first text in the volume, however, is a work entitled *Shumūs al-anwār wa-kunūz al-asrār*. It is the the Malay translation of the Arabic work by that title by Ibn al-Ḥājj al-Tilimsānī al-Maghribī,[47] which contains mystical calculations, diagrams, *da'irah* etc., and is divided into thirty chapters.[48] This work does indeed offer some features which can also be encountered in works by al-Būnī, including a sequence on *ḥurūf*, *asmāʾ* and *āyāt* in the first three chapters: f. 2b. *Bab* 1. *Fī sirr al-ḥurūf*; f. 14a. *Bab* 2. *Pada menyatakan sekali khāṣṣiyyat Asmāʾ Allāh yang Ḥusnā*; f. 49a. *Bab* 3. *Fī khawāṣṣ baʿḍ al-āyāt*. Al-Tilimsānī's authorship of this text, whether genuine or spurious, demonstrates again that this type of magic is considered to come from the West, as is the case with al-Būnī, whose *nisba* refers to Bône, now ʿAnnāba, in Eastern Algeria.

3. *Printed editions of works by al-Būnī*

Būnī, Aḥmad b. ʿAlī al-, *Manbaʿ uṣūl al-ḥikma* (Cairo, 1951). This volume contains four works by al-Būnī and two related works by other authors. The works by al-Būnī in the volume are: *al-Uṣūl wa ʾl-ḍawābiṭ al-muḥkama, Bughyat al-mushtāq fī maʿrifat waḍʿ al-Awfāq, Sharḥ al-birhatiyya al-maʿrūf bi-Sharḥ al-ʿahd al-qadīm*, and *Sharḥ al-Jaljalūtiyya al-kubrā*. The two other works are: *al-Sirr al-maẓrūf fī ʿilm basṭ al-ḥurūf*, by Muḥammad al-Shāfiʿī al-Khalwatī al-Ḥanafī, and *al-Durra al-bahiyya fī jawāmiʿ al-asrār al-rūḥāniyya*, by ʿAlī b. Muḥammad al-Ṭandatāʾī al-Qārī. I used an undated reprint by al-Maktaba al-Thaqāfiyya of Beirut, made after the edition of Muṣṭafā al-Bābī al-Ḥalabī published in Cairo in 1370/1951 (a copy of the original edition is extant in the Leiden University Library, class-mark 8096 B 17).

Būnī, Aḥmad b. ʿAlī al-, *Shams al-maʿārif al-kubrā*. I used three editions, two from Cairo, the third published in India. 1. Edition published Muṣṭafā al-Bābī al-Ḥalabī (Cairo, 1 Rajab 1345/1927) [Leiden 8192 A 15, olim 862 D 90]. Apart from the *Shams al-maʿārif*, this volume contains four treatises by ʿAbd al-Qādir al-Ḥusaynī al-Adhamī: *Risālat mīzān al-ʿadl fī maqāṣid aḥkām al-raml, Risālat fawātiḥ al-raghāʾib fī khuṣūṣiyyāt awqāt al-kawākib, Risālat zahr al-Murūj fī dalāʾil al-burūj*, and *Risālat laṭāʾif al-ishāra fī khaṣāʾiṣ al-kawākib al-sayyāra*; 2. Edition published by Maktabat al-Jumhūriyya al-Miṣriyya, edited by ʿAbd al-Fattāḥ ʿAbd al-Ḥamīd Murād, Cairo 1380/1960. 3. Indian lithographed edition, possibly published in Bombay or Calcutta, dated 1291/1874 [Leiden University Library class-mark 8196 C 21].

Būnī, Aḥmad b. ʿAlī al-, *Sharḥ ism Allāh al-aʿẓam fī ʾl-rūḥānī, wa-yalīhi Kitāb al-lamʿa fī ʾl-fawāʾid al-rūḥāniyya ʿazīzat al-sumʿa* (Cairo: Maḥmūd ʿAlī Ṣubayḥ, n.d.).

---

[46] Teuku Iskandar, *Catalogue of Malay, Minangkabau, and South Sumatran manuscripts in the Netherlands* (Leiden, 1999), I, pp. 400–401 (No. 878).

[47] C. Brockelmann, *GAL* G II, p. 83 mentions this author as Muḥammad b. Muḥammad b. Muḥammad b. al-Ḥājj al-Fāsī al-ʿAbdarī al-Qayrawānī al-Tilimsānī, a scholar of *maghribī* origin, who died in Cairo in 737/1336. In *GAL* S II, p. 95 arguments are given for a different authorship of this popular magical compilation, of which many editions exist.

[48] See G.W.J. Drewes, *Directions for travellers on the mystic path. Zakariyya al-Ansari's* Kitāb Fatḥ al-Raḥmān *and its Indonesian adaptations with an appendix on Palembang manuscripts and authors* (The Hague, 1977), pp. 207–208.

Būnī, Aḥmad b. ʿAlī al-, *Shams al-maʿārif al-ṣughrā al-maʿrūf bi-Shams al-maʿārif wa-laṭāʾif al-ʿawārif al-ṣughrā*, ed. Abū Salāma al-Farīdī al-Falakī (Al-Dār al-Bayḍā: Maktabat al-Waḥda al-ʿArabiyya, 1424/2003). The editor signs in the introduction (p. 3) without the addition *al-Falakī* to his name. He adds to it: *al-Azhar al-Qāhira*. In his list of writings of al-Būnī (18 items) he postulates the existence of a work *Shams al-maʿārif al-wusṭā*, but he adds to it that he has never come across a manuscript of the text. The books proves to a newly typeset edition of a manuscript version, which had on its titlepage, in clear *naskh* script: *Kitāb Shams al-maʿārif wa-laṭāʾif al-ʿawārif lil-Shaykh al-Imām al-ʿAllāma al-Qudwa Muḥyī ʾl-Dīn Abī ʾl-ʿAbbās Aḥmad b. al-Shaykh al-Ajall al-Muqrī b. ʿAlī b. Yūsuf al-Būni raḥimahu ʾLlāhu taʿālā wa-naffaʿanā bihī*. From this is evident that the double *al-Ṣughrā* addition on the title-page is of the making of the editor, to whom it had not escaped that this was indeed a version of more modest size than the large editions. Copyist of the manuscript (or is it in fact a lithograph?) is Ḥasan al-ʿInānī al-Saʿdī. The manuscript seems to bear the date 1280 (1863–1864, see the reproductions on pp. 7–9).

A CURIOUS ARABIC TALISMAN[1]

Petra M. Sijpesteijn

In this paper I discuss an Islamic stone amulet that came into my possession as a present for my twenty-fifth birthday. Of unknown provenance, it was bought from a Dutch dealer in 1976 and is datable to the early twentieth century. I hereby dedicate it to Remke, who helped to introduce me to the special magic of the Arabic language.

The rectangular piece of cornelian is 3 cm wide, 2.5 cm high and 2 mm thick. Across its face concentric 'waves' ranging from whitish light yellow to dark reddish brown run from left to right, moving generally from lighter to darker colouring. The four sides of the stone have been cut so that it has a slight trapezoidal shape. Traces of some kind of adhesive material appear on the sides of the amulet and the stone is slightly damaged at the right bottom corner. There is a positive engraving on one face of the stone only.[2] The writing is cursive and most of the diacritical dots, the *shadda* and *hamza*, are written. The last two words on line 2 are repeated at the beginning of line 3 and repetition occurs again in lines 6 and 7, suggesting the engraver was copying his text from a transcript in front of him (Fig. 1, p. 210).

The carefully executed text on this small amulet consists of seven lines containing the *basmala* and the 'throne verse' (Qur'an 2:255). Under this text several 'figurae magicae' appear intended to fortify the power of the Qur'anic verse, as well as words, some written with numbers, and some individual letters written in isolation. Such a combination of different symbols is unusual in magical objects of this type, attributable probably to its late date.[3]

[1] I would like to thank Alexander Fodor, Emilie Savage-Smith, Gideon Bohak, Venetia Porter and Alexander Schubert for their comments. Needless to say, any errors that remain are mine.

[2] As opposed to a negative engraving used in seals which sometimes use the same symbolism and phrases (V. Porter, 'Islamic seals. Magical or practical?,' *University lectures in Islamic studies* 2 (1998), p. 135).

[3] Porter, pp. 142–143.

Moving from right to left, we first find two lunette sigla written under the Qur'anic verse, with two series of numbers beneath them: 5, 1, 1, 1, 5 and 5, 1, 2, 1, 1. Next appears a group of individual letters: the letter *sīn* five times and the letter *hāʾ* eight times. At the end of the line two more series of numbers, 5, 1, 1, 5 and 5, 1, 1, 5 under and separated by a double line appear.

On the register below, again starting from the right, there is a five-pointed star, or pentagram, under the lunette sigla. In each of the points of the star the number 5 is inscribed, and in the spaces between the points single Arabic letters appear; in the centre what seems to be the name 'Muḥammad' is repeated in a very cursive manner. Next to it is a six-pointed star, or hexagram, filled with numerals. In the centre the number 7 and in the points the numbers 1 through to 6 are written in such a way as to make each of the two opposing points add up to seven. At the end of the register we see a magic square (*wafq al-ʿadad*), divided into nine small squares filled with numbers 1 to 9 (except 4), with the number 5 in the middle.[4] The numerals 1, 3, 2, 6 occur twice written horizontally over the top left with two heart-shaped figures engraved above them and vertically adjacent to the bottom right corner of the magic square.

Some phrases and words are written under the pentagram and hexagram: *Makka mukarrama* (sic!), *mā shāʾa ʾllāh* and *tabāraka ʾllāh.*

Before delving into what all of this might mean, let us take a closer look at the amulet's text and drawings.

Edition of the Amulet

Text

1. بسم الله الر حمن الرحيم الله لا اله الا هو الحىّ
2. القيوم لا تاخذه سنة ولا نوم له ما فى
3. ما فى آلسموات وما فى الارض من ذا الذى
4. ليشفع عنده الا باذنه يعلم ما بين ايديهم

[4] The number five is often written in this manner, especially in texts from Northern India and Pakistan. It is also the way it is written in Urdu (Alexander Fodor, e-mail communication).

5. وما خلفهم ولا يحيطون بشى من علمه
6. الا ما شاءه وسع كرسيه السماوات وا لارض
7. وا لارض ولا يعوده حفظهما وهو العلى العظيم

8. مكة مكرمة ما شا الله تبارك الله

Translation

1. In the name of God, the Compassionate, the Merciful. God: there is no god but Him, the Living,
2. the Eternal One. Neither slumber nor sleep overtakes Him. His is what is in
3. what is in the heavens and the earth. Who can intercede with Him
4. except by His permission? He knows what is before
5. and behind them. They can grasp only that part of His knowledge
6. which He wills. His throne is as vast as the heavens and the earth,
7. and the earth and the preservation of both does not weary Him. He is the Exalted, the Immense One.

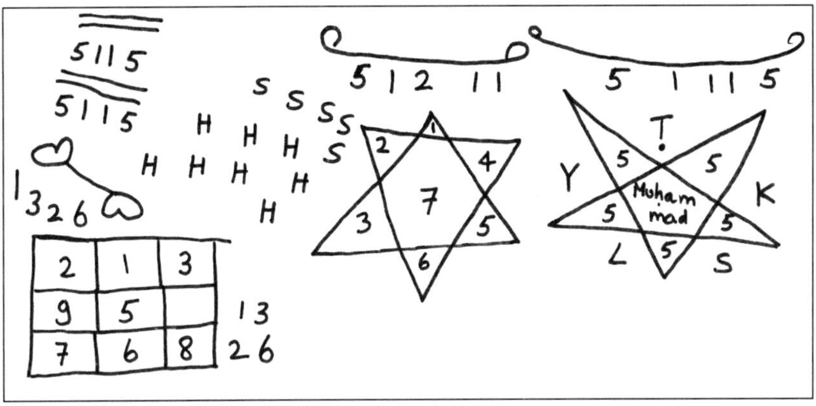

8. Venerable Mecca. Whatever God intend. God bless!

Commentary

2.–3. The last two words on line 2, *mā fī*, are repeated at the beginning of line 3.
3. and 6. *Al-samāwāt* is written with an *alif* rather than *alif superscriptum* as according to Qur'anic ductus.
6.–7. The last three words on line 6, *wa 'l-arḍ*, are repeated at the beginning of line 7.
7. *Ya'ūduhū* is written with *'ayn* instead of *hamza*.
8. Usually, the epithet of the city of Mecca has an article: مكة المكرمة.

Use and Meaning

On the sides of the stone we can still detect traces of a glue residue, presumably dating from when it was made, used to fasten the amulet into some kind of setting. The damage in the right bottom corner of the stone probably occurred when the stone was removed from this setting. The precious nature of the cornelian, the careful and precise stonecutting, and the meticulous execution of the writing suggest that this was an object of some value. The setting was probably made of precious metal such as silver, allowing the amulet to be worn both as adornment and apotropaic protection, in keeping with the many known examples of such amuletic jewellery.[5] Such objects functioned as a kind of never-ending prayer, a continuous repetition of invocations that conferred upon their wearers a defence against evil and misfortune that lasted as long as the material it was written on. The power of this automatic repetition of the prayer was reinforced by the magical nature of the symbols.

Magical drawings and many of the symbols that appear on Islamic amulets can be traced to pre-Islamic traditions. But unlike pre-Islamic amulets, which include demons and other 'dark' forces among their roster of invokees, Islamic amulets appeal exclusively to God, for aid and succour, for good fortune, and for general protection against the evil eye, devils and assorted demons.[6] Islamic amulets, moreover, are generally

[5] See for example L. Kalus, *Catalogue of Islamic seals and talismans* (Oxford, 1986), nos. UU 1.3; 1.18; 1.23; 1.28; 1.30; 1.32; 1.35; 1.36; 1.39; 1.42; A. Fodor, 'Amulets from the Islamic World,' *The Arabist. Budapest studies in Arabic* 2 (1990), no. 98.

[6] E. Savage-Smith, 'Introduction,' in E. Savage-Smith (ed.), *Magic and divination in early Islam*. The formation of the classical Islamic World; 42 (Aldershot, 2004), p. xvii.

aniconic, having as their sole decoration plant or geometrical motifs.[7] Their focus is their text, with a textual staple made up of Qur'anic quotations, prayers and invocations to God, the names of angels, the 99 'Beautiful Names of God' (*al-asmāʾ al-ḥusnā*), and other pious phrases, sometimes combined with magical symbols.

Turning to the different elements on the amulet and the magical and invocatory powers attributed to them, we should first discuss the *basmala* which was considered to be the Qur'an's most significant *āya* and to possess special powers.[8] The Qur'anic verse (2: 255) depicted on the amulet is also called *āyat al-mustaʿīdhīn* ('the verse of those seeking refuge'), *āyat al-mustaʿīnīn*, ('verse of those seeking help'), *al-āya al-muḥaṣṣina*, ('the fortifying verse'), *al-āya al-mukhrija lil-shayṭān*, ('the verse driving out Satan') and *sayyidatu āyi ʾl-qurʾān*, ('the mistress of the verses of the Qur'an').[9] It is found particularly often on amulets from all over the Islamic world, and because of its clear and simple wording, as well as its resonant message about the omnipotence of God, it was considered to have special protective powers.[10]

The hexagram and pentagram stars were called 'Seal of Solomon.' Often used to represent God's name in a logogram with the symbols which are written inside the stars, they were interchangeable and do not usually appear together on the same magical object.[11] The number five that is written in the five points of the pentagram was a favourite charm against the Evil Eye.[12] The letters surrounding the star's points are, clockwise starting from the right upper corner: *ṭāʾ*, *kāf*, *sīn*, *lām*, and *yāʾ*. Four

---

[7] Exceptions are ninth-century amulets on which a long-horned stag or oryx appears whose symbolism is unknown as well as early amulets and magic-medicinal bowls which depict animals such as scorpions, wild dogs and snakes, considered to cause sudden and inexplicable deaths (Savage-Smith 2004, pp. xix, xxiii; E. Savage-Smith, 'Magic and Islam,' in F. Maddison & E. Savage-Smith (eds.), *Science, tools and magic* (Oxford, 1997), I, pp. 135–137). From the modern period *zār* amulets have depictions of men and women, fish, soldiers, and mermaids, etc. (R. Kriss & H. Kriss-Heinrich, *Volksglaube im Bereich des Islam* (Wiesbaden, 1962), pp. 150–155; depictions pp. 115–122, 124; Fodor, nos. 288–294.

[8] For the special blessings and power attributed to the *basmala*, see W.A. Graham, 'Basmala,' in J. Dammen McAuliffe (ed.), *Encyclopedia of the Qurʾān*, I (Leiden, 2001), p. 211.

[9] T. Canaan, 'The decipherment of Arabic talismans,' *Berytus. Archaeological studies* 5 (1937), p. 75.

[10] Kalus, p. 48; E. Savage-Smith (1997), I, p. 59; Kriss & Kriss-Heinrich, pp. 61–62.

[11] H.A. Winkler, *Siegel und Charaktere in der muhammedanischen Zauberei* (Berlin [etc.], 1930), p. 119.

[12] E. Westermarck, *Ritual and belief in Morocco* (London, 1926), p. 142.

of these letters form the favourite demon name ṬYKaL, the 'lord of the sea.' As one of the 'four heads' (*al-arbaʿ ruʾūs*) this demon's name is normally found together with the other three 'heads' surrounding a magic square, but it appears also individually as in our text. Ṭykal is supposed to help one take revenge on one's enemies and was also considered to be one of the throne bearers.[13] The letters forming this demon's name can also be read as the numbers 9, 10, 20 and 30.[14] The *sīn* might symbolise the blessing *ʿalayhi ʾl-salām* following the mention of Muḥammad in the centre of the star.[15] Seven, figuring in the hexagram, has a special place in Islamic numerology (i.e. seven heavens, seven divisions of hell with seven gates, seven days of the week, etc.).[16]

The magic square is a simple three-fold grid, i.e. divided 3x3, making it the simplest of its kind and suggesting it is a so-called *budūḥ* square in which the sum of every horizontal, vertical and diagonal line is 15.[17] Already discussed extensively by al-Ghazzalī (d. 1111) as an amulet to help in childbirth, especially in difficult labour, this simplest of the magic squares was soon ascribed wider amuletic qualities and became also the most popular one. Associated with the planet Saturn, it was thought, among its many other properties, to relieve headaches, temporary impotence and menorrhagia, and to ensure the safe arrival of letters and packages. It could also make you invisible.[18] It has retained its popularity, appearing on amulets from the earliest to the latest period.[19]

The numbers 1 to 9 as read in our square, however, do not coincide with one of the eight ways in which these numbers can be arranged to add up to 15.[20] Only in the two diagonals, the numbers do add up to 15. The horizontal and vertical rows do not add up to 15 and the middle square in the right vertical is left blank. Did the amulet engraver detect his mistake and decide not to finish the square? Squares like ours in which the numbers do not seem to have an obvious order or meaning do appear on amulets, but whether these are corrupted forms of the true

[13] Canaan, pp. 84–85 and n. 93.
[14] Canaan, p. 159.
[15] I would like to thank Alexander Fodor for his help in interpreting these letters.
[16] Westermarck, p. 142.
[17] D.B. Macdonald, 'Budūḥ,' in *Encyclopaedia of Islam* (New ed.), Supplement (Leiden 2004), pp. 153–154; Savage-Smith 1997, p. 106.
[18] Macdonald, pp. 153–154.
[19] Kriss & Kriss-Heinrich, pp. 84–87; W. Ahrens, 'Studien über die magischen Quadrate der Araber,' *Der Islam* 7 (1917), pp. 198–221.
[20] Canaan, pp. 100–103. W. Ahrens, pp. 188–194.

magic squares, random orderings of numbers or have some other significance to which we have no access is impossible to determine.[21]

The other symbols appearing on our amulet, the individual letters and numbers under the lunette sigla, are harder to interpret. Islamic magic drew extensively upon the supposed occult properties of letters and the divine and angelic names they form. The *sīn* and *hāʾ* both belong to the 'fire letters,' which were supposed to ward off evils connected with cold and, where necessary, generate heat.[22] Individual letters like these appear in groups on amulets mostly written on paper and, in a continuation of the pre-Islamic practice, need to be read out loud to take effect.[23] The four series of numbers stubbornly resist decipherment, but might be a simplified form of the seven signs appearing together, sometimes in combination with lunette figures and probably represent names of God or other pious words.[24] It does not seem possible to read any words in them using magical alphabets.[25]

Besides the Qur'anic verses and magical symbols, the writing material itself, in this case the stone used for the amulet, was thought to have specific functions and address particular problems. Cornelian or agate (*ʿaqīq*) came to the Middle East from the Yemen and India and was especially valued for its magical properties. Even the prophet himself is said to have approved its use in rings, and the fourth caliph and son-in-law of the prophet ʿAlī (r. 35–40/656–61) allegedly had a ring with an agate stone.[26] In general the stone is thought to protect against poverty and give happiness, and more specifically to guard travellers from

[21] E.g. Fodor, no. 162. Cf. Ahrens, pp. 246–248. Canaan considers these 'errors in copying' (p. 160). Cf. Winkler, p. 150.

[22] T. Fahd, 'Ḥurūf ('ilm al-),' in *Encyclopaedia of Islam* (New ed.), III (Leiden, 1971), pp. 595–596.

[23] E.g. Kriss & Kriss-Heinrich, p. 103, depiction 99; Fodor, no. 133. A magical formula including a series of individual *hā*'s connected on the left side as on our amulet appears also in al-Būnī as cited by Winkler (p. 120). For pre-Islamic *voces magicae*, where the 's' and 'h' can, however, not be found, see W.M. Brashear, 'The Greek magical papyri. An introduction and survey. Annotated bibliography (1928–1994),' *Aufsteig und Niedergang der römischen Welt* II 18/5 (1995), pp. 3380–3684. I would like to thank Chris Faraone for referring me to this article.

[24] Canaan, p. 144. See note 22 for Arabic magical alphabets. For an example of a modern paper amulet containing also a series of number fives and ones, see e.g. Fodor nos. 133, 149.

[25] Canaan, pp. 102–108; M. Casanova, 'Alphabets magiques arabes,' *Journal asiatique* 19 (1922), pp. 250–262.

[26] Porter, p. 145.

the annoyance of beggars and dervishes, ward off accidents, shut the mouths of false witnesses, protect one in battle and heal eye diseases.[27] Cornelian is still a popular component of jewellery, used for beads, ring stones and other jewellery and it appears especially often in amulets.

Finally, it remains to address the dating and provenance of our object. Pentagrams and hexagrams, which did not become popular until the twelfth and thirteenth century respectively, give us a rough date *post quem*.[28] More can be gleaned from the fact that in our amulet these depictions appear collectively, together with the Qur'anic verses and individual letters and numbers used on magical objects. This mix-and-match approach to magic elements usually appears only on later objects.[29] Finally, we can probably interpret the numbers appearing twice around the magic square as a date, 1326, bringing us into the twentieth century to 1906. This seems to be supported by the style and writing of the amulet. While it is not common for amulets to carry dates, it does occasionally occur.[30]

The words appearing at the bottom of the amulet, suggest that this was a pilgrimage souvenir, ordered to commemorate a visit to Mecca, since this is an inscription shared by other amuletic pilgrimage souvenirs (*makka al-mukarrama*).[31] The ductus of the writing, however, points to a Pakistani or Northern Indian origin for the amulet.[32] The lack of the article in the expression *makka mukarrama* confirms a Persian-speaking background for the engraver. This supposition that the engraver was not a native Arabic speaker is strengthened by the mistake made in the Qur'anic inscription in line 7. How can we reconcile these two contradicting geographical references? On our amulet the words referring to

---

[27] B.A. Donaldson, *The wild rue. A study of Muhammadan magic and folklore in Iran* (London, 1938), p. 152; J. Hell, "Aqīq,' in *Encyclopaedia of Islam* (New ed.), I (Leiden, 1960), p. 336.

[28] But they do already appear on ninth-century amulets. See for example J. Karabacek, *Führer durch die Ausstellung Papyrus Erzherzog Rainer* (Vienna, 1894), no. 840.

[29] Kriss & Kriss-Heinrich, p. 59.

[30] Kalus, depictions II 1.26 (dated 1121/1709–10) and II 2.3 (III) (dated 1078/1667–8).

[31] Kriss & Kriss-Heinrich, p. 53; depiction 53 figure 2. See also one mentioning Medīna in figure 1. Also in Fodor no. 93. For a possible Shi'ite example, see the 20th-century glass amulet with a depiction of ʿAlī's mosque in Najaf (Fodor, no. 247).

[32] See the inscription on the magic bowl published by A. Fodor, 'A magic bowl from India in the Tareq Rajab Museum,' in P.L. Baker & B. Brend (eds.), *Sifting sands, reading signs. Studies in honour of Professor Géza Fehérvári* (London, 2006) pp. 187–198.

Mecca and the extra religious phrases at the bottom of the stone are written in a cramped way, and were perhaps added as something of an afterthought after the rest of the inscription was completed. We can thus posit an amulet brought by a pilgrim from India or Pakistan that was made into a pilgrimage souvenir by adding the phrase referring to Mecca, either executed by an engraver in Mecca with whom the pilgrim could communicate in his mother tongue or completed after his return to his home country. The amulet's efficacy would have been greatly enhanced by the visit to the holy city.

One final question remains: is it possible to uncover the real meaning behind the magic square, the numerical series, the pentagram and the hexagram, or were these figures drawn for their symbolic and visual properties only? And this leads us to the larger question of how we should understand the magical and invocatory power of magic symbols when they do not seem to make sense? The same question has been raised for amulets covered in pseudo-writing. How did such objects obtain their invocatory force; were they made by or for analphabetic users? In other words, do we have to do here with a kind of graphical version of 'abracadabra' – a formula whose magical qualities have far outlived its original sense, or perhaps whose power has even been enhanced by the loss of its original verbal moorings? Or was this amulet, more prosaically, merely produced by an incompetent engraver for a customer who did not know any better him or herself? For the moment any answers to these questions must remain speculation. It is certain, however, that someone assigned this object power enough to merit it being mounted and worn, and that power continues to beguile.

Fig. 1 Cornelian amulet with inscription.

HISTORY

THE VICISSITUDES OF TIME
A CONTEST BETWEEN BLACK AND WHITE

Pieter Smoor

On the evening preceding his death, Ḥusayn the grandson of the Prophet was in a very black mood following his unfortunate journey to the city of Kūfa, where he had been stopped by the generals and supporters of the Dynasty of Muʿāwiya in the vicinity of a relatively unimportant village named Karbalāʾ. As a result of this incident, Ḥusayn and his travelling entourage had been unable to quench their terrible thirst with the waters of the River Euphrates. Ḥusayn's feeling of melancholy was expressed in a few lines of poetry, the full significance of which penetrated with sharp clarity into the sensitive mind of Ḥusayn's sister Zaynab:

> 1. Woe unto you, oh Dahr, oh Destiny! What kind of friend are you? In the prime of the morning and the lateness of the evening how many have been killed
> 2. From among the companions and the seekers of knowledge? For the Dahr is not content with a substitute.
> 3. The Elevated One is especially suited to the issuing of commands; and every living being must walk his appointed path.

Hearing this, Zaynab jumped up, ripped open the neck of her gown and began to lament. As the *Tārīkh* of al-Yaʿqūbī informs us, she cried: 'Woe unto those who cause the death of my children and family. Oh, would Death but take away my life from me! Now Fāṭima has died and ʿAlī [her parents], and my brother Ḥasan, the son of ʿAlī!'[1]

---

[1] See Aḥmad b. Abī Yaʿqūb [...] al-maʿrūf bi ʾl-Yaʿqūbī, *Tārīkh al-Yaʿqūbī* (Beirut, 1960), II, p. 244; Abu ʾl-Fidāʾ Ibn Kathīr, *al-Bidāya wa ʾl-nihāya* (Riyad [etc.], 1966), VIII, p. 177. Al-Yaʿqūbī died in 292 AH, see M.Q. Zaman, 'al-Yaʿḳūbī,' in *Encyclopaedia of Islam* (New. ed.), XI (Leiden, 2001), p. 257. ʿImād al-Dīn Ismāʿīl b. ʿUmar Ibn Kathīr died in Damascus in 774 AH, see H. Laoust, 'Ibn Kat͟hīr,' in *Encyclopaedia of Islam* (New ed.), III (Leiden, 1971), p. 817.

This article is chiefly concerned with the literary and historical view of the contest between black and white as given by al-Yaʿqūbī and Ibn Kathīr; more details from other sources are also given, for example part of J. Wellhausen, *Das arabische Reich und sein Sturz*. 2nd unchanged edition [pref. R. Hartmann] (Berlin, 1960), pp. 306–352, ch. 9; also W. Madelung, 'The Sufyānī between tradition and history,' *Studia Islamica*

To learn what else happened to Ḥusayn, the reader is referred to the numerous relevant historical reports. Only a few points are of interest to us now, in connection with the subject of this study. The prognostication that the result of Ḥusayn's expedition would be disastrous was foretold by an archangel to Ḥusayn's grandfather, the Messenger of Allah. He then informed Umm Salama, one of his wives, about this future development and handed over to her a particularly strange implement, a jug full of sand. He explained that his instructions had come from the Archangel Gabriel himself:

> Gabriel informed me that Ḥusayn would be killed by my followers, my *umma*. The angel gave me this sand whilst saying: "When this changes into fresh blood, then you will know that Ḥusayn has already been killed." The jug filled with sand remained at her home; later, when the moment had come, she started looking at the contents of the jug and behold, the sand had already turned into blood. She cried out lamenting: "Oh Ḥusayn! Oh son of the Messenger!" Thereupon all of the town of Kūfa resounded with the cries and lamentations of the women.[2]

Umm Salama saw that the (yellow) colour of the sand had been changed to red. This implied that innocent blood had been shed in Karbalāʾ, the village which would later become notorious, for the colour red would not remain restricted to the contents of Umm Salama's jug.

Centuries later Shīʿites would remember what had happened in Karbalāʾ with feelings of intense sorrow and the colours of dawn and dusk would become a daily reminder of Ḥusayn's death. This can be deduced from a passage in a dirge dedicated to a deceased member of the descendants of the family of the Prophet, the *sharīf* Abū Ibrāhīm, who died in the 4th century or at the beginning of the 5th century AH. Even at this relatively advanced moment in time, the poet was able to portray the moment with remarkable poignancy:

> 21. Because of the blood of the two martyrs, ʿAlī and his son, two witnesses will appear to the end of time.

---

63 (1984), pp. 5–48. However, I was unfortunately not able to lay hands on M. Sharon, *Black banners from the East* (Jerusalem, 1983). For subsequent changes of fortune between Black and White, Sunna and Shīʿa, and the struggle for power between the viziers, see Pieter Smoor, *Wazāra the killer of many husbands*. Publications de l'Institut Français d'Archéologie Orientale du Caire (Cairo, [*c.* 2007]).

[2] *Tārīkh al-Yaʿqūbī*, II, pp. 245–246; Ibn Kathīr, *Bidāya*, VIII, pp. 199–201.

22. For they are the two aspects of dawn, the real one and the false one at the end of the night; at nightfall however, they are the two aspects of dusk, the red one and the white.
23. These have both been embroidered upon the shirt of time, until time shall attain the day of Resurrection, while beseeching protection from the Merciful One.[3]

After the death of ʿAlī and his son Ḥusayn, the Umayyad clan began their reign. The relations which the Umayyad princes entertained towards the members of the clan of the Prophet, vacillated between fierce enmity and a desire to repair the injustices perpetrated against them.

For instance, the first Umayyad to reign, Muʿāwiya, did not feel anything like repentance; but he did have a sense of grief when he felt that the end of his life had almost come. However, it had nothing to do with the many victims among his opponents from the clans of ʿAlī and of ʿAbbās, both of whom were related to the Prophet. It is interesting to note that the colour which Muʿāwiya used to manifest his feeling of grief was the colour black. Thus we can read the following passage in the *Tārīkh al-Yaʿqūbī*:

> I saw Muʿāwiya wearing a black turban, which he had let down till just above his mouth. He did so, whilst speaking the following words to the assembled crowd: "Oh you people, my age has now become advanced and my strength is seriously weakened. I have been stricken in the best part of my life, may Allah have compassion upon the one who shall pray on my behalf." Having said this, he wept.
>
> Not long after this, a servant of Muʿāwiya laid out his funeral shrouds over his preaching chair. He said: "Muʿāwiya was the Canine Tooth of the Arabs and he was their Strong Rope, but now he has died. These are his funerary shrouds in which we have to wrap him and we shall make him descend into his grave. That will be the last encounter."[4]

As the years went by the situation changed, although the colour black remained of importance. Indeed, a certain sense of loss and mourning was to prevail again, but this time it overcame the people of the clan of ʿAlī b. Abī Ṭālib and those of the clan of ʿAbbās b. ʿAbd al-Muṭṭalib.

---

[3] From the poem to rhyme *bi-fānī*, composed by Aḥmad b. ʿAbd Allāh Abū ʾl-ʿAlāʾ al-Maʿarrī, *Shurūḥ Saqṭ al-zand*, ed. Ṭāhā Ḥusayn (Cairo, 1945–1948), I, pp. 441–442.

[4] *Tārīkh al-Yaʿqūbī*, II, p. 239; Ibn Kathīr, *Bidāya*, VIII, pp. 142–143: 'Muʿāwiya was the Wall and Support of the Arabs and their Good Fortune; and Allah has cut off civil war through him, has made him king over His servants and has opened the lands through him; but now he has died.'

The sufferings of some of the members of the ʿAlī clan were more apparent in the eyes of the wider public and therefore we must go into detail concerning the clan of ʿAlī and that of ʿAbbās. Due to the uncouth and merciless behaviour of the Umayyad princes and governors, certain descendants of the Prophet's family would end their lives in miserable circumstances, and this would awaken a sense of grief and even revenge amongst the two factions.

As far as the ʿAlī clan is concerned, we find quite a profusion of forced endings to their lives described in the *Tārīkh al-Yaʿqūbī*. For instance, there is a report about the evil deeds of Sulaymān b. ʿAbd al-Malik who organised his secret agents to set up a line of tents in the desert, with the sole purpose of killing a member of the ʿAlī clan offering him a cool draught of poisoned milk! The tired desert traveller whom they knew would ride past their tents was none other than the eminent ʿAbd Allah Abū Hāshim b. Muḥammad b. ʿAlī b. Abī Ṭālib:

> Having at first refused politely: "May Allah give you recompense, thank you," he travelled farther and farther. But finally overwhelmed by the desert heat, he gratefully accepted the offered drink. However, after the milk had anchored itself in his stomach, he started to cry out repeatedly: "I am as good as dead (*anā mayyit*)!"[5]

His troubled condition did not prevent him from organising his testament and empowering an uncle of his, a member of the rival clan of the ʿAbbās family, to dispose of all his assets pertaining to the ʿAlī clan, the direct descendants of the Prophet. Having handed to him the written document, the *ʿahd*, which he owned from his forefathers, he even

[5] According to a short notice by Ibn Kathīr, *Bidāya*, X, p. 40, the poisoned milk had been offered to Ibrāhīm b. Muḥammad, a member of the ʿAbbās clan. See further, *Tārīkh al-Yaʿqūbī*, I, p. 296. Another, less graphic account is found in Abū ʾl-Faraj al-Iṣfahānī, *Maqātil al-Ṭālibiyyīn*, ed. Aḥmad Ṣaqr (Beirut, n.d.), p. 126, under the heading: ʿAbd Allāh b. Muḥammad b. ʿAlī; also see for Abū Hāshim H. Brentjes, 'Die Imamatslehren im Islam nach der Darstellung des Aschʿarī,' *Abhandlungen der Sächsischen Akademie der Wissenschaften zu Leipzig. Phil. hist. klasse*, Bd. 54, Heft 5 (Berlin, 1964), p. 20 and the reference to Abū ʾl-Ḥasan ʿAlī b. Ismāʿīl al-Ashʿarī, *Kitāb Maqālāt al-islāmiyyīn wa ʾkhtilāf al-muṣallīn*, ed. H. Ritter (Wiesbaden, 1963), p. 20. Ibn Kathīr, *Bidāya*, X, p. 5, states that it was ʿAbd Allah b. Muḥammad the son of the Ḥanafiyya (i.e. Khawla bint Jaʿfar, whom ʿAlī had married in addition to his wife Fāṭima) who had handed over the full rights pertaining to the ʿAlī clan to the representative of the ʿAbbās clan (who was incidentally the future first ʿAbbāsid Caliph) al-Saffāḥ Abū ʾl-ʿAbbās, ʿAbd Allāh b. Muḥammad, to whom he announced the following fortunate message: 'The Caliphate will inherit into the possession of your descendants.'

found time to explain by word of mouth the stipulations in it, dying only after all the details had been considered.

The enmity between the Umayyads and the Prophet's descendants had other, less obvious effects. Some victims are mentioned again, such as the wife of al-Mukhtār the Propagandist of certain rights of the ʿAlī clan. For this see the text of al-Yaʿqūbī, which we quote here:

> 'Asmā' the daughter of al-Nuʿmān b. Bashīr was arrested. She was the wife of al-Mukhtār. Thereupon the Umayyad Governor Muṣʿab b. Zubayr said to her: "What do you say concerning al-Mukhtār?" She answered: "I say that he was godfearing, pure and someone who fasted frequently." He said upon this: "Oh you enemy of Allah, you belong to the people who dare to declare him pure!" Thereupon he gave an order to someone to cut her neck and so she became the first woman whose head was severed without so much as a trial. The poet ʿUmar b. Abī Rabīʿa said the following lines concerning this:
>
> 1. In my opinion, the most startling thing was the killing of a fair long-necked woman (*ʿuṭbūl*), noble and free,
> 2. She was killed without having committed any crime. Even as a victim she was still an excellent person.
> 3. For us, killing and fighting is ordained but the only obligation of a beautiful woman is the wearing of a train of long skirts.[6]

Perhaps the decease of another descendant from the ʿAlī clan must have made more impression upon people's minds, especially in view of the Umayyads' treatment of the victim's dead body. The corpse of Zayd b. ʿAlī, the grandson of Ḥusayn, was first burned, then the ash was spread over the agricultural lands and the waters of the river Euphrates. This was organised by an Umayyad dignitary in order to humiliate the Shīʿa population of Kūfa, who were forced into eating and drinking the powdered remains of an almost holy member of the ʿAlī clan. See the text of the *Tārīkh al-Yaʿqūbī*:

> Zayd b. ʿAlī had been killed. He was transported, borne upon the back of a donkey. His head was set up on top of a long pole. After this he was gathered together, head and rump, and burnt and his ashes strewn around, one half of him over the Euphrates and the other half of him spread out

---

[6] *Tārīkh al-Yaʿqūbī*, II, p. 264, also discussed in H.D. van Gelder, *Mohtar de valsche profeet* (Leiden, 1888), pp. 140–141. According to Ibn Kathīr, *Bidāya*, VIII, p. 289, her name was ʿAmra bint al-Nuʿmān.

> over agricultural land. The Umayyad Governor Yūsuf b. ʿUmar said: "By Allah, Oh inhabitants of Kūfa, I shall certainly ensure that you will eat him in your food and drink him in your water."

The murder perpetrated on Zayd b. ʿAlī occurred in the year 121 AH.

> When Zayd was killed as happened in the course of things, the Shīʿa partisans started to move. Their affair became visible and many people came towards them and were attracted to them. They began to mention the deeds of the Banū Umayya and what they had perpetrated against the clan of the Messenger of Allah. This continued until not a single town remained except that this information was spread inside it. The Dāʿī Propagandists started to come out in the open and dream visions were seen and the books of the legendary battles (*malāḥim*) became subjects of study.[7]

In the meantime the rival ʿAbbās clan could boast of its own martyrs. The violent death of Ibrāhīm b. Muḥammad b. ʿAlī b. ʿAbd Allah b. ʿAbbās made an especially deep impression on the minds of the anti-Umayyad population. Marwān, al-Ḥimār, the last Wild Ass at the end of the Umayyad line of kings, was responsible for the bloodless way of killing Ibrāhīm. He wrapped a velvet piece of cloth around his head or inserted his head into a bag made from sheep's leather (*jirāb*). Such bags, incidentally, were intended for the storage of finely powdered charcoal (*nūra*) which was kept for medicinal purposes. A comment on this sad event is given by a certain poet called Ibn Harma:

> 1. I considered myself a hardened person; what weakened me was a grave in Ḥarrān where the Sinlessness of Religion was laid.
> 2. It robbed everyone, both rich and poor; therein lies the Imām whose misfortune was all-embracing.[8]

---

[7] *Tārīkh al-Yaʿqūbī*, II, p. 326. But according to Ibn Kathīr, *Bidāya*, IX, pp. 329–331, it happened in 122 AH that Zayd b. ʿAlī was killed. For Zayd b. ʿAlī, see also C. van Arendonk, *De opkomst van het Zaiditisch imamaat in Yemen* (Leiden, 1919), pp. 31–33; Brentjes, 'Darstellung,' p. 13 and the reference to *Maqālāt*, ed. Ritter, p. 65 and p. 78.

[8] For the circumstantial context of the poem quoted, see *Tārīkh al-Yaʿqūbī*, II, p. 342 and see Ibn Kathīr, *Bidāya*, X, pp. 39–40, where one finds that Ibrāhīm b. Muḥammad had been killed by the Umayyad agents, who covered his face with a *muraqqaqa*, a thin piece of dough i.e. a thin sheet of uncooked pancake. Or perhaps he was the one whom they had killed by offering the poisoned milk mentioned earlier (The reader is reminded of the Biblical tale in II Kings 8, 15: 'And it came to pass on the morrow, that he [Hazael] took a thick cloth, and dipped it in water, and spread it on his [Ben-hadad's] face, so that he died: and Hazael reigned in his stead.' – thanks to Juut Bruins Slot). The poem was most probably composed by the contemporary Ibrāhīm b. ʿAlī b. Salama b. Harma Ibn Hudhayl who died about 150 AH; for his poetry and biography see Abū ʾl-Faraj al-Iṣbahānī, *Kitāb al-Aghānī*. 2nd pr. (Cairo, 1963), IV, pp. 367–397; *Das biographische*

The time had come for a concerted and effective action against the Umayyads who had dealt so evilly with the Prophet's family, albeit with a few exceptions.

The revolutionary action started successfully in the farflung province of Khurāsān, where the Propagandist and supreme leader Abū Muslim had chosen the colour black to express his sorrow and feeling of mourning about the misfortune which had again struck the person of the Imām. This can be ascertained from reading Dīnawarī's *al-Akhbār al-ṭiwāl*, where we find the following:

> The moment had come which Abū Muslim had appointed with his followers. All of them set out towards him on one definite day from all the villages of Khurāsān until they had come to him. They had put on black clothes in order to comfort themselves for the loss of Ibrāhīm b. ʿAlī b. ʿAbbās (from the ʿAbbās clan not from the ʿAlī clan).[9]

Marwān the last of the Umayyad line had been responsible for this murder.

In this incident and in the other incidents reported in the different historical sources we very frequently come across the verb *sawwada* i.e. to make black and the substantive form *sawād* with meanings like black clothes, flags or banners, next to the name for the region of this same name, the region called Sawād which surrounds Kūfa.

In al-Ṭabarī, *Tārīkh al-rusul wa ʾl-mulūk*, we find an interpretation of the meaning of the colour black which goes beyond its signification of mourning, *ḥidād*, as al-Ṭabarī informs us of its importance as an indication of a pro-ʿAbbāsid view, explaining:

> When it had become Thursday 25 Ramaḍān in the year 129 AH, they took the flag (*liwāʿ*) which the Imām [Ibrāhīm, or his father Muḥammad, both of whom were descended from ʿAbbās] had sent to them and they fastened it to a long pole or lance of 14 ells. They called the flag by the name of Shadow (*al-ẓill*). The standard (*rāya*) which the Imām had likewise sent received from them the name of Cloud (*al-saḥāb*) and it was fastened to another pole somewhat smaller in length, being only 13 ells long. Open war was declared against the Umayyads when Abū Muslim started to recite a verse from the Qur'an, namely Sūrat al-Ḥajj 22 verse 39, which says:

Lexikon des Ṣalāḥaddīn Ḫalīl ibn Aibak aṣ-Ṣafadī, al-Wāfī bi ʾl-wafayāt, VI, ed. S. Dedering (Wiesbaden, 1972), pp. 59–60. For Ibrāhīm b. Muḥammad, see also Brentjes, 'Darstellung,' p. 21 and the reference to *Maqālāt*, ed. Ritter, p. 21.

[9] Abū Ḥanīfa Aḥmad b. Dāwud al-Dīnawarī, *al-Akhbār al-ṭiwāl*, ed. ʿAbd al-Munʿim ʿĀmir & Jamāl al-Dīn al-Shayyāl (Cairo, 1960), p. 360.

> Permission is granted to those who fight because they have suffered wrong. Verily Allah is able to help them.[10]

Whilst he was reciting this verse, Abū Muslim had put on black clothing (*sawād*). But of more interest to us here is the esoteric meaning of the two words naming the flag and the standard which are suddenly mentioned by al-Ṭabarī in the course of this passage:

> The esoteric interpretation (*ta'wīl*) of The Shadow and The Cloud is as follows: The Cloud covers the earth and this is brought about by the Propaganda of the Banū ʾl-ʿAbbās; the implication of The Shadow is that the earth will never be empty of an ʿAbbāsid Caliph, through all eternity till the end of Time, and of the Dahr.[11]

A change of fashion: from ʿAbbāsids to Fatimids

Fashions change, however. The eternity of the black Shadow soon appeared to be confined within a relatively short period of time, because in the year 358 AH, the contrasting colour white appeared in Egypt which manifested itself in flags and standards. Nevertheless, this colour had actually made an earlier appearance during the time of Abū Muslim's arrival from the East. There had been some sporadic resistance from pro-Umayyad factions against himself and his ʿAbbāsid Imām Abū ʾl-ʿAbbās 'Ibn al-Ḥārithiyya'. In the *Tārīkh al-Ṭabarī* for instance we read about one of the last generals of Marwān who was known as Abū ʾl-Ward (Majzaʾa b. al-Kawthar b. Zufar b. al-Ḥārith al-Kilābī). He offered fierce resistance to the black banners and standards which came like a flood from all over the East. Al-Ṭabarī explains in a vague way why the colour white was preferred by people like Abū ʾl-Ward. In his *Tārīkh* he says:

> The reason that he practised wearing white clothes and flags stemmed from a fear of losing his life and the lives of the members of his clan.[12]

---

[10] Abū Jaʿfar Muḥammad b. Jarīr al-Ṭabarī, *Tārīkh al-rusul wa ʾl-mulūk [Tārīkh al-Ṭabarī]*, ed. Muḥammad Abū ʾl-Faḍl Ibrāhīm (Cairo, 1977), VII, p. 356; Ibn Kathīr, *Bidāya*, X, p. 30.

[11] *Tārīkh al-Ṭabarī*, VII, p. 356; the same esoteric interpretation is found in Ibn Kathīr, *Bidāya*, X, 30. Compare Sūrat al-Baqara II, verse 210: 'What prospect have they but that Allah will come to them in the shadow of the clouds with the angels; the affair is decided, and to Allah return (all) affairs.' (transl. R. Bell)

[12] *Tārīkh al-Ṭabarī*, VII, p. 443.

There appears no clearer argument for the choice of White by the opposition parties who wanted the Umayyad reign upon the throne of the Caliphate to continue.

The Shīʿa preference for White became apparent when the Fatimid al-Muʿizz ordered his Chancellor, a Slav *mamlūk* who served as a general of his, namely Abū ʾl-Ḥusayn Jawhar al-Ṣiqillī, to attack Egypt on behalf of the Fatimids and to do so under the aegis of white banners and standards and flags. The reason for this is not immediately clear but it might have been a reaction to the contrasting colour black.

Another explanation for the importance of White is that it was the colour of shrouds worn by the dead. This was already pointed out by Henri Pérès, *La poésie andalouse en arabe classique*, when he refers to a remarkable passage in a poem by the Andalusian pro-Fatimid eulogist of the Fatimid Imām al-Muʿizz, the commander and master of Jawhar.[13] The writer of this poem, Muḥammad Ibn Hāniʾ, had his origins both in Spanish Seville and in the old Fatimid capital on the seashore of present day Tunis, the city of al-Mahdiyya. In the poem to rhyme *rikābā* the poet describes his love for some ladies travelling in camel chairs. He says their departure makes him so sad that he tries to find suitable colours to express his sadness and grief because of their disappearance. Here we find the following lines about mourning and the colour of White which was used in al-Andalus and particularly in the city of Cordoba:

> 6. You have now gone away. Had I not tried in vain to alter my hair colour and had I not encountered you in a moment when you were filled with anger towards me,
> 7. Then I would have painted the colour white over my side whiskers in a deceitful way and I would thus have wiped away the black hairs of youth.
> 8. And then I would have removed that colour just as a sword belt is tossed aside, in a manner which people disapprove of; and then I would have exchanged the mantle of youth for another mantle, another *jilbāb*.
> 9. And then I would have painted, using the white colour of mourning – provided that I had been able to find such a thing as the colour white.

---

[13] H. Pérès, *La poésie andalouse en arabe classique* (Paris, 1953), pp. 289–303: 'Mais l'on est surpris de relever de nombreuses attestations qui prouvent que le costume de deuil en Espagne, à des époques et en des régions que nous allons essayer de préciser, était blanc.' In 19th-century Egypt, the colour white was still preferred for mourning, compare E.W. Lane, *An account of the manners and customs of the modern Egyptians. The definitive 1860 edition*, intr. J. Thompson (Cairo, 2003), p. 513: 'The colours most approved for the grave-clothes are white and green; but any colour is used, except blue, or what approaches to blue.'

> 10. Whenever you desire to attain the colour grey you should take the centuries as your steeds on the road towards it.
> 11. Then, from Time, you should take a white dove; and to Time, you should then send your black raven [your hair].
> 12. But what shall I say to Time in respect of its treachery in uniting my enemies and dispersing my friends?[14]

However beautiful these lines may sound, the use of white in the practice of army life was more important. In 358 AH when the Fatimid general Jawhar succeeded in occupying Egypt in a relatively peaceful manner, we suddenly see White on the flags, standards, and clothing.[15] More impressive is how quickly the colour black was dismissed from the official cloaks of the *faqīh*s and *qāḍī*s in Egypt. The disappearance of Black is exemplified in the story which refers to Jawhar and how he laughed when Black was humiliated in an official meeting. In al-Maqrīzī, *Itti'āẓ al-ḥunafā* we read the following:

> And the qāḍī ʿAbd Allah b. Ṭāhir al-Ḥusaynī entered the hall of Jawhar, while he was clothed in a darkish official cape (*ṭaylasān kuḥlī*). In the Salon there were judges, scholars and assistants of the judicial court (*qāḍīs*, *ʿulamāʾ* and *shuhūd*), thereupon he, Jawhar, disapproved of the darkish cape and stretched out his hand and tore the cloak off. Thereupon Ibn Ṭāhir grew angry and spoke angry words, but Jawhar gave the order to shred the cloak to pieces and it was indeed torn to pieces, whilst Jawhar laughed. The victim remained without any covering for his head and

---

[14] *Dīwān d' Ibn Hānī al-Andalusī (augmenté de pièces inédites)*, ed. Muḥammad al-Yaʿlāwī (Beirut, 1995), pp. 48–49, which is a passage from the poem to rhyme *rikābā* in praise of Jaʿfar b. ʿAlī al-Andalusī, one of the Banū Ḥamdūn (the Amīr was ruler in the town of Masīla, of the Zāb region in present-day Algeria). It is also found in the ed. Zāhid ʿAlī, *Tabyīn al-maʿānī fī sharḥ Dīwān Ibn Hāniʾ al-Andalusī al-Maghribī* (Cairo, 1353 [1932]), No. 6, pp. 107–108. The poet Ibn Hāniʾ died 23 Rajab 362 AH, according to Mohammed Yalaoui, *Un poète chiite d'Occident au IVème / Xème siècle. Ibn Hāniʾ al-Andalusī*. Publications de l'Université de Tunis (Tunis, 1976), p. 122.

[15] Taqī al-Dīn Aḥmad b. ʿAlī al-Maqrīzī, *Ittiʿāẓ al-ḥunafā bi-akhbār al-aʾimma al-Fāṭimiyyīn al-khulafā*, ed. Jamāl al-Dīn al-Shayyāl (Cairo, 1967), I, p. 97; al-Ṣafadī, *al-Wāfī bi ʾl-wafayāt*, XI, ed. Shukrī Fayṣal (Wiesbaden, 1981), p. 224.

For Jawhar's assumption of power and his agreement of *amān*, see M. Brett, *The rise of the Fatimids. The world of the Mediterranean & the Middle East in the tenth century C.E.* (Leiden, 2001), pp. 299–303.

The official Friday preacher, the *khaṭīb* ʿAbd al-Samīʿ al-ʿAbbāsī was in hiding; thereupon his substitute officiated in White instead of Black during the Friday sermon. See H. Halm, *Das Reich des Mahdi. Der Aufstieg der Fatimiden* (München, 1991), p. 366.

On the Fatimids' rise to power in Egypt see also Thierry Bianquis, 'La prise du pouvoir par les Fatimides en Égypte (357–363/968–974)', *Annales islamologiques* 11 (1972), pp. 49–108.

> without his cloak. But Jawhar rose up and caused a turban to be brought forth for him and a green cloak. Then he clothed him and provided him with a turban with his own hand.[16]

From that moment the black colour of the ʿAbbāsid-appointed cleric was removed from public view, and we can only be surprised by the choosing of a green colour, since it was not white.

For the rest, it seems that the population of Egypt felt a nostalgic longing for the good old times of the Umayyads of the distant past, who had also been known to use white flags against their ʿAbbāsid opponents from Khurāsān.

If we believe al-Maqrīzī in his *Ittiʿāẓ*, several curious incidents took place in Egypt, such as occurs in his story of an old woman. She apparently felt antipathetic towards the existing situation, which had started with Jawhar, as she tried to associate his regime with that of the Umayyads. Being both blind and deaf however, she was perhaps not familiar with the significance of colours:

> In the month of Ramaḍān an old woman was arrested on the public road, whilst she was shouting aloud and she was put into jail, because a group of the populace agreed with her and clamoured together with her with slogans mentioning the Prophet's Companions, the *ṣaḥāba* and they shouted: "Muʿāwiya is the uncle, *khāl*, of the Believers and he is the uncle of ʿAlī!"
>
> Thereupon Jawhar sent town-criers whom he caused to announce the following communication in Old Cairo inside the Old Mosque of ʿAmr b. al-ʿĀṣ: "Oh people, do not speak too much and set curiosity aside. For we have arrested the old woman in order to protect her. Therefore, let no one speak about this in a disapproving manner, or a painful punishment will befall him." And indeed later on the woman was released.[17]

According to al-Maqrīzī, the slogan shouted by the old woman was not exceptional as nearly the same words were being declaimed on the square of the moneychangers in Cairo. Thus their tumultuous demonstration also tried to make a connection between the ʿAlī clan and the Umayyads.[18]

But Jawhar the general of al-Muʿizz and faithful follower of the Ismāʿīlī doctrine kept his loyalty to these special descendants of the Prophet's daughter Fāṭima, and continued his policy of displaying white standards

---

[16] Al-Maqrīzī, *Ittiʿāẓ*, I, p. 132 under the year 362 AH.
[17] Al-Maqrīzī, *Ittiʿāẓ*, I, pp. 130–131.
[18] Al-Maqrīzī, *Ittiʿāẓ*, I, p. 132.

and other items including the white costumes of the official preachers in the Friday mosque.

In this connection the text of *Ittiʿāẓ al-ḥunafā* is of interest where it says:

> Hibat Allāh b. Aḥmad spoke his sermon, his *khuṭba*, on behalf of them, i.e. the Fatimids, especially al-Muʿizz, and he did so whilst clothed in White. When he had reached the subject of the prayer his *duʿāʾ* on behalf of the reigning authority in Egypt al-Muʿizz, he said:
>
> "Oh Allah, raise his status on high and cause his word to be elevated and make his argument clear and unequivocal. May You assemble the congregation, the Umma, in obedience to him and the hearts in loyalty to him, and incline them towards companionship with him.
>
> "Let him inherit the lands to the East and the West.
>
> "Let the origin and result of these things offer him their praise. This is so because You speak and Your word is the Truth."
>
> [Sūrat al-Anbiyāʾ XXI, verse 105 says:]
>
> "And We have written in the Psalms, after the reminder: "My servants the righteous shall inherit the earth." (…)
>
> "Oh Allah, grant Your support to the armies which he has caused to set out, and to the expeditions which he has dispatched in order to fight against the Polytheists; and to the holy war, the *jihād*, against the Unbelievers; and to the Hallowed Area, the *ḥaram*; and to cause injustice and suspicion and desire to end. May justice spread amongst the nations.
>
> "Oh Allah, cause his standards (*rāyāt*) to be held high and to be widely known, and his armies to prevail and to conquer. Bring reparation through him and through his hands, and on our behalf from Your side make a raised enclosure (*waqiyya*)."[19]

This moment of triumph was remarkable and is well reported by the poets of the time, for example by Ibn Hāniʾ, the eulogist of the Fatimid Caliph al-Muʿizz, who would so tragically die in his prime. A few lines of his masterful poem to rhyme *al-amru* depict the mood of optimism which prevailed at the court of al-Muʿizz even when the latter had not yet taken his residence in the newly founded city of al-Qāhira:

> 5. Do not keep mentioning the Time which has passed for it is a time which has gone; it was another age.
> 6. Did you doubt the capacity of the army? Be silent, for the lance quivers and the army is huge.
> 7. The riders of the Godhead [the Angels] were looking down from on high at Religion and the World, just as dawn does when it breaks.

[19] Al-Maqrīzī, *Ittiʿāẓ*, I, pp. 114–115.

The Angels were the allies of the Fatimid warriors on the earthly battlefield.

> 8. This is the Son of the Prophet of Allah, who asks for the revenge to which he is entitled. And He [Allah] is clever and does not allow the loss of such an entitlement.
> 9. Permit his riders to descend to the waters of the Euphrates, for you [the opponents of al-Muʿizz] can hold him back from neither drought nor flood.
> 10. And as for the Sun, is there any doubt that it is the real sun, once it has revealed itself before you without any actual act of unveiling?
> 11. It is nothing but a miraculous sign, following upon an earlier sign, and a warning to you – if indeed a warning can still be of any use to you.
> 12. [Oh enemies,] may you be granted a harvest of discouraging fatigue; may you be converted into a King in whose hand both Death and Resurrection are trapped.
> 13. Show obedience to an Imām who surpasses other Imāms, just as pious deeds surpass other deeds.
> 14. Make your way down towards someone who is offering a drink and whose pools you will never be able to exhaust even in the fulness of their flood, just as tiny ants can never drink the ocean dry.[20]

From Fatimids to Ayyubids

As hinted at by the preacher Hibat Allāh in the prayer for the reigning Sultan, the 'raised enclosure' protecting the Fatimids would not remain of such an inaccessible height in later ages, because the Sunnite dynasty of the Ayyūbids would gradually take the place of the Fatimids and vest its own power on the shores of the Nile.

Whilst the Fatimid Imām remained for some short time in his post, the generals of the new Ayyūbid family were appointed as viziers with all the paraphernalia of the colour White. In this manner the short-lived reign of Asad al-Dīn Shīrkūh (who was a general in the service of Nūr al-Dīn Maḥmūd b. Zankī, lord of Aleppo and Damascus) started on 27 of the month of Rabīʿ II, 564 AH. And though Shīrkūh was a *mamlūk* of Nūr al-Dīn, he was immediately appointed by the appreciative Fatimid Imām al-ʿĀḍid. The Imām was grateful because Shīrkūh had helped him to get rid of an evil vizier, Shīrkūh's predecessor Shāwar, the latter being

[20] See *Dīwān Ibn Hāni'*, ed. al-Yaʿlāwī, pp. 136–143 = edition Zāhid ʿAlī No. 22, pp. 335–337.

beheaded at the orders of Shīrkūh after a special request from al-ʿĀḍid. The decapitation occurred by means of the Imām's 'Sword of Blood'.[21]

After his nomination, the new vizier Shīrkūh could be seen in the hall of the Palace behind the window specially reserved for the Imām's audiences, the *shubbāk al-īwān* in al-Qāhira.[22] After Shīrkūh's death, his nephew (*ibn ʿamm*) Ṣalāḥ al-Dīn was appointed as vizier of the Fatimid Imām, whilst simultaneously remaining in the service of Nūr al-Dīn. History knows him as the staunch founder of orthodox religious schools, Sultan Saladin of the Ayyūbid Dynasty. However, at the moment of his installation Ṣalāḥ al-Dīn was observed wearing a white turban of gold-brocaded material from the town of Tinnīs, where this sort of material was usually fabricated. Moreover the edict of his appointment on 25 Jumādā II, year 564 AH, was inserted into a white satin case.[23]

Only after the decease of the last Fatimid Imām in 567 AH, did the superiority of the colour white finally come to an end. Even before this, the Sultan Nūr al-Dīn b. Zankī must have abhorred the custom of wearing white as he himself was an adherent of the Majestic Black (*al-sawād al-aʿẓam*) of the ʿAbbāsid Dynasty in Baghdad. Ṣalāḥ al-Dīn did not respond positively to the requests coming from Nūr al-Dīn in Damascus asking him to change the colours in Egypt. It seems that Ṣalāḥ al-Dīn wanted to maintain his relative distance from his master in Damascus by keeping his independence.[24]

---

[21] See for Shāwar and his eulogiser, P. Smoor, "Umāra's poetical views of Shāwar, Ḍirghām, Shīrkūh and Ṣalāḥ al-Dīn as viziers of the Fatimid caliphs,' in F. Daftary & J.W. Meri (eds.), *Culture and memory in medieval Islam. Essays in honour of Wilferd Madelung* (London, 2003), pp. 410–432.

[22] For the installation of Asad al-Dīn Shīrkūh on the date of 27 Rabīʿ II 564 AH, see al-Maqrīzī, *Ittiʿāẓ*, III, pp. 302–303 and see Shihāb al-Dīn ʿAbd al-Raḥmān b. Ismāʿīl b. Ibrāhīm al-Maqdisī al-Dimashqī Abū Shāma, *Kitāb al-Rawḍatayn fī akhbār al-dawlatayn al-Nūriyya wa ʾl-Ṣalāḥiyya*, ed. Ibrāhīm al-Zībaq (Beirut, 1997), II, p. 111 = edition Muḥammad Ḥilmī Muḥammad Aḥmad (Cairo, 1998), I, pt. 2, pp. 436–437.

Another source is Sāwīrus Ibn al-Muqaffaʿ, Bishop of al-Ashmūnīn, *Tārīkh baṭārikat al-kanīsa al-Miṣriyya al-maʿrūf bi-siyar al-bayʿa al-muqaddasa*, ed. A. Khater & O. Burmester (Cairo, 1970), III, pt. 2, Arabic text p. 63 and English translation p. 106: Shīrkūh reportedly slaughtered Shāwar with the Imām's Sword of Blood on 2 Rabīʿ I, and he entered the city of al-Qāhira on 4 Rabīʿ I when he was installed as vizier [Note the different dates assigned, PS].

[23] Al-Maqrīzī, *Ittiʿāẓ*, III, p. 309; Abū Shāma, *Kitāb al-Rawḍatayn*, ed. Zībaq (Beirut, 1997), II, pp. 115–116 = ed. Aḥmad (Cairo, 1998), I, pt 2, p. 439; al-Fatḥ b. ʿAlī al-Bundārī, *Sanā ʾl-barq al-shāmī ikhtiṣār min kitāb al-Barq al-shāmī li l-ʿImād al-Iṣfahānī*, ed. Fatḥiyya al-Nabrāwī (Cairo, 1979), p. 42. According to *Sanā ʾl-Barq*, Shīrkūh died Sunday, 22 Jumādā II 564 AH.

[24] The *khuṭba ʿabbāsiyya* was introduced for the first time in Egypt on the second Friday of the month Muḥarram of the year 567 AH. According to Ibn Saʿīd al-Andalusī,

Another change of fashion

In the month of Shawwāl 566 AH, Nūr al-Dīn b. Zankī had already noted in his correspondence that he was not happy with the appointment of two of his generals as viziers of the Fatimids in Egypt. There was also a question in the correspondence about the prayer on behalf of the reigning Imām or Caliph. He indicated that the person to be mentioned in the prayer should be the Caliph of Baghdad, not the Imām of the Fatimids.

After the appointment in Baghdad of a new Caliph who was called by the honorific al-Mustaḍī' bi-Nūr Allah b. al-Mustanjid b. al-Muqtafī, the pressure from Baghdad became stronger. In order to accelerate a solution to the problems of colour and prayer, a very elevated functionary, a certain eunuch called Ṣandal ʿImād al-Dīn, arrived in Damascus from Baghdad in the course of the year 567 AH. This dignitary belonged to the so-called Ustādhiyya, of the Muqtafawiyya court circles.[25]

In Damascus at the court of Nūr al-Dīn, Ṣandal was received with great pomp and circumstance. On this occasion more than one Chancellor accepted the gifts from Baghdad, including the well-known literary figure Chancellor ʿImād al-Dīn al-Kātib al-Iṣfahānī, who later composed a poem of praise in connection with this important visit.

Among the presents to be given to Nūr al-Dīn were a number of black standards (*rāyāt sūd*), and splendid ʿAbbāsid official capes (*uhbas*) which were intended for the preachers, the *khaṭībs*, to wear while declaiming their sermons in the Friday Mosques. For the ʿAbbāsid presence in Egypt, black would mean a new start. Among the presents given there, was also a pair of brilliant swords. When the Chancellor of Nūr al-Dīn enquired about the two swords: 'What is the meaning of these two swords and the two sword belts?' he was informed that they symbolised

---

al-Nujūm al-zāhira fī ḥulā ḥaḍrat al-Qāhira. Al-qism al-khāṣṣ bi 'l-Qāhira , part of the *Kitāb al-Mughrib fī ḥulā 'l-Maghrib*, ed. Ḥusayn Naṣṣār (Cairo, 2000), p. 97, Ibn al-Athīr said: 'Ṣalāḥ al-Dīn wished to leave the Daʿwa ʿĀḍidiyya as it was, so as to prevent Nūr al-Dīn from acquiring the lands of Egypt should he wish to increase his possessions. He frequently offered this as an excuse for not annulling the prayer on behalf of al-ʿĀḍid. Only when he could no longer excuse himself, did he annul it.'

[25] Al-Fatḥ b. ʿAlī al-Bundārī, *Sanā 'l-Barq*, p. 53 and pp. 60–61; Abū Shāma, *Kitāb al-Rawḍatayn* (ed. Zībaq), II, pp. 205 and 207 (= ed. Aḥmad, I, pt. 2, p. 503). According to Badr al-Dīn Maḥmūd al-ʿAynī, *ʿIqd al-jumān fī tārīkh ahl al-zamān al-ʿaṣr al-Ayyūbī*, ed. Maḥmūd Rizq Maḥmūd (Cairo, 2000), I, p. 55, the inauguration of the ʿAbbāsid Caliph al-Mustaḍī' occurred in Baghdad, on Sunday, 9 Rabīʿ II in the year 566 AH.

the union of Syria and Egypt in the person of Nūr al-Dīn. In the historical report of this event we read that Nūr al-Dīn on one occasion had to ride out in full pomp wearing several black mantles and swords as he sat upon a horse of noble ancestry which trotted over the Maydān Akhḍar, the Green Hippodrome in Damascus, right up till the end of that huge parade ground.

A lesser mantle, again in black, was sent via Damascus towards Ṣalāḥ al-Dīn at his residence in al-Qāhira, along with garments such as black capes destined for the Cairo preachers to use during their Friday sermons.[26]

The Chancellor of Nūr al-Dīn, the littérateur al-Kātib al-Iṣfahānī, was enthusiastic at the impending change in Egypt and expressed all of this in his poem to rhyme *al-ʿaṣri*:[27]

1. In Egypt we have already given a sermon on behalf of al-Mustaḍīʾ, the Substitute of al-Muṣṭafā [i.e. the Prophet Muḥammad] and the Imām of the Age.
(…)
10. The population of Egypt acknowledged the Truth, which had previously been denied by some and acknowledged by others.
11. To the Propagandist of the Pretender, say: "This is sufficient for you because Allah, as the Best One to apportion rights, has now apportioned them [to the praised one, the Caliph of Baghdad]."
12. This is the very first opening up [of the land of Egypt] for Allah has designated us, above all other creatures, to deflower this virgin.
13. We have earned praise, reward, victory, a sweet-scented eulogy and a good reputation.
14. With superior strength, we have unfurled our black Flags (*aʿlāmanā ʾl-sūda*) against the Blue enemies, transporting to them Red destinies of Death.
15. From the Pretenders we have retrieved the Rights which used to be required by them for all and sundry.
16. And he who pretended to the Imamate of al-Qāhira has collapsed into a deep trough of despair.

---

[26] On day 21 of the month of Rajab 567 AH, Ṣalāḥ al-Dīn took part in a procession through the city of al-Qāhira, wearing the black mantle of honour. See al-Fatḥ b. al-Bundārī, *Sanā ʾl-Barq*, p. 61.

[27] For the poem to rhyme *al-ʿaṣri* and its circumstantial context, see ʿImād al-Dīn al-Kātib al-Iṣfahānī as quoted in Abū Shāma, *Kitāb al-Rawḍatayn* (ed. Zībaq), II, pp. 206–209 (= ed. Aḥmad, I, pt. 2, pp. 503–506); Nāṣir al-Dīn Muḥammad b. ʿAbd al-Raḥīm Ibn al-Furāt, *Tārīkh Ibn al-Furāt*, IV, pt. 1 (years 563–588 AH), ed. Ḥasan Muḥammad al-Shammāʿ (Basra, 1967), pp. 182–183; al-ʿAynī, *ʿIqd al-jumān*, I, pp. 71–72.

17. As to his desires, the Dahr deceived him; let no-one with insight ever attempt to trust the Dahr.
18. The (new) Imām is only appointed by right, since a shapely woman [i.e. Egypt] is not accepted unless she receives a bride price.
(...)
25. May the Victory of Right Guidance continue throughout the kingship of the Sons of ʿAbbās until such time as the Day of Resurrection shall arrive.

The passage quoted above shows how the circle had finally become closed again, as the black colours of the ʿAbbāsids returned complete with their shadows. The regime of White for official clothing and flags was now definitely dead and buried. Interestingly, it is precisely this aspect which is expressed in a less well-known poem by Ibn al-Taʿāwīdhī, a contemporary of the Caliph al-Mustaḍīʾ in Baghdad, when he reminisces the poet Ibn Hāniʾ and the latter's panegyrics upon the Fatimid Imām al-Muʿizz:[28]

10. How can he [al-Mustaḍīʾ] be congratulated upon his Time when, on the contrary, it is the days, and the year, and the age, for which congratulations are due to him?
11. But for the Imām al-Mustaḍīʾ and his opinion, the powers of Islam would have collapsed, and a breach would have appeared in the frontier.
12. After the disease of untruth had grown serious and after evil had spread widely, Allah made use of him to help the Caliphate.
13. Who will inform Ibn Hāniʾ as he lies beneath the dust, and who will tell the tomb of al-Muʿizz – if indeed it is capable of hearing –
14. That in his time the rights have been retrieved in spite of someone who was deceitful towards him, and that Egypt has been opened up again,
15. And that, as for the nights, after them his brilliant days glowed with righteousness?

The rebirth, the victorious procession of the black flags and standards of the ʿAbbāsid reign had now been realised. This joyful display had already been visualised even before it occurred in reality. An ascetic

[28] Poem to rhyme *al-ḍarru* by Abū ʾl-Fatḥ Muḥammad b. ʿUbayd Allah b. ʿAbd Allah, who was a Chancellor known as Ibn al-Taʿāwīdhī. See al-ʿAynī, *ʿIqd al-jumān*, I, pp. 55–58. This Chancellor in the 'Department for the feudal lands', the *Dīwān al-Aqṭāʿ* (the register of land divided into lots) died in Baghdad in the year 583 or 584 AH. For this poet see other sources where, however, this special poem is not found, such as the *tarjamas* in Abū ʾl-ʿAbbās Shams al-Dīn Aḥmad b. Muḥammad b. Abī Bakr Ibn Khallikān, *Wafayāt al-aʿyān*, ed. Iḥsān ʿAbbās (Beirut, 1971), IV, pp. 466–473, No. 680; Abū ʿAbd Allah Yāqūt b. ʿAbd Allah al-Rūmī al-Ḥamawī, *Muʿjam al-udabāʾ* (Beirut, 1991), V, pp. 365–374, No. 884; al-Ṣafadī, *al-Wāfī bi ʾl-wafayāt*, IV, ed. S. Dedering (Stuttgart, 1992), pp. 11–16.

jurisprudent had forecast that this time would surely come, as he had seen it in a dream of the type which usually occurs when great changes threaten:

> I saw as it were two men, of whom one was more brilliant than the other. The one who was shining more, turned his face towards the direction of prayer, the *qibla*. He had a very long black beard. The slightest Zephyr breeze caused that beard to move, and its movement and its shadow cast their shadowy effect over the earth. The man who saw this was surprised about it. And it was as if he heard the voices of people who were reciting in a manner and with sounds which were melodious, such as he had never heard before. And it was as if someone who was present on this occasion asked a question saying: "What is this?" They answered him and said: "The people have caused their Imām to be replaced."[29]

The vision in the dream, seen in 552 or 555 AH, was concerned with a cosmic black beard, which would cast its black shadow over the earth up to our own times, if not to all eternity.

[29] *Tārīkh Ibn al-Furāt*, IV, pt. 1, p. 156 and *Kitāb al-Rawḍatayn* (ed. Zībaq), II, pp. 200–202. According to the editor of *Kitāb al-Rawḍatayn*, II, p. 200, n. 3, there is a gloss in the manuscript which states: 'I saw in the relevant Biography that the person who saw this dream was no other than the ascetic *faqīh* Muḥammad 'Afīf b. al-Mubārak b. Maḥmūd al-Aḥmadī, in the year 552 AH, but Allah knows best.'

IBN BAṬṬŪṬA ON PUBLIC VIOLENCE IN THE DELHI SULTANATE

David Waines*

Introduction

In her short, insightful book *On Violence*, Hannah Arendt commented that George Sorel's reflection in 1906 that 'The problems of violence still remain very obscure' was as true when she was writing sixty-three years later. More recent attempts to define the phenomenon of violence in modern societies have led social scientists to acknowledge the problems and perils involved in doing so.[1] The anthropologist, David Riches, has cautioned that the Anglo-Saxon notion of violence is unhelpful as it popularly implies such acts are illegitimate and irrational. Elizabeth Stanko adds the point that among her colleagues, apart from lacking a single, precise meaning of violence, there is even disagreement over what might constitute a hierarchy of harm.

The analysis of violence in pre-modern societies, too, poses problems, if for different reasons. Consider the case of China. In an essay entitled, 'Rethinking "violence" in Chinese culture.'[2] the author observes that 'The place of violence – however defined – in Chinese culture has hardly

---

* The author would like to express his appreciation to colleagues, Drs Maurits van den Boogert, Han den Heijer and Gautier Juynboll for reading early drafts of this paper and making valuable comments.

[1] H. Arendt, *On violence* (New York, 1969); See G. Aijmer & J. Abbink (eds.), *Meanings of violence. A cross cultural perspective* (Oxford, 2000); E. Stanko (ed.), *The meanings of violence* (London, 2003); D. Riches (ed.), *The anthropology of violence* (Oxford, 1986). Crime writer Ruth Rendell expresses the elusive nature of violence in the opening passage of her 1977 novel *A Judgement in stone*: 'Eunice Parchman killed the Coverdale family because she could not read or write. There was no real motive and no premeditation; no money was gained and no security... She accomplished nothing by it but disaster for herself, and all along, somewhere in her strange mind, she knew she would accomplish nothing. And yet, although her companion and partner was mad, Eunice was not. She had the awful practical sanity of the atavistic ape disguised as twentieth-century woman... But there was more to it than that.' I owe this reference to Remke Kruk's passion for the crime/thriller genre, second only perhaps to her passion for bitterballen.

[2] B.J. ter Haar, in Aijmer and Abbink, pp. 123–140.

been studied in a systematic way. I suspect that at least one reason for this lack of attention is a feeling that this is not an interesting question, since we 'know' that violence did not play a large role in Chinese culture anyhow.'[3] One could scarcely imagine similar scholarly neglect of any other major bloc of medieval societies. Problems remain especially when violence is associated with religious values. The historian of the European crusades, Jonathan Riley-Smith has wryly observed that the contemporary Western 'consensus' on the use of force is conceptually weak, impractical, and close to collapse. Part of that consensual view, he notes, is that violence is regarded as 'intrinsically evil, but can be allowable in the last resort when confronted by a greater evil.'[4]

The academic study of violence in pre-modern Islamic societies has been sporadic and relatively recent. It has tended to focus particularly upon the ideology and practice of jihad or Holy War[5] or inter-communal relations between Muslim rulers and Christian and Jewish minorities.[6] There is now a major contribution to the study of violence from a broader perspective in the collection of essays on religion, politics and violence in al-Andalus.[7] The present essay deals with the account of the fourteenth century Delhi Sultanate and the reign of Muḥammad b. Tughluq (1324–1351) written by the Moroccan traveller and sojourner Ibn Baṭṭūṭa. Peter Hardy has already covered some of the ground in his wide ranging essay on 'Force and violence in Indo-Persian writing on history and government in medieval South Asia.'[8] Hardy concedes that it is ultimately untenable to distinguish between 'force' as 'controlled coercion for a known and relatively clearly-defined purpose' and the term 'violence' to denote 'physical coercion expressive of wild emotion and with no clearly-formulated purpose in view…'[9] In other words, the

[3] Ibid., p. 123.

[4] 'Religious violence,' in A. Sapir-Abulafia (ed.), *Religious violence between Christians and Jews* (Basingstoke, 2002), p. 184.

[5] Exceptionally, Albrecht Noth's study, *Heiliger Krieg und Heiliger Kampf in Islam und Christentum. Beiträge zur Vorgeschichte und Geschichte der Kreuzzüge*, was published as early as 1966. There is now a burgeoning literature.

[6] D. Nirenberg, *Communities of violence. Persecution of minorities in the Middle Ages* (Princeton, 1996).

[7] M. Fierro (ed.), *De muerte violenta. Política, religión y violencia en al-Andalus* (Madrid, 2004).

[8] P. Hardy in M. Israel & N.K. Wagle (eds.), *Islamic society and culture. Essays in honour of Professor Aziz Ahmed* (Manohar, 1983), pp. 165–208.

[9] Ibid., p. 166.

problem of violence is expressed here as the difficulty in discriminating between purposeful and irrational acts.

Depicting public violence in Sultan Ibn Tughluq's reign, Ibn Baṭṭūṭa wrote as a reasonably well informed observer, if not as a native insider. He spent more time in India than anywhere else on his travels and was employed for several years by the Sultan as a qadi in Delhi; on several occasions he attended his private dinners. His account of numerous episodes accords closely with Riches' understanding of violence as 'contested legitimacy' between performer and victim. Among the many possible aims of violence, he argues, there exists a 'core purpose' that is, 'the means by which its legitimacy will ultimately be argued' which constitutes both 'the *substance* of violence's legitimacy' and amounts to a 'necessary condition of violence.'[10] Violence can serve both practical (instrumental) and symbolic (expressive) purposes so that 'the desire to achieve a very wide variety of goals and ambitions, is a *sufficient condition* for acts of violence to be performed.'[11] Hence as a means of communication and used as a 'strategically, consciously employed resource'[12] violence need not be regarded as either intrinsically evil or irrational.[13] In addition to Ibn Baṭṭūṭa's description of public violence we are interested as well in his attitude to or evaluation of the phenomenon which suggests he also perceived a hierarchy of harm. This in itself reflects another aspect of the phenomenon's complexity.

---

[10] Riches, p. 7 (emphasis in text).

[11] Ibid., p. 11 (emphasis in text).

[12] Ibid., p. 12. Riches recognises his approach is at odds with those who accept violent tendencies as somehow 'biologically innate or as imprinted in the human subconsciousness.' In the Preface (p. xi) to the Aijmer and Abbink volume, Jon Abbink cites the minimally defining elements of violence by paraphrasing Riches' position as: 'the contested use of damaging physical force against other humans with possible fatal consequences and with purposeful humiliation of other humans. Usually this use of force – or its threat – is pre-emptive and aimed at gaining dominance over others. This is affected by physically and symbolically communicating these intentions and threats to others.'

[13] In a similar manner, Nirenberg, p. 245, concludes his study of communal violence in the contemporary 14th century Kingdom of Aragon reminding us that where Jewish and Muslim minorities lived under a Christian majority, '*Convivencia* was predicated upon violence; it was not its peaceful antithesis. Violence drew its meaning from coexistence, not in opposition to it.'

Part One

Ibn Baṭṭūṭa arrived at the Indian frontier, the River Sind or Banj Āb meaning the 'five waters', in September, 1333.[14] Muḥammad b. Tughluq, had already been in power for nearly a decade. The Tughluqs were the latest in a succession of Muslim dynasties comprising the Delhi Sultanate established in 1191. Muḥammad b. Tughluq's father and founder of the dynasty was Ghiyāth al-Dīn who had ruled briefly from 1320. Ibn Baṭṭūṭa's chief source for the events of Ghiyāth al-Dīn's rule was the pious and learned imam, shaykh Rukn al-Dīn al-Multānī who died in 1335 and who was an eye witness to certain of the episodes he recounted for his Moroccan guest.

As Ibn Baṭṭūṭa begins his story of the Tughluqs, father and son, they are united in revolt (*khilāf*) against the usurper of the Khaljī throne and its incumbent, Sultan Quṭb al-Dīn Mubārak Shāh (1316–1320). The usurper proclaimed himself Sultan and assumed the title of Nāṣir al-Dīn Khusraw Shāh. He then effectively eliminated the Khaljī dynasty by slaying all male members of Quṭb al-Dīn's family. Battle with the Tughluqs was now inevitable although its outcome, at first indecisive, by good fortune swung in the Tughluqs favour. Nāṣir al-Dīn was forced into hiding and the elder Tughluq was in turn declared Sultan Ghiyāth al-Dīn Tughluq Shāh. When the fugitive was finally caught and brought to the palace, he said to Ghiyāth al-Dīn, 'O Tughluq! Treat me according to the code of kings and do not dishonour me.' ('*Yā Tughluq! If'al ma'ī fi'l al-mulūk wa-lā tafḍaḥnī.*')[15] According to Ibn Baṭṭūṭa, the request was granted. The usurper was despatched on the same spot and in the same manner as Quṭb al-Dīn, namely, he was beheaded first then both head and body thrown from the palace roof. The body was washed, wrapped in a shroud and buried. Thus began Ghiyāth al-Dīn's four year reign (1320–1324).

The conflict had manifestly concerned a direct struggle for the Sultanate and, in Riches' terms, was one of 'contested legitimacy.' Even the Khaljī Sultan, Quṭb al-Dīn, had experienced the impulse to violence, first to win and then to protect his throne. He had deposed his own brother

---

[14] Ibn Baṭṭūṭa, pp. 93–94. The Arabic edition used here is that of ʿAbd al-Hādī al-Tāzī, 5 vols. (Rabat, 1997) who provides marginal pagination references to the edition of Defrémery and Sanguinetti (thus 75/3, is volume 3, page 75). Only the page number is used here and all references, except in footnotes 40, 41 are to volume three.

[15] Ibid., p. 207.

to usurp the Sultanate and when he heard of a plot to replace himself with his ten year old nephew, Quṭb al-Dīn 'seized [the boy] by the feet and dashed his brains out on the stones.' [16] For his part, the usurper Nāṣir al-Dīn had successfully dealt with any threat from the Khaljīs. Political reality dictated that as long as an enemy existed *in potens*, he existed everywhere, both inside and beyond the ruler's own immediate circle of family and close supporters. On the other hand, Nāṣir al-Dīn's appeal to a 'code of kings' suggests that contenders for power within the political elite operated by certain rules of the game which possibly was intended to help check violence descending into chaos. Here Ibn Baṭṭūṭa provides no clear picture as to the norms of the code. Other episodes, however, provide clues to their clarification as we shall see below.

As the story continues, Muḥammad b. Tughluq attempts to revolt (*qiyām*) against his father and fails. Sultan Ghiyāth al-Dīn had sent Muḥammad with a large army to restore the Sultanate's authority over Telingana in the Deccan, a three months journey from Delhi. Muḥammad was accompanied by some of the principle amirs, including the Keeper of the Seal. Muḥammad planned his revolt (*mukhālafa*) and ordered his boon companion (*nadīm*), the poet-jurist Ubayd, to spread the rumour of the Sultan's death; he imagined the army would swiftly pay allegiance to him but he gravely miscalculated. The amirs refused to be drawn and proposed killing him, but were restrained by one of their number, allowing Muḥammad to escape with a handful of men to Delhi. The Sultan furnished him again with money, troops and orders to return to Telingana. When he learned what Muḥammad's real intentions had been, the Sultan ordered Ubayd killed and the Keeper of the Seal was executed by skewering him on a sharpened stake, head downwards and left on display. The remaining amirs who had in effect remained loyal to the Sultan, fled in fear of being suspected of collusion, and sought refuge at the court of the independent Muslim ruler of Lakhnawti in Bengal.[17] While Ubayd's execution perhaps suited a minor troublemaker, The Keeper of the Seal's exemplary punishment served its purpose as a warning to all potential rebels, including his son Muḥammad.

The next episode recounted by Ibn Baṭṭūṭa describes Sultan Ghiyāth al-Dīn's campaign to restore order in Bengal by supporting its deposed ruler against his usurping younger brother. During the Sultan's absence,

---

[16] Ibid., pp. 191–193.
[17] Ibid., pp. 208–210.

Muḥammad b. Tughluq was left in the capital as his deputy (*nāʾib*). News of disturbing developments reached the Sultan from Delhi. His son had developed a close attachment to the famous Sufi shaykh Niẓām al-Dīn Awliyā who, while in a trance, had foretold Muḥammad's succession to the Sultanate.[18] When the shaykh died, Muḥammad assisted shouldering the bier at the funeral. In addition, Ghiyāth al-Dīn's suspicions and anger were further aroused by reports that Muḥammad was spending lavishly on the purchase of mamluks, gifts, and on efforts to captivate the hearts of the people. The Sultan even threatened some astrologers who had reportedly predicted he would never again enter Delhi. Returning from campaign he ordered Muḥammad to build a (temporary) palace (*kushk*) at Afghanpur near Delhi. It was constructed in three days of wood resting above the ground on wooden columns and was designed to collapse when elephants passed a particular side of it. Ibn Baṭṭūṭa's informant, the above-mentioned shaykh Rukn al-Dīn, was present with the Sultan and his favourite son (*al-muʾthar*), Maḥmūd. While the shaykh left to attend his prayers, Muḥammad paraded the elephants past the kiosk and the edifice fell, as planned, upon those left inside. The bodies were recovered from the ruins – slowly – following Muḥammad's orders, and not until after sunset was the Sultan found bent over his son as if to protect him. Some said that they were already dead, others that they were alive and then quietly finished off. The Sultan was buried in the mausoleum he had built for himself outside his palace at Tughluqabad. The palace contained vast treasure from which Muḥammad benefited when he succeeded as Sultan and the designer of the kiosk, Khwāja Jahān, enjoyed an unrivalled position among wazirs and others during Muḥammad's Sultanate.[19]

Muḥammad's initial plan of revolt had failed due to hasty or inept planning. His second attempt indicated a more careful and cautious preparation. Muḥammad employed the tactics of winning over (or, sowing sedition among) the Sultan's local support through gifts or promises, the purchase of mamluks loyal to him and by exploiting a religious/magical dimension through his cultivation of Niẓām al-Dīn and astrologers. If access to power was chiefly a matter of forging alliances, Muḥammad was methodically preparing a 'home base' which ultimately would yield

[18] Gibb's translation, p. 654, has 'We give him the kingdom' while the Arabic, p. 211 reads 'We give you the kingdom.'

[19] Ibn Baṭṭūṭa, pp. 210–216.

him both his father's mortal remains and his vast treasure. He may also have been concerned by a potential danger from his younger brother Maḥmūd whom, Ibn Baṭṭūṭa notes in passing, was his father's favourite son. The method of the Sultan's death caused disquiet within court circles. Providing a proper burial for the former Sultan, however, may have muted the criticism of potential rivals. If this were so, the matter of the burial may be an important component of the 'code of kings' noted by Ibn Baṭṭūṭa above.

Part Two

Ibn Baṭṭūṭa reports that Muḥammad b. Tughluq assumed control of the Sultanate without competitor (*munāziʿ*) or opponent (*mukhālif*). He then proceeds to describe the new ruler's character. 'This king is the most dedicated of persons to conferring gifts and to shedding blood; the gate (of the palace) is never free of some poor man enriched or of some living soul executed (*ḥayy yuqtal*).'[20] Stories abounded of his generosity and courage as well as of his 'cruelty and violence towards criminals' (*wa-shuhirat… ḥikāyātuhu fī 'l fatk wa 'l baṭsh bi-dhawī 'l-jināyāt*).[21] Ibn Baṭṭūṭa concludes that of all men, he was the most humble and most ready to display justice and acknowledge right. But above all, he was generous.

Recalling his adventures in India some years later while secure on his native soil, Ibn Baṭṭūṭa's description of Muḥammad b. Tughluq's apparently conflicting qualities is likely a fair assessment drawn from his own experience.[22] He himself richly benefited from the Sultan's generosity and also fell under the shadow of his wrath.[23] Yet he was clearly distressed by instances of Ibn Tughluq's violence. 'The Sultan used to punish small faults and great without respect of persons, whether men of learning or piety or noble descent…, may God deliver us from misfortune.' (*aʿādhanā Allāh min al-balāʾ*)[24] For example, he recounts his horror at the sight of mutilated bodies and stuffed skins of executed

[20] Ibid., p. 216 and again at 290–291.
[21] Ibid., p. 216.
[22] Remke Kruk offers further insight into Ibn Baṭṭūṭa's character in her article 'Ibn Baṭṭūṭa: travel, family life and chronology,' *Al-Qantara* 16/2 (1995), pp. 369–384.
[23] Ibn Baṭṭūṭa, pp. 395, 402 for benefits acquired.
[24] Ibid., p. 291.

prisoners exhibited on a city wall, noting its purpose was to spread terror among the public.[25] He once witnessed such stuffed skins being devoured by dogs and remarked 'God preserve us.'[26] Ibn Baṭṭūṭa's illustrations of the Sultan's darker side deal chiefly with victims from among the ulama class, pious sufi shaykhs like Shihāb al-Dīn, several jurists, a qadi, a market inspector (*muḥtasib*) and the chief preacher of Delhi. Concluding his account of each victim Ibn Baṭṭūṭa expresses a blessing of sympathy, 'May God Most High have mercy upon him.' In the remaining cases involving one of the Sultan's relatives or other of the notables (*kibār*) who ran afoul of the royal temper, no blessing is mentioned. This cannot be accidental. As a member of the ulama class, albeit a foreigner, he was especially sensitive to attacks upon them; his attitude towards the ruling elite was more aloof than indicated by his general sentiment noted above.

The case of the Sufi shaykh Shihāb al-Dīn is notable.[27] Muḥammad b. Tughluq attempted to bring the respected Sufi shaykh into his employ, but Shihāb al-Dīn publicly refused the offer. This brought about his banishment to Dawlat Abad for seven years. Once restored to favour, the shaykh was appointed to a government post and continued to rise in rank until there was no one higher in the palace. Later he secured the Sultan's permission to leave the capital for the nearby hills where he housed in an immense cave a slave workforce to irrigate and cultivate the land. The enterprise earned the shaykh a considerable fortune. When the Sultan summoned him to the capital, Shihāb al-Dīn proclaimed he would never again serve an oppressor (*ẓālim*). Forcibly brought to Delhi in irons, the shaykh declined food and drink for two weeks, all the while being pressed to recant by fellow shaykhs and jurists. He continued to resist, even rejecting food sent by the Sultan and stoutly declaring his readiness to die. The Sultan ordered him force fed with human excrement and when Shihāb al-Dīn stubbornly refused to yield, he was beheaded. Ibn Baṭṭūṭa perhaps felt a special affinity with Shihāb al-Dīn. It was his visit to the shaykh's cave that caused royal suspicion also to fall upon him. Although detained, Ibn Baṭṭūṭa suffered nothing worse than the terrifying thought that his life was coming to an end. In another instance, two jurists who protested their innocence against an accusa-

[25] Ibid., p. 108.
[26] Ibid., p. 331.
[27] Ibid., pp. 293–299 and 444–445.

tion by the Sultan were tortured. They then wrote their confessions, stating they did so of their own free will, for to have claimed otherwise would have led to more painful torture. Both were beheaded. In each of these cases involving ulama, Ibn Baṭṭūṭa concludes with the expression of condolence, 'May God Most High have mercy upon him.'

Ibn Baṭṭūṭa describes as *thawras* several revolts against the Sultan. Two of them involved the governor of Ma'bar in the distant south, sharif Aḥsan Shāh, who declared himself independent ruler there and that of his son, sharif Ibrāhīm who harboured ambitions on the Sultanate. Ibn Baṭṭūṭa had a particular interest in their story. Their claim to belong to the *shurafā'* (sg. *sharīf*), descendents of the Prophet, gave them special religious status in society, and Ibrāhīm, moreover, was Ibn Baṭṭūṭa's brother-in-law. Aḥsan Shāh managed to remain secure in his local power base. The Sultan's army, led against Aḥsan Shāh by Sultan Muḥammad himself, only managed to reach the province of Telingana, three months march from Ma'bar, when pestilence (*waba'*) broke out decimating the troops thereby ending the campaign. Ibrahim was less fortunate. Initially a court favourite, he aroused the Sultan's suspicions and, in fear of torture, he confessed his intentions and was executed. His body, notes Ibn Baṭṭūṭa, was left in the open for three days 'according to custom'. On the third night, Hindus employed by the authorities removed the corpse to a trench outside the city and guarded it to prevent relatives retrieving it. Money, however, changed hands and the body was released to the family who gave it a proper interment. Ibn Baṭṭūṭa concludes the tale with the familiar words of condolence. He also conveys the impression that in this case, albeit by subterfuge, the 'code of kings' had been honoured by the victim's kin.

Not so where others were involved. For example, the Sultan suspected his half brother, Mas'ūd Khān, of plotting a revolt (*qiyām*) and had him questioned. He readily confessed for fear of torture which, in the case of the jurists and the sharif Ibrāhīm above, was customarily applied when suspects denied charges against them. Execution, says Ibn Baṭṭūṭa, was considered lighter punishment than torture, observing what he clearly perceived as a hierarchy of harm. Mas'ūd Khān was subsequently beheaded in the market place and Ibn Baṭṭūṭa remarks, his body was left for three days 'according to custom.' By this last phrase, and without further comment, Ibn Baṭṭūṭa seems to imply that Mas'ūd Khān was not awarded a proper burial.

Two other men, including the Sultan's cousin, openly defied a royal command, were hunted down, captured and killed. The cousin was treated in a particularly brutal manner. The Sultan first ordered him humiliated in front of his own family, then 'flayed alive (*bi-qayd al-ḥayāt*), and when he had been flayed his flesh was cooked with rice and sent to the sons of his family.'[28] Both victims were treated in customary fashion. Their skins were stuffed with straw and sent on tour of the provinces as an example and warning to others against disobedience. The stuffed remains (*jildayn*) eventually arrived at Multan in the northern province of Sind. There the amir, Kishlū Khān, whom Muḥammad b. Tughluq addressed as 'uncle', had the remains buried. The Sultan was enraged and Kishlū Khān fell in the subsequent campaign against him, his severed head which Ibn Baṭṭūṭa later witnessed, placed over the city gate. Aware of the symbolic value of the bodies' public display, Kishlū Khan had removed that value with their burial. Possibly he had relied on his relationship with the Sultan to avoid punishment but was nonetheless prepared to face the consequences of his action. In doing so, he seemed to express the sentiment that the mutilated remains deserved better treatment, a decent burial. It was a blatant challenge to the Sultan's authority and perhaps a reminder of the failure to adhere to the 'code of kings'. The Sultan's spy system touched the lives of all amirs great or small. A mamluk attached to each one acted as the Sultan's 'eye'. Slave girls placed in the amirs' homes passed on information to female sweepers (*kannāsāt*) who entered homes uninvited and transmitted news to the chief of intelligence (*malik al-mukhbirīn*) who reported to the Sultan.[29] Information transmitted along these channels proved fatal for one amir. He desired intercourse with his wife, rejecting her protest, 'by the head of the Sultan', to desist. Summoned to the Sultan's presence the following day, the amir could not deny the incident and suffered the ultimate penalty. Spies uncovered a plot (*qiyām*) involving a governor, one ʿAyn al-Mulk and his brothers. ʿAyn al-Mulk was captured but his wife, instead of escaping with the brothers, declared, 'Shall I not do as the wives of infidels do who burn themselves with their husbands? I too shall die for my husband's death and live for his life.'[30] Her words reached

[28] Ibid., p. 321.

[29] Ibid., p. 343.

[30] Ibid., p. 353. See also, pp. 319–320. Muḥammad b. Tughluq's nephew, Bahā al-Dīn, who had refused to acknowledge him as Sultan, fled and was protected by a Hindu king, Ray Kanbila. When the king could hold out no longer against the Sultan's forces, he burnt his possessions and his wives and daughters, and those of his amirs and wazirs cast themselves onto the pyre. The king and those who sought to die with him went into battle lightly armed.

the Sultan and he was overcome by compassion (*riqqa*) for her and she was allowed to visit her husband. Those of ʿAyn al-Mulk's followers from among the rabble (*lafīf al-nās*) and of no account were ordered released. ʿAyn al-Mulk himself had to witness the execution by trained elephants of some sixty of his associates. He was, however, subsequently pardoned along with the leader of another revolt and both set the task of overseeing the Sultan's gardens each with a daily ration of flour and meat.

In Ibn Baṭṭūṭa's account, the politics of pardon or reconciliation was employed sparingly by the Sultan. In this connection the figure of the Sultan's preceptor, Quṭlū Khān, appears. He is described as a man of his word (*ṣāḥib ʿahd*) in whom people placed their confidence. He enjoyed the Sultan's respect and held a high position in his court. He only entered his presence on invitation so that the Sultan should not tire in rising to greet him. He was liberal in giving alms and helping the disadvantaged. Quṭlū Khān settled several conflicts by negotiation with rebel leaders, arranging their safe conduct to the Sultan who pardoned each of them. Another rebel, Amīr Bakht (Sharaf al-Mulk) who had fled to Sind with some followers was apprehended, punished with one hundred lashes daily, but then pardoned and banished to an outlying province. He later married the Sultan's sister and was appointed to a post of provincial governor. Ibn Baṭṭūṭa remarked on this turn of events, 'Glorified be the Turner of hearts and Changer of conditions.'[31]

Of the remaining rebels recorded by Ibn Baṭṭūṭa, mention of the qadi Jalāl al-Dīn will suffice. He and a body of Afghan followers rose up and seized the city of Kanbaya (Cambay) on the Indian coast. He claimed the Sultanate, successfully defended his position against troops sent to crush him, and attracted to his standard in the process members of the 'turbulent and criminal classes' (*ahl al-fasād wa 'l-jarā'im*). One of his supporters was the highly respected shaykh ʿAlī al-Ḥaydarī who was arrested and charged with praying for Jalāl al-Dīn's success. His execution by beheading (accomplished on the second stroke of the sword) is curiously the only instance mentioned by Ibn Baṭṭūṭa of the Sultan's order for a judgement to be made by jurists according to the shari'a.[32]

---

[31] Ibid., p. 361.
[32] Ibid., pp. 310–311, 368.

Part Three

Ibn Baṭṭūṭa's account can be supplemented here by reference to another written by the historian Ḍiyā al-Dīn Baranī (d. 1357), a long time confidant (*nadīm*) of the Sultan. Although he and Ibn Baṭṭūṭa were contemporaries, there is no firm evidence they ever met. Conceivably they might have done so on the occasions of the Sultan's small private dinners to which Ibn Baṭṭūṭa was invited.[33] The two men differed in background. Baranī was Indian born and his family was well connected to Delhi's ruling circle of indigenous aristocrats. His view of political violence appears similar to that of Ibn Baṭṭūṭa inasmuch as he concedes that 'the affairs of kingship involve (a combination of) cruelty and benevolence; and kingship subsists in (such) opposite qualities.'[34] But Baranī was primarily concerned with the long term interests of his own, the ruling class, that lay in its unity achieved by despotic power: 'Royalty is nothing but terror, power and the claim to unshared authority', he stated.[35] Baranī records what purports to be a conversation with the Sultan who spoke of his policy of severe punishment for the slightest whisper of sedition or trifling act of disobedience. Times had changed, he declared, from the days of the ancient kings of Persia. 'In these days a large number of wicked and mischievous persons have been born. I inflict capital punishments on the basis of suspicion and presumption of rebellion, disorder and conspiracy. I put people to death for every slight disobedience that I see in them, and I will keep inflicting capital punishments in this way till either I perish or the people are set right and give up rebellion and disobedience. I have no wazir who can frame such laws for my kingdom as did the Persians that it may become unnecessary for me to smear my hands with blood... I have distributed so much trea-

---

[33] Ibid., pp. 238–239.

[34] I. Habib, 'Baranī's theory of the history of the Delhi sultanate,' *The Indian historical review* 7 (1980–81), pp. 99–115, cited at 105 from Baranī's *Ta'rīkh-i Fīrūz Shāhī*. A. Black, *The history of Islamic political thought* (Edinburgh, 2001), pp. 161–162, argues that Baranī was the first Muslim theorist to attempt to resolve the contradictions between Islamic and Iranian views on government; that a king must combine mercy and severity. The Andalusi wazir, Ibn al-Khaṭīb (d. 1375) also reflected this view in his writings, although Black claims this echoed Christian rather than Muslim sentiment of the time. Ibn Baṭṭūṭa was no political theorist but his narrative suggests agreement with Baranī (and Ibn al-Khaṭīb) on this score and it may be that other jurists of the day thought along similar lines, so that Baranī's views may not be as unique as Black claims.

[35] Ibid., p. 104, cited from Baranī's *Ta'rīkh*.

sure among the people, but no one has become my sincere well-wisher. The temper of the people has been clearly revealed to me; they are my enemies and opponents.'[36] The modern Indian historian, Irfan Habib, has observed that the medieval ruler faced an insoluble dilemma when 'the possession of power required a proportionate exercise of terror; but the latter once initiated became itself a major element of instability. Once instability set in, terror in the next round of the cycle could only be greater.'[37]

As we have seen, Ibn Baṭṭūṭa also acknowledged the contrary qualities possessed by the Sultan. Like Baranī, he accepted the ruler's violence as a fact of political life, as much as it distressed him. Each man disapproved of political violence most vigorously where it was perceived to affect the interests of their respective segments of the population, whether the ruling elite (Baranī) or the ulama (Ibn Baṭṭūṭa). For his part, the Sultan justified violence for both instrumental and symbolic purposes as instances of contested legitimacy according to Riches' analysis.

There is, however, another aspect of violence about which Ibn Baṭṭūṭa displays no squeamishness. Among the Sultan's positive qualities he notes his being strict (*shadīd*) about prayer observance, especially Friday congregational prayers, with grave penalties meted out for their omission. In one day alone he had nine persons put to death for not fulfilling their religious duty. Anyone found in the markets after prayer commenced was punished; he gave orders that people acquire knowledge of the basic rituals and beliefs of Islam and they were punished if, when questioned, they could not answer correctly.[38] These measures likely only applied in Delhi and surrounding area where close supervision could be exercised. Ibn Baṭṭūṭa relates with approval another, albeit less severe, punishment related to prayer that he had witnessed only in Khwarizm. The muezzin would announce the approaching prayer time passing through the streets in the neighbourhood of his mosque. Anyone who failed to appear was

---

[36] K.A. Nizami, 'Sultan Muhammad bin Tughluq (1324–51),' in M. Habib & K.A. Nizami (eds.), *A comprehensive history of India*, V, *The Delhi Sultanate*, cited at pp. 551–552. See also R.C. Majumdar (ed./author), '*The Delhi Sultanate*,' in *The history and culture of the Indian people* (Bombay, 1964), p. 85; W. Haig (ed., author), *Turks and Afghans. The Cambridge history of India*, III (Cambridge, 1928), pp. 162–163. On the historical use of these reported 'conversations' of Baranī, see P. Hardy, 'The *oratio recta* of Barani's *Ta'rikh-i-Firuz Shahi* – fact or fiction?,' *BSOAS* 20 (1957), pp. 315–321.

[37] Habib, p. 106.

[38] Ibn Baṭṭūṭa, pp. 286–287. For these other cases Ibn Baṭṭūṭa does not spell out the nature of the punishment inflicted.

whipped by the imam in front of the congregation and fined a sum of money put towards the mosque's upkeep.[39] Later, while acting as qadi in the Maldive Islands, Ibn Baṭṭūṭa once ordered a thief's hand severed and was astonished that some locals present in the court, all upright and pious persons, swooned at the sight.[40] On the other hand, while in Mali he was surprised to see the children of a qadi in leg chains who would not be released until they had memorised the Qur'an.[41] Severity 'in the name of God', could have been explained by Ibn Baṭṭūṭa as either fulfilling the divine will (punishment of a thief), or as a practical means of securing and sustaining Muslim commitment to the faith, especially in the precarious Indian context of Muslims living as a minority, albeit a ruling minority, in an overwhelming Hindu environment.

Ibn Baṭṭūṭa left Delhi in 1342,[42] some nine years before the death of his patron the Sultan. At the commencement of his reign, Muḥammad b. Tughluq was acknowledged to possess authority over a larger area of the continent than any of his predecessors. By his death in 1351 'Bengal and every tract south of the Vindhyas had declared their independence and none of these provinces was ever recovered.'[43] Whether these rebellions and uprisings were, according to one view cloaked in 'opportunistic responses to a prolonged crisis' or whether, according to another view, the regime had become entrapped by its own 'reputation for harshness',[44] Ibn Baṭṭūṭa's narrative illustrates both the sufficient and necessary conditions for violence as proposed by Riches. Exercised by the Sultan, violence was practical and symbolic, at least in the short term. As much as anything, that mirrored the prevailing conditions of Muslim rule from Delhi. Ibn Baṭṭūṭa describes the Hindu and Muslim territories as physically side by side. He says, '...their lands are contiguous, but though the Muslims have the upper hand over them yet the infidels maintain themselves in inaccessible mountains and rugged places, and they have

---

[39] Ibid., pp. 4–5.

[40] Ibid., IV, p. 114. In a reverse situation, Ibn Baṭṭūṭa (3: 136–141) nearly tumbled from his horse in a faint witnessing a Hindu ritual of widows preparing to join their deceased husbands on the funeral pyre.

[41] Ibid., IV, p. 423.

[42] For a chronology of his travels, see A. Miquel, 'Ibn Baṭṭūṭa,' in *Encyclopaedia of Islam* (New ed.), III (Leiden, 1971), pp. 735–736. He spent several months travelling to southern India along the west coast on his way to the Maldives, Ceylon and China, returning briefly to India in 1347 on his way back to Morocco.

[43] P. Jackson, *The Delhi sultanate. A military and political history,* (Cambridge, 1998), p. 255, and for his account of the troubles during the reign, pp. 255–277.

[44] Ibid., pp. 269, 270.

forests of reeds... which are for them as good as city walls...'[45] Jackson has observed of this realistic approach that 'Obeisance, like tribute, was intermittent... The Muslim population of the Sultanate largely resided in its fortified towns and cities, and even there they were not unusually a minority...In India, the 'war zone', peopled by *ḥarbī*s, was never far away.'[46] The *ḥarbī*s in practice, as Ibn Baṭṭūṭa describes them, were chiefly Muslims who presented a formidable challenge to the stability of the Sultanate. The Sultan, despite the harshness of the declaration attributed to him by Baranī, also practiced the politics of reconciliation and pardon following the counsel of his preceptor Quṭlū Khān. The general political situation is reflected in Hannah Arendt's summary distinction between power and violence: 'Power and violence are opposites; where the one rules absolutely, the other is absent. Violence appears where power is in jeopardy, but left to its own course it ends in power's disappearance...Violence can destroy power; it is utterly incapable of creating it.'[47]

We have already suggested that Ibn Baṭṭūṭa maintained a certain reserve towards violence within ruling political circles. The language of political contestation illustrates the point. He most frequently employs the terms *thawra*, *khilāf* and *qiyām* for political revolts resolved by resort to arms. The single use of the term *fitna* referred to widespread but unspecified disturbances fortuitously resolved by God. In his survey of Indo-Persian historical writing, Peter Hardy cites the Arabic terms employed for 'wrongful violence' against kings such as *ṭughyān*, *ʿiṣyān*, *maʿṣiya*, *baghy* and *fasād*.[48] These acts are deemed 'wrongful' inasmuch as the user of force is seen as having broken faith with his master (and therefore God), the terms themselves suggesting, in their Qur'anic context, unbelief or treachery. Ibn Baṭṭūṭa does not appear to share this paradigm. Neither the terms he uses for revolt, nor his accounts of them, hint at movements that should be described or judged conceptually in religious terms. Moreover, in the one instance where the lists correspond, Ibn Baṭṭūṭa uses the word *fasād* of anonymous groups (*ahl al-fasād*, including criminals, or more simply, *mufsidūn*, disturbers of the peace) who were attracted by the success of certain revolts (*khilāf*). In

---

[45] Ibn Baṭṭūṭa, p. 389.
[46] Jackson, p. 126.
[47] Arendt, p. 56.
[48] Hardy 1983, p. 182ff. See also Hardy's essay on 'The authority of Muslim kings,' in M. Gaborieau (ed.), *Islam et société en Asie du Sud* (Paris, 1986), p. 46.

other words, they were 'hangers-on', 'fellow travellers' not leaders challenging for access to power.

We have also seen that Ibn Baṭṭūṭa's attitude to violence was ambivalent. He was personally disturbed by some manifestations, political or otherwise. Yet, he could approve violence, including capital punishment, 'in the name of God'. On the other hand, his religious commitment had a deeper vein that touched the core of the human condition. A striking episode is recounted later in the India narrative during his visit to the Sultan of Ma'bar (Madura), Ghiyāth al-Dīn Dāmaghānī. The kingdom of the 'infidel' sultan, Balal Deo, was nearby whose forces greatly outnumbered those of Dāmaghānī. Indeed they included a sizable number of Muslim 'scoundrels, criminals and run away slaves.' Despite the odds, Dāmaghānī scored 'one of the greatest victories of Islam' and Balal Deo was captured. Dāmaghānī promised to release him in return for his riches, but when these were secured he was executed, his skin filled with straw and hung on the wall of Madura where Ibn Baṭṭūṭa saw it suspended. He makes no further comment. Earlier, Ibn Baṭṭūṭa had been in the same Sultan's presence when a *kāfir* was brought in with his wife and young son. Dāmaghānī ordered them beheaded and while this was done, Ibn Baṭṭūṭa had to avert his gaze. On another occasion in similar circumstances, Ibn Baṭṭūṭa rose and excused himself to go and pray much to the Sultan's amusement. When he saw 'infidel' prisoners, men, women and children, put to death on sharpened stakes, Ibn Baṭṭūṭa, writes, 'That was an abomination ('*amr shanī'*') which I had never known a king to do. For this, God hastened [Dāmaghānī's] death.'[49]

Political reality perceived by Ibn Baṭṭūṭa was a circle of deadly conflict. Or rather, he experienced violence as part of the human condition, disturbing yet unavoidable. He could justify its legitimacy in some cases, adopt a more neutral position in others, or contest its exercise from a sense of human compassion. Violence is indeed, according to Ibn Baṭṭūṭa's narrative, a perplexing phenomenon.[50]

---

[49] Ibn Baṭṭūṭa, IV, pp. 192–198.

[50] A different approach to the same period, utilising the insights of Weber's patrimonial polity, is the excellent study by S. Conermann, *Die Beschreibung Indiens in der 'Riḥla' des Ibn Baṭṭūṭa. Aspekte einer herrschaftssoziologischen Einordnung des Delhi-Sultanates unter Muḥammad b. Tuġluq* (Berlin, 1993).

IBN KHALDŪN, A CRITICAL HISTORIAN AT WORK

THE *MUQADDIMA* ON SECRETARIES AND SECRETARIAL WRITING[1]

Maaike van Berkel

Ibn Khaldūn belongs to a small group of medieval Arab men of learning who enjoy great fame in the West. Discovered and translated by European Orientalists in the nineteenth century, he soon gained the reputation of being a universal scholarly hero and his reputation spread outside a small group of Arabists. Since then, he has been extensively studied and been labelled alternately the father of sociology, the founder of scientific history and a forerunner of historic materialism. Indeed comparisons have been made with almost every great thinker of Western civilization, from Machiavelli to Marx.

Ibn Khaldūn has even achieved the honour of becoming the favourite Arab quoted by a twentieth-century president of the United States, who has not exactly gone down in history for his scholarly mind or interest in Arab-Islamic culture. On many occasions, Ronald Reagan praised Ibn Khaldūn for his plea for low taxes. He qualified him as a forerunner of the supply-side economics, quoting from the third chapter of Ibn Khaldūn's most famous work, the *Muqaddima*: 'It should be known that at the beginning of the dynasty, taxation yields a large revenue from small assessments. At the end of the dynasty, taxation yields a small revenue from large assessments.'[2]

Indubitably, heroic descriptions or, even more so, a-historic appropriations can easily obscure a clear-sighted view of Ibn Khaldūn and his work. It is indeed doubtful whether the man and his work can live up to

[1] Many of the arguments of this article have profited from discussions I had with Rudi Künzel. I would like to thank him and Remco Raben for reading earlier versions of this text. Also I am grateful to Remke Kruk, who – without knowing that I was writing an article on Ibn Khaldūn for her festschrift – took off a morning in her busy schedule to discuss this intriguing scholar with me.

[2] Ibn Khaldūn, *al-Muqaddima* (Tunis, 1991), p. 154. Ibn Khaldūn, *The Muqaddimah. An introduction to history*, tr. F. Rosenthal (Princeton, 1967), II, pp. 89–91. All citations from the *Muqaddima* in this article are taken from Rosenthal's translation.

the ascriptions of later eras, a debate which has been intensified by the appearance of Aziz al-Azmeh's studies on Ibn Khaldūn.[3] It goes without saying that this short article will not provide a final judgement on the merits of this great scholar who has intrigued generations of historians and sociologists. Nor will it deal with the *Muqaddima*, the first book of his *Universal History* (*Kitāb al-ʿIbar*), as a whole. In the *Muqaddima*, Ibn Khaldūn set forth a historical approach he proclaimed to be new. By testing transmitted historical information against the general laws underlying human civilizations in their various stages, he asserted he had developed a new method for sifting reliable from unreliable historical data. For this reason he has often been labelled the inventor of scientific history. The novelty or traditionalism of his reflections on historical science in itself – the topic of some of the prominent studies on Ibn Khaldūn – will not be discussed in the following pages. Rather, this article will analyse – on a very modest scale – the manner in which Ibn Khaldūn actually applied his historical methodology and how he composed his text.

To provide tentative answers to these questions, I will take a closer look at a few specific paragraphs of the *Muqaddima*, namely those dealing with secretaries (*kuttāb*, sing. *kātib*) and secretarial writing. Clerks were very familiar to Ibn Khaldūn. During his lengthy official career, he was employed in many secretarial jobs. Before he had even reached the age of twenty he was appointed secretary to the office of the *ʿalāma* (the Signet) in Tunis and a year later he started working for Abū ʿInān, the new Marīnid ruler in Fez, as secretary in charge of the royal petitions (*tawqīʿ*). He was in his proper place. Ibn Khaldūn descended from a family which had brought forth many influential *kuttāb*. His great-grandfather, Abū Bakr, had even written a treatise on secretary-ship, with which, however, his famous great-grandson does not seem to have been familiar.[4] Setting out on this brief journey will be an interesting exercise to see whether Ibn Khaldūn applied the critical historical attitude he promotes so vividly in the introductory parts of his *Muqaddima* to this familiar field of knowledge. Furthermore, an analysis of his views on clerks is especially intriguing since it may reveal a potential

[3] Aziz al-Azmeh, *Ibn Khaldūn in modern scholarship. A study in Orientalism* (London, 1980) and *Ibn Khaldūn. An essay in reinterpretation* (London, 1982).

[4] Lévi-Provençal, 'Le traité d'*adab al-kātib* d'Abū Bakr Ibn Khaldūn,' *Arabica* 2 (1955), pp. 280–288.

tension between the self-proclaimed novelty of his approach and the firmly rooted tradition of writing about secretaries in manuals and histories, a tradition with which Ibn Khaldūn was very well acquainted. To gain some understanding of his method of working, three specific questions will be asked. What kind of information about secretaries did Ibn Khaldūn select? How did he analyse the selected material? And did his analysis and (possibly) critical methodology produce new views and theories on secretaries and their position in society?

The people of the pen in the Muqaddima

The first step is to identify the places in the *Muqaddima* where Ibn Khaldūn refers to civil servants and secretarial writing and analyse what precisely he has to say about the topic. *Kuttāb* figure throughout the text, but most prominently in a few paragraphs in the third chapter and only modestly in others. In the section entitled 'The ranks of royal and governmental authority and the titles befitting these ranks', Ibn Khaldūn provides the most extensive information.[5] Here he introduces the various types of governmental institutions and their employees: the vizierate (*wizāra*); the office of doorkeeper (*ḥijāba*); the ministry of financial operations and taxation (*dīwān al-aʿmāl wa ʾl-jibāyāt*); the ministry of official correspondence and writing (*dīwān al-rasāʾil wa ʾl-kitāba*); the police (*shurṭa*); and the admiralty (*qiyādat al-asāṭīl*). He contrasts his own historical perspective with studies on the legal aspects of these institutions. Royal and governmental positions, he argues in the introduction to this paragraph, will not be discussed with reference to their legal status – this has been done thoroughly in books on the ordinances of government (*al-aḥkām al-sulṭāniyya*), such as that by al-Māwardī (d. 1058) –, but as something required by 'the nature of civilization in human existence'.

The section on the vizierate discusses the main governmental tasks and the chronological and geographical development of its corresponding institutions from the early Islamic state until the author's own day.[6] Although the vizierate and its various guises are obviously the focal

[5] Ibn Khaldūn, pp. 130–142. Paragraph 32 of chapter three of Rosenthal's translation, II, pp. 3–46.

[6] The vizierate: Ibn Khaldūn, pp. 131–133; Rosenthal, II, pp. 6–14.

point here, Ibn Khaldūn often refers to other, generally subordinate, governmental institutions such as those concerned with secretarial writing. On account of the 'simplicity of Islam' and the illiteracy of the early Muslims, he argues, offices involved in tax collection, expenditure, bookkeeping, correspondence and the writing down of orders – all of which had existed in states before the advent of Islam – disappeared from the early Islamic state. For bookkeeping, *ahl al-kitāb* (people of the book, Jews and Christians) or *mawālī* (non-Arab, newly converted Muslims) were employed. The caliph appointed occasional writers for the writing of state papers. Afterwards – Ibn Khaldūn does not provide any precise chronological indication, but from the context it is clear that he is referring to the Umayyad era – a special professional secretary was appointed in an effort to prevent the ruler's secrets from becoming public knowledge. Unlike the vizier, this official's sphere of activity was not concerned with oral matters, but was restricted to writing.[7] At that time, bookkeeping still remained the work of *mawālī* and non-Muslims.

During the 'Abbāsid era, the vizier's influence reached its apex. He supervised both the bookkeeping and 'the pen', the latter task not only for the protection of the ruler's secrets, but also to preserve good style. At that time, the Arabic language had become corrupted, according to Ibn Khaldūn. After this era, with the seizure of power by non-Arab rulers, the title of vizier was used to describe a fairly powerless official in the private retinue of the caliph. In the course of this period, Ibn Khaldūn continues, language decayed even more and secretary-ship became a craft which was the preserve of a group of specialized people. And, since viziers were often non-Arabs and eloquence could not be expected of them, it was these specialists – Ibn Khaldūn does not mention titles or ranks – who were responsible for good style. Later, under the Mamlūks, the term of vizier was used for the person in charge of the tax collection. Quite different was the application of this term under the Umayyads in Spain. The Umayyads had several viziers, each one of them concerned with a specialized function: one for correspondence, another for accountancy and yet another for supervising the border regions.

---

[7] Generally, the introduction of the institution of the vizierate is situated among the early ʿAbbāsids, whereas Ibn Khaldūn already identifies the vizier as the highest official rank in the Umayyad state. Cf. D. Sourdel, *Le vizirat ʿAbbāside de 749 à 936 (132 à 324 de l'Hégire)* (Damascus, 1959–1960), pp. 41–73.

'The ministry of taxation,' Ibn Khaldūn argues in one of the next sections, 'is an office that is necessary to the royal authority'.[8] In a nutshell, its task was the administration of the state revenues and expenses. Its employees had to be specialists in accounting. The argument continues with an analysis of the stages during which dynasties would be in need of a tax office. This is followed by the introduction of the taxation bureau under Caliph 'Umar (r. 634–644). The languages used in the tax office remained as they had been before the advent of Islam, Ibn Khaldūn adduces. Persian was used in Iraq; while Byzantine Greek was used in Syria. The *kuttāb* of both countries were appointed from among the *ahl al-'ahd* (conquered people). In the reign of Caliph 'Abd al-Malik (r. 685–705), Arabic was introduced in the Iraqi and Syrian tax bureaus. At this point in his narration Ibn Khaldūn again relates the objectives of his historical approach to the legal aspects of governmental institutions. In the *Muqaddima* his intention is to deal with royal authority. There are three pillars of royal authority, he continues, namely: soldiers, money and the means to communicate with those who are absent. Therefore rulers need specialists to deal with the sword, finances and the pen. Detailed geographical and chronological variations in the financial pillar in Spain, the Maghrib, and Egypt conclude the section on the tax bureau.

Unlike a ministry of finance, Ibn Khaldūn does not consider a bureau for correspondence and writing an absolute requirement for royal authority.[9] Indeed, he opens the relevant section with the statement that: 'Many dynasties were able to dispense with it completely'. This is especially true for dynasties which were 'rooted in the desert and not (yet) affected by the refinements of sedentary culture and the high development of the crafts.' In the early Islamic state, the *Muqaddima* continues, a writer was generally a relative of the person for whom he worked, whether this latter was the caliph or one of the leading men in Syria and Iraq. Later, the writing of official correspondence became a specialized craft, not entrusted to a sibling by chance well-versed in the art, but practised by a corps of experts trained in the composition of state documents. In the 'Abbāsid era, the secretaries of the correspondence

---

[8] The ministry of financial operations and taxation: Ibn Khaldūn, pp. 134–136; Rosenthal, II, pp. 19–26.

[9] The ministry of official correspondence and writing: Ibn Khaldūn, pp. 136–139; Rosenthal, II, pp. 26–35.

bureau manage to climb high on the social ladder. As a consequence, they were even allowed to authorize documents with their own signature. However, in later periods – a more detailed chronological specification is missing – the writing office lost most of its prestige. Other types of officials gained prominence. The situation in the later Ḥafṣid dynasty is presented as an example of this. At that time, the doorkeeper became the highest state official, eventually controlling even the ruler himself.

One of the secretary's main tasks, Ibn Khaldūn continues in the section on the correspondence bureau, was recording decisions on petitions. This job required a wealth of stylistic talents. Even more demanding were the qualities required from the person in charge of the bureau. His qualifications were defined not only in terms of his specialized skills, but they also took account of his general cultural baggage and education and his good manners. At this point, Ibn Khaldūn draws attention to the possibility of a dynasty under which the chief secretary could be a military man or under which a military companion of the ruler could be given authority over the head of the correspondence bureau. This is what happened in Mamlūk Egypt, where the head of the chancery, the *ṣāḥib al-inshā*ʾ, was supervised by an *amīr*. The section concludes with the integral recording of a treatise in which the Umayyad secretary ʿAbd al-Ḥamīd b. Yaḥyā (d. 750) addresses his fellow secretaries. This treatise is – according to Ibn Khaldūn – the most complete account of the qualifications that rulers have to consider when choosing their secretaries.

In 'The different importance of the ranks of the sword and the pen in the various dynasties',[10] Ibn Khaldūn discusses the position of civil servants in relation to the other important group of governmental officials, the military. Both at the beginning and at the end of a dynasty, a ruler is more in need of the sword than the pen, he argues. During these stages 'the pen is merely a servant and agent of the ruler's authority, whereas the sword contributes active assistance.' At the beginning, the military plays a major role in establishing the dynasty. At the end it is needed to avoid its collapse. The middle period, on the other hand, witnesses an increase in esteem for the pen. During this stage 'the swords stay unused in their scabbards.' After royal authority has been firmly established, the ruler is in need of the pen to administer his realm. Consequently, the people of the pen are closest to him during this period.

---

[10] Ibn Khaldūn, p. 142. Paragraph 33 of chapter three of Rosenthal's translation, II, pp. 46–48.

The position of the people of the pen in the various stages of a dynasty is also discussed in the paragraph: 'The ruler and his entourage are wealthy only in the middle (period) of the dynasty'.[11] Here Ibn Khaldūn draws attention to the connection between the social status of secretaries and their financial rewards. At the beginning of a dynasty, the author argues, state revenues are distributed mainly among those who 'share in the ruler's group feeling' and on whom he depends in the establishment of the dynasty. Secretaries and the rest of the entourage have to be satisfied with a very small remuneration. The situation changes in the middle period of the dynasty. Commensurate with their increasing usefulness to the dynasty, the civil servants in the ruler's entourage are able to acquire property. During this period they enrich themselves. Then, in the last phase of a dynasty, the ruler again requires military supporters, this time from outside the group which shared his group feeling. Consequently, his revenues are bestowed on these military men. There is a constant lack of money and therefore the ruler starts to confiscate the riches of his officials. The loyalty of these civil servants, in their turn, grows fragile. As a consequence, the dynasty suffers seriously from the loss of its servants and men of distinction and wealth. This happened, for example, to some of the ʿAbbāsid vizier families and some of the prominent families in Spain at the end of the Umayyad dynasty. Ibn Khaldūn concludes this paragraph with a recommendation to officials in this precarious last phase of a dynasty not to abandon their jobs or to flee to another region, especially should they not take along their acquired riches. This – so he warns them – will eventually ruin their lives and fortunes.

More recommendations for governmental officials are given in numerous miscellaneous paragraphs in other parts of the *Muqaddima*. For example, from the great historian al-Ṭabarī (d. 923), Ibn Khaldūn copied a letter that Ṭāhir b. al-Ḥusayn (d. 822) wrote to his son.[12] The letter was intended to advise the son about government and administration on the latter's appointment as governor. The father exhorts his son to arrange his agenda to see each one of his secretaries for a specific time of the day in order to be able to deal quietly and deliberately with the matters they present. And elsewhere Ibn Khaldūn speaks of the

[11] Ibn Khaldūn, pp. 156–158. Paragraph 39 of chapter three of Rosenthal's translation, II, pp. 97–102.
[12] Ṭāhir's letter to his son: Ibn Khaldūn, pp. 168–172; Rosenthal, II, pp. 139–156.

advantages of obsequiousness and flattery in the obtaining of profitable positions.[13] Here, he admonishes skilled scribes to refrain from haughtiness and arrogance.

Finally, the peculiarities of the secretarial craft figure in some paragraphs on writing in general. For example, in his extensive section on calligraphy in Chapter Five – discussing scripts from the Tubbaʿ dynasty (late 3rd–early 6th centuries) up to his own day –, Ibn Khaldūn adds an excursus on coded script used by governmental officials to conceal information.[14] He refers to their use of, for example, the names of perfumes, fruits, birds or flowers to denote specific letters of the alphabet. The codes are meant to be decipherable only by a small group of colleagues. The rub is that sometimes, it transpires, an intelligent scribe from outside is able to solve the puzzle. Another excursus on secretarial writing is found in the paragraph dealing with 'The division of speech in poetry and prose'.[15] Here Ibn Khaldūn notes his strong disapproval of the mixing up of poetry and prose styles in official correspondence. Present-day secretaries, he says, use poetical idioms in their governmental documents. Such a mannered style is incompatible with the function of these documents, which is 'the encouraging or frightening of the masses'. The officials who use such a pompous style are often of non-Arab origin. By adding excessive embellishments, they try to 'cover up their inability to make their speech conform to the things they want to say and to the requirements of the particular situation.'

A critical historian at work

Before analysing Ibn Khaldūn's characterization of the *kātib* in more detail, I would like to pay attention to the way in which he worked. Or, more precisely to look at the way in which the application or the misapplication of his critical methods emerges in the selection and presentation of his historical information on secretaries. Such an analysis helps

---

[13] Obsequiousness and flattery: Ibn Khaldūn, pp. 215–217; Rosenthal, II, pp. 328–334.

[14] Code writing: Rosenthal, II, pp. 390–391. This section is an addition, not found in all manuscripts and absent in the Arabic edition used for this article, cf. Rosenthal, II, p. 388, n. 192.

[15] Ibn Khaldūn, pp. 326–327. Paragraph 52 of chapter six of Rosenthal's translation, III, pp. 368–371.

to identify the new and more traditional elements in his *kātib*-image, precisely because it is in his critical evaluation of historical facts that we can discern Ibn Khaldūn's additions to the then existing literary image of the *kātib*.

Much has been said about Ibn Khaldūn's historical criticism. For an understanding of the rest of my argument, a very short summary will suffice. In the introductory sections of the *Muqaddima,* Ibn Khaldūn complains about the absence of a critical attitude among his predecessor-historians.[16] He criticizes them for trusting historical information as it was transmitted, neither 'checking it with the principles underlying the corresponding historical situations, nor comparing it with similar material.' He therefore argues in favour of the verification of transmitted historical anecdotes against the historical evolution of societies. How then is the historian able to detect the characteristics of the development of human civilization? To accomplish this, Ibn Khaldūn introduces a new science, *ʿilm al-ʿumrān*, which investigates human social organization. It studies the patterns underlying and determining historical evolution. Knowledge of these patterns will provide historians with 'a normative method for distinguishing right from wrong and truth from falsehood in historical information....'

With this promise in mind, we shall look once more at Ibn Khaldūn's discussion of secretaries and try to identify the application of this critical methodology in the presentation of his material. In most of the sections discussing secretaries and their tasks, Ibn Khaldūn interlards his text with theoretical explanations on the laws governing various societies at different stages in their development. Often these theoretical clauses – sometimes no more than small notes – provide detailed elaborations of the general models of the *Muqaddima* as, for example, the rise and decline of civilizations. And they are generally followed (or occasionally preceded) by one or more concrete historical examples. Having critically evaluated these against the general laws of human social organization, Ibn Khaldūn therefore presents his historical anecdotes as reliable historical facts.

The section on 'The ministry of financial operations and taxation' may be taken as an example to demonstrate this alternation of general patterns and specific historical examples. Ibn Khaldūn commences with

---

[16] Ibn Khaldūn, pp. 4–26. Rosenthal, I, pp. 6–85.

an indication of the phase in which a tax bureau may develop by stating a general law: 'It should be known that the office (of tax collections) originates in dynasties only when their power and superiority and their interest in the different aspects of royal authority and in the ways of efficient administration have become firmly established.'[17] Then the specific historical example follows with the discussion of the first *dīwān* during the reign of Caliph ʿUmar (634–644). Further on in the section we find the next example. Dealing with the introduction of Arabic instead of Greek and Persian in the financial departments of Syria and Iraq during the reign of the Umayyad Caliph ʿAbd al-Malik (r. 685–705), Ibn Khaldūn again inserts a general law. He argues that this linguistic shift coincided with a phase in which 'people turned from the low standard of desert life to the splendour of sedentary culture and from the simplicity of illiteracy to the sophistication of literacy.'[18] In the other paragraphs dealing with secretaries, the alternation between model and historical anecdote is also clearly distinguishable. For example, the development from occasional writers employed by the early Islamic state to a permanent staff later is linked to general patterns of development in the direction of civilization and urbanization and the theoretical excursus on the ruler's abandonment of civil servants in the last stage of a dynasty is followed by the historical example of the fall of some ʿAbbāsid vizier families.

These are classical examples of Ibn Khaldūn's methodology. However, the author does not obey his own rules consistently. Sometimes information seems to elude verification against the evolutionary laws of human civilization. Or worse, Ibn Khaldūn's description remains prescriptive. An illustrative example is the introduction to the section on the tax bureau in which he characterizes the office by its tasks and procedures without applying any chronological or geographical differentiation.[19] Again in the section on the ministry of official correspondence, the person in charge of the bureau is defined in terms of ideal qualifications. His required education, cultural baggage and moral attitudes are mentioned without testing this ideal against the historical evolution of societies. This gap is even more evident in the description of the requirements for

[17] Ibn Khaldūn, pp. 134–135. Rosenthal, II, pp. 20–21.
[18] Ibn Khaldūn, p. 135. Rosenthal, II, p. 22.
[19] Ibn Khaldūn, p. 134. Rosenthal, II, pp. 19–20.

secretaries in general, civil servants of all types and ranks. At this point, Ibn Khaldūn refers to the ideal qualifications put forward by the eighth-century Umayyad secretary ʿAbd al-Ḥamīd b. Yaḥyā in his famous letter to his fellow *kuttāb*. Here the self-proclaimed critical historian pays no attention at all to any sort of historical differentiation or evolution.

The kātib*-image: something old, something new, something borrowed...*

Ibn Khaldūn's plea for a critical historical approach brings us to the next and last question of this article: Does his methodology lead to a critical evaluation of previous images of the *kātib* and, in exchange, does he provide new ideas and views on this important official? Does he openly criticize the then existing literary images by testing them against the changing laws underlying societies? And does he add unique and original elements to them? Comparison between Ibn Khaldūn's views and those by his predecessors is especially interesting in the light of the tradition of writings about secretaries. Many images of the *kātib* handed down by Ibn Khaldūn's predecessors have a rather static, a-historical, character. Meant to sketch a normative ideal, they lack chronological, geographical or hierarchical differentiations. This is particularly true of the administrative literature, prescriptive manuals intended to provide examples of good administration. Even in historical writings normative descriptions are present, often inserted as the presentation of historical examples of good and bad secretary-ship. At the same time, as may have been expected, historical texts pay attention to particular historical events and situations, and therefore they also provide a broad and differentiated range of *kātib*-images. The *Muqaddima* carries elements of both traditions, the normative and the descriptive, but it also extends beyond them.

Let us start with the latter, 'the extra', which contains no doubt the most exciting elements of Ibn Khaldūn's *kātib*-image. In the paragraphs in which Ibn Khaldūn presents the patterns underlying historical evolution, he has also new and provocative things to say about secretaries. These are new visions, at least as far as can be reconstructed from the surviving literature. These new images are not concerned with particular, unique historical situations but, unlike the pre-existing historical texts, address general models. Moreover, in contrast to the normative descriptions of the administrative literature, these models are not static, but pay attention to the alteration in the positions of various officials in changing societies of varying dimensions.

One of the most telling examples of such a new image is the paragraph on the pen and the sword.[20] Here Ibn Khaldūn is able to specify his model of various types and stages of dynastic rule for the positions of civil and military servants. In sharp contrast to the administrative literature – always harping on the absolute importance of *kuttāb* in any given situation – he refreshingly puts the people of the pen in perspective by pointing out their naturally inferior position to the people of the sword during the early and late stages of a dynasty.[21] Less useful means lower on the hierarchical ladder, is his argument. Protection and defence are needed at the beginning and the end of a dynasty, a well-oiled administration is important in the mid-term. Therefore military officials are the ruler's main pillars in the early and later phases of a dynasty, while civil servants play this role in the middle period. He elaborates on this theme in one of the next paragraphs arguing that, as a consequence of this status, the financial rewards to secretaries are also less in the beginning and end of a dynasty.[22] Cogently, from this line of argument also follow his remarks on the waning loyalty of secretaries in the end phase of a dynasty. Another model he adduces is the influence of territorial size on the differentiation in the bureaucratic organization of a dynasty, as discussed in the section on the Mamlūk financial administration.[23] The same is true of the observation that the extent of urbanization influences the organization of governmental administration and the disposition of its secretaries.[24]

These are all examples of specified models, detailed elaborations of the general patterns underlying social organization. However, alongside these daringly innovative theories, Ibn Khaldūn also adds numerous

---

[20] Ibn Khaldūn, p. 142. Paragraph 33 of chapter three of Rosenthal's translation, II, pp. 46–48.

[21] Also in literary debates between pen and sword an evolutionary or cyclic description of the relation between the two is absent. Cf. G.J. van Gelder, 'The conceit of pen and sword: on an Arabic literary debate,' *Journal of Semitic studies* 32/2 (1987), pp. 329–360.

[22] 'The ruler and his entourage are wealthy only in the middle (period) of the dynasty': Ibn Khaldūn, pp. 156–158; Rosenthal, II, pp. 97–102.

[23] Ibn Khaldūn, p. 136; Rosenthal, II, p. 25.

[24] For example, his conclusion about the absence of a correspondence bureau in dynasties rooted in the desert. Ibn Khaldūn, p. 136. Rosenthal, II, p. 26.

historical examples, anecdotes of unique historical events. Many of these are nestled among a description of the general laws in their society, this being the way they are subjected to the author's critical methodology. Incontrovertibly, while their function as building blocks for the general 'theories' is highly innovative, in themselves many of the stories do not differ greatly from the way they are told in earlier sources. In this sense, Ibn Khaldūn also clearly demonstrates he has his place in the Arab-Islamic tradition of historical writing. He simply filtered out those anecdotes which seemed unreliable when tested against the general laws of historical evolution. Consequently, the stories which found their way into the *Muqaddima* are those anecdotes which had stood the test. These seem to stem mainly from pre-existing texts and perhaps partially from the author's own experiences and observations at the various courts in which he worked. The examples copied from previous texts – either vaguely from recollection or exactly to the letter – refer, although not always explicitly, to their original sources.[25] Large parts of the section on the financial bureau, for example, are based on al-Māwardī's *al-Aḥkām al-sulṭāniyya* without this source always being mentioned specifically by Ibn Khaldūn.[26] On the other hand, those historical details we cannot retrace to a specific source do not necessarily spring from the author's own experiences or from hear-say. Even when it concerns Ibn Khaldūn's own period and area, caution is advised since a relatively small amount of the literature from the Maghrib has survived. As Rosenthal aptly cautions us: 'Under these circumstances, we should perhaps be justified in assuming that practically every matter of detail found in the *Muqaddima* was probably not original with Ibn Khaldūn, but had been previously expressed elsewhere.'[27] Even so, some of the untraceable details dealing with specific secretarial functions in North Africa and Spain contain many unique *kātib*-images which are – with the situation as it stands – new to us and therefore add to our knowledge of secretarial writing in the West.

Finally, the most unexpected element in Ibn Khaldūn's secretary-image is the normative ideal copied from the administrative literature. This is most unexpected because its static, prescriptive and a-historical character seems to be in conflict with Ibn Khaldūn's explicit emphasis

---

[25] See Rosenthal's clear analysis of Ibn Khaldūn's scholarly craftsmanship in the introduction to the translation, pp. lxx–lxxi and for the sources, pp. lxxxiv–lxxxvii.

[26] Cf. al-Māwardī, *al-Aḥkām al-sulṭāniyya* (Beirut, [ca. 1980]), p. 249ff.

[27] Rosenthal, I, p. lxxxv.

on historical evolution. The most striking examples are the sections about the ideal qualifications required of secretaries. Here the author – unlike in many other instances – makes no chronological, geographical or hierarchical differentiation at all. On the contrary, he presents a norm applicable to all types of secretaries in all kinds of societies throughout the centuries. In a more subtle way this is also true of other fragments. In the section on poetry and prose, for example, Ibn Khaldūn complained about the unsuitable and pompous style used by his contemporaries in official correspondence, which he attributes to their non-Arab background.[28] Pertinently, criticism of *kuttāb* regretting their failure to handle the Arabic language correctly was not a novel topic in Arabic literature. Indeed, it could be said to have been one of the *topoi* of the administrative literature in the early days. On more than one occasion, such authors as al-Jāḥiẓ (d. 869) and Ibn Qutayba (d. 889), for example, ridiculed their contemporary *kuttāb* for their clumsy use of the Arabic language, citing precisely their non-Arab origin.[29] The same can be said about code writing, described as necessary for secretaries wishing to conceal governmental secrets.[30] Knowledge of codes and concomitantly of their decipherment was already defined as a prime requisite of secretaries in the ʿAbbāsid period and many administrative manuals have sections devoted to this topic.[31]

These were the three elements of the *kātib*-image presented in Ibn Khaldūn's *Muqaddima*: the general pattern, the historical anecdote and the normative ideal. What remains are questions about his choices in the selection of material about secretaries. The query which looms largest is why a critical historian, who insisted on the importance of historical evolution and was very confident about his new methodology, still inserted so many non-historical, normative descriptions. Was he upholding a tradition? Was he deliberately providing recommendations for secretaries and rulers of his own age and by so doing combining historical models and anecdotes with political advice? Or did he believe

---

[28] Ibn Khaldūn, pp. 326–327. Rosenthal, III, pp. 368–371.

[29] See, for example, al-Jāḥiẓ, *Dhamm akhlāq al-kuttāb* in: J. Finkel (ed.), *Three essays* (Cairo, 1926), pp. 40–51 and Ibn Qutayba, *Adab al-kātib* (Beirut, 1986), especially the introduction.

[30] Rosenthal, II, pp. 390–391.

[31] C.E. Bosworth, 'The section on codes and their decipherment in Qalqashandī's "Ṣubḥ al-Aʿshā",' *Journal of semitic studies* 8 (1963), pp. 17–33.

that, regardless of the actual historical situation, norms did not change and secretaries from all ranks, areas and times should be obliged to live up to this high standard? Unfortunately, here Ibn Khaldūn leaves us with these questions unanswered. But then, perhaps it is a comfortable conclusion to assume that even such a formidable man as Ibn Khaldūn, who proposed such a novel historical approach, is still recognizably a child of his time. A writer, after all, always cherishes his traditions.

DIE LITERARISIERUNG DER MAMLUKISCHEN HISTORIOGRAFIE

VERSUCH EINER SELBSTKRITIK

Bernd Radtke

Der titel zeigt, dass dies nicht mein erster versuch ist, um über die 'literarisierung' der mamlukischen historiografie klarheit zu gewinnen. Vorangegangen sind fünf weitere. Sie begannen mit der einleitung zur edition des 1. band der weltchronik des Ibn al Dawādārī.[1] Ausführlicher konnte ich in *Weltgeschichte und Weltbeschreibung* werden,[2] weiteres brachte ich in drei kürzeren artikeln vor.[3]

Worum ging es? In aller kürze: Ulrich Haarmann hatte durch das studium mamlukischer chroniken – vor allem der des Ibn al-Dawādārī (gest. nach 736/1335) – den eindruck gewonnen, dass hier eine art von geschichtsschreibung vorliege, die sich von einer älteren unterscheide. Im gegensatz zu dieser, wissenschaftlich orientierten, geschichtsschreibung liege eine 'literarisierte' vor, die nicht mehr als rein wissenschaftlich zu bewerten ist. Die gründe sind das eindringen von unterhaltenden stoffen, von 'volksliteratur', von berichten über mirabilia, weiter der sorglose umgang mit den historischen 'fakten'.[4]

In den genannten arbeiten wies ich darauf hin, dass die von Haarmann beobachteten fänomene für die mamlukenzeit nichts eigentlich neues darstellen, sondern samt und sonders fast schon zu beginn der islamischen geschichtsschreibung vorhanden sind. Wichtig war mir vor allem, dass sie nicht als beweis für eine mögliche dekadenz der islamischen kultur herhalten können. Solche bewertung projiziere ein modernes wissenschafts- und wirklichkeitverständnis auf die vergangenheit.

---

[1] *Die Chronik des Ibn al-Dawādārī*, I, *Kosmografie*, hrg. von Bernd Radtke (Kairo [etc.], 1982).

[2] Bernd Radtke, *Weltgeschichte und Weltbeshreibung im mittelalterlichen Islam* (Beirut [etc.], 1992).

[3] Bernd Radtke, 'Das Wirklichkeitsverständnis der islamischen Universalhistoriker,' *Der Islam* 62 (1985), S. 59–70; *id.*, 'Die älteste islamische Kosmographie. Muḥammad-i Ṭūsīs ʿAǧāʾib ul-maḫlūqāt,' *Der Islam* 64 (1987), S. 279–288; *id.*, 'Zur literarisierten Volkschronik der Mamlukenzeit,' *Saeculum* 41 (1990), S. 44–52.

[4] Zusammenfassung in *Weltgeschichte*, S. 186 f.

Ich musste feststellen, dass eine eingehende auseinandersetzung mit meinen positionen ausbliebt,[5] zumindest was die mamlukische historiografie angeht. In letzter zeit zeichnet sich eine veränderung ab. Hinzuweisen ist auf den aufsatz von Stefan Leder 'Post-klassisch und prä-modern: Beobachtungen zum Kulturwandel in der Mamlukenzeit',[6] der einmal meinen ergebnissen zustimmt und zum anderen neue gesichtspunkte bringt. Auch andere arbeiten zeigen eine fortgeschrittene reflexion über das thema.[7]

Und doch: Es beschlichen mich im laufe der zeit zweifel, ob ich nicht zu weit gegangen war und das kind mit dem bade ausgeschüttet hatte. Wenn ich auch meine ergebnisse im allgemeinen nicht zu korrigieren hatte, so vielleicht doch im detail. Gab es womöglich entwicklungen, die ich übersehen hatte?

Ich möchte dieser frage nachgehen, indem ich über den begriff der *fiktion* in der arabischen literatur nachdenke.

Was ist fiktion – literarische fiktion? Selbst eine grobe begriffsbestimmung ist nur möglich, wenn die begriffe literatur und wirklichkeit annähernd bestimmt werden. Wenn es auch trivial klingen mag, so muss doch zunächst festgestellt werden, dass literatur die schriftliche, sprachliche wiedergabe von wahrnehmungen ist, die in ein sinnbestimmtes ganzes gefasst werden. Bei der wahrnehmung mag es sich entweder um sinnliche oder um innere handeln, die sich auf durch menschliche eigenaktivität hervorgebrachtes bezieht: auf produkte der vorstellungstätigkeit also. Realität, wirklichkeit besitzt zunächst im gleichen grad das objekt der sinnlichen wahrnehmung und das objekt der inneren wahrnehmung, das durch menschliche vorstellungstätigkeit erzeugt wird – man kann hier den begriff *fantasie* einführen.

Das verhältnis, das zu beiden wahrnehmungsobjekten gewonnen werden kann, ist variabel. Ständig wird unter den objekten der äusseren wahrnehmung, die uns jeweils zugänglich sind, eine grösstensteils unbewusst bleibende auswahl getroffen. Dieser tatbestand ist der psychologie bekannt. Andrerseits wird die äussere wahrnehmung bzw. das äussere wahrnehmungsobjekt sehr unterschiedlich interpretiert, d.h.

[5] So selbst in Haarmanns besprechung, *JAOS* 115 (1995), S. 133–135.

[6] In: Stephan Conermann & Anja Pistor-Hatam (hrg.), *Die Mamluken. Studien zu ihrer Geschichte und Kultur. Zum Gedenken an Ulrich Haarmann (1942–1999)* (Schenefeld, 2003), S. 292 f.

[7] Thomas Bauer, 'Ibrāhīm al-Miʿmār. Ein dichtender Handwerker aus Ägyptens Mamlukenzeit,' *ZDMG* 152 (2002), S. 63 ff.

mit sinn versehen, indem nach den gründen der erscheinungen gefragt wird. Nicht zuletzt wird die interpretation der äusseren wahrnehmung von dem verhältnis abhängen, das zu den hervorbringungen der inneren vorstellungstätigkeit gewonnen wird. Solange sprachliche reproduktion der äusseren wahrnehmungswirklichkeit von sprachlicher reproduktion der inneren vorstellungsproduktion geschieden wird, kann man von einer bedingt sauberen wirklichkeitsreproduktion sprechen. Die bewusste vermischung von den beiden reproduktionsebenen nenne ich *fiktion*.

Ohne weiteres ist einzusehen, dass fiktion bewusst mit durchaus unterschiedlichen intentionen eingesetzt werden kann. Es kommen, wie mir scheinen will, vor allem zwei mögliche intentionen in betracht: Fiktion kann benutzt werden, um die sogenannte wirklichkeit der sinne verändert wiederzugeben, damit seelische wirkungen beim hörer oder publikum erzeugt werden, die wir aesthetisch-künstlerisch nennen. Diese mit fiktion durchsetzte reproduktion von wirklichkeit wird nicht unbedingt einen absoluten wahrheitsanspruch für das erzählte geltend machen wollen.

Anders, wenn fiktion bewusst eingesetzt wird, indem sie wirklichkeit so verändert reproduziert, dass sie einer schon vorgegebenen, vorgefassten sinngebung entspricht. Diese vorgefasste sinngebung können wir *ideologie* nennen. Darunter kann auch die religion fallen. Die durch diese art der fiktion beabsichtigte wirkung ist nun nicht künstlerisch, sondern zielt auf das sozial-politische verhalten.

Sehen wir, wie sich unsere theoretischen versuche auf die literarischen verhältnisse des islamischen mittelalters anwenden lassen. Es gab in der islamischen mittelalterlichen gesellschaft zwei bildungstraditionen, bzw. literaturtraditionen. Die mitglieder der einen werden *ʿulamāʾ*, genannt, die anderen *udabāʾ*. Die kultur des *ʿālim* ist von der theologie in ihren unterschiedlichen ausformungen geprägt. Die wissenschaft des *ʿālim* ist *scientia sacra*. Kultur und wissenschaft des *ʿālim* dienen dem gottesdienst, haben darin ihren sinn. Ziel ist theologisch bestimmte wahrheit, die zum rechten handeln führt, und dies wiederum zur ewigen seligkeit.

Selbstverständlich enthält die kultur des *adīb* diese theologische komponente ebenfalls – etwas anderes ist in der theokratisch verfassten islamischen gesellschaft des mittelalters nicht vorstellbar. Man sollte deshalb die dichotomie *ʿilm* (wissenschaft des *ʿālim*) – *adab* (wissenschaft des *adīb*) nicht überstrapazieren. Es handelt sich nicht um gegensätze, sondern um sich gegenseitig ergänzende aspekte derselben bildungstradition. *Adab*, die bildung des *adīb*, ist vielleicht recht gut mit den *artes liberales* des europäischen mittelalters zu vergleichen. Auch ist sie ja kein gegensatz zur *sacra theologia*, sondern vorbereitung und ergänzung. So mag man die bildung des *adīb*, den *adab*, eine auf theologischer grundlage beruhende weltbildung nennen.

Betrachten wir etwas näher die arabische literatur selbst, und zwar zunächst die historiografische, zu der wir auch die biografische zu rechnen haben. Beim studium dieser literatur drängt sich ein eindruck geradezu auf: Man hat es mit einem ungeheuren wust von einzelnachrichten zu tun, ja geradezu mit einer detailverliebtheit. Diese wird auch noch durch gewisse charakteristika der arabischen sprache gefördert. Sowohl die entwicklung des vokabulars als auch der grammatischen formen zielen auf eine scharfe beschreibung punktueller elemente. So hätte man es mit einer literatur zu tun, die es sich zum ziel gesetzt hätte, eine detailtreue abschilderung der wahrnehmung vorzunehmen. Dieses bild täuscht jedoch. Denn es ist durch rezente untersuchungen erkannt worden, dass die arabische historiografie und biografik in der darstellung historischer abläufe und individueller ereignisse überwiegend festgelegte schemata und topoi verwendet. Die eroberung einer stadt, der ablauf einer schlacht, das leben eines heiligen mannes – alles das wird in immer wiederkehrenden topoi und schemata geschildert, so dass das individuelle ganz hinter dem allgemeinen schema, einer allgemeinen konzeption zu verschwinden droht.[8]

Woher nun diese tendenz zum schematisieren und generalisieren, der die schilderung des individuellen historischen ablaufes fast geopfert wird? Einmal ist zu sagen, dass schemata und topoi literarisch-handwerkliche hilfsmittel sind, durch deren gebrauch die darstellung der fast unendlichen individuellen ereignisse erleichtert wird. Zum anderen ist die theologische ausrichtung islamisch-arabischer, besonders frühislamischer historiografie in betracht zu ziehen. Da sie von der setzung ausgeht, dass das eigentliche handlungssubjekt der geschichte Gott und nicht die individuelle menschliche persönlichkeit ist, hat das einzelne, individuelle nur insofern bedeutung und wert, als es ausdruck eines allgemeinen, der heilsgeschichtlichen bestimmtheit von geschichte, ist.

---

[8] Dazu Bernd Radtke, 'Versuch einer Grundsatzbetrachtung über das Allgemeine und das Individuelle in der islamischen Historiographie,' in *XXIV. Deutscher Orientalistentag* (Köln, 1990), S. 265.

Hier hätten wir es dann mit der fiktion zweiter art zu tun, mit der ideologisch bedingten.

Die verfasser historiografischer literatur stammen sowohl aus den kreisen der *ʿulamāʾ* und der *udabāʾ*, und auch die verwendung heilsgeschichtlich bedingter topoi und schemata ist nicht etwa nur auf die gruppe der *ʿulamāʾ* beschränkt, wenn auch hier vielleicht besonders verbreitet.

Betrachten wir eine andere literaturgattung: die geografisch-kosmografische literatur. Ihre entstehung verdankt sie einmal rein praktischen erwägungen. Verwaltungsbeamte des kalifenreiches brauchten informationen über ausdehnung, bevölkerungsdichte, produktion, landwirtschaftliche erzeugnisse usw. des reiches, und so ist es kein wunder, wenn wir viele staatsbeamte – also *kuttāb* – als verfasser dieser literatur finden. Diese beamten standen dem bildungsideal des *adab* nahe. Ein *kātib* hatte ein *adīb zu sein.*

Die geografisch-kosmografische literatur diente jedoch nicht allein der trockenen information, sondern auch, und hier spielt das bildungsideal des *adab* hinein, der unterhaltung. Sie sollte und wollte *prodesse et delectare.* Das gilt besonders für die schilderung ferner länder und fremder sitten. Man kann feststellen, dass autoren geografischer literatur zu mancherlei literarischen mitteln greifen, um ihren berichten eine erhöhte wirkung auf das interesse des lesers zu verleihen.

Von diesen mitteln seien zwei behandelt: 1. der erfundene augenzeugenbericht, und 2. das erfundene zeugnis einer autorität. Erfundene augenzeugenberichte werden fast immer mit der arabischen formel *raʾaytuhū* oder *shahidtuhū ʿiyānan* (ich habe es mit den eigenen augen gesehen) eingeleitet.[9] Wie lässt sich bei solchen anscheinend nicht zu bezweifelnden wahrheitsbeteuerungen die erfindung beweisen? Oft recht einfach durch quellenkenntnis. Man sieht dann, dass der betreffende bericht aus einer früheren quelle abgeschrieben ist. Es muss jedoch eingeräumt werden, dass ein plagiat noch kein zwingender beweis dafür ist, dass der autor im 'Ich habe es selbst gesehen' die unwahrheit sagt. Es besteht ja durchaus die möglichkeit, dass er die eigene erfahrung mit den worten einer früheren quelle wiedergibt, die eine grössere autorität als er besitzt. In diesem fall hätte man nicht gerade eine individuell gestaltete abschilderung von wirklichkeit, aber auch keine erlogene. Oft lässt

---

[9] Radtke, *Weltgeschichte*, S. 171 ff.

sich jedoch durch die aufdeckung von werkimmanenten widersprüchen beweisen, dass der autor die unwahrheit sagt, zum beispiel zum angegebenen zeitpunkt nicht am vorgebenen ort gewesen sein *kann*. Bevor wir eine erklärung für diesen hang zur fiktion in der geografisch-kosmografischen literatur suchen, sei das zweite erwähnte literarische mittel besprochen: Die erfundenen autorität.

Manchmal tritt ein angeblich uralter, weiser mann auf, der wundersames, ja unglaubliches über die vergangenheit zu berichten weiss. Durch die scheinbare autorität dieses zeugen wird unglaubliches wahr. Und wieder wäre nach den gründen für diese art der fiktion zu fragen.

Es können verschiedene angeführt werden. Einmal erinnert gerade das beispiel des erfundenen zeugnisses eines uralten weisen daran, dass die arabische geografisch-kosmografische literatur von autoren stammt, die unter dem einfluss der antiken tradition standen. Das zeugnis erfundener autorität, aber auch der erfundene augenzeugenbericht galt schon der antiken rhetorik als ein beliebtes mittel, um der *narratio* einen unwiderlegbaren wahrheitsanspruch zu verschaffen. Das motiv mag womöglich aus der antiken tradition stammen – die intendierte wirkung ist die unterhaltung und spannung des lesers. Demnach wird hier fiktion in der ersten erwähnten funktion gebraucht.

Neben diesen sind weitere gründe anzuführen: Islamische scholastische erkenntnistheorie, die ihren ausgangspunkt von rechtlich-theologischen überlegungen nahm, legt auf das augenzeugnis, den sinnesschein, allergrössten wert. Es ist davon auszugehen, dass unsere autoren von diesen erkenntnistheorien wussten. So konnten sie sich sagen, dass selbst erfundener augenschein besser als gar keiner ist.

Ferner ist zu beachten, dass auch die ganz theologisch orientierte literatur, wie frühislamische historiografie in grossem mass, den wahrheitsgehalt allein durch die person des tradenten der narratio gesichert sah. Das zeigt die methodik des *ḥadīth* überdeutlich. Wahrheitskriterium war nicht so sehr der inhalt, der wortlaut des *matn*, sondern fast ausschliesslich die persönliche integrität des tradenten. Dass es trotzdem gang und gäbe war, persönliche autorität zu erfinden, wird nicht erstaunen. Denn alle politisch-juristisch-staatliche autorität wurde aus der autorität profetischer setzungen abgeleitet.

Werfen wir noch einen blick auf die dichtung. Ich möchte eine gattung der arabischen dichtung herausgreifen: die weindichtung (*khamriyyāt*). Weinverse gibt es bereits in der dichtung der vorislamischen zeit, aber unverändert auch in islamischer zeit. Das ist eigentlich etwas, was nicht sein sollte, denn es ist bekannt, dass das islamische gesetz den weingenuss verbietet. Die verherrlichung und besingung durch dichtung konnte und kann daher wohl schwerlich vom religiösen gesetz erlaubt sein.

Wenn man nun den absolutheitsanspruch des islamischen religionsgesetzes nimmt, das vorgibt, göttlichen ursprungs zu sein, und dann die ungeheure menge der in islamischer zeit entstandenen weindichtung betrachtet, dann steht man vor einem erstaunlichen widerspruch, der zu erklären versucht werden will.

Zuerst ist zu sagen, dass der islamische dichter eine gewisse narrenfreiheit besass. Sie war durch einen vers des korans abgesichert, wo es heisst, dass die dichter nicht täten, was sie sagten (koran 2:226). Die aussagen von weingedichten mussten also nicht ernst genommen werden, als dichterische aussagen waren sie nicht dem gesetz unterworfen.

Dazu kommt ein weiteres: Ein charakterzug arabischer dichtung ist die ständige verwendung derselben topoi und themen – wir trafen ähnliches in der historiografie. Man kann vielleicht sogar soweit gehen und behaupten, dass islamische poesie oft – ich meine hier vor allem auch die persische – nicht aus dem bestreben eines lyrischen ichs lebt, sich individuellen ausdruck zu verschaffen, vielmehr danach strebt, einmal geschaffene topoi und themen geistreich und geschickt abzuwandeln. Als genial gilt der dichter, der ein neues concetto, eine neue variation der schon vorhandenen topoi erfindet. Und gerade weil das so ist, darf der einzelne vers eines dichters über weingenuss etwa nicht von vornherein als schilderung persönlichen erlebens betrachtet werden.

Und das bringt uns zu einem weiteren punkt: Der islam als religion ist ernst, ja streng, puritanisch, nüchtern, vernunft bestimmt. Ähnlich die mittelalterliche islamische gesellschaft in vielen zügen: puritanisch, autoritär, nüchtern, traditionsgebunden, alles verhalten ist reglementiert, von vielen tabus bestimmt. Das ausleben und öffentliche zurschaustellen persönlicher, individueller gefühle ist verpönt. Einen freiraum konnte in dieser reglementierten gesellschaft fast nur im innenleben, in der fantasie gefunden werden. Weindichtung kann somit als eine form der kompensation durch die fantasie betrachtet werden.

Versuchen wir eine konklusion. Fiktion tritt in arabischer literatur u.a. in historiografie, biografik, geografie-kosmografie und dichtung auf. Als gründe konnten wir zum einen theologische anführen. Sie sind am besten an der art der abfassung der historischen narratio zu studieren. Da geschichte ein sinnvoller, von Gott gewollter und geleiteter prozess ist, muss das in der historiografischen wiedergabe zur erscheinung kommen. Man führt daher ausserweltliche triebkräfte des historischen prozesses an, die nach unserem modernen verstehen fingiert sind. Zum anderen muss der inhalt der narratio durch die autorität des tradenten gesichert sein – und selbstverständlich kann diese autorität erfunden sein.

An zweiter stelle sind ästhetische gründe zu nennen. Die narratio sollte nicht nur der erkenntnis dienen, sondern auch erfreuen und unterhalten. Das konnte u.a. der erfundene augenzeugenbericht in der kosmografisch-geografischen liteartur. Diese diente zwar der weltbildung, wusste sich letztlich jedoch auch in einem theokratisch bestimmten kosmos aufgehoben.

Ich möchte eine frage zwischenschalten: Gibt es grundsätzliche unterschiede im gebrauch der literarischen fiktion im islamischen und europäischen mittelalter einerseits und in islamischer und europäischer neuzeit andrerseits? Anders gefragt: Ist die rolle eines heutigen islamischen literaten, der noch in einer relativ theokratisch bestimmten gesellschaft lebt, grundsätzlich von der eines europäischen literaten verschieden – gesetzt, es gibt überhaupt *den europäischen* literaten?

Nun einige textbeispiele. Als erstes habe ich eine kurze passage aus der weltchronik Ṭabarīs (gest. 310/923) ausgewählt. Der autor war historiker, jurist und theologe. Seine monumentale weltgeschichte reicht von der weltschöpfung bis in seine zeit. Der als beispiel gewählte text behandelt ereignisse im jahr 13/635–36 während der muslimischen eroberung von Palästina und Syrien. Er lautet in übersetzung:

> Es überliefert uns Ibn Ḥumayd von Salama von Ibn Isḥāq. Der erzählt: Als die muslime mit der eroberung von Ajnādayn fertig waren, zogen sie nach Fiḥl im Jordanland. Dort hatten sich nämlich die führerlosen heerreste der byzantiner gesammelt. Die muslimische führung jedoch war intakt, und Khālid b. al-Walīd war führer der vorhut. – Als die byzantiner in Baysān lagerten, stachen sie die dämme der dortigen flüsse durch, so dass sie aus den ufern traten. Baysān ist salzartiges marschland und wurde jetzt zu morast. – Die muslime also lagerten sich in Fiḥl. Baysān befindet sich zwischen Palästina und dem Jordan. Als nun die muslime über Baysān kamen und nicht erkannten, was die byzantiner gemacht hatten, versanken ihre pferde im schlamm und sie gerieten in not. Dann aber errettete sie Gott. Und Baysān wurde daraufhin "die mit dem schlamm" genannt wegen der dinge, die den muslimen dort widerfuhren. Die muslime wandten sich nun eilends gegen die byzantiner, die in Fiḥl lagerten. Die heere gerieten in streit. Die byzantiner wurden überwunden und die muslime besetzten Fiḥl. Die zersprengten haufen der byzantiner zogen sich nach Damaskus zurück. Das treffen bei Fiḥl war im jahr 13 der hijra, nachdem sechs monate des kalifats ʿUmars vergangen waren. In diesem jahr leitete ʿAbd al-Raḥmān b. ʿAwf die wallfahrt für die menschen. – Von Fiḥl aus rückten die muslime auf Damaskus, mit Khālid b. al-Walīd als

> führer der vorhut. – Die byzantiner hatten sich in Damaskus um einen mann namens Bāhān gescharrt, und der kalif ʿUmar hatte Khālid b. al-Walīd durch Abū ʿUbayda als oberbefehlshaber ersetzt. – Die muslime und die byzantiner trafen in der umgebung von Damaskus aufeinander und gerieten heftig aneinander. Schliesslich zernichtete Gott die byzantiner, und die muslime gewannen die oberhand über sie. Die byzantiner zogen sich in die stadt zurück und verschlossen die tore, während die muslime die stadt belagerten und bestürmten. Schliesslich wurde Damaskus erobert und die bewohner bezahlten die kopfsteuer.[10]

Soweit der text. Dies kann auffallen: Einerseits werden eine menge von einzelnachrichten auf verhältnismässig engem raum mitgeteilt. Sie bleiben jedoch merkwürdig blass. Man erfährt gar nicht, wie z.b. die kämpfe abliefen. Die einzelnachrichten stehen fast ohne rechte kausale verknüpfung nebeneinander. Wichtig ist dem erzähler, dass es um gottgewollte geschichte geht: Gott besiegt die feinde, er errettet die muslime usw. Davor verschwinden alle einzeltatsachen, sie werden unwichtig. Fiktion ist hier also nicht der eigentliche historische prozess, sondern die erklärung des prozesses. Man könnte von fantastischer ätiologie sprechen.

Der zweite text stammt aus den *Murūj al-dhahab* Masʿūdīs. Masʿūdī starb 956–57 in Ägypten.[11] Er war, anders als Ṭabarī, nicht theologe sondern *adīb*. Sein hauptwerk *Murūj al-dhahab* besteht aus weltgeschichte und weltbeschreibung, *historia* und *imago mundi*, vereinigt also historiografie und kosmografie.[12]

Im rahmen der behandlung von Ägypten erzählt er:

> Aḥmad b. Ṭūlūn in Ägypten hörte ungefähr im jahr 260/875 von einem mann in Oberägypten unter den kopten, der 300 jahre alt sein sollte. Er sei von jugend an eine autorität in wissenschaft, in der theologischen spekulation und der kenntnis der verschiedenen filosofischen schulen und religionssysteme gewesen. Er kenne zudem land, gewässer, geschichte und dynastien Ägyptens überaus gut. Auch habe er die erde bereist, die verschiedenen reiche besucht und die völker der weissen und schwarzen gesehen; auch besitze er kenntnis der astronomie und astrologie. – Aḥmad b. Ṭūlūn sandte einen seiner offiziere mit mannschaft, der den alten auf dem weg des Nils zu ihm brachte, obwohl dieser es ungern mit sich

---

[10] Der text ist übernommen aus Brünnow-Fischer, *Arabische Chresthomatie*. 6. rev. aufl. (Wiesbaden, 1984), S. 73 ff.; verkürzt übernommen aus Muḥammad b. Jaʿfar al-Ṭabarī, *Taʾrīkh al-rusul waʾl-mulūk*, ed. M. de Goeje, IV, ereignisse des jahres 13 AH.

[11] *Weltgeschichte*, S. 26 f.

[12] Ibid., S. 143 f.

> geschehen liess, denn er hatte sich von den menschen zurückgezogen und lebte auf der spitze eines bauwerkes, das er sich selbst errichtet hatte. Er hatte schon die 14. generation seiner enkel erlebt.
>
> Als er nun vor Aḥmad b. Ṭūlūn stand, da erwies er sich als ein mann, der zwar spuren höchsten alters zeigte, aber durchaus noch gesund an sinnen, urteil und vernunft war. Er verstand die leute, die ihn anredeten, wusste beredt zu antworten und von sich zu sprechen. Aḥmad b. Ṭūlūn liess nun seine gelehrten herbeiholen. Er beschäftigte sich intensiv mit dem alten, verbrachte viele tage und nächte mit ihm allein und lauschte seiner rede, seinen erzählungen und seinen fragebeantwortungen.
>
> Der alte wurde u.a. über den see von Tinnīs und Dimyāṭ gefragt. Seine antwort war: "Dieser see war einst land, und zwar ein land, wie es keines ähnlicher qualität, güte und fruchtbarkeit in Ägypten gab. Es bestand aus gärten, palmen und weinbergen."
>
> (Weiter wusste er zu erzählen): "Zwischen al-ʿArīsh und der insel Zypern gab es einst einen weg, der von tieren auf dem trockenen begangen werden konnte. Zwischen al-ʿArīsh und der insel Zypern gab es nur eine furt, heute ist der weg zur see zwischen Zypern und al-ʿArīsh weit, genauso zwischen Zypern und dem land der byzantiner."

Der bericht des alten setzt sich noch über etliche seiten fort, behandelt u.a. die entstehung der pyramiden. Die pyramidenerzählung wird dann speziell von späteren autoren übernommen.[13]

Ausgangspukt ist die suche nach der erklärung für kultur- und naturfänomene. Dafür wird eine autorität erfunden. Der alte Masʿūdī wird mit zügen ausgestattet, die denjenigen eines antiken weisen und filosofen ähneln: Er hat die welt durchwandert, kennt alle lehren der filosofie und religionen, er weiss um die geheimnisse des himmels, denn astronomie und astrologie sind ihm geläufig. Unwillkürlich fallen hier Pythagoras und Platon als vorbilder ein. Auch das motiv: 'Alter weiser gibt auskunft vor herrscher über die vorzeit' ist ein bekannter topos antiker literatur.[14] Selbstverständlich hat es jedoch auch seinen islamischen platz im leben, denn zweifellos haben islamische herrscher weise zu sich gerufen und sich von ihnen belehren lassen, wenn diese auch schwerlich über dreihundert jahre alt waren.[15] Ein detail mag auf christliches hinweisen: Der alte lebt zurückgezogen von der welt und den menschen auf der spitze

---

[13] *Weltgeschichte*, S. 430 ff.

[14] Ibid., S. 178.

[15] Ibid., S. 48 und anm. 205; jetzt auch Leder, 'Heraklios erkennt den Propheten. Ein Beispiel für Form und Entstehungsweise narrativer Geschichtskonstruktionen,' *ZDMG* 151 (2001), S. 1–42, siehe S. 4 ff.

eines bauwerkes. Hier kann man an die christlichen einsiedlermönche denken, die bisweilen auf säulen lebten.

Die wahrheit unseres berichtes ist also einmal bezeugt durch das hohe alter des zeugen, der zudem über umfassende welterfahrung verfügt – also auch wieder durch autorität – andrerseits durch die persönlichen qualitäten des tradenten: Er ist durchaus bei sinnen und verstand. Masʿūdī versteht es, einem sonst völlig unglaubwürdigen bericht den schein der wahrhaftigkeit zu geben. Dadurch kann er auch mit dem interesse des lesers rechnen. Hier ist also der inhalt der *narratio* selbst fiktion.

Zum schluss zwei verse aus einer anthologie, die der anfangs genannte mamlukische welthistoriker Ibn al-Dawādārī dem 1. band seiner weltchronik einfügte.

Die verse lauten:

> *Eine 'tränkung' einem orte, an dem mich, sooft ich hier eingeschlafen war, der morgenwind und das schlagen der holzklappern weckten.*
> *Seine lilien auf jener anhöhe über den feldern waren gleichsam schwänze von pfauen.*[16]

Der zweite vers soll nicht weiter beachtet werden. Das letzte wort des ersten verses zeigt, worum es sich handelt: *nāqūs*, pl. *nawāqīs*, ist eine holzklapper, die in kirchen und klöstern anstelle der verbotenen glocken benutzt wurde.[17] Der dichter ist nach reichlichem weingenusse eingeschlafen – wein wurde in oder in der nähe von klöstern ausgeschenkt[18] – und wird nun mehr oder weniger sanft geweckt.

Die beiden verse werden meistens einem dichter der frühen abbasidenzeit namens Ukhayṭil al-Ahwāzī[19] zugeschrieben. Im belegstellenapparat der edition findet man jedoch ausser diesem dichter auch die zuschreibung an die dichter Abū Nuwās,[20] Ibn al-Muʿtazz[21] und Muharrim b. Khālid al-ʿAbdī.[22] Daher darf die folgerung erlaubt sein, dass es

---

[16] Auch übersetzt bei Gregor Schoeler, *Arabische Naturdichtung* (Beirut, 1974), S. 126.

[17] Bei Schoeler a.a.o. falsch *Glocken*.

[18] Dazu Peter Heine, *Weinstudien. Untersuchungen zu Anbau, Produktion und Konsum des Weins im arabisch-islamischen Mittelalter* (Wiesbaden, 1982), S. 54 f.

[19] Mir nicht näher zu identifizieren. Schoeler bringt keine daten.

[20] Gest. 814–815. Zu ihm Fuat Sezgin, *Geschichte des arabischen Schrifttums (GAS)*, II (Leiden, 1975), S. 543 ff.

[21] Gest. 908. *GAS*, II, S. 569 ff.

[22] Mir nicht näher zu identifizieren.

sich *nicht* um die schilderung eines individuellen erlebnisses des dichters Ukhayṭil oder eines der drei anderen handelt, sondern um einen topos: 'Dichter berauscht sich im christlichen kloster und wird vom klappern des *nāqūs* geweckt'. Selbstverständlich, so muss man hinzufügen, kann und wird dieses erlebnis irgendwann einmal als wirkliches, individuelles stattgefunden haben, denn auch ein islamischer dichter wird sich nicht nur an seiner fantasie berauscht haben.

Man sieht: Fiktionalität ist keine erfindung mamlukischer historiografie. Zu untersuchen wäre allein das *wie*: Gibt es einen wandel von früher zu späterer. Und: Gibt es, wie gefragt worden ist,[23] eine vermehrung dieser fiktionalen elemente, sprich der unhistoriozität. Das zu untersuchen wäre eine aufgabe.

---

[23] Besprechung Haarmann von *Weltgeschichte,* a.a.o., S. 134 f.; *Weltgeschichte*, S. 193–195.

LITERATURE

SOME REMARKS ON THE WOMEN'S STORIES IN THE JUDEO-ARABIC *AL-FARAJ BA'D AL-SHIDDA* BY NISSIM IBN SHĀHĪN (990–1062)

Arie Schippers

The *Kitāb al-faraj ba'd al-shidda* ('Book of relief after adversity') by the eleventh-century North African Jewish author Nissim Ibn Shāhīn is an interesting example of a Judeo-Arabic narrative work. The *Faraj* belongs to a literary genre that has religious and secular Arabic parallels in, among others, the works of Abū 'Alī al-Muḥassin al-Tanūkhī (941–994). The 'Relief after adversity' genre consists of collections of stories in which God grants relief to someone who finds himself in a difficult situation.[1] The author, Rabbi Nissim ben Yacob ben Nissim Ibn Shāhīn, had a difficult life himself. He was born in Kairouan (present-day Tunisia), where his father was Head of the Academy, a learned assembly and institution for the study of the Torah and the Talmud. As a scholar, Ibn Shāhīn became widely known in the Jewish world. His later years were troubled by unhappy events, which were caused by the destructive invasion of the Banū Hilāl and Banū Sulaymān Bedouins. Ibn Shāhīn's text is written in Judeo-Arabic and Middle Arabic, although an intellectual of his stature could have written in Classical Arabic had he wished to do so.

The first printed edition of Nissim Ibn Shāhīn's *al-Faraj ba'd al-shidda*, Julian Obermann's edition of 1933,[2] was printed in Arabic script. Obermann called the book *Ta'līf ḥasan fī 'l-faraj* [A Beautiful Composition about Relief]; the title was a translation of the widespread Hebrew

---

[1] Nissim ben Jacob Ibn Shāhīn, *An elegant composition concerning relief after adversity*, translated from the Arabic with introduction and notes by W.M. Brinner (New Haven [etc.], 1977)], p. xxvii.

[2] J. Obermann, *Studies in Islam and Judaism. The Arabic original of Ibn Shâhîn's Book of comfort known as the Ḥibbûr yaphê of R. Nissîm B. Ya'aqobh*, edited from the unique manuscript by Julian Obermann. Yale Oriental series. Researches; 17 (New Haven, 1933). The MS that Obermann used is preserved in the library of the Jewish Theological Seminary of America, MS no. 2472. See E.N. Adler, *Catalogue of Hebrew manuscripts in the collection of Elkan Nathan Adler* (Cambridge, 1921), p. 59 No. 4026.

version of the book *Ḥibbur Yafeh me-ha-Yeshuʿah*. The edition also featured a facsimile of the Harkavy manuscript in Hebrew script. In his edition, Obermann frequently 'corrected' the Middle Arabic of the text, often relying on the Hebrew translation of the text[3] in his reconstruction of the 'original' text.

The contents of Nissim's work are completely different from those of al-Tanūkhī's *Faraj* book: while the relief in al-Tanūkhī's stories often consists of the escape of a government official from a difficult situation or an execution, or of a traveller from robbers or wild animals, the comfort offered by Nissim Ibn Shāhīn's *Faraj* is exclusively of a moral nature. In principle, all of the stories have a pious aim.[4] Compared to other Jewish storytelling (e.g., Ibn Zabara's *Sefer Shaʿashuʿim*), the stories in this work are more religious and less mundane. This aspect is also manifested in the types of women that appear in Nissim's *Faraj*. Most of Ibn Zabara's women are either wicked or stupid, but some are intellectually superior to their husbands or play the role of a clever girl, but all in a worldly way. The women in Nissim's book, however, are all embedded within a religious society. Rather than occurring in contemporary times, most of the stories are set in the period around the first century CE, a time in which rabbis like Rabbi Eleazar, Joshua and Meʾir (who were important for the Mishna and the Talmud) played a prominent role. The moral ideals of this era were different; for instance, the sacrifice of a woman for her husband occurs more than once in Nissim's work, while such incidents are completely absent from the works of Ibn Zabara and al-Tanūkhī.

Not all the women in Nissim's book, however, are characterized by morally exemplary behaviour, for the stories also contain examples of the opposite. In the following survey of the most characteristic of Nissim's stories we will attempt to define the various types of women that appear in these stories. The numbers between round brackets refer to the folio numbers of the Harkavy MS in Obermann's edition.

1. (f. 38b–42b) An interesting story about the constancy of women relates how Rabbi Eleazar and Rabbi Joshua went to Jerusalem. While

[3] He was criticized for this by D.H. Baneth in his review in *Kirjath sepher* 11 (1934), pp. 349–357.

[4] The Arabic *Faraj* genre is probably of Jewish origin. The first book of the Arabic genre (e.g., by Ibn Abī ʾl-Dunyā) is also based on pious inspiration, as are the beginning chapters of Tanūkhī's *Faraj*. Cf. Brinner, p. xxvi; Abū ʿAlī al-Muḥassin al-Tanūkhī, *Kitāb al-Faraj baʿd al-shidda*, edited by ʿAbbūd Shāljī (Beyrouth, 1978).

they are in the Temple on the eve of the Day of Atonement, an angel appears to them, holding in his hand a brilliant robe, which shines like the sun, but which lacks a collar. When they ask for whom the robe is intended, they are told that it is meant for a person in Ascalon, whose name is Yosef the Gardener.[5]

After their pilgrimage the two rabbis go to Ascalon, where the notables offer them their hospitality. The two rabbis, however, wish to be the guests of Yosef the Gardener. They are thus brought to Yosef in his garden, where he is busy picking herbs. The rabbis ask for his hospitality, but he has nothing to offer them except two loaves of bread. After the meal the rabbis ask Yosef to tell them about his occupation. He tells them that he has become poor only recently. His father's wealth was lost after his death, so he was forced to leave the town and live outside it. Yosef says that he uses one half of the crop from his garden to nourish himself and his wife, while giving the rest away as charity for the poor.

The two rabbis tell Yosef about the shining robe in the hands of the angel, and that it is meant for him on the condition that he provide the missing collar. They then tell him how to become worthy of doing so. Yosef's wife replies that he should sell her as a slave in the market place and give the proceeds as charity for the poor. This is what Yosef does. After many adventures in which his wife remains chaste, a voice comes from Heaven, revealing the location of a treasure that has been hidden by his father. Both Yosef and his wife continue to perform acts of benevolence and charity, and they live on a far grander scale than before.

This story emphasizes the gardener's wife, who is the first to suggest a solution, thereby sacrificing herself by becoming a slave and even remaining chaste in spite of the advances of her new masters.

2. (f. 43b–45b) The virtues of women are also stressed in other stories, including one in which the daughter of the executed rabbi Ḥananyah ben Ṭeradion appears. This daughter remains chaste even after being confined to a brothel or whores' pavilion, and is therefore rescued by rabbi Meʾir.[6]

---

[5] Brinner, pp. 48–52; Obermann, pp. 44–47; See the facsimile of the Harkavy MS in Obermann, f. 38b–42b; R. Nissim Gaon, *Libelli quinque*, edition, introduction and notes by S. Abramson (Jerusalem, 1965), pp. 512–513. In this article, Brinner's translation was consulted for the summary of the stories.

[6] Brinner, pp. 53–54; Abramson, pp. 454–455; Obermann, pp. 48–50.

3. (f. 45b–49b) Another story,[7] which is also known from the later Hebrew work *Sefer Shaʿashuʿim* (Book of Delight, ca. 1170) by Joseph Ibn Zabara,[8] provides an unfavourable image of women. In this story a king wonders whether a single woman can be found in his city who is completely virtuous (*i.e.* who is both chaste and wise). His viziers start searching and they find a chaste and wise woman, as beautiful as the sun, who is the wife of a wealthy merchant. The king puts her husband to the test by asking him to kill his wife and children and to marry his only daughter. The husband, however, remains faithful to his wife. The merchant's reluctance to kill his wife and children prompts the king to say, 'You are not a man; your heart is that of a woman'. His wife is then put to the test. If she kills her husband, so the king promises, she will be queen tomorrow – but she is given a sword of tin. That night, when she strikes her husband in order to kill him, the sword proves ineffective. Both husband and wife must then relate their stories in the palace. The king's wise men hear the stories and are convinced of the king's wisdom concerning the faithlessness of women.

4. (f. 52b–53a) Another story concerns Yoḥani bat Reṭibi, who at first sight appears very pious and God-fearing. In secret, however, she is a witch, who casts spells upon pregnant women to impede their delivery, thus working against, rather than assisting with, the birth of children. She then revokes her spell, creating the illusion that the birth of a child depends on her magical powers. The women consider Yoḥani righteous and they believe that her prayers are answered in heaven. However, her witchcraft is discovered when a boy is left to watch her house. Upon hearing a sudden movement, he looks around and finds a covered jar in a corner of the house. This jar contains Yoḥani's magic, which becomes ineffective after having been exposed. From that moment pregnant women are able to give birth without her assistance. Once Yoḥani's situation becomes known she is sent into exile.

5. (f. 54a–b) Another story concerns a woman who appears pious at first sight, but who ultimately proves to be calculating. The story goes as

[7] Brinner, pp. 54–58; Abramson, pp. 455–458; Obermann, pp. 51–54.
[8] Ibn Zabara [= Joseph ben Meir ibn Zabara, *Sepher Shaʿashuʿim. A book of mediaeval lore*, ed. by I. Davidson. (New York, 1914)], pp. 26–30; Davidson, in his introduction to Ibn Zabara, *Sepher Shaʿashuʿim*, pp. l–lii.

follows: 'Another time I saw a widow who went to the synagogue in my neighbourhood at nightfall to pray. Because I knew that there was a synagogue in her own neighbourhood, I asked her, "Why did you leave the synagogue in your neighbourhood and come here?" She replied, "My lord, in order to gain recompense and earn reward for the additional steps from that synagogue to this one [*li-an anāla ajran wa-aktasiba thawāban fī ziyādati ʾl-khuṭā ʾllatī min tilka ʾl-kanīsati ilā hādhihi*]".'

The notions of recompense and reward in this story are characteristic of the Jewish religious character of the book. They rarely appear in other narratives.

6. (f. 55a–58a) One remarkable story concerns a girl who chooses to marry her father's servant. After years of study this servant becomes the famous Rabbi ʿAqiba. One unusual feature of this story is that the daughter is apparently free to choose her husband independently from her father. The introduction to the book exhorts men to marry learned women, or at least the daughters of scholars.[9] The story has a slightly different angle, however, since it is a woman – the daughter of a famous rabbi – who chooses her own husband, whom she encourages to acquire knowledge despite her father's resistance.

The father in this story, Ben Kalba Sebuʿah ['The one who is so generous that he even satisfies the hunger of the dogs'], is a rich benefactor who at first does not agree with his daughter's choice, the poor shepherd who watches his flock. He swears he will cut off her allowance and ʿAqiba lives in poverty with his wife, owning nothing but the straw on which he lies. He even gives away some of the straw to a poor man whose wife will soon give birth.

ʿAqiba's wife now tells her husband that he must study the Torah for twelve years. To do so, he leaves to study with the famous rabbis Eleazar and Joshua. After twelve years he returns home, but departs again to study for another twelve years, an endeavour strongly encouraged by his wife. Through his studies he achieves the utmost glory. Once, when his wife attempts to pay him a visit, his entourage nearly pushes her away. He intervenes, however, saying that he owes everything to her.

---

[9] Cf. the Harkavy MS in Obermann, f. 54b–55a. A similar passage can also be found in Ibn Zabara (chapter 12), who was perhaps influenced by Nissim Ibn Shāhīn's present work.

In the meantime her father, Ben Kalba Sebuʿah, has heard of ʿAqiba's reputation and wishes to ask his advice about the ban he put on his daughter, whom he now knows is living in poverty [*Qāla: Amḍi wa-astaftīhi laʿalla an ajida lī rukhṣatan fī-mā andhartuhū wa-uʿṭiya ʾbnatī shayʾan taʿīshu bihī fa-qad aḥraqa fuʾādī mā ushāhiduhā fīhi min ḍīqi ʾl-ḥāl*]. He does this without realizing that ʿAqiba is her husband.

He approaches [R. ʿAqiba] and tells him about the vow he made. R. ʿAqiba asks him, 'For what reason did you do this [*li-ayyi sababin faʿalta dhālika*]?' The father replies, 'Because she married someone who cannot read and has neither knowledge [of the Torah] nor wealth [*li-annahā tazawwajat bi-man laysa yaqraʾu wa-lā fīhi ʿilmun wa-lā dhū mālin*]'. R. 'Aqiba then asks, 'What if he had done as I have done?' The father replies, 'My lord, if he could read only one chapter I would give her most of my wealth [*law kāna yā mawlayya yaqraʾu wa-law firqan wāḥidan la-aʿṭaytuhā akthara mālī*]'. At this, R. ʿAqiba says to him, 'I am her husband!' The father then rises, kisses R. ʿAqiba upon the head and gives his daughter half of his wealth. R. ʿAqiba makes a crown of gold for her, as he promised. They are thus delivered from poverty, their hardships have ended and their well-being has increased.

7. (f. 74a–75b) In contrast to the active role of the woman in the preceding story, the woman in the following story does not choose her own husband, but remains as passive as possible. The story gives her no active role. A butcher [*qaṣṣāb*] tells a pious man why he figures among the chosen pious men, despite his being a butcher. One day, so the butcher relates, a caravan passes with several captives, one of whom is a girl who is weeping bitterly [*Kuntu fī baʿḍi ʾl-ayyāmi wa-anā qāʾimun fī shughlī, ḥattā ʾjtāza bī rafaqatun wa-fīhā qawmun masbiyyīn fa-taʾakhkharat minhum ṣabiyyatun wa-hiya tabkī bukāʾan shadīdan*]. The butcher asks her, 'Why are you weeping when no one else in the caravan weeps?' The girl replies, 'My lord, I am a Jewess. Because I have been taken captive I fear for my religious way of life at the hands of these unbelievers. I wish I would arrive at a place where there are Jews who might redeem me.' [*Qālat: yā sayyidī anā yahūdiyyatun wa-qad subītu wa-akhshā ʿalā madhhabī min hāʾulāʾi ʾl-kuffār wa-kuntu ashtahī law waṣaltu ilā mawḍiʿin fīhi yahūdun yafdūnī*]. The butcher decides to redeem her so that his twenty-year old son may marry her. The son obeys. The marriage contract is drawn up and all the necessary preparations are made.

The butcher invites everyone, poor and rich, to the wedding. All the guests eat and rejoice, with the exception of those at one table, who have eaten nothing. When asked about this, they reply, 'This poor man whom you have placed among us – ever since he sat down he has been weeping bitterly'. The butcher asks why, and the man replies, 'I weep for the girl whom you have chosen as a bride for your son. She is from such-and-such city and I am from there as well. I once had her pledged to me [*qad qaddastuhā lī*], such that she was 'put into my possession' [*mumlakatun*; betrothed]; she is betrothed to me. She was taken captive, however, as was I some time later. I have the document of her betrothal here in my hand [*wa-hādhā kitāb taqdīsihā bi-yadī*]'.

The butcher asks him to describe some characteristic mark on the girl [*'laka fī hādhihi 'l-ṣabiyyati 'alāmatun?'*], and the man affirms that 'she has a mole on her toes and its appearance is such-and-such' [*'Na'am, ra'aytu muddatan fī bayti abīhā wa-lahā shāmatun fi ba'ḍi aṣābī'ihā wa-ḥilyatuhā kadhī wa-kadhī'*]. Thereupon, the butcher asks his son to halt the marriage, saying 'This girl is betrothed to another man, and her husband is present here. I have examined her marriage contract, and she is now indeed forbidden for you [*hādihi 'l-ṣabiyyatu mumlakatun li-ghayrika wa-huwadhā ba'luhā ḥāḍirun wa-qad waqaftu 'alā kutaybatihā wa-hiya 'l-āna ḥarāmun 'alayka*]'.

Thereupon the reunited couple stays with the butcher for some time, joyful after the distress that has befallen them.

Upon hearing this story the scholar says to the butcher, 'You have dispelled my sorrow. I do not doubt the greatness of God Most High and my rank in the hereafter, as you are one of my companions'.

A striking feature of this story is that in neither case does the girl oppose the marriage that is arranged for her. However, she is less passive at the beginning of the story, when, as a captive, she weeps because she is afraid to fall into the hands of the unbelievers. Her 'Jewishness' is the most important thing to her.

8. (f. 82b–84b) The story of 'The young woman and the oppressor' is about a woman in a besieged town who slips out to the enemy's camp and succeeds in seducing their king, getting him drunk and cutting off his head, which she brings to her own besieged town. Brinner (p. 92, n. 1) is aware of the origin of the story as well as the differences with the Story of Judith and Holofernes: 'This story, based on the apocryphal book of Judith, differs in many details from it.'

Some key episodes are as follows: 'The king asked her, "Who are you, and whence do you come?" She replied, "Master, I am one of the daughters of the prophets, and I have heard from my father that you will capture this city and take possession of it. I have therefore come before you to ask you to protect me and my family once you have captured the city. My heart will then be at ease with the knowledge that you will support us and do us no harm"'.[10]

The king wants to fulfil her desire, but only on the condition that she will marry him. She then tells the king that she is unclean [*'Lākin uʿarrifu ʾl-malika anna ʾl-jāriyata nājisatun wa ʾl-yawma kamālu ḥālī najāsatihā'*] and that she needs time. The king gives orders for a dinner with his viziers and generals. As the evening draws near they notice that the king has become drunk and they all leave. The girl is then alone with the king, who is asleep in a drunken stupor. She declares her sincere intentions to her Creator [*Ḥīnaʾidhin akhlaṣat niyyatahā quddāma khāliqihā*], she takes the king's sword, cuts off his head, puts it inside her wrap and departs with her maid towards the fountain [*wa-akhadhat sayfahū, ḍarabat bihī wa-ḥamalat raʾsahū fī izārihā wa-maḍat maʿa ʾl-khādima ilā ʾl-ʿayn*]. None of the king's servants opposes her, as the king has forbidden them to approach the two women, until they reach the city gate.

The girl cries out, 'Open the gate for me, for God Most High has helped me against this unbeliever. I have killed him, and here is his head', but the guards do not believe her. One of the officers, however, a refugee who had once almost been crucified by the king whose head is now cut off, is able to identify the head of his former master. Thereupon the gate is opened.

News of the dead king spreads immediately throughout the city, and the young men of the people assemble, seizing their swords and shouting, '*Hear, O Israel the Lord our God, the Lord is One!*' (Deut. 6: 4).

When the soldiers of the king see this, they go to the king's tent and find that he has been killed. They all leave their possessions and flee, but they are pursued until they reach Antioch, where they are either killed or taken captive.

9. (f. 90a–95b) Two stories about deposits follow, in which the protagonists are travellers who wish to consign purses or other possessions to

---

[10] Brinner, pp. 92–93.

someone they do not know and whom they must trust. In both cases the person to whom the deposit is given refuses to return it, saying that he knows nothing about it. In both cases the duped traveller turns to the swindler's wife in her husband's absence and uses trickery to recover his deposit. In one story, a secret Christian who feigns to be a Jew plays the most important role; in another story, the treacherous person is named Kidor.

10. (f. 104b–106b) The story of 'R. Meʾir and the perfidious woman' tells of R. Meʾir, who used to go from Syria on a pilgrimage to Jerusalem. His route would take him through Tiberias, where he would stay with a man named Judah, who had a righteous and chaste wife. When R. Meʾir stayed with them, Judah's wife would serve him food. One year she dies and Judah marries another woman. When the time for R. Meʾir's arrival approaches, the man instructs his new wife to serve R. Meʾir when he passes by during his absence. Thus, the new wife receives R. Meʾir when Judah is not yet returned home. Judah arrives later, and both men eat, drink and fall asleep at their places, having become somewhat tipsy.

The woman goes to R. Meʾir, removes his clothes and lies down close to him in the hope that he will help her fulfil her desire, but he remains unaware of her [*fa-qāmati ʾl-imraʾatu ilā rabbi Meʾir anzaʿathū thiyābahū ʿanhū wa-nāmat mulāṣiqatahū an yusāʿida ʿalā shahwatihā wa-lam yashʿur bihā*]. The following morning she awakens the men and brings them water to wash their faces and feet, whereupon they say their prayers. Judah leaves for his work, while R. Meʾir remains seated in the house. She then rises, serves him food and begins to flirt with him, but he does not raise his eyes to look at her. The wife then suggests that he 'did such-and-such a thing yesterday'. As proof, she tells him that she knows that he has a particular mark on his body. At this, R. Meʾir becomes suspicious of himself, rises, flees from the place and returns to his home, bewildered. A sage with whom he takes counsel in this matter remarks, 'Perhaps the woman merely noticed this mark on you while you were asleep, and you did nothing, and are guilty of neither an unintentional lapse nor an intentional sin. Why then have you brought this great distress upon yourself?'

After several days of fasting, R. Meʾir eventually has a dream in which someone says to him, 'You have committed neither an unintentional lapse nor an intentional sin or transgression. The wicked woman has lied'. He awakes from his sleep full of joy, free of all the doubts that he has felt, and returns to Jerusalem for the pilgrimage.

In this story, the wife's role is similar to the ones that can be recognized in the Koranic story of Yūsuf and Zulaykha, or the Biblical story of Joseph and the wife of Potiphar: a woman whose advances are ignored tries to take revenge by accusing the man of having harassed her.

11. (f. 112a–116a) The next story concerns a wife, Ḥannah, who does not surrender and remains chaste to the end, resisting the pressure of a rich man who is desperately in love with her, and that of her own husband, a debtor who is imprisoned. Hannah's loyalty and virtue are tested by the rich man, Nathan the Halo Bearer, who desires to meet with her. At the same time the woman's husband beseeches her to go to Nathan 'to implore him to give you money to pay off at least some of my debts, so that you can settle my affairs with it and help me out of this prison'. She refuses, as she considers Nathan's feelings illegitimate. She also reproaches her husband as follows: 'The length of your stay in prison has caused you to lose your senses'.

Ḥannah does not visit her husband in prison for three days, and on the fourth day she finds him near death. Her husband accuses her of desiring to marry Nathan and leaving him to die. She says to herself, 'This husband of mine bids me, "Go and commit adultery to help me out of prison!"'

Ḥannah implores God Most High to preserve her from error, and then goes to Nathan the Halo Bearer. When his slaves see her they tell their master, 'Ḥannah is at the door!' She enters and says, 'My husband is imprisoned for debt, and I would like you to advance me one hundred dinars to redeem him. You will attain a high degree of merit by this good deed'. He immediately orders several purses to be brought to her and bids her to take out of them as much as she wishes. He then says to her, 'It is not hidden from you how deeply I am in love with you. Do now accede to my entreaty, and restore me to life.' Ḥannah replies, 'Do with me as you wish, but I must tell you that here and now is the moment you can earn for yourself the hereafter'.

When Nathan hears her wish, he humbles himself before God Most High, asking Him to subdue his passion. Nathan then says to Ḥannah, 'Depart from me in peace to your home, and redeem your husband with this sum'. Ḥannah thus returns to her husband, sets him free and tells him what happened between them. Her husband becomes suspicious, however, thinking that Nathan has had his way with Ḥannah and that she is hiding this from him.

Seeing Nathan outside riding a horse, Rabbi ʿAqiba calls out to ask some of his disciples who he is. On his head, Nathan has a halo that shines like a light on his head, as bright as the shining sun. His disciples reply that the man is 'Nathan the Halo Bearer, the whoremonger [*Natan de-susitā roʿeh ha-zonot*]'. His disciples do not see the halo, but they know that it is connected with his great distinction.

One day, Ḥannah's husband is passing R. ʿAqiba's academy. Upon seeing Nathan the Halo Bearer sitting in the same seat with the rabbi, the husband asks one of the disciples what brought Nathan to this high rank. The disciple recounts all that happened to Nathan; the man's heart is reassured, and he no longer doubts his wife. He returns to Ḥannah, kisses her, imploring, 'Forgive me for having doubted you. [..] Today, I saw this man sitting with R. ʿAqiba in his seat'.

12. (f. 151b–152b) Nahum, the man of Gamzu, is dismissed by the king as a judge. He then goes to his wife and tells her to return to her father's house. When she inquires about his reason for divorcing her, Nahum replies that he no longer has employment. Because his wife has a very learned father, he asks whether she has some of her father's wisdom, and proceeds to recount the details of the matter. The wife says that she will go the next day to have him reinstated in his office, and that she will see to it that the man who has been appointed in his place will be beheaded. Pretending to be a widow she succeeds in introducing herself to the king, and she tells him a parable about a servant – in reality, a palm tree – from which she and her children used to receive food and clothing. The 'servant' has been 'beheaded', but she has not received compensation. The king summons his new judge to provide an explanation: 'Examine the record of this case and declare judgment upon it'. When the judge reads the case, he replies, 'There is no such thing in the world. It must be one of the tricks of the Jews'. The king asks him, 'If we call the Jewish judge and he rules otherwise, what shall we do?' The judge replies, 'Have me beheaded'.

The king then summons Nahum, the man of Gamzu, and tells him the parable. Nahum explains it as follows, 'The widow had a date palm; that is the servant. She would take its produce and fruit and store it to feed her children and [...] clothe them with its garments. She is now demanding [...] its price, as it has been cut down'. The king thereupon orders, 'Let this judge be beheaded, and let Nahum, the man of Gamzu, be restored to his office'.

In these twelve stories we see a great variety of different women. Although the examples presented here do not include all of the stories, they are presented in their order of appearance in the book. As in *adab* collections, stories of a similar type are gathered together. The first and the second stories are concerned with the chastity of wives. The first story also features a wife's self-sacrifice for the sake of her husband. This self-sacrifice is illustrated clearly by the wife's willingness to be sold as a slave in the market place so that the proceeds can be used as charity for the poor, in order to earn glory for her husband. In the second story, the daughter of Ḥananyah ben Ṭeradion remains chaste, despite being confined to a brothel or whores' pavilion.

The third story stresses the contrast between a husband and his wife. It is the famous story of the king who tempts both, a story that features in Ibn Zabara's collection as well. This story is meant as a contrast with the first two tales, just as the tenth story, about the perfidious woman, forms a contrast with the surrounding tales.

The fourth and fifth stories share the theme of women who at first sight appear to be devoutly pious at first sight, but who ultimately prove to be devious. The fourth story concerns a seemingly God-fearing woman who is a witch in secret. The fifth story involves a woman who at first sight appears to be pious, but who later proves to be calculating because of the extra steps she prefers to make by going to another synagogue further away instead of the one in her own neighbourhood.

The sixth and the seventh stories offer a contrast as well. The sixth story tells of a woman who chooses her own husband against the will of her father. She also endures self-sacrifice and suffers for her husband. The wife in this story reflects the tone of the introduction, in which men are exhorted to choose learned women, or at least daughters of scholars. The seventh story, in contrast, concerns a woman who does not choose her own husband, but remains as passive as possible. She is given in marriage to a passive son by his father, while apparently failing to reveal that she had previously been engaged to someone else.

Stories 8, 9a, 9b, and 10 provide different types of contrast. Stories 9a and 9b are both about women whose lack of intelligence results in the deposited goods that their husbands want to steal being restored to their rightful owners. Stories 8 and 10 are of a different calibre. Story 8, about a woman who beheads the foreign king after introducing herself to him under false pretexts, is reminiscent of such fierce Biblical women as Jael who killed Sisera and Judith who killed Holofernes. In the tenth story a perfidious woman, who is married to someone else, tries to seduce a famous rabbi, fails and then accuses the rabbi of sexual harassment. This is reminiscent of the stories of Potiphar's wife or Zulaykha.

The eleventh story presents the image of a woman who remains chaste despite her husband's attitude. The twelfth story tells of a clever woman who, by telling a story, persuades a king to reinstate her husband in his job and have his rival beheaded.

The character of the women's stories by Nissim Ibn Shāhīn is primarily religious. Although the *Sefer Sha'ashu'im* of Ibn Zabara, which appeared a century later, has one story in common with the present book, the stories of the latter are far more secular and are more similar to medieval, ancient classical and oriental narrative in general.[11] The stories that are contained in another contemporary narrative work, the *Disciplina Clericalis* by the Jewish convert Petrus Alphonsi (1062–1121), are secular and of Arabic origin; several are even derived from Indian sources. Examples include stories about a wife who deceives her husband, or about a woman who uses a stratagem to recover a deposit.[12]

Nearly all of the stories in Nissim's book are set at the beginning of the Common Era. This was the period of rabbis like Rabbi Me'ir, Rabbi 'Aqiba, Rabbi Joshua and Rabbi Eleazar. It was an idealized period, in which virtues were taught and, supposedly, largely practised. More than the stereotype of the perfidious woman, therefore, we tend to remember the images of women who sacrifice themselves for their husbands, or of women who are exclusively concerned about remaining true to their religious way of life.

[11] Cf. H. Schwarzbaum, 'The value of Ibn Zabara's 12th-century *Sepher Sha'ashu'im* (Book of delight) for the comparative study of folklore,' in V.J. Newall (ed.), *Folklore studies in the twentieth century. Proceedings of the centenary conference of the British Folklore Society* (Woodbridge [etc.], 1980), pp. 391–397; A. Schippers, 'Ibn Zabara's *Book of Delight* (Barcelona, 1170) and the transmission of wisdom from East to West,' *Frankfurter judaistische Beiträge* Heft 26 (1999), pp. 149–161.

[12] Cf. H. Schwarzbaum, 'International folklore motifs in Petrus Alphonsi's "Disciplina Clericalis",' *Sefarad* 21 (1961), pp. 267–299; 22 (1962), pp. 17–59; 22 (1962), pp. 321–344; 23 (1963), pp. 54–73.

THE DAWĀDĀR'S HUNTING PARTY

A MAMLUK *MUZDAWIJA ṬARDIYYA*, PROBABLY BY SHIHĀB AL-DĪN IBN FAḌL ALLĀH

Thomas Bauer

Mamluk hunting literature

Since the dedicatee of the present volume has contributed so essentially to our knowledge about the *birth* and the *generation* of animals,[1] I feel compelled to apologize for presenting a text that is so fundamentally preoccupied with the *killing* of animals: 'They emptied the sky from birds, and this provided for the utmost joy!' However, since Remke Kruk combines an interest in animals with an interest in the Mamluk period, I hope that despite the blood that will be shed in the following pages, this contribution on hunting poetry in the Mamluk period will find favour in her eyes.

The main purpose of the following pages is to bring Mamluk hunting literature to the attention of historians of Arabic literature. So far, not even its mere existence has ever been recorded in the modern studies of the *ṭardiyya*. No article or monograph goes so far as to even mention what happened to this genre in the time after the fourth/tenth century.[2] But Kushājim and Abū Firās were not the last hunting poets in Arabic literature. Rather, after the Umayyad and early Abbasid period (from Shamardal to Abū Nuwās) and the middle Abbasid period (from Ibn al-Muʿtazz to the poets of the entourage of Sayf al-Dawla), the Baḥrī Mamluk period is a third heyday of Arabic hunting literature. During the seventh and eighth/thirteenth and fourteenth centuries, a considerable number of hunting texts, both in prose and poetry, were composed that are in no way inferior or less interesting than those of the earlier

---

[1] To mention only the articles *Takwīn*, *Tawallud* and *Waḥam* in vol. X and XI of *Encyclopaedia of Islam* (New ed.) and the literature quoted there.

[2] See T. Seidensticker, 'Ṭardiyya,' in *Encyclopaedia of Islam* (New ed.), X, pp. 223–224 and J.E. Montgomery, '*ṭardiyya* (hunting poem),' in J.S. Meisami & P. Starkey (eds.), *Encyclopedia of Arabic literature* (London [etc.], 1998), II, pp. 759–760, and the titles mentioned in the bibliographies of these articles.

periods, whereas the two preceding centuries, more or less coinciding with the period of the so-called 'Sunni revival,' constitute a crisis within hunting poetry, if not of Arabic poetry in general.

But this situation changed fundamentally after the Ayyubids and Mamluks had established a stable rule over Syria and Egypt, where the civilian elite had thus appropriated the literary culture of the Abbasid *kuttāb* and where countless madrasas provided for a mass education that made it possible for popular literature to thrive with a hitherto unknown intensity. This development also brought about a new (last?) blossoming of Arabic hunting literature, which may have been initiated by the Ayyubid dynasty itself,[3] but reached its apogee a few decades later. During the short period examined here, the end of the seventh/thirteenth and the first half of the eighth/fourteenth century, a considerable number of texts on hunting were produced, covering all forms of the *ṭardiyya* known so far and even forms hardly used for this purpose before. Especially remarkable is the great number of prose text in the form of a *risāla* 'epistle,' either called *risāla ṭardiyya* or *ṣaydiyya* 'hunting epistle' or, with a more specific subject, *risāla bunduqiyya* 'epistle on hunting (birds) with the crossbow.' The hunting *risāla* reached the height of its popularity in the Mamluk period, but it had been created much earlier. The earliest hunting epistle known so far is by ʿAbd al-Ḥamīd al-Kātib (d. after 132/750).[4] Later, epistles of this kind were written by Ibrāhīm b. Hilāl al-Ṣābiʾ (313–384/925–994)[5] and Hilāl b. al-Muḥassin al-Ṣābiʾ (359–448/969–1056).[6] Another author of a *risāla bunduqiyya* from Ayyubid times is Ḍiyāʾ al-Dīn Ibn al-Athīr (558–637/1163–1239).[7] Further texts are mentioned by Hämeen-Anttila.[8]

The following literary hunting texts date from the second half of the seventh and the first half of the eighth century:

---

[3] A chapter of *ṭardiyyāt* comprising two qasidas can be found among the poetry of the Ayyubid Sultan al-Malik al-Nāṣir Ṣalāḥ al-Dīn Dāwūd (603–657/1205–1259), see ʿAbd al-Ḥasanayn al-Khaḍir, *al-Shuʿarāʾ al-ayyūbiyyīn*, II: *al-Fawāʾid al-jaliyya fī ʾl-farāʾid al-nāṣiriyya* (Damascus, 1996), pp. 313–315.

[4] See Jaakko Hämeen-Anttila, *Maqama. A history of a genre* (Wiesbaden, 2002), pp. 213–215.

[5] Shihāb al-Dīn Aḥmad Ibn Faḍl Allāh al-ʿUmarī, *Masālik al-abṣār fī mamālik al-amṣār*, XII, ed. Ibrāhīm Ṣāliḥ (Abū Ẓabī, 2002), pp. 11–15.

[6] Hilāl b. al-Muḥassin aṣ-Ṣābiʾ, *Kitāb Ghurar al-balāgha*, ed. Muḥammad al-Dībājī. 2nd ed. (Beirut, 2000), pp. 390–398.

[7] Ibn Faḍl Allāh, *Masālik*, XII, pp. 303–305, see also al-Nuwayrī, *Nihāyat al-arab fī funūn al-adab* (Cairo, 1923–1998), X, p. 327.

[8] Hämeen-Anttila, *Maqama*, pp. 214–215.

1. Sirāj al-Dīn al-Warrāq (615–695/1218–1296): A *muzdawija ṭardiyya.*[9]
2. Muḥyī 'l-Dīn Ibn 'Abd al-Ẓāhir (620–692/1223–1292): A *risāla ṣaydiyya.*[10]
3. Kamāl al-Dīn Ibn al-'Aṭṭār (626–702/1228 or 1229–1303): A *risāla bunduqiyya.*[11]
4. Shihāb al-Dīn Maḥmūd (644–725/1246–1325): A long and famous *risāla bunduqiyya*[12] and a *risāla ṭardiyya.*[13]
5. Shihāb al-Dīn Muḥammad Ibn al-Ṣā'igh (645–ca. 722/1247–ca. 1322): A *risāla bunduqiyya* and a *mukhammasa* on the same subject, addressed to Ṣalāḥ al-Dīn b. Muḥyī 'l-Dīn Ibn Faḍl Allāh.[14]
6. Ṣafī al-Dīn al-Ḥillī (677–750/1275–1349 or 1350): A *mukhammasa* in praise of the Artuqid Sultan al-Malik al-Manṣūr Najm al-Dīn Ghāzī, describing a hunt with the *bunduq*;[15] another *mukhammasa* (without a panegyric section) on the same subject,[16] and a series of poems in the form of the qasida, the *urjūza*, and the epigram, which form a chapter of his *Dīwān.*[17]
7. Ibn Nubāta (686–768/1287–1366): The longest and most ambitious *muzdawija ṭardiyya* of Arabic literature, entitled *Farā'id al-Sulūk fī Maṣā'id al-Mulūk.*[18]
8. Tāj al-Dīn al-Bārinbārī (696–756/1296–1355): A *risāla ṭardiyya.*[19]

---

[9] Text in Shihāb al-Dīn Ibn Faḍl Allāh, *Masālik*, XIX, ed. Yūnus Aḥmad al-Sāmarrā'ī (Abū Ẓabī, 1424/2002), pp. 15–306; the *ṭardiyya* is no. 454 (p. 253–258).

[10] Al-Nuwayrī, *Nihāyat al-arab*, VIII, pp. 105–106. On Muḥyī 'l-Dīn see J. Pedersen, 'Ibn 'Abd al-Ẓāhir,' in *Encyclopaedia of Islam* (New ed.), III, pp. 679–680.

[11] Ibn Faḍl Allāh, *Masālik*, XII, pp. 381–385. On Ibn al-'Aṭṭār see Khalīl b. Aybak al-Ṣafadī, *al-Wāfī bi 'l-wafayāt*, ed. Helmut Ritter [et al.]. Bibliotheca islamica; 6 (Wiesbaden [etc], 1962–), VIII, pp. 167–172.

[12] Shihāb al-Dīn Maḥmūd b. Salmān b. Fahd al-Ḥalabī, *Ḥusn al-tawassul ilā ṣinā'at al-tarassul*, ed. Akram 'Uthmān Yūsuf (Baghdad, 1980), pp. 353–367 = al-Qalqashandī, *Ṣubḥ al-a'shā fī kitābat al-inshā*, (Cairo, 1913–18), XIV, pp. 288–299 = Ibn Faḍl Allāh, *Masālik*, XII, pp. 414–429 (including a letter about the text). On the author see also al-Ṣafadī, *A'yān al-'aṣr wa-a'wān al-naṣr*, ed. 'Alī Abū Zayd [et al.] (Damascus, 1418–9/1998), V, pp. 372–399.

[13] *Ḥusn al-tawassul*, pp. 347–349 = Ibn Faḍl Allāh, *Masālik*, XII, pp. 402–404.

[14] Al-Qalqashandī, *Ṣubḥ*, XIV, pp. 282–288.

[15] *Dīwān Ṣafī al-Dīn al-Ḥillī*, ed. Muḥammad Ḥawar, ('Ammān, 2000), I, pp. 195–202.

[16] *Dīwān Ṣafī al-Dīn al-Ḥillī*, I, pp. 407–414.

[17] *Dīwān Ṣafī al-Dīn al-Ḥillī*, I, pp. 407–440.

[18] Abbreviated version in *Dīwān Ibn Nubāta al-Miṣrī*, ed. Muḥammad al-Qalqīlī (Cairo, 1905), pp. 585–592 (169 lines) and ed. As'ad Ṭalas in *Majallat al-Majma' al-'Ilmī al-'Irāqī* 2 (1951), pp. 302–310 (177 lines). The probably complete version of 193 lines can be found in the manuscripts Berlin MS 8400, f. 47b–51a (the same MS that contains the *muzdawija* edited here), and in Shihāb al-Dīn al-Ḥijāzī, *Rawḍ al-ādāb*, Vienna, MS 400, f. 59b–62b.

[19] Al-Qalqashandī, *Ṣubḥ*, XIV, pp. 165–173. On the author see al-Ṣafadī, *A'yān*, V, pp. 170–180. A selection of Mamluk and older texts on hunting is given in al-Jazūlī, *Maṭāli' al-budūr fī manāzil al-surūr* (Cairo, 1419/2000), pp. 527–555.

9. The text to be presented here can be attributed to Shihāb al-Dīn Aḥmad Ibn Faḍl Allāh (700–749/1301–1349) with a sufficient amount of probability. It is an *urjūza muzdawija* in the tradition of Sirāj al-Dīn al-Warrāq.

The dawādār *Nāṣir al-Dīn, the governor Tankiz, and the Syrian hunt*

A hunting poem or a hunting *risāla* could not be created until several conditions were fulfilled. (1) A rich and mighty personality arranges sumptuous hunting expeditions that are worthy of immortalization in the form of a literary work. (2) This person or another equally important person from his entourage, who also takes part in these excursions, has a good command of Arabic and is interested in Arabic literature. (3) A writer of poems and/or prose literature is interested in dedicating a literary portrayal of the hunt to this person, and even more cogently if the writer is allowed to take part in the hunt himself. (4) There is a broader general public that is interested in literary texts about hunting. Otherwise the genre of the *ṭardiyya* loses its relevance and neither patron nor poet will expend any effort on the creation of a new text. The absence of one of these preconditions helps to explain the discontinuities in the history of Arabic hunting literature.

At the time of the creation of Shihāb al-Dīn's *muzdawija*, spectacular hunting safaris were conducted by the Sultan himself, by his governors or, more rarely, by other leading Mamluk amirs.[20] For the Turkish Mamluks, the hunt offered an opportunity to practise military skills in a time of peace. But contrary to Ibn Ṭughj, Sayf al-Dawla and ʿAḍud al-Dawla, to whom al-Mutanabbī, Abū Firās and their contemporaries directed their hunting poems, these Mamluks had little if any interest in Arabic poetry. By and large, architecture had taken the representative function that poetry had fulfilled during the Abbasid period, and many amirs did not have enough command of the Arabic language to enjoy the subtleties of contemporary literature. On the other hand, poetry had acquired new functions as a privileged means of communication between the *ʿulamāʾ* and an opportunity to distinguish oneself, to show one's mastery of the Arabic language and the aesthetic code of literature,

[20] See Nabīl Muḥammad ʿAbd al-ʿAzīz, *Riyāḍat al-ṣayd fī ʿaṣr salāṭīn al-Mamālīk* (Cairo, 1999), pp. 127–203.

and thus to prove worthy to be included among the crème of the cultivated civilian elite. As a consequence, the time-honoured genre of panegyric poetry *(madīḥ)* could survive, but in a more bourgeois form as literature directed by one *ʿālim* to another *ʿālim* in a communicative exchange on the same social and intellectual level. For hunting texts, the situation was more complicated, since more often than not (1) the leader of the hunting expedition, and (2) the person interested in hunting literature were not one and the same person. A major exception is al-Malik al-Afḍal of Ḥamāh. His father Abū ʾl-Fidāʾ, the famous author of a book on geography and another on history, was a scion of the Ayyubid dynasty and governed Ḥamāh for the Mamluks under the title al-Malik al-Muʾayyad.[21] His son, who bore the title al-Malik al-Manṣūr before he became the successor of his father as al-Malik al-Afḍal in 732/1331, was a great Nimrod before the Lord and used to go on hunting expeditions with the *nāʾib al-Shām* Tankiz.[22] At the same time, he was a patron of literature in the footsteps of his father. Ibn Nubāta addressed him with about twenty panegyric qasidas. Around the year 728, when the future al-Afḍal was still the heir-apparent al-Manṣūr, Ibn Nubāta composed for him his *muzdawija Farāʾid al-Sulūk*, the masterpiece of hunting literature of the period.

But al-Manṣūr/al-Afḍal was an exception.[23] The most enthusiastic hunter Tankiz, the powerful governor of Syria between 712/1312 and 740/1340,[24] had no pronounced interest in Arabic poetry, nor did most of the sultans in Cairo. Great literary competence, however, was assembled in the chancellery of state, in the *dīwān al-inshāʾ* of Cairo, Damascus, and Aleppo. It is no accident, therefore, that most authors of Mamluk hunting texts stood in close contact to the *dīwān al-inshāʾ*. Ibn al-ʿAṭṭār, al-Bārinbārī and Ibn Nubāta were *kuttāb* of renown, and al-Shihāb Maḥmūd and his disciple al-Shihāb Ibn Faḍl Allāh held the office of the head of the chancellery, *kātib al-sirr*, of Damascus for many years.

---

[21] See H.A.R. Gibb, 'Abū ʾl-Fidāʾ,' in *Encyclopaedia of Islam* (New ed.), I, pp. 118–119.

[22] See ʿAbd al-ʿAzīz, *Riyāḍat al-ṣayd*, p. 198.

[23] Other possible exceptions are the Artuqid princes to whom al-Ḥillī addressed panegyric poems.

[24] See ʿAbd al-ʿAzīz, *Riyāḍat al-ṣayd*, pp. 197–199, and S. Conermann, 'Tankiz,' in *Encyclopaedia of Islam* (New ed.), X, pp. 185–186.

Given the fact that the hunt was led by a person who had little interest in literature and that the most significant literary competence had been assembled in the *dīwān al-inshāʾ*, a crucial role came to be played by persons who held an office that combined competence in the realm of the civil elite with power in the realm of the military elite. An office in this intermediate position was the vizier, in this time identical with the *nāẓir al-dīwān*, 'supervisor of the chancellery.' Another office even closer to the military was the *dawādār*, the 'bearer of the royal inkwell.' His duties and his importance varied a great deal over time. The holders of these offices were close enough to the amirs to take part in their hunting expeditions side by side with the amirs and close enough to the *kuttāb* to be interested in sophisticated literature and pay attention to communication by means of poetry and artistic prose.

The first *urjūza muzdawija* of the Mamluk period was composed by al-Warrāq for the vizier al-Ṣāḥib Tāj al-Dīn Ibn Ḥinnā and his brother.[25] Shihāb al-Dīn Ibn Faḍl Allāh composed his *urjūza* for the *dawādār* Nāṣir al-Dīn Muḥammad b. Kawandak. From 712/1312–1313 to 734/1333–1334, Nāṣir al-Dīn was an intimate friend of the governor Tankiz and the most powerful man in the office of the *dawādār*. 'People never saw a *dawādār* like him.'[26] When Nāṣir al-Dīn came to office, Shihāb al-Dīn's uncle Sharaf al-Dīn was *kātib al-sirr* of Damascus (712–717). Ten years later, his father Muḥyī ʾl-Dīn was appointed to the same office, which he held between 727–729 and again between 732–733. In the years between and after 733, he held the same office in Cairo. Muḥyī ʾl-Dīn, born 645, was already aged at that time and assisted by his son, who was the éminence grise of the *dīwān* even before he came to hold the office himself between 741 and 743.[27] It is quite natural that al-Shihāb Ibn Faḍl Allāh should have an interest in the mighty and well educated *dawādār*. But the network of relations was even more complicated as a result of the presence of Ibn Nubāta as well as the prince of Ḥamāh. The picture thus presented is much like a tableau: Tankiz, the governor of Syria, liked to go on hunting expeditions with the prince of Ḥamāh. Ibn Nubāta had close relations to the Ḥamawī court and dedicated many poems and other works to al-Muʾayyad and his son. Among these works

[25] Ibn Faḍl Allāh, *Masālik*, XII, p. 258; see also al-Ṣafadī, *Wāfī*, I, pp. 217–228.

[26] Al-Ṣafadī, *Aʿyān al-ʿaṣr*, V, p. 105.

[27] See K.S. Salibi, 'Faḍl Allāh,' *Encyclopaedia of Islam* (New ed.), II, pp. 732–733; idem, 'Ibn Faḍl Allāh al-ʿUmarī,' *Encyclopaedia of Islam* (New ed.), III, pp. 758–759; al-Ṣafadī, *Wāfī*, VIII, pp. 252–70, XXVIII, pp. 267–272; idem, *Aʿyān*, IV, pp. 417–434, V, pp. 571–576.

is a *Mufākhara bayna ʾl-Sayf wa ʾl-Qalam* 'Debate between the Sword and the Pen,' first dedicated to al-Muʾayyad. He accompanied his son al-Manṣūr on one of his hunting safaris and immortalized it in his hunting *urjūza*. However, Ibn Nubāta was also in close contact with the *dīwān al-inshāʾ* of Damascus, especially with the Banū Faḍl Allāh. He addressed poems to Muḥyī ʾl-Dīn and exchanged poems and letters with Shihāb al-Dīn and his brothers.[28] But he also approached the *dawādār* himself. We know of an epigram addressed to him,[29] and the Berlin manuscript, which contains both al-Shihāb's hunting *muzdawija* and Ibn Nubāta's *mufākhara*, shows that Ibn Nubāta dedicated his *mufākhara* not only to al-Muʾayyad, but also to the *dawādār* Nāṣir al-Dīn.[30] The *muzdawija* that is presented here is therefore much like a knot within a rather complex network that can be visualized as follows:

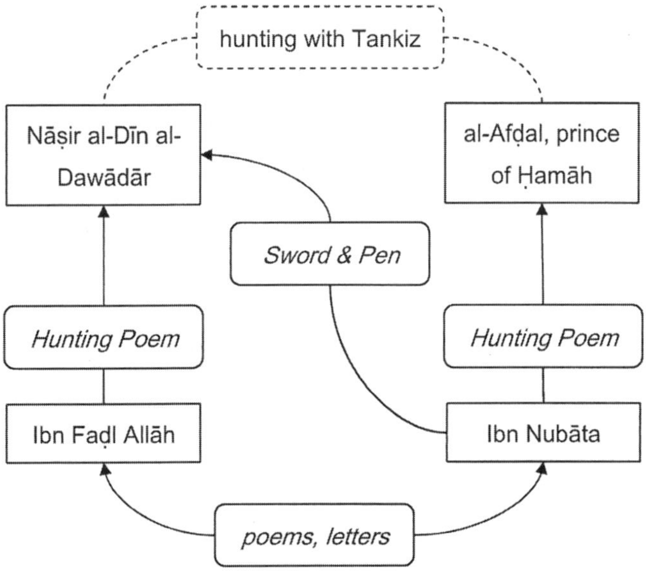

[28] See Bauer, 'Ibn Nubātah al-Miṣrī (686–768/1287–1366). Life and works,' in *Mamlūk studies review* (forthc.).

[29] Ibn Nubāta, *Dīwān*, p. 253 (Munṣariḥ, *-āṣirī*), see also Paris, Bibliothèque Nationale MS arabe 2234, f. 162b.

[30] See Bauer, *Ibn Nubātah* (forthcoming).

Not visible in the chart is the general public, which is nonetheless always present in the mind of both poet and addressee. As the many sources for Ibn Nubāta's hunting *urjūza* show, the literary public of the time had a vivid interest in hunting literature of this kind. Ibn Nubāta's *muzdawija* was praised as a model for *insijām* 'fluency.'[31] This fluency is without doubt also prominent in the poem by Ibn Faḍl Allāh.

The manuscript Berlin 8400 and the authorship of the urjūza

The only source of the *urjūza* known so far (at least to me) is a Berlin manuscript that bears the title *Tadhkirat al-Nawājī*.[32] This manuscript is an important, if not the only, source for several literary texts, among them a satirical criticism of Abū Tammām's poetry by one of the Khālidī brothers (*Ẓulāmat Abī Tammām*), Ibn Nubāta's *Dīwān* of epigrams, entitled *al-Qaṭr al-Nubātī*, his *Mufākhara bayna 'l-Sayf wa 'l-Qalam* and, finally, his hunting poem *Farāʾid al-Sulūk fī Maṣāʾid al-Mulūk*. The author of the compilation, however, is clearly not al-Nawājī. The title, which is written by a third hand, is nothing but the guess of a bookseller. Instead, several remarks make it clear that its author is al-Ṣafadī, and that we probably have a volume of the *Tadhkirat al-Ṣafadī* before us.[33]

The text of the *urjūza muzdawija* published below can be found on f. 42b–43b. The preceding page, f. 42a, starts with the headline *al-Qāḍī Shihāb al-Dīn Aḥmad b. al-Mawlā al-Makhdūm al-Qāḍī Muḥyī 'l-Dīn Ibn Faḍl Allāh*. Several poems and prose passages by this author fol-

---

[31] See Ibn Ḥijja, *Khizānat al-adab*, III, pp. 102–115 and G.J. van Gelder, 'Poetry for easy listening,' *Mamlūk studies review* 7/1 (2003), pp. 31–48, esp. p. 41.

[32] Berlin MS 8400; see also Wilhelm Ahlwardt, *Verzeichniss der arabischen Handschriften*, (Berlin, 1887–1899), VII, p. 383.

[33] F. 6a: The author quotes Ibn Sayyid al-Nās (d. 734/1334) and adds: '*wa-samiʿtu dhālika min lafẓihī*'; al-Ṣafadī mentions several titles of Ibn Sayyid al-Nās he had heard from him (*Wāfī*, I, p. 292). f. 8a: *wa-anshadanī al-shaykh... Ibn Muṣaddaq... li-nafsihī... fī 15 dhī 'l-qaʿda sanata 728...*; the same verses are cited also in al-Ṣafadī, *Wāfī*, XIII, pp. 24–25, introduced by the words *anshadanī jumlatan min shiʿrihī, min dhālik....* The author entered Cairo in the year 727 (f. 55b), he met Abū Ḥayyān al-Gharnāṭī in Cairo in the year 728 (f. 27b), and he met Ibn Nubāta in Damascus (f. 52a). He heard Ibn Nubāta's *Mufākhara bayna 'l-sayf wa 'l-qalam* from the author himself in the year 729 in the Umayyad Mosque in Damascus. All these events of al-Ṣafadī's life can be corroborated by other sources. Since on f. 73a Shihāb al-Dīn Ibn Faḍl Allāh is called *kātib al-sirr*, the work may have been written between 740 and 743.

low. At the very end of the page, the writer added another headline, saying: *al-Mawlā al-Shaykh al-Imam al-Adīb al-Fāḍil Jamāl al-Dīn Abū ʿAbd Allāh b. al-Shaykh al-Imām al-Muḥaddith Muḥammad Ibn Nubāta.* The words *'Ibn Nubāta'* are crossed out and substituted by the words *'Muḥyī 'l-Dīn Faḍl Allāh'*. This error can be explained by the following hypothesis: The *urjūza* is still part of the section comprising texts by al-Shihāb Ibn Faḍl Allāh. The writer, however, realised that the following text is a hunting *urjūza*. At this time, the most famous hunting *urjūza*, was, of course, Ibn Nubāta's poem, a text that is also included in this very collection a few pages later. Further, the writer must have realised that several leaves of his source were in disorder. This is very obvious in the text of *al-Qaṭr al-Nubātī*. Obviously, he thought that Ibn Nubāta's hunting *urjūza* had already started even in these earlier pages. Therefore he added the headline appropriate for this text. Later, perhaps when he reached f. 47b, he realised his error. On that page, Ibn Nubāta's *urjūza* begins, and it is introduced with exactly the same headline as the bottom of f. 42a. Having realised his error, the writer crossed out the words 'Ibn Nubāta' and substituted them with the last words of the headline at the top of the page. Even so, he still did not realise that not only the end of the headline was wrong, but also its beginning. I conjecture that the original headline at this place was either a simple *wa-lahū (ayḍan)* or a repetition of the headline at the top of the page. The writer of the title page, which contains a list of the principle texts included in the *Tadhkira*, refers to our text as *Urjūzat Ibn Faḍl Allāh fī 'l-Ṣayd.*

This is corroborated by internal criticism. The poem presented here cannot be a work of Ibn Nubāta. A comparison between this poem and Ibn Nubāta's *Farāʾid al-Sulūk* reveals a clear stylistic contrast. Whereas Ibn Nubāta repeatedly uses the *tawriya* (double entendre), this stylistic device hardly plays a role in the *urjūza* published here. Instead, its author makes ample use of all kinds of *jinās*. This corresponds exactly to what we are told by Ibn Ḥijja al-Ḥamawī, who mentions al-Shihāb Ibn Faḍl Allāh among the poets who possess no virtuosity in the field of the *tawriya* and make only sporadic use of it.[34] In favouring *jinās* instead of the *tawriya*, his style is closer to that of al-Ḥillī than to that of Ibn Nubāta. Other sources corroborate that the *urjūza* is a literary

---

[34] Ibn Ḥijja, *Khizānat al-adab*, III, p. 522.

form that was practiced by Shihāb al-Dīn. Though he mentions no titles, al-Ṣafadī explicitly states that al-Shihāb is the author of 'many qasidas, *rajaz* poems, epigrams and *dūbayts*.'[35]

Finally, Ibn Faḍl Allāh's relation to the *dawādār* that resulted from his position in the *dīwān al-inshāʾ* and his place in the literary system of his period, as sketched above, make him the most likely candidate for the authorship of the *urjūza*. The most probable dates for its creation would be the years 727–729 and 732–733, when Shihāb al-Dīn acted as a deputy for his aging father, the official *kātib al-sirr* of Damascus.

Shihāb al-Dīn Ibn Faḍl Allāh as a poet

In Western sources, the author to whom the *muzdawija* can be attributed is normally called 'Ibn Faḍl Allāh al-ʿUmarī.' But there were more famous members of the Banū Faḍl Allāh, including his father Muḥyī ʾl-Dīn and his younger brother ʿAlāʾ al-Dīn. Therefore, it is appropriate to use no other form of his name than the form used in all contemporary sources, Shihāb al-Dīn (or al-Shihāb) Ibn Faḍl Allāh. Today, Shihāb al-Dīn's manual on penmanship (*al-Taʿrīf bi ʾl-Muṣṭalaḥ al-Sharīf*) and his encyclopaedia *Masālik al-Abṣār* are used as essential sources for the history of the period. But his contemporaries saw Shihāb al-Dīn principally as a man of letters, an author of poetry and prose literature. Paradoxically, his achievements in the field of literature are not even mentioned in the entry on him in the *Encyclopedia of Arabic Literature*.[36] For al-Ṣafadī, however, our author was the *imām ahl al-ādāb*.[37] According to this statement, al-Shihāb Ibn Faḍl Allāh reached the absolute summit as a prose author (i.e. as an author of *inshāʾ*). No *munshiʾ* after al-Qāḍi al-Fāḍil could ever reach his level, despite such prominent writers as Ibn al-Athīr, Muḥyī ʾl-Dīn Ibn ʿAbd al-Ẓāhir, al-Shihāb Maḥmūd and Ibn al-ʿAṭṭār.[38] Note that three of them also appear in our list of Mamluk authors of hunting literature! Al-Ṣafadī also finds warm words for his

---

[35] Al-Ṣafadī, *Aʿyān*, I, p. 419.

[36] C.E. Bosworth, 'Ibn Faḍl Allāh al-ʿUmarī,' in *Encyclopedia of Arabic literature*, I, pp. 322–323.

[37] This and the following according to al-Ṣafadī, *Wāfī*, VIII, pp. 252–253.

[38] Al-Ṣafadī, *Wāfī*, VIII, p. 253.

poetry, though there is a light reserve here. Whereas in prose nobody can ever attain his level, he says, in his poetry 'only single poets' (*al-afrād*) can equal him.[39] Another contemporary, the scholar Tāj al-Dīn al-Subkī, leaves no doubt that Ibn Nubāta is unassailable both in poetry and prose.[40] But in another place he tries to promote Ibn Qāḍī 'l-ʿAskar by calling him 'one of the three *udabāʾ* of their time in the realm of prose, in addition to Ibn Nubāta and Ibn Faḍl Allāh, being superior to both in scholarship (*al-ʿulūm*), whereas they are superior to him in poetry.'[41] Ibn Ḥajar al-ʿAsqalānī, himself a poet of renown and a great admirer of Ibn Nubāta, contradicts vehemently, saying that Ibn Nubāta and Ibn Faḍl Allāh do not belong in the same category of poets.[42] Though he speaks highly of Shihāb al-Dīn's prose, he is less enthusiastic about his poetry and remarks that al-Shihāb 'composed very much poetry, but of an average quality (*wasaṭ*).'[43] Nevertheless, al-Shihāb Ibn Faḍl Allāh remains one of the most respected *udabāʾ* of his time, and this should be reason enough to turn attention to his poetic production as well.

Al-Shihāb Ibn Faḍl Allāh belongs to the sphere of the chancellery, and his literary production reflects the aesthetics of the chancellery rather than that of other milieus of poetic activity. It may be no accident, therefore, that one of the few major poetic works of Ibn Faḍl Allāh that has been preserved is a hunting *urjūza*. The above list of Mamluk literary hunting texts demonstrates quite clearly that the proper domain of Mamluk hunting literature was the chancellery. The texts are either *rasāʾil*, the chancery genre *par excellence*, and/or they are composed by authors who were professional secretaries or held a leading position of the chancellery, if only temporarily. A major exception is Ṣafī al-Dīn al-Ḥillī, who, despite his intensive engagement in hunting literature, conspicuously wrote neither a narrative *urjūza* nor a *risāla* on this subject.

---

[39] Al-Ṣafadī, *Wāfī*, VIII, p. 253.

[40] See Tāj al-Dīn al-Subkī, *Ṭabaqāt al-shāfiʿiyya al-kubrā*, (Beirut, 1420/1999), V, pp. 111, 153.

[41] See Ibn Ḥajar al-ʿAsqalānī, *al-Durar al-kāmina fī aʿyān al-miʾa al-thāmina*, (Ḥaydarābād, 1929–1931), IV, p. 49.

[42] Ibn Ḥajar, *Durar*, IV, p. 49.

[43] Ibn Ḥajar, *Durar*, I, p. 395.

Structure and content of Shihāb al-Dīn's muzdawija ṭardiyya

The structure of the poem is clear-cut and well-balanced. The poem may be said to consist of three parts, each part divided into two subsections. The first part starts with an introductory section of eleven couplets, primarily of a descriptive nature. The second section, the 'parade of the horses,' is again a descriptive section and again comprises eleven lines. The central section of the poem is constituted by two hunting episodes that combine both descriptive and narrative elements. Each of these sections has the length of twelve couplets. A transition of five lines, the *takhallus*, leads up to the concluding panegyric section, which with its thirteen lines is again one couplet longer than the central sections. A chart may illustrate this structure:

Lines	Subject	Length
1–11	1–3: frame: address to the audience; 4–11: description of nature	11
12–22	the huntsmen and their horses: 'parade of the horses'	11
23–34	*episode 1: hunting gazelles and onagers with arrows, dogs and cheetahs*	*12*
35–46	*episode 2: hunting birds with falcons and the crossbow*	*12*
47–51	*takhalluṣ*: return from the hunt; transition to the panegyric section	5
52–64	praise *(madīḥ)* of the *dawādār*	13

(1) Introduction and description of nature. With its reference to *surūr* 'joy' the first couplet may allude to the *urjūza muzdawija* by Abū Firās al-Ḥamdānī (320–357/932–968), in which we read *al-ʿumru mā tamma bihī l-surūru* 'life is when joys are complete.'[44] But there are only few similarities between both poems, apart from their form. Abū Firās's poem, for instance, is more than twice as long as Shihāb al-Dīn's. Consequently, Shihāb al-Dīn cannot give as detailed a record of the events

[44] *Dīwān Abī Firās*, ed. Sāmī al-Dahhān, (Beirut, 1944), III, p. 435. See also James E. Montgomery, 'Abū Firās's Veneric *Urjūzah Muzdawijah*,' *Arabic and Middle Eastern literatures* 2 (1999), pp. 61–74.

of the hunt as Abū Firās. Further, Mamluk authors seem interested less in a narration of events than in an endeavour to transform the events into a sequence of literary images. In any case, the aesthetics of Mamluk literature were less dependent on their Abbasid predecessors than is often assumed. Shihāb al-Dīn's principal model as regards structure and style was not that of Abū Firās, but Sirāj al-Dīn al-Warrāq. This is clearly demonstrated by the lines that follow the frame motive of the first three couplets, the description of nature. Nearly all Mamluk hunting texts, both poetry and prose, start with a description of an idyllic landscape in the early morning light before sunrise.

(2) The parade of the horses. The background of time and place is set now and the attention turns to the protagonists, the huntsmen and their horses. The hunters are introduced only briefly. The greater part of the paragraph is dedicated to a very popular motive of Ayyubid and Mamluk literature, which I call 'the parade of the horses.' It occurs in hunting poems, in *rasāʾil ṭardiyya* (but not in *rasāʾil bunduqiyya*), but also in *rasāʾil* on horses and other texts, in which horses are treated in a literary fashion. In these 'parades,' the author stages horses of different colours, one after the other, and provides a (shorter or longer) description for each of them.

The principal colour terms for Arabian horses that were used at that time were *ashhab* 'white,' *adham* 'black,' *akhḍar* 'seal brown,' *aḥmar* = *kumayt* 'bay,' *ashqar* 'chestnut,' *aṣfar* 'dun,' and *ablaq* 'dappled.' Horse colours are treated in several works of the Mamluk period.[45] At that time, the meaning of these terms was not necessarily exactly the same as it is today or even as it was in ancient Arabia. The terminology of horse colours was very detailed. All authors of theoretical texts know several subtypes of every basic colour, but these subtypes do not play a role in the literary texts. One of the theorists considers *akhḍar* a subtype of *adham*.[46] This may be the reason for the fact that only four of the nine texts mentioned below differentiate between *adham* and *akhḍar*. If both are differentiated, *adham* is equated with *aswad*, whereas al-Nuwayrī's description of the type *akhḍar adgham* is that of the English 'seal brown':

---

[45] See al-Nuwayrī, *Nihāyat al-arab*, X, pp. 5–12; ʿAlī b. ʿĪsā al-Fahrī al-Basṭī, *Zahrat al-ādāb wa-tuḥfat ulī ʾl-albāb*, Alexandria, al-Maktaba al-Baladiyya MS 1676 ب, pp. 33–36; al-Malik al-Ashraf ʿUmar b. Yūsuf, *al-Mughnī fī ʾl-bayṭara*, ed. Muḥammad Altūnjī (Abū Ẓabī, 1425/2004), pp. 33–42. A good modern study on horses, shooting, and hunting in classical times is ʿAbd al-Ḥalīm Salāma, *al-Riyāḍa al-badaniyya ʿinda ʾl-ʿArab*. 2nd ed. (Tunis, *c.* 2000).

[46] Al-Basṭī, *Zahrat al-ādāb*, p. 34.

wa-huwa al-akhṭabu lawni wajhihī wa-udhunayhi wa-manākhirihī.[47] Other shades of dark may also be subsumed under the term *adham*. The difference between *ashqar* and *aḥmar/kumayt* is explained by al-Basṭī as follows:

والفرق بين الكميت والأشقر بالعرف والذنب فان كانا أحمرين أو أصهبين فهو أشقر وان كانا أسودين فهو كميت.[48]

This is exactly the difference between the bay with its black mane, tail, and points, and the chestnut. Five of our nine authors differentiate between both. Dappled horses may be reckoned among the one or the other colours according to the dominating colour. This is the reason why only three of our authors treat *ablaq* as a category of its own.

The following chart lists the horse colours (note that *kumayt* = *aḥmar*) portrayed in nine 'parades' of the late Ayyubid and the early Mamluk period. The lines are arranged according to the date of birth of the author. The numbers indicate the sequence in which the different colours are treated.

	ashhab	adham	akhḍar	aḥmar	ashqar	aṣfar	ablaq
Ḍiyā' al-Dīn al-Qurṭubī (*602)[49]	4	3		1	2		
Sirāj al-Dīn al-Warrāq (*615)[50]	2	1		3		4	
Ibn al-ʿAṭṭār (*626)[51]	2	4			1	3	
al-Shihāb Maḥmūd (*644)[52]	1	2	6	4	3	5	7
al-Makhzūmī (*680)[53]	1	5		3	2	4	

[47] al-Nuwayrī, *Nihāyat al-arab*, X, p. 7.

[48] Al-Basṭī, *Zahrat al-ādāb*, p. 33.

[49] A letter in al-Nuwayrī, *Nihāyat al-arab*, VIII, pp. 67–68. On al-Qurṭubī see *Wāfī*, VII, pp. 339–346.

[50] In his *muzdawija*, Ibn Faḍl Allāh, *Masālik*, XIX, p. 253.

[51] In his *risāla* on the postal service, *Rasf al-farīd fī wasf al-barīd*, ibid., XII, p. 368.

[52] Not in his hunting letters, but in a *Risāla fī awṣāf al-khayl*, see Shihāb al-Dīn Maḥmūd, *Ḥusn al-tawassul*, pp. 343–347 (the 'parade' on p. 344) and al-Nuwayrī, *Nihāyat al-arab*, X, pp. 70–75.

[53] Tāj al-Dīn ʿAbd al-Bāqī b. ʿAbd al-Majīd al-Makhzūmī, see *Wāfī*, XVIII, pp. 23–28, is the author of a *risāla* on horses, see al-Nuwayrī, *Nihāyat al-arab*, X, pp. 75–78.

Table (*cont.*)

	ashhab	adham	akhḍar	aḥmar	ashqar	aṣfar	ablaq
Ibn Nubāta (*686)[54]	3	5	4	1		2	
al-Bārinbārī (*696)[55]	1	3		2			
*al-Shihāb Ibn Faḍl Allāh (*700)*	*1*	*2*	*3*	*4*	*5*	*6*	*7*
Badr al-Dīn Ibn Ḥabīb (*710)[56]	1	2	6	4	3	5	7

A comparison between these texts would give interesting insight into the descriptive styles of the period. Here it must suffice to note that obviously al-Shihāb Maḥmūd was the first to arrange a parade of horses of seven different colours. The author of our *muzdawija* emulated him in poetic form, and Ibn Ḥabīb followed this model even in the sequence of the colours. Shortly afterwards, Ibn Ḥijja al-Ḥamawī included four 'parades of the horses' in his *Qahwat al-inshā'*: the well-known text by al-Shihāb Maḥmūd and three *muʿāraḍāt* by Ibn Nubāta, Ibn Faḍl Allāh and Ibn Ḥijja himself (see ed. R. Veselý, Beirut 2005, pp. 86–108).

(3) The first hunting episode. Different forms of the hunt were combined in a single hunting excursion. The first episode narrates a hunt with bows, dogs and cheetahs on gazelles and onagers. Our author creates a succinct text by combining narrative and descriptive elements in a compact and dense way. The episode may be read as a narrative sequence, but it is at the same time a sequence of descriptions of actions and movements. A general narrative line (line 23) introduces lines on (1) the hunters, their bows and arrows (lines 24–26), (2) the dogs (lines 27–28), which are in their turn compared with arrows, and (3) the cheetahs (lines 29–30). Another general narrative line on the quarry in general (line 31) introduces lines on the (4) gazelles (lines 32–33) and on the (5) onagers (line 34). The Syrian onager, *Equus hemionus hemippus*, is now extinct.[57]

[54] Ibn Nubāta, *Dīwān*, pp. 588–589.

[55] In his *risāla ṭardiyya*, al-Qalqashandī, *Ṣubḥ*, XIV, p. 169.

[56] Badr al-Dīn Ibn Ḥabīb, *Nasīm al-ṣabā*, (Alexandria, 1289/1873), pp. 65–66 = Paris, Bibliothèque Nationale MS arabe 3362, f. 52a–53b.

[57] See Thomas Bauer, *Altarabische Dichtkunst. Eine Untersuchung ihrer Struktur und Entwicklung am Beispiel der Onagerepisode* (Wiesbaden, 1992), I, pp. 13–14.

(4) The second hunting episode. A hunt with the crossbow and hunting birds (sakers and goshawks) follows. This form of the hunt was also the subject of the *rasāʾil bunduqiyya* that were so popular in these days. The episode in this *muzdawija* is of a predominantly narrative character. Descriptive lines are dedicated to the crossbow (line 37) and the hunting birds (lines 41–42). The hunted birds, which are described in detail in the *risāla bunduqiyya* by Ibn al-ʿAṭṭār, are mentioned only in the act of fleeing (line 40), and no further description is dedicated to them. The first verse of couplet 45 is a near verbatim quotation of al-Warrāq, line 17a: وأخلى بها الجوَّ من طيورٍ ....[58] The reference to the sunset in line 46 closes the time frame that was opened in the first paragraph with the description of the daybreak. This does not necessarily mean that the hunt lasted a single day only. Instead, it is quite probable that hunting expeditions of this kind extended over more than one day and included bivouacking, as the following line shows.

(5) The takhallus̱. The hunting party returns to their bivouac full of joy and starts to praise the *dawādār*, who appears as the master of the hunt, though probably its real master was the governor Tankiz. Just as the *muzdawija* poems by al-Warrāq and Ibn Nubāta, this is a panegyric poem. *Ḥusn al-takhalluṣ*, a good transition between the first part and the panegyric section, is one of the main requirements of panegyric texts. In this field, Ibn Ḥijja considered the achievements of 'modern' poets like Ibn Nubāta much superior to that of Abbasid and older authors.[59] Shihāb al-Dīn's *takhalluṣ* creates an atmosphere of joy and enthusiasm after a successful hunt. In this situation, the desire to thank and praise its master comes in quite naturally, most typically within the frame of a concluding banquet.

(6) The panegyric section. The concluding *madīḥ* rounds off the poem. Despite its new subject, it is not felt like an alien element. This is mainly due to the fact that it is a perfect echo of the description of nature in the first paragraph. In the introduction, dew, rain, bounty and beauty set the mood of the hunting expedition. In the conclusion, we learn the true nature of nature: it is nothing but the *dawādār* himself. Nearly all key words from the first section are repeated in the last one or substituted by near synonyms: compare *muzn*, *ghamāma* (lines 5, 10) – *saḥāb*,

---

[58] Ibn Faḍl Allāh, *Masālik*, XIX, p. 255.

[59] See Ibn Ḥijja, *Khizānat al-adab*, II, pp. 399–433.

ghamām (lines 54, 61); *ṭall* (line 6) – *nadan* (line 54); *nūr*, *shams* (line 7) – *shams*, *nūr* (lines 55, 56); *rīḥ*, *ṣabā* (lines 8, 10) – *nasīm* (line 58); *tabʿath* (line 8) – *tuḥyā* (line 53); *ghayth* (line 9) – *ghayth* (line 61); *al-najm fī āfāqihī* (line 11) – *al-mirrīkh fī samāʾihī* (line 57).

The last lines characterize the position of a *dawādār* between the civil and military elite. He is praised for his bounty and his good governance, yet he may also behave as angry as a lion –the hunter par excellence– against his enemies. But he is also part of the world of the *ʿulamāʾ*. This is stressed in the two last couplets. Therefore, he is a bearer of both sword and pen. Shihāb al-Dīn's concluding words may be understood as a reference to the *Mufākhara bayna ʾl-Sayf wa ʾl-Qalam* that Ibn Nubāta dedicated to the *dawādār*. This *ṭardiyya* is Shihāb al-Dīn's gift to the 'bearer of the royal inkwell.'

Edition

Notes to the edition: Although I do not reproduce all vowel signs, I also do not add any vowel signs that are not already in the manuscript. I add a few *shadda* and *hamza* signs where appropriate. I differentiate between ى and ي (the manuscript has always ى). In the following cases, I add dots that have been omitted in the manuscript: Line 10: أذياله instead of اذياله, line 29: ثارت instead of تارت, line 33: قرنيه without dots under the ي, line 49 يمدحه without dots under the ي. In line 62 لكل من is written لكلمن. Three emendations (lines 9, 19 and 24) are explained in the annotations to the translation.

1 – إن شئْتَ يا مقتَطفَ السُرُور * ومجُتبي لطائفِ الحُبُور
2 – أنّك تَروي خَبرَ الصُيُود * بالطير والكلاب والفُهُود
3 – فاسمعْ حَديث مَن رأى وجَرّبَا * كيفَ تخاتل الطيور والظبا
4 – لمّا خَرَجْنا والرياضُ تُجْلَى * وكلُّ غُصنٍ جِيدُهُ مُحلَّى
5 – قَد نَثَر المُزنُ عليها جَيْبَهُ * وأسبلَ الزَهْرُ عليها ثَوبَهُ
6 – وكلّلِ الطلُّ على رُباها * جواهراً تَزيدُ في حُلاها
7 – تخالُ نُور بُرْدها الموشّع * شمسَ ضحىً قد وقفت لِيُوشَعِ
8 – والريح في عقْدِ الزُهورَ تعبَثُ * تُميتُهُ طوراً وطوراً تبعَثُ
9 – كم غادرِ الغيثُ بها غَديرا * صافي النطاف يخرج الضميرا
10 – قد صَقَلتْ أذيالهُ الغمامَهْ * وقّركتْ ريحُ الصَّبا أكمامَه
11 – والنجمُ في آفاقه حَيرانُ * والصُّبحُ في مَضْجَعِه وَسْنانُ

12 – حتىّ إذا أسْفر وجهُ الشرق * قمنا جميعاً لابتغاء الرزق

13 – في فتيةٍ كأنجمِ الظَلامِ * ألحاظهم أمضَى من السِهام

14 – قدْ ركبوا الجيادَ للطِّرادِ * كمِثْل أطوادٍ على أطواد

15 – من كلّ سلهبٍ جميل الوصْفِ * يَسبَقُ في الجرْي ارتدادَ الطرف

16 – فأشهبٌ يُقبلُ كالنهار * كأنّه طيرٌ مِنَ الأطيار

17 – وأدهمٌ مُبتَسِمُ المُحيَّا * كالليل في جبْهتِهِ الثُريَّا

18 – وأخضر كمُسقط الغمام * قدْ خلَطَ الصباحَ بالظلام

19 – وأحمرٌ يَفيضُ كالمُدام * ملتهب [الغرّة] كالضرام

20 – وأشقرٌ كقَدْحَةِ البُروقِ * لسبقه رضّخ بالخلوق

21 – وأصفرٌ مُنتَسِبٌ أصيلُ * كأنّما ذابَ به الأصِيلُ

22 – وأبلقٌ يجمَعُ في ألوانه * ما يجمعُ الأزهار في بُسْتانِه

23 – فلم يقِفْ حتّى أتانا القنصُ * سانحةً لكلّ مَن يقتَنِصُ

24 – فابْتدَر الرُماةُ بالقسيِّ * ترمي بها شاكلَةَ الرميِّ

25 – كم قمر في يَده هلال * طُلُوعهُ تُقْضَى به الآجال

26 – وانقضَّتِ السهامُ كالنجوم * وفاضتِ الدماءُ كالغُيوم

27 – وأطلقوا الكلابَ من أطواقها * تبتدرُ السُبَّاقَ من أسباقها

28 – من كلّ كلبٍ راشقٍ كالسهم * أو كوكب مُنْحَدر للرَجم

29 – وثارتِ الفِتْنةُ بالفهود * تنتهبُ الأرضَ إلى الصَيود

30 – من كلّ فهد فيه يُضرَبُ المثلْ * قد افتدتْ منه الظباءُ بالمُقَلْ

31 – فلم يَدَعْ صَيْدًا يُصادُ إلاّ * في قبضةِ السكّين أو في المَقلى

32 – فكم غَزال عن كناسه كَنِس * وخلّفوهُ في الفلاة محتبَسْ

33 – فكم به ذُرِع دَم تُحيطُ * إبرة قرنيه له تَخيطُ

34 – وكم رمَينا من حمار وَحْشِ * كمِثْلِ دينار بغير نقْشِ

35 – ثمّ انْثنتْ إلى مَرامي الطيْرِ * طَالبةً منها وُفورَ الخيرِ

36 – فكم أصابتْ من قسيّ البُندقِ * ما تشتهي في مغرب أو مشرق

37 – كأنّها في شكلها عقاربُ * تَسقُط من قبضتِها الكواكبُ

38 – وزهزهتْ ما بيننا الرُماةُ * حتّى تعالى الصَّوتُ والصيحاتُ

39 – فأشرقَتْ برَمْينا الوجوهُ * وانصرفَ المرء بما يرجُوهُ

40 – وطارتِ الطيورُ كيما تجفلا * فأرسَلوا لها بَلاءً مُرْسَلا

41 – من كلّ صَقرٍ خافق الجناح * مُصوَّر من عُنْصُر الرياح

42 – ومن بُزاةٍ فائقاتٍ في الرتبِ * فضيّةِ الجسم بعين من ذَهب

43 – تهتزّ وسطَ كفّها المناسرُ * كأنّها في فعلها بواتِرُ

44 – وكلّ جَفّةٍ بها تنقضّ * كمثْل سلك لؤلؤ يرفضُّ

45 – فأخْلتِ الجوَّ من الطيور * وكان هذا مُنتهى السرور

46 – وأصغت الشمسُ إلى الغُروبِ * مصفرّةً كوجنةِ المريبِ

47 – وبعد ذا عُدْنا إلى الخِيامِ * بأوفر الحُظوظ والأقسام
48 – نَلهجُ بالشكر وبالدعاء * وأفخر المديح والثناء
49 – لخير مَن يمدحه الأنامُ * وترتدي بفضله الأيّامُ
50 – لناصر الدين الذي يَطيبُ * بذكره المديحُ والنسيبُ
51 – هو الدوادارُ الذي كم داروا * فما روتْ مثْلاً له الأخبارُ
52 – يا واقفاً أمامَهُ أما تَرَى * في واحدٍ يُجمَعُ مجموعُ الورى
53 – أوْحَدُ أهل العصر في الصِفاتِ * تحيىَ به المنى من الرُفات
54 – أجدى من السحاب في نداهُ * ومن وُضوحِ البدر في هُداهُ
55 – كأنما الشمس التي تنيرُ * مِن فهمه شَرارةً تَطيرُ
56 – كأنما البدرُ الذي تكمّلا * نُورُ محيّاهُ إذا ما سُئِلا
57 – كأنما المرّيخ في سمائه * قَطرةَ سيفٍ من دِما أعدائه
58 – أخلاقه كهيئة النسيم * وهزّة المدامِ للنديم
59 – حتّى إذاما هجته أو غضبَا * تحسبُهُ ليثَ الشرَى قد وثبا
60 – قد أتقن المُلك بما يَسُوسُ * سياسةً طابت بها النفوسُ
61 – تدبيرُه أنفع للشام * مِن مُسْقطِ الغَيث من الغمام
62 – فكم غدا يفيء أوَ يفيدُ * لكلّ من يطلبُ ما يُريدُ
63 – يفيضُ بالعلوم والرغائب * فلا خلتْ أبوابُه من طالب
64 – لا برحتْ في طوعِه الأحكامُ * وفي يدَيه السيفُ والأقلامُ

Translation

1. Oh gatherer of joy, oh collector of subtle delight: if you wish
2. to tell stories about hunting with birds, dogs and cheetahs,
3. then listen to the report of one who witnessed and experienced how birds and gazelles outwit each other!
4. As we started off – at a time when the meadows lay disclosed and the necks of all twigs were adorned (with dewdrops) –,
5. a cloud had just emptied its pockets on them, and blossoms had made their clothes hang down over them,
6. and dew decked their hills with jewels that added to their ornament.
7. You would take the glow of the hill's variegated garments for the morning sun, which stopped its course to bring forth flowers.[60]
8. And the wind plays among the necklace of the flowers, once destroying, once enlivening it.

[60] The MS explicitly gives the vocalisation *nūr*, not *nawr*. To the different forms of *wshʿ* see Ibn Manẓūr, *Lisān al-ʿArab*, (Beirut, 1388/1968), VIII, p. 394: *washshaʿa ʾl-thawba: raqamahū bi-ʿalam wa-naḥwihī… awshaʿa ʾl-shajara wa ʾl-baqla: akhraja zahrahū…*

9. How many a pond has been left by the rain, pure its water, so that it cannot keep a secret.[61]
10. The cloud had polished its trails, and the east wind's breeze had rubbed its sleeves,
11. and the star stood confused at the horizon while the morning was still sleepy on its bed.
12. But when the face of the east became bright, we all set out to seek the boon,
13. in the midst of a valiant band of men like stars in the darkness, whose glances are more penetrating than arrows.
14. They mounted their coursers to ride in a row, and it seems as if mountains had mounted mountains.
15. All of them long-bodied, of good quality. When they run, they beat the twinkling of an eye.[62]
16. There is a white horse, which is like the light of the day when it approaches, and which is as if it were kin to the birds,
17. and a black horse with a smiling face, as if the Pleiades shone on the forehead of the night,[63]
18. and a seal brown horse like deep hanging clouds that blend the morning light with darkness,
19. and a bay horse that pours forth like an old (red) wine, its blaze glowing like a flame of fire (?),[64]
20. and a chestnut as if lightning would strike fire…?…,
21. and a purebred dun with noble pedigree, which appears as if the evening sun had melted in him,[65]
22. and a dappled horse that combines the same colours as the flowers in its grazing ground.[66]
23. Hardly had they started to follow the tracks when the game came to us, presenting its right side to the hunters,

---

[61] I propose to read (صافي النطاف) instead of (صاف النطاف). *Nuṭfa* is explained *al-māʾ al-ṣāfī*, and the phrase *ṣafā l-niṭāf* is attested in an often quoted line, see *Lisān al-ʿArab*, IX, p. 335 (s.r. *nṭf*) and XII, p. 377 (s.r. *ẓlm*), resp.

[62] Note the *ṭibāq* (antithesis) between *sabaqa* 'antecede/to go ahead' and *irtadda* 'fall back,' which cannot be expressed in the translation.

[63] The horse is black and has a blaze that extends to the muzzle or even encompasses it. This facial markings are compared with the Pleiades. If also certain legs of the horse bear a white marking, it fits the Prophet's description of the ideal horse, see Ibn Mufliḥ al-Ḥanbalī, *al-Ādāb al-sharʿiyya wa ʾl-minaḥ al-marʿiyya*, ed. ʿIṣām Fāris al-Ḥarastānī (Beirut, 1417/1997), III, p. 167: خير الخيل الأدهم الأقرح الأرثم المحجّل طلق اليمين فإن لم يكنْ أدهم فكمَيتٌ على هذه الشية

[64] I propose to read (الغرّة) or (القرحة) instead of (الغرمه), which does not make sense to me. Note that the Prophet's favoured bay (*kumayt* = *aḥmar*) has the same markings as his favoured black, see the preceding note. *mudām* may also be a 'continuous rain,' but the association with red wine is more obvious, cf. the title of al-Nawājī's anthology of wine poetry, *Ḥalbat al-kumayt* 'The Racecourse of the Bay/Red Wine.'

[65] The comparison of the dun with the evening before sunset (*aṣīl*) already in Ibn al-ʿAṭṭār's *Risāla*, see Ibn Faḍl Allāh, *Masālik*, XII, p. 368. The word is here combined with *aṣīl* 'purebred' to form a *jinās tāmm*.

[66] *Bustān* here obviously means 'pasture' rather than 'garden.'

24. and every archer hastened to grasp an arrow to shoot with it at the flank of the hunted animal.[67]
25. How many a full moon could you see, a crescent in his hand:[68] a moon that rises to bring death!
26. And arrows poured down like shooting stars, and streams of blood poured forth like clouds,
27. and they released the dogs from their collars and let them strive to be the first to win the stake.
28. All of them are dogs, <u>slender</u>/*shooting*[69] like arrows, or shooting stars coming down (from heaven) to stone (the rebel Satans).
29. And cheetahs, which plunder the earth to hunt, stirred up turmoil.
30. Every cheetah is such that it can well be said: The gazelles will sacrifice their eyes to escape from it;
31. and it lets every animal that is hunted end up in the grip of the knife and in the frying pan.
32. How many a gazelle did these cheetahs sweep away from its covert and left to be captured in the open country!
33. How often was it covered by a bloody coat of mail, which the needle of its horn had sewn for it!
34. And how many onagers, which resemble the pure gold of unstruck coins, did we shoot!
35. Then (the company?) turned to aim at the birds to seek from them a surplus of benefit.
36. And what did they gain from the crossbows from that which they desired in the west and in the east!
37. Their shape is like a scorpion from whose grip the stars fall down.
38. The marksmen among us applauded until the noise and the shouting rose high.
39. This bag of us made faces beaming, and everyone went away with what he desired.
40. And off flew the birds, only to take fright and flee away, for the hunters had sent to them a 'sent' affliction
41. in form of all the sakers with quivering wings, created from the element of wind,
42. in form of all the goshawks of outstanding rank, silver their body, golden their eyes.
43. Their beaks move back and forth between their claws, and in the action they look like sharp swords.
44. And they make every flock swoop down, so that it seems as if a string of pearls was broken, scattering its beads.

---

[67] Instead of (فابْتَدَر), the MS reads (فابْدَر).

[68] The same simile in Ḍiyāʾ al-Dīn Ibn al-Athīr's epistle, Ibn Faḍl Allāh, *Masālik*, XII, p. 304: *ahillatun ṭāliʿatun min akuffi aqmār*.

[69] A *tawriya* (metalepsis, double entendre) is noted in my translations in the following way: The primarily intended meaning is underlined, the secondarily suggested meaning is written in italics.

45. They emptied the sky from birds, and this provided for the utmost joy!
46. Now the sun inclined towards setting, pale like the cheek of a disconcerted man.
47. Thereafter we returned to our tents, and most plentiful was our fortune and our share,
48. in order to devote ourselves to thank and to bless, to extol and to eulogize most splendidly
49. the best man whom mankind can praise, a man whose lavishness is the clothing of fate:
50. Nāṣir al-Dīn! How pleasant becomes all panegyric and amatory poetry when he is mentioned in the poem!
51. However far one may roam, he will never hear such reports similar to what can be said about this *dawādār*!
52. Oh you who stand before him: Do you not see how in a single person the whole of mankind is summarized?
53. Due to his qualities, he is unequalled among the people of his time; he is the reviver of desires that have decayed.
54. In his magnanimity he is more generous than the clouds. In his guidance he is more pure than the full moon.
55. The shining sun seems to be a floating spark of his intellect.
56. The moon, when it is a full moon, seems to be the light of his countenance when he is asked for a favour.
57. Mars in its orbit seems to be a drop of the blood of his enemies on a sword.
58. His temper is like a gentle breeze and like the briskness, which wine arouses in a drinking companion.
59. But if you provoke him or when he is filled with anger, you will take him for the lion of al-Sharā when it leaps.
60. Thanks to his leadership, the empire establishes governance that brings happiness to the people.
61. His direction brings more benefit to Syria than the clouds' falling rain.
62. How much did he bestow and how much benefit did he bring to everyone who has asked for what he desired!
63. He pours forth knowledge and presents, so that his doors never lack a student/supplicant![70]
64. May the judgments never cease to be obedient to him, and may the sword and the pen never cease to be in his hands!

[70] The line contains an *istikhdām*, a form of the *tawriya* in which both possible meanings are intended at the same time. Shihāb al-Dīn, who was no expert in this field, 'borrowed' the whole line from an epigram that Ibn Nubāta once had dedicated to al-Muʾayyad of Ḥamāh, see Ibn Nubāta, *Dīwān* (Sarīʿ, *-āxibī*) and f. 33a of our MS (Ibn Nubāta, *al-Qaṭr al-nubātī*).

PRECIOUS STONES, PRECIOUS WORDS
AL-SUYŪṬĪ'S *AL-MAQĀMA AL-YĀQŪTIYYA*

Translated and Annotated by Geert Jan van Gelder

Introduction

'Many good wishes and good luck / to the recipient of this article, Professor Remke Kruk, / a jewel in Leiden's crown / and the envy of many a university town, / a unique pearl not randomly strung / and on the necklace of Oriental Studies hung, / for whose scholarship and learning / all lovers of Arabic studies are perpetually yearning, / since she excels not only in the field of medieval Arab zoology, biology, *et cetera*, / but also is distinguished in the area of Arabic *belles lettres*...'

This is obviously a pastiche of the style in which innumerable Arabic books, treatises and letters used to be written for many centuries. The floweriness and the rhyming of this prose (since being unmeasured it would never count as poetry to the Arabs) have long been rather repellent to modern tastes, of Arabs and Arabists alike. A reappraisal is on its way: currently several scholars are aiming at a re-evaluation of what is known of the Arabic literature of the Mamluk and Ottoman eras, and the exploration of the vastly greater quantity of what is as yet unknown.

The present contribution does not claim to discover anything new: al-Suyūṭī's text has been edited several times and a German translation by Rescher was published in 1918.[1] A much shortened Latin translation with annotation by a Maronite scholar from Lebanon, Ibrāhīm al-Ḥāqilānī (alias Abraham Ecchellensis) was printed in Paris as early as 1647.[2] A recent monograph by ʿAbd al-Malik Murtāḍ, somewhat

[1] The text used is that of *Sharḥ maqāmāt Jalāl al-Dīn al-Suyūṭī*, ed. Samīr Maḥmūd al-Durūbī, 2 vols. (Beirut, 1989), II, pp. 1140–1171, which is much superior to that of *Maqāmāt al-Suyūṭī*, ed. ʿAbd al-Ghaffār al-Bundārī & Muḥammad al-Saʿīd Basyūnī Zaghlūl (Beirut, 1986), pp. 68–80 (I refer to them as D and B&Z, respectively). Al-Durūbī has also helpfully identified many lines of verse quoted anonymously. For a German translation (unrhymed), with minimal annotation, see *Die Maqâmen des Sojûṭî (I–VI)*, aus dem Arabischen übersetzt von O. Rescher (Kirchhain N.-L., 1918), pp. 67–79.

[2] *De proprietatibus ac virtutibus medicis animalium, plantarum ac gemmarum. Auctore*

misleadingly entitled *Maqāmāt al-Suyūṭī: Dirāsa* (Al-Suyūṭī's Maqāmas: A Study), is in fact wholly devoted to *al-Maqāma al-yāqūtiyya*. It is a semiotic analysis of the text, with chapters on isotopy (*tashākul*), colour, space, and rhythm.[3] The present article is intended as an experiment: using the quaint medium (*saj'*), rhymed prose, in English. Whether or not it is deemed successful, it should at least give a flavour of the original. The choice of the text was partly inspired by the knowledge of the recipient's interest in the natural world, be it animal, vegetable or mineral (including precious and semi-precious stones), as well as in Arabic literary texts. At the same time this article is, in a sense, also a sequel to my translation (in ordinary prose) of a similar short composition by the same author, on perfumes.[4]

The author and the genre

It is difficult to avoid the word 'polymath' when discussing al-Suyūṭī (Egypt, 849–911/1445–1505), who wrote on virtually any subject in the realms of Islamic religion, history, philology and literature. The exact and natural sciences were not among the fields in which he excelled, but the natural world dominates some of his lesser works: his *maqāmāt* on perfumes, flowers, vegetables, fruits, nuts, and precious stones. Rather than being botanical or mineralogical treatises, they combine several

---

Habdarrahmano Asiutensis Aegyptio, nunc primùm ex Arabico idiomate Latinitate donatus ab Abrahamo Ecchellensi (Parisiis, 1647). The *Maqāma yāqūtiyya* is translated as *Disputatio Iacutina* (pp. 147–53); it comprises only five of the seven sections (*De Iacut, sive Carbunculo, De Margaritis, De Smaragdis, De Corialio, De Cyano, sive Turchina*), the peridot and cornelian being left out. The references to Qur'an and Hadith as well as virtually all the poetry are omitted, so that only the passages on the stones' properties remain. The annotation (pp. 154–179), which also discusses the two other stones (*De Topazio, De Achate*) includes some more material from the Arabic.

[3] 'Abd al-Malik Murtāḍ, *Maqāmāt al-Suyūṭī. Dirāsa* (Damascus, 1996). Although it makes some good points, much of it is a verbose stating of the obvious. Nine pages (45–54), including a complicated diagram, deal with the four words from the opening: *wa-qaṣadū* [*taṣaddaw* in the text used by me] *li l-mufākhara lā li l-mufājara* ('to embark on a mutual boasting, rather than a mutual roasting') and a further eight pages with the thirteen words of passage *ayyuhā fī 'l-rutba a'lā wa... ajlā* ('to see who would rank higher... told' in my translation).

[4] 'Four perfumes of Arabia. A translation of al-Suyūṭī's *al-Maqāma al-miskiyya*', in R. Gyselen (ed.), *Parfums d'orient* (Louvain, 1991) = *Res Orientales* 11 (1999), pp. 203–212.

domains: religious knowledge (Qur'an and Hadith, some Islamic law), philology, and literature, with perhaps some popular science and medicine. Several of these shortish texts are cast in the form of an imaginary colloquium, or rather a series of monologues, put into the mouths of the personified objects, each arguing its superiority over the others, if not in absolute terms then at least in some aspects. It is therefore closely related to the ancient genre called in Arabic *munāẓara adabiyya* or 'literary debate', and one would be tempted to use that term here, were it not that there is no true debate, nor, in this case, a final judgment by an arbiter. Such literary debates were usually composed in rhymed prose, interspersed with suitable quotations from poetry. Al-Suyūṭī called his compositions *maqāma*s, for which he has been criticised by some modern scholars who think that *risāla*, for instance, would have been a better term, in the belief that the term *maqāma* should be restricted to the more clearly narrative types made famous by al-Hamadhānī and al-Ḥarīrī.[5] But it is rather presumptuous to pretend one knows better than al-Suyuṭī. The term *maqāma* easily accommodates any text written in ornate prose which has a fictional element and in which one or more speeches are uttered. Al-Suyūṭī's aim with these texts was to instruct and entertain at the same time, quoting appropriate Qur'anic verses, traditions of the Prophet, and elegant epigrams by poets, mostly relatively late, Ibn al-Muʿtazz (d. 296/908) being the oldest one as far as can be ascertained.

Altogether, it is a rather light-weight composition, as the author would have been the first to acknowledge. Modern scholarship does not greatly like this. The erudition and labour needed for the edition and study of such texts has to be justified with more seriousness, and efforts are often made to find a deeper significance. Samīr Maḥmūd al-Durūbī, the very erudite editor of al-Suyūṭī's *Maqāmāt*, is not the first to seek and find (he thinks) a hidden allegory in texts of the literary debate genre. He believes (see his introduction, pp. 64, 68) that it is likely that the seven jewels stand for *amīr*s, military commanders, who often have names such as Lu'lu' or Yāqūt, and that the *Maqāma* 'clearly deals with political criticism' (p. 77). He assumes (p. 86) it was written 'in the period of political unrest that prevailed in Egypt after the death of sultan Qāyitbāy in 901[/1496] until the deposition of Ṭūmān Bāy by al-Ghūrī'. All this is

[5] *See* J. Hämeen-Anttila, *Maqama. A history of a genre* (Wiesbaden, 2002), p. 342.

speculation; there is not a shred of evidence and the rather un-polemical nature of the text argues against the hypothesis: the precious speakers in the text do not attack or even address one another. It is easy and tempting to suppose there is some hidden message or meaning in any text: anyone can do it. Let us consider, for instance, the initial letters of the Arabic words for the seven speaking stones in our text: these spell *Y-L-Z-M-Z-ʿ-F*, in this order; surely this acrostic stands for the sinister injunction *yalzamu zaʿfu*…, 'It is incumbent to kill X on the spot'? The Stones Have Spoken! And the missing name at the end of this incomplete acrostic surely is that of the Sultan, who is conspicuously absent at the end of the text, where he should have been the arbiter in the dispute according to a common convention of the genre? Of course not: if you believe this you will believe anything.

Precious stones in Arabic literature

There is something exceptional about precious stones (which here include the materials of 'organic' origin, pearls and coral) in Arabic literary contexts. Natural objects are described in innumerable passages in poetry and prose, usually in simple or elaborate comparisons or metaphors. Almost any natural object may occur as the *primum* or the *secundum comparationis*: eyes are compared to narcissi, and *vice versa*; Laylā's smile is like lightning, or is it the other way round? Yet jewels rarely occur 'poetically' except as *secundum comparationis*. There is a tendency to descend the natural 'chain of being', from human to vegetable to mineral: a boy's rosy cheek may be likened to roses or apples, roses or apples are compared to rubies or emeralds, but there it stops. Flowers are regularly compared to jewels, but jewels are not compared to flowers; wine is always a ruby, but to my knowledge one never reads that a ruby is like a drop of wine. This is also seen in the poetic quotations in the present *maqāma*; but since the stones adduce these images to stress their own beauty and excellence, the effect is as if they compared themselves to flowers and other beautiful things, thus reversing the normal process.

The selection of seven stones is not explained by the author. Seven is a popular number, of course, as are ten and four. Al-Suyūṭī selected seven vegetables in al-*Maqāma al-zumurrudiyya* ('The Emerald *Maqāma*', incidentally illustrating the phenomenon hinted at in the preceding paragraph), seven nuts in *al-Maqāma al-fustuqiyya* ('The Pistachio *Maqāma*'), and seven fruits in *al-Maqāma al-tuffāḥiyya* ('The

Apple *Maqāma*'). His flower *maqāma* lists ten participants and his perfume *maqāma* has four contestants. It would have been easy for al-Suyūṭī to include more precious stones, but the point of *adab* is to be neither exhaustive nor boring. Some stones, one imagines, are disqualified because of their un-Arabic and cacophonous names, to Arab ears at least (*balakhsh* 'spinel', *bāzahr* 'bezoar'). In fact, the majority of the seven jewels that are present also have names, more euphonious ones, of non-Arabic origin: *yāqūt* (ὑάκινθος) and *marjān* (μαργαρίτης) are ultimately from Greek, *fīrūzaj* is Persian, and *zumurrud* and *zabarjad*, whatever their origin (see below), are not Arabic. By modern standards, the conspicuous absentee is the diamond (*almās* or *al-mās*), which was known more for its unsurpassed hardness and its use for cutting other stones than for serving as a jewel, apparently because it was rarely found except as small stones.[6]

Translating the text

Rhymed prose, the normal medium of the *maqāma* and related medieval Arabic literary prose forms, is no longer used by Arab writers except in parodies and pastiches. It is precisely the quaintness of this form that I have tried to convey by using it in my translation, which, if berated for being 'precious', could be considered appropriate to the subject, at least. Rhyme-prose does not seem to be very common in world literature, or at least is rarely discussed.[7] The choice of this medium has resulted in some

---

[6] For more information on stones in classical Arabic literature, see the various references in my annotation. By far the best medieval text, from the scientific point of view but also interesting for students of literature, is that by the great scholar Abū Rayḥān al-Bīrūnī (d. *c.* 442/1050), *al-Jamāhir fī 'l-jawāhir*, ed. Yūsuf al-Hādī (Tehran, 1995). For a modern survey, see the entry 'D̲j̲awhar,' in the *Encyclopaedia of Islam* (New ed.), Supplement (Leiden, 2004), pp. 250–262, by M. Keene and M. Jenkins. Informative on the use of jewels is Ghāda al-Ḥijjāwī al-Qaddūmī's annotated translation of the anonymous *Kitāb al-hadāyā wa 'l-tuḥaf: Book of gifts and rarities* (Cambridge, Mass., 1996). On minerals in general see e.g. F. Sezgin (with E. Neubauer), *Wissenschaft und Technik im Islam*, IV (Frankfurt am Main, 2003), Kap. 9, 'Mineralien und fossile Substanzen,' pp. 155–210.

[7] See the short entry 'Rhyme-prose', mostly on Latin (by O.B. Hardison) and Arabic (by R.M.A. Allen), in A. Preminger & T.V.F. Brogan (eds.), *The new Princeton encyclopedia of poetry and poetics* (Princeton, NJ, 1993), p. 1065. English rhyme prose is widespread these days, alas, in the form of 'rap'. I have predecessors among Arabists, however, notably German scholars, who have repeatedly produced prose-rhyme in their translations of *saj'*: e.g. Rückert in his al-Ḥarīrī and Qur'an adaptations, Littmann in

liberties with the meaning and the syntax: I have used many 'poetic' inversions in the English. In Arabic there are scores of words rhyming with e.g. *yāqūt* or *ʿaqīq*; in English there are very few, if any, suitable rhymes with 'ruby' or 'cornelian'. Even al-Suyūṭī seems hard pressed at times to find a fitting rhyme. Naturally, in this context, the translations of the poems had to rhyme, too. In Arabic editions of rhymed prose the rhyming clauses or phrases are often separated by means of punctuation, either commas or full stops. In my translation I have used 'ordinary' punctuation, rather than a surfeit of commas or slashes (as in the introductory flourish, above), in order not to clutter the text, leaving it to the reader to spot the rhymes. The annotation includes some information intended for the general reader rather than specialists.

Translation

In the name of God, the Merciful, the Compassionate:
Seven precious stones[8] gathered at some point in time, in order to embark on a mutual boasting, rather than a mutual roasting, with noble vying, not arrogant lying, to see who would rank higher, be more precious to a buyer, be sweeter to behold, and have better reports about him told. So they sat down and each formed a ring, and lauded Him who beautifully created everything. For each of them a dais was raised in a circle, all alone, and fingers pointed to where he had become the eye of the signet ring and its central stone. Each one of them had [1141] a story – and what story! – of his own, and made it known.

---

the *Thousand and one nights*, Rescher and Rotter in al-Hamadhānī's *Maqāmāt*, and Stefan Wild in a *maqāma* by Ibn Nāqiyā (in W. Heinrichs & G. Schoeler (eds.), *Festschrift Ewald Wagner zum 65. Geburtstag*, II, *Studien zur arabischen Dichtung* (Beirut [etc.], 1994), pp. 427–438). I have attempted it in Dutch in three *maqāmas* by al-Hamadhānī and selections from al-Maʿarrī's *al-Ghufrān*, in A. Vrolijk (ed.), *De taal der engelen. 1250 jaar klassiek Arabisch proza* (Amsterdam, 2002), pp. 437–468. In English, R.A. Nicholson used rhyme in two *maqāmas* by al-Ḥarīri, see his *Translations of Eastern poetry and prose* (London, 1987), first published in 1922, pp. 115–124; one of them reprinted in R. Irwin, *Nights and horses and the desert. An anthology of classical Arabic literature* (London, 1999), pp. 188–193. See also B.B. Lawrence, 'Approximating *sajʿ* in English renditions of the Qur'an. A close reading of Sura 93 (*al-Ḍuḥā*) and the *basmala*,' *Journal of Qur'anic studies* 7 (2005), pp. 64–80.

[8] *yawāqīt*: the author uses *yāqūt*, here and in the title, in the general sense of 'precious stone', rather than its more precise meaning ('ruby' or 'sapphire'). The *yāqūt*, generally deemed the most precious stone, is the first speaker.

The RUBY (*al-yāqūt*)[9] said:
Praise be to God, Who created me as such a well-formed thing, more splendid to the eye than pearls on a string! Above many peers He raised my fame, since in the Qur'an, the Sura 'The Compassionate', He mentions me by name: "[maidens] as if they were rubies and coral" (Q 55:58). He mentioned me first, which shows that I am above corals, in renown and rank, in mould and morals.

How often am I not mentioned in Traditions[10] that are good or sound, and among the fine features of God's Paradise that may be found! Among them is a tradition going back to him to whom God has abundantly given noble characteristics:[11] 'God has built the Garden of Eden; of red ruby, green peridot, and white pearls are its bricks'. A tradition transmitted by an eminent transmitter who is revered a lot: [1142] 'In the third rank of Paradise, the houses, rooms, doors, couches and hooks(?)[12] are made of ruby, pearls, and peridot.' We hear in a tradition of the soundest grade: 'Of pearls and rubies are its pebbles made'. A tradition that is 'good' rather than 'sound': 'Its ranks are pearls and rubies, its gravel are pearls, and saffron is its ground.' In a tradition transmitted by al-Bayhaqī,[13] promising a reward to him who prays, it is said: 'For every believing man who prays at night in Ramadan God will build a house in Paradise of rubies red.' We are told, in traditions both 'good' and 'sound': 'In Paradise there are horses of ruby that have two wings of gold, and that with their riders fly around'. Whenever I am mentioned by way of incitement and stimulation, it is a matter of pride for me, of glory and exaltation. About me there are traditions whereby my honour and pride are upheld: 'Use ruby signet rings, and poverty will be dispelled.'

[1143] The special properties lodged in me are exalted, my highly useful qualities cannot be faulted. Thus, using me in signet rings, as pendants[14] and the like, will give certain protection against the plague

---

[9] Strictly speaking, *yāqūt* could be 'sapphire' as well as 'ruby'. In poetry and literature, it is always red and 'ruby' is the obvious translation. See Ghada al-Hijjawi al-Qaddumi, 'Yāḳūt,' in *Encyclopaedia of Islam* (New ed.), XI (Leiden, 2002), pp. 262–263.

[10] When referring to a hadith (a saying or anecdote going back to the Prophet or his contemporaries), 'sound' (*ṣaḥīḥ*) is the term used for the most reliable category; 'good' (*ḥasan*) is the next best.

[11] *viz.* the Prophet.

[12] The meaning of *maʿālīq* is not clear. Rescher, reading *maghālīq*, has '(Tür)-Schlösser(?)'.

[13] Abū Bakr Aḥmad b. al-Ḥusayn (d. 458/1066).

[14] *taʿālīq*; perhaps amulets are intended, as Rescher suggests.

that may strike.[15] Cheer and joy I do impart, I fortify a wounded heart, to venom I am an antidote, grief and worry I keep remote: for all this I am renowned and famous all around. Among my properties is that by files I am not affected, that if I am heated in a fire no traces on me can be detected. Suffice it to quote the poet's words in which this is reflected:

> Oft rubies are roasted on fire, euphorbia its fuel:
> wood turns to dying ember, ruby is still a ruby jewel.[16]

Or another poet:[17] [1144]

> – Why is he so hard-hearted? Mildness goes
> with a mild face, say all those on the planet!
> – Do not be fooled by tender rosy cheeks:
> a ruby is by nature hard as granite.

Poets have compared to me all things of which one may boast, and that are valued most. A poet[18] said,

> Don't you see the roses on their stalks
> in the garden,[19] pleasant to behold:
> Ruby platters that have been inlaid
> in the middle part with yellow gold.

[1145] Another poet:[20]

> One day in a garden, with food and with friends,
> I spent there a few of my pleasantest hours.
> In it, I was clothed in a mantle of green,
> with buttons of ruby made from the red flowers.

---

[15] Here (and on similar occasions) al-Bundārī and Zaghlūl add a lengthy editorial footnote calling for the 'pruning of Islamic books from these superstitions'.

[16] A line by, or at least quoted by, al-Ḥarīrī (in his *maqāma* no. 46, 'al-Ḥajriyya'). The euphorbia tree (*ghaḍā*) is proverbial for the heat it produces. This line is, exceptionally, translated by Abraham Ecchellensis (*De proprietatibus*, p. 147): *Ardentiores extinguuntur carbones / Sed carbunculus semper remanet carbunculus.*

[17] Identified by al-Durūbī as Ibn Miknasa al-Iskandarī, an Egyptian poet from the Fāṭimid period.

[18] Unidentified.

[19] Reading, with B&Z, *al-bustān* instead of *al-insān*.

[20] Unidentified.

Another poet:[21]

> Is there a finer sight than narcissus eyes
> exchanging glances, gathered in the glade?
> Pearls bursting forth from rubies[22] perched on stalks
> of peridot on carpets of brocade.

Another poet:[23] [1146]

> Look: the narcissus, in a pretty garden
> lush, with all kinds of sundry flowers dressed,
> Is like a yellow ruby; on its stalk,
> around it, are six petal-pearls impressed.[24]

[1147] The PEARL (*al-lu'lu'*) said:
Praise be to God, Who has honoured me with a robe that is white; Who made me among precious stones like the garden's shining light! He has granted me praise and veneration; He has mentioned me repeatedly in the Revelation, giving me precedence in the Qur'an, when He says in the Sura called 'al-Rahman': "From them[25] come forth pearls and coral" (Q 55:22). He has compared to me the paradisial youths and beautiful girls, saying in His Book: "Wide-eyed houris, like hidden pearls" (Q 56:22–23). He, the exalted, has said, inciting the believers and warning them not to obey sinners and infidel deceivers: «Immortal youths are going round among them: when you see them you deem them pearls unstrung» (Q 76:19). He says, informing about the great bounty shown

[21] al-Ṣanawbarī (d. 334/945–6), cf. his *Dīwān*, ed. Iḥsān ʿAbbās (Beirut, 1970), p. 180. For a translation and discussion of the poem, see G. Schoeler, *Arabische Naturdichting. Die zahrīyāt, rabīʿīyāt und rauḍīyāt von ihren Anfängen bis aṣ-Ṣanaubarī* (Beirut, 1974), pp. 295–297. I apologise for 'glade', chosen for the sake of rhyme and alliteration, which evokes an English landscape rather than an Arab garden.

[22] Schoeler has 'Hyazinthen', which reflects the etymology but not the sense of Arabic *yāqūt* (from Greek ὑάκινθος); the context demands an inorganic precious material (ruby or sapphire) rather than a flower.

[23] The second verse is found in Abū 'l-Walīd al-Ḥimyarī, *al-Badīʿ fī waṣf al-rabīʿ*, ed. H. Pérès (n. pl., 1989), p. 81, in an epigram extemporized by Abū ʿĀmir Muḥammad b. Maslama, the vizier (Seville, 5th/11th century). See also al-Nuwayri, *Nihāyat al-arab* (Cairo, 1923–), XI, p. 232, al-Suyūṭī, *Ḥusn al-muḥāḍara* (Cairo, 1968), II, p. 410; for a modern discussion, see A. Hamori, *On the art of medieval Arabic literature* (Princeton, 1974), pp. 78–79.

[24] At this point D adds, in a footnote, a passage found in one manuscript with more sayings on the properties of the ruby. As the editor says, it is unlikely that it is part of al-Suyūṭī's original text (it is not written in *sajʿ*, for one thing) and is quoted from *Nukhab al-dhakhā'ir fī aḥwāl al-jawāhir* by Ibn al-Akfānī (d. 749/1348), see the edition by Mārī al-Kirmilī (Cairo, 1939), pp. 2–13.

[25] From the 'two seas that meet', often explained as 'the salty and the sweet'.

to the people of Paradise and those of that ilk: «There they are adorned with golden bracelets and pearls and their clothes are made of silk» (Q 22:23).

[1148] I have been mentioned often in the Tradition, in the description of Paradise by mouth of him who was sent as bringer of good tidings and warning admonition. Thus, in a tradition about those who are given the river Abundance[26] as their own: 'In Paradise there are chambers made of various kinds of precious stone.' In a tradition transmitted by the specialists in Prophetic lore: 'The lowliest of the people of Paradise will have a house hewn of one pearl, including its every room and door.' In a tradition transmitted by Abū Nuʿaym, of prodigious memory, excellent:[27] 'The rivers of Paradise flow on the face of the earth; they are bordered by tents of pearl and their mud is musk of purest scent.' In a tradition spoken by him who had the idol overthrown:[28] 'The two banks of al-Kawthar are made of pearls, peridot and ruby stone.' In a tradition on the explanation of the verse of adornment, in clear Arabic expressed:[29] 'They wear crowns, the least pearl of which lights up what is between the East and the West.'

[1149] It is reported by al-Bukhārī and Muslim, on whose transmission one can rely:[30] 'The tent is a pearl hollowed out, stretching into the sky sixty miles high.' One of the leading theologians, Mujāhid,[31] vouches: 'Of pearls and rubies are the couches.'[32] In a tradition deemed sound as the chain of authorities goes: 'The music heard by the people of Paradise comes from beds of reeds made of moist pearls in which the wind

---

[26] Cf. Q 108:1 "We have given you *al-kawthar*"; the word seems to mean 'abundance' but it is often interpreted as the name of a river in Paradise, as also in a tradition quoted in what follows.

[27] Abū Nuʿaym al-Iṣfahānī (d. 430/1038), who quotes this in his *Ḥilyat al-awliyāʾ* (Cairo, 1932–1938), VI, p. 205.

[28] The Prophet Muḥammad destroyed the idols in Mecca after the conquest in AD 630.

[29] The editor refers to Q 18:31, 22:23 and 76:21, similar verses that mention the inhabitants of Paradise as being adorned with bracelets of gold or silver and with pearls.

[30] On a tent found in Paradise; see e.g. Muslim, *al-Ṣaḥīḥ*, ch. *al-Janna*, al-Bukhārī, *al-Ṣaḥīḥ*, ch. *Tafsīr* (on sura 55).

[31] Early commentator of the Qurʾan, d. between 100/718 and 104/722.

[32] *al-arāʾik*, mentioned five times in the Qurʾan.

blows.'[33] On ʿIkrima's[34] authority: 'Every raindrop that God sends down from heaven He turns into a plant on earth or a pearl in the sea.'

How many useful properties did the Merciful to me impart! I fortify the human heart. I am useful [1150] against melancholy's terrors and trepidations, as well as against palpitations: these I abolish. The teeth I polish. I whiten the eyes, clearing them of any dark speck, cleansing them of any dirty fleck. Any eye disease I ease, I serve to strengthen their nerve, I dry their moisture, I retain the blood and give relief from grief. For every occurrence I give beneficial aid, I am profitable in trade, for him who wants an ornament, or wants to ward off any dire event. Poets, with comparisons involving me, have filled a brimming sea. A poet[35] said: [1151]

> A branch atop a hillock tortured me,
> that is both soft and firm. Alas!
> I'm jealous when a wineglass nears his mouth
> and pearls are being kissed by glass.

Another poet:[36]

> How pretty, almond trees
> drenched by what rainclouds bring them:
> The light is scattered from them
> like pearls from hands that string them.

Another:

> Cucumbers! Nice to eat, and nice to get
> (If only getting them could make you rich!)
> Like bars of peridot in which are set
> Almond-like pearly seeds (the pearls are wet).

[1152] The EMERALD (*al-zumurrud*)[37] said:
Praise be to God, Who with a high status has blessed me, and Who with a green mantle did invest me! He clothed with my hue the sky's

[33] Though called 'sound', the hadith is not found in al-Būkhārī or Muslim. B&Z will have nothing of this and the preceding traditions: 'All these reports are based on speculation about the Unseen and are wholly unsound'.

[34] d. 105/723–4.

[35] Kushājim (d. *c.* 350/961), see *Dīwān*, ed. Khayriyya Muḥammad Maḥfūẓ (Baghdad, 1970), p. 93. The lines employ the metaphors common in love lyrics, of branch (slender body), hillock (ample hips), and pearls (white teeth).

[36] This and the following quotation have not been identified.

[37] The word, Persian in form and often spelled *zumurrudh*, is etymologically related

azure,[38] and made me more than water pure. A remedy for any pain I will procure, and any sickness I will cure. A large share of merit did I inherit. I am mentioned in reports and sayings of the Prophet, and there is much of it! Thus al-Bayhaqī relates in his *Faith's Branches*, that book of glory and much profit,[39] on the authority of Anas ibn Mālik, one of the early Supporters,[40] who was one of the Chosen Prophet's reporters: 'Whoever fasts on Wednesday, Thursday, and Friday, for him will God prepare in Paradise [1153] a palace of pearl, ruby, and emerald built, and He will save him from hellfire, erasing his guilt.' As authorities in an unbroken chain record, the Prophet explained the 'goodly mansions' (Q 9:72) which are stored as the believers' reward, as 'a palace made of pearls in which seventy buildings made of ruby can be seen, in each building seventy rooms made of emerald green'. And, as Ibn 'Abbās transmits, he made it known in general, that 'The palm-tree trunks in Paradise are made of green emerald.'

I possess lofty useful qualities and not a few special properties. One may determine that I help against poison and bites of any vermin. Scrape off me seven grains in weight, [1154] then let a poisoned person drink it and it will save him from a fatal fate. His hair will not fall out and of his skin he will not be peeled, but through this he will be healed. If you fix your gaze on me,[41] cured of tired eyes you will be. If in a necklace or a signet ring I am held one will not by a fit be felled. Therefore physicians tell princes that I[42] as an amulet should be worn by their children as soon as they are born. If someone loses blood I, whether drunk as a potion or worn as a charm, will stop the flood. If a viper looks at me

---

to 'emerald', Old French 'esmeralde', Greek σμάραγδος or μαραγδος. Semitic and Indian origins have been suggested (Akkadian *barraqtu*, Biblical Hebrew בָּרֶקֶת *bāreqeth*, Sanskrit *□marakata*), see E. Klein, *A comprehensive etymological dictionary of the English language* (Amsterdam, 1971), p. 244, L. Koehler & W. Baumgartner, *Lexicon in veteris testamenti libros*, 3. Aufl., Lieferung 1 (Leiden, 1967), p. 153, Samar Najm Abul Huda, *Arab roots of gemology. Ahmad ibn Yusuf Al Tifaschi's* [*sic*] Best thoughts on the best stones (London, 1998), p. 192. See also below, n. 51, and Ghada al-Hijjawi al-Qaddumi, 'Zumurrud,' in *Encyclopaedia of Islam* (New ed.), XI (Leiden, 2002), pp. 569–571.

38 The word *akhḍar*, normally corresponding to our 'green' (as in the previous clause), also covers a range of colours and hues varying from blackish to blue; the feminine form, *al-khaḍrā'*, is used as an epithet of the sky. The word *azraq*, which often corresponds to our 'blue', has a more limited use and often has negative connotations.

39 On al-Bayhaqī, see above (n. 13).

40 *al-Anṣār*, those Medinans who supported the Prophet after his migration from Mecca in AD 622.

41 Both editions have 'at him' (*ilayhi*), but al-Durūbī mentions in a note two earlier editions that have *ilayya*, as one might expect here.

42 Again the Arabic text lapses into the third person singular.

with his eye, the eye will presently melt and liquefy.[43] They have used me in similes for everything that is highly esteemed and as something precious deemed. Thus a proficient poet has said,[44] [1155]

> Behold! The roses' army, armed with lances
> yellow and green, takes to the field!
> Armoured with thorns, each blade an emerald,
> and made of gold is every shield.

Another:[45]

> See, in the garden, the red willow tree[46]
> on which the rain has battered:
> [1156] Red rubies, leaves of emerald, and dew
> like pearls that have been scattered.

The CORAL (*al-marjān*) said:
Praise be to God Who with a robe of red has framed me, and in His holy book has named me! He has mentioned me by my name twice,[47] in the Sura 'The Compassionate' comparing to me the black-eyed damsels in Paradise. He made the seas my native base, and necklaces my dwelling-place. I come next, being the third of the jewels attested in the Holy Text, which are pronounced superior, and to which pure gold is inferior.

As Traditions from the Prophet about me relate, thereby enhancing my state, 'The house of a believer in the Garden is made of a pearl, in the midst of which stands a tree from which grow robes. He takes with two fingers seventy robes with belts of pearls and coral'. Another Tradition says: 'In the Garden is a river called Well-Watered. On the river is

---

[43] Here one MS again inserts an unrhymed passage on the properties of the stone, taken from Ibn al-Akfānī (cf. supra, n. 24). Its author reports about experiments with snakes, but without the promised result (*Nukhab al-dhakhā'ir*, pp. 51–52), unlike al-Tīfāshī, who says, 'I tried it myself and found it to be correct' and then describes his experiment in some detail; see al-Tīfāshī, *Azhār al-afkār fī jawāhir al-aḥjār*, in Abul-Huda, *Arab roots of gemology*, Arabic facsimile text f. 26a–b, transl. p. 106, and al-Tīfāshī, *Azhār al-afkār fī jawāhir al-aḥjār*, ed. Muḥammad Yūsuf Ḥasan and Maḥmūd Basyūnī Khafājī (Cairo, 1977), pp. 84–85; also quoted in al-Ghuzūlī, *Maṭāli' al-budūr* (Cairo, AH 1299–1300), II, pp. 149–150. Al-Bīrūnī (*al-Jamāhir fī 'l-jawāhir*, pp. 272–273) says that he has gone to great lengths experimenting, but that the snake's eyes did not change one bit ('unless their sight became even sharper', he adds sarcastically).

[44] Identified by al-Durūbī as al-Ṭughrā'ī (d. 514/1120). The lines are also found in al-Suyūṭī's *Ḥusn al-muḥāḍara*, II, p. 405.

[45] Identified by al-Durūbī as Abū Sa'd al-Kanjarūdhī (or Ganzarūdhī), see al-Tha'ālibī, *Tatimmat al-Yatīma*, ed. Mufīd Qumayḥa (Beirut, 1983), pp. 187–188.

[46] *ṣafṣāf* (*Salix safsaf*).

[47] Q 55:22 and 55:58; both have been cited above.

a town of [1157] coral which has a thousand gates of gold and silver, for any man who holds a Qur'an.'

Many a useful property, given by my Creator, in me lies. I serve as a salve for sore and tearful eyes. Any human heart that is sad I will make glad; against the heart's palpitation I provide fortification. Should any limb lose blood, I will staunch its flood. Brushing one's teeth with my powder strengthens the gum, and the teeth will not to decay succumb. Pulverised and mixed with oil of balm, dripped into ears I protect against deafness and harm and serve as a charm. I am astringent and dehydrating, to any moistness desiccating. When hung on the neck of him who suffers from a fit I am full of benefit; when I am put on someone's gouty foot I drive out the gout. When drunk with water I have been seen to dissolve tumours of the spleen; if someone [1158] by strangury is grieved, by me he will be relieved.

To anything fair do poets me compare. A poet said:[48]

> See, the sweet basil! It revives us with
> its fragrant gift so fair and floral.
> You think it looks, beshadowed and bedewed,
> like emerald that carries bits of coral.

Another poet:[49]

> Look at the garden wonderful, the flowers
> in strings and rows, each in its plot,
> The pomegranate's blossom on their twigs
> like coral beads on top of peridot.

[1159] Another poet:[50]

> She is as pure as ruby or as coral,
> a pearl preserved from every enemy,
> Or a white rose in a bouquet of flowers,
> tinged with the hue of red anemone.

---

[48] Identified by al-Durūbī as Abū ʾl-Qāsim Ibn al-ʿAṭṭār; anonymously in al-Nuwayrī, *Nihāyat al-arab*, XI, p. 254. Also in al-Suyūṭī, *Ḥusn al-muḥāḍara*, II, p. 421 (with *ṭall*, 'dew' instead of *ẓill*, 'shade').

[49] Identified by al-Durūbī as ʿAlī b. Muḥammad b. ʿAbd al-Raḥmān al-ʿUbyī (d. 790/1388), see Ibn Ḥajar al-ʿAsqalānī, *al-Durar al-kāmina* (Hyderabad, AH 1348–1350), III, pp. 105–106.

[50] Identified by al-Durūbī as ʿAbd Allāh [*recte* ʿUbayd Allāh] b. ʿAbd Allāh b. Ṭāhir (d. 300/913).

The PERIDOT (*al-zabarjad*)[51] said:
Praise be to God, Who has made me to the emerald a brother, and has forever strung me on a thread together with the other! My name is mentioned in Prophetic reports and in sound traditions of all sorts, such as a saying which a full chain of authorities supports, one that is sound: 'In Paradise there are columns of ruby on top of which chambers of peridot are found.' Another: 'That chamber is a red ruby, or a green peridot, or a white pearl all round'. A tradition that al-Ṭabarānī[52] has recorded: 'Whoever fasts a day in [1160] Ramadan quietly in silence, will with a house built of red ruby or green peridot be rewarded.' As many detailed traditions tell, with me are covered the wings and feet of Gabriel. And if I possessed only one honourable trait, one that would raise me to the highest state, then my elevation, taking me up unto the highest station, is ensured by this alone: that in the Chosen Prophet's signet-ring I was the stone; from authentic reports this fact is well-known. It is an achievement that none of the other precious stones can boast: on this road I have beaten them all at the post! Who can vie with me since the touch of the Chosen One's hand he gave me, and [1161] with his own name and

---

51 I have chosen 'peridot' (to rhyme with dot, as the *Oxford English dictionary* recommends, not with doe). It is the clear-green gemstone variety of olivine (called chrysolite when yellow). It is said to be found especially in Egypt, see al-Thaʿālibī, *Thimār al-qulūb* (Cairo, 1985), p. 540, idem, *Laṭāʾif al-maʿārif* (Leiden, 1867), p. 116, al-Tīfāshī, *Azhār al-afkār*, in Abul Huda, *Arab roots of gemology*, Arabic text, f. 30a–b, transl. p. 110, comm. p. 197. The word *zabarjad* is also sometimes translated as beryl or topaz, and even emerald, to the extent that *zabarjad* and *zumurrud* are treated as synonyms by some. See e.g. Ibn Manẓūr, *Lisān al-ʿarab*, roots *ZBRJD* and *ZMRDh*, al-Qazwīnī, *Mufīd al-ʿulūm wa-mubīd al-humūm* (Beirut, 1985), p. 163, Anon., *The sea of precious virtues*, tr. J. Scott Meisami (Salt Lake City, 1991), pp. 202 and 360 – where *zabarjad* is rendered as beryl –, al-Ibshīhī, *al-Mustaṭraf* (Cairo, 1952), II, p. 164, al-Tīfāshī, *Azhār al-afkār*, in Abul Huda, *Arab roots of gemology*, Arabic text, f. 23a, transl. p. 103, comm. p. 197, Dāwūd al-Anṭākī, *Tadhkirat ulī ʾl-albāb*, ed. ʿAlī Shīrī (Beirut, 1991), p. 248, J. Ruska (ed. & transl.), *Das Steinbuch des Aristoteles* (Heidelberg, 1912), p. 98 (Arabic), 134–135 (German), al-Bīrūnī, *al-Jamāhir fī ʾl-jawāhir*, p. 262. The word is sometimes erroneously said to be derived from Zabargad or Zebirget, the name of an island in the Red Sea off the Egyptian coast near the border with Sudan (also known as St John's Island, called Topazios by the ancient Greeks), the crystalline parts of which contains peridotite masses. Obviously, it is the other way round: the island is called after the mineral. Etymologically, the word *zabarjad* is related to *smaragdos/zumurrud* (see above, n. 37); intermediate forms such as *zᵉmargad* and *izmargᵉdā* are attested in Targumic Hebrew and Aramaic and in Syriac. Al-Suyūṭī's Peridot, in what follows, also stresses his close relationship with the emerald. The word 'peridot' itself has been connected tentatively but unconvincingly with Arabic *farīda* 'unique, single' (often used for large pearls or jewels).

52 Sulaymān b. Ayyūb al-Ṭabarānī (d. 360/971), author of several important collections of Hadith.

title, 'Muḥammad, Messenger of God', did engrave me? This is enough to raise me and praise me. When from 'Uthmān's hand I fell into the well of Arīs all the trouble began and gone was the peace.[53] The people of the Faith fell upon one another with swords and great zeal: it was because in me had been contained all the secret power of Solomon's seal.

Because the emerald and I are of one kind, the same useful properties in us one will find. One of the qualities commonly ascribed to me, is that my scrapings in a potion are good against leprosy. When poets want to praise a thing to the utmost degree they compare it in their similes to me. A poet said:[54]

> The red anemones sway up
> and downwards in this windy spot,
> Like ruby banners that have been
> unfurled on spears of peridot.

[1162] Another poet:[55]

> You'd think the fresh and pretty daffodils
> were languid looks of big-eyed oryx does;
> Peridot stalks with golden pupils charged,
> that eyes as white as camphor do enclose.

Another:[56]

> Along the surface of your rosy cheeks
> your pretty cheek-down crawls[57]

[53] The reign of the third caliph, 'Uthmān, 23–35/644–655, is traditionally divided into six 'good' and six 'bad' years, with his murder marking the beginning of the first great civil war in Islam. The accidental loss of the Prophet's ring (never retrieved) halfway through his reign in AH 30 is seen as a symbolical beginning of the troubles.

[54] The often-quoted lines are by al-Ṣanawbarī (d. 334/945), see his *Dīwān*, ed, Iḥsān 'Abbās (Beirut, 1970), p. 477. For discussions, see 'Abd al-Qāhir al-Jurjānī, *Asrār al-balāgha*, ed. H. Ritter (İstanbul, 1954), pp. 145–146, 158 (German translation by Ritter in 'Abd al-qāhir al-Curcānī, *Die Geheimnisse der Wortkunst* (Wiesbaden, 1959), pp. 180, 192–193), Hamori, *On the art of medieval Arabic literature*, pp. 81–83, Schoeler, *Arabische Naturdichtung*, pp. 285–286. My 'windy spot', prompted by the rhyme, is not in the Arabic original but implied by the context.

[55] Unidentified.

[56] Identified by al-Durūbī as 'Alī b. Muḥammad al-Akhfash: not one of the more famous grammarians called al-Akhfash, but a minor one, *fl.* 452/1060. See 'Imād al-Dīn al-Kātib al-Iṣfahānī, *Kharīdat al-qaṣr, Qism shu'arā' Miṣr*, ed. Aḥmad Amīn, Shawqī Ḍayf [et al.] (Cairo, [1951]), I, pp. 238–240 (the verses on p. 240), al-Suyūṭī, *Bughyat al-wu'āt*, ed. Muḥammad Abū 'l-Faḍl Ibrāhīm (Beirut, 1979), II, p. 202.

[57] Al-Iṣfahānī, who admires the conceit, condemns the repetition of 'cheek'. With 'surface' and 'rosy' I have adopted both *ṣafḥa* of the present text and its variant *ḥumra*, found in *Kharīdat al-qaṣr* and *Bughyat al-wu'āt*.

> [1163] Just like a polo-stick of peridot,
> curving round ruby balls.

Another:[58]

> Look at the date-palms, shedding fruits,
> glad tidings of King Fresh Date's reign!
> Kohl bottles carved from peridot,
> while golden caps their tops contain.

The CORNELIAN (*al-ʿaqīq*)[59] said,
[1164] Praise be to God, Who to the elite admitted me, Who in a most splendid vestment fitted me, and Who with the best characteristics benefited me! He blesses through me the wearer. On me said he who is of Truth the receiver and the bearer: 'The number of cornelians exceeds all other Paradise-dwellers' gemstone beads'. A report that any harm will repel: 'He who uses cornelian in signet-rings will always be well.' Other reports attest: 'Use cornelian in a signet-ring for it is blessed'; and: 'Use cornelian in signet-rings for by this poverty is banned.' It is reported in a hadith that is well-supported: 'He who uses cornelian in signet-rings, his good luck will by God never be thwarted.' And a final important hadith: 'He who uses cornelian in signet-rings will have success and the two (guardian) angels will love him, here beneath.'[60]

[1165] Whoever in a dispute feels alarmed, by wearing me in his signet-ring he will be becalmed. I will also staunch the blood that may flow from him, no matter from which limb, and this is especially so for women suffering from excessive menstrual flow. Rubbing one's

[58] As al-Durūbī points out, the lines are ascribed to Ibn Wakīʿ al-Tinnīsī (d. 393/1003) in Ibn Ẓāfir al-Azdī, *Gharāʾib al-tanbīhāt ʿalā ʿajāʾib al-tashbīhāt*, ed. Muḥammad Zaghlūl Sallām & Muṣṭafā al-Ṣāwī al-Juwaynī (Cairo, 1971), p. 112 and al-Nuwayrī, *Nihāyat al-arab*, XI, pp. 126–127; and to Ibn Ḥamdīs (d. 527/1133) in Abū ʾl-Baqāʾ al-Badrī *Nuzhat al-anām fī maḥāsin al-Shām* (Cairo, 1980), p. 195. They are quoted anonymously (not ascribed to Ẓāfir al-Ḥaddād, *pace* al-Durūbī) in al-Suyūṭī, *Ḥusn al-muḥāḍara*, II, p. 435, as well as in al-Nawājī, *Ḥalbat al-kumayt* (Cairo, 1938), p. 261.

[59] Also translated, less correctly, as 'agate' (e.g. H. Wehr, *A dictionary of modern written Arabic*).

[60] As the editors point out, all these hadiths are deemed unauthentic by the Muslim specialists. It is interesting to note that one ninth-century authority, Ibrāhīm al-Ḥarbī, quoted in Ibn Manẓūr's dictionary *Lisān al-ʿarab* (*s.v.* ʿQQ), believed that a negative counterpart, *lā takhattamū bi ʾl-ʿaqīq* ('do not use cornelian in signet-rings'), is a misreading of *lā tukhayyimū bi ʾl-ʿAqīq* 'do not pitch your tents in [one among the many places in Arabia called] al-ʿAqīq'.

teeth with my ashes[61] or chips means caries' end and plaque's eclipse; it also prevents the wobbling of teeth and keeps them in place below and beneath.

Poets have often used me when things are to be compared and their great value has to be declared. A poet said:[62]

[1166] He is all jewels; to describe him well
  the mind is ill-equipped:
Moustached with peridot,[63] front-teethed with pearl,
  and with cornelian lipped.

Another poet:[64]

Look at this carrot with its leaves!
  A flame it seems, a candle;
A fly-whisk made of sarsenet
  with a cornelian handle.

Another poet:[65]

[1167] Like red anemones: look at
  the colour of the unripe dates!
The palm leaves are like emerald
  that bears cornelian fruit in spates.

Another poet:[66]

Spring has a carpet for us spread
  where wonderful embroidery has grown.
On it marsh mallows rosy red
  appear like cups cut from cornelian stone.

---

[61] *ḥurāqatī*; the variant *kh.zāfatī* (e.g. in B&Z) is rejected by D. Other texts (e.g. al-Anṭākī, *Tadhkira*, p. 341, which has *ramāduhū*) confirm D's reading, although I do not know how the stone could be burned to 'ashes'.

[62] Identified by al-Durūbī as Ibn Wakīʿ al-Tinnīsī (393/1003); al-Thaʿālibī, *Yatīmat al-dahr*, ed. Muḥammad Muḥyī 'l-Dīn ʿAbd al-Ḥamīd (Cairo, 1947), I, 379, Ibn Khallikān, *Wafayāt al-aʿyān*, ed. Iḥsān ʿAbbās (Beirut, 1972), VII, pp. 50–51.

[63] This does not mean, of course, that the moustache is green: it is dark (cf. above, n. 38; *akhḍar* often refers to hair). Instead of peridot (*zabarjad*), Ibn Khallikān has emerald (*zumurrudh*).

[64] Identified by al-Durūbī as Ibn al-Muʿtazz (d. 296/908); see his *Dīwān* (Beirut: Dār Ṣādir, n.d.), p. 344, but not found in *Dīwān*, ed. Bernhard Lewin (İstanbul, 1945–1950) or ed. Muḥammad Badīʿ Sharīf (Cairo, 1977–1978); anonymously in al-Nuwayrī, *Nihāyat al-arab*, XI, p. 57, al-Suyūṭī, *Ḥusn al-muḥāḍara*, II, p. 445.

[65] Identified by al-Durūbī as Abū 'l-Qāsim Ẓāfir al-Ḥaddād (d. 529/1134). Anonymously in al-Nawājī, *Ḥalbat al-kumayt*, p. 261, al-Nuwayrī, *Nihāyat al-arab*, XI, p. 127, al-Suyūṭī, *Ḥusn al-muḥāḍara*, II, p. 436.

[66] Unidentified; the lines are quoted anonymously in al-Nawājī, *Ḥalbat al-kumayt*, p. 254.

Another:[67]

[1168] Bunch of red roses, freshly plucked and tender,
by a gazelle with magic eyes and slender:
Held in his hand, they looked to me like gems,
cornelian salvers held on peridot stems.[68]

The TURQUOISE (*al-fayrūzaj*) said:
Praise be to God! A superior jewel with colours dual He made me, and in two raiments He arrayed me.[69] He made me part of Alchemy and for eye-disease a remedy. My essence is so subtle that my colour moves like a shuttle: when the air is clear, so do I appear; and when the sky turns murky, then so do I, the 'turkie'! God gave me exclusively to the mountains of Nishapur: from nowhere else do they me procure. Whoever drinks me, to a powder ground, I will greatly benefit all round. I am useful against ulcers [1169] internal and for the sting of dreaded scorpions infernal. I will act against eye cataract, the swelling of the pupil I contract, split eyebrows I render intact.[70]

Poets have used me in similes for things they highly rate, which they wish to promulgate[71] and propagate. A poet says:[72]

---

[67] Unidentified; the lines are quoted anonymously in al-Nawājī, *Ḥalbat al-kumayt*, p. 240.

[68] In the epigram by Ibn Wakīʿ quoted above (p. 1163 of the Arabic text), *muqammaʿāt* seems to mean 'contained, encased', but here I have taken *qummiʿat* to be derived from *qamʿ* ('stem or stalk of a fruit or flower'). Rescher has 'eingefaßt(?) mit Chrysolith', apparently taking *zabarjad* to refer to the boy's hand, which at first sight looks plausible; but *zabarjad* always refers to something green (which is why I have not used 'chrysolite'). For the verb *qammaʿa* see also the penultimate epigram of the *maqāma*.

[69] The 'two colours' may refer to the fact that turquoise changes when affected by its environment, as explained in what follows; or it may be because the colour verges between blue and green (e.g. Ps.-Aristoteles, in Ruska, *Das Steinbuch*, p. 106: *ḥajar akhḍar mashūb bi-zurqa*, al-Anṭākī, *Tadhkira*, p. 363: *yatarakkabu min khuḍra wa-zurqa*).

[70] *ajmaʿu ḥujub al-ʿayn al-munkhariqa* (with variants such as *munḥarifa* in B&Z, which does not rhyme, and *muḥtaraqa*, *munḥariqa*, *mutaḥarriqa*, and *mutakharriqa*, culled from parallel sources in D). Rescher (p. 79) has '[ich] (habe die Eigenschaft), auseinanderstehende [wörtl.: a. gerissene] Augenbraunen [*sic*, for 'Augenbrauen', vG] zusammen zubringen', adding in a note that he thinks of a hair tonic and mentioning the 'oriental' habit of joining the line of the eyebrows with kohl. Huda, translating the same words in al-Tīfāshī's *Azhār*, has 'It strengthens drooping eyelids' (*Arab roots of gemology*, p. 137, see Arabic text f. 57a). More likely, some astringent power is ascribed to the stone.

[71] The verb *asarra* has two opposite meanings; 'to keep secret' seems less likely here.

[72] Identified by al-Durūbī as Abū 'l-Maʿālī Saʿd b. ʿAlī al-Ḥaẓīrī, 'Dallāl al-kutub' ('the Book Hawker'; poet and anthologist, d. 568/1172). The epigram is found in ʿImād al-Dīn al-Iṣfahānī, *Kharīdat al-qaṣr wa-jarīdat al-ʿaṣr, al-Qism al-ʿIrāqī*, ed. Muḥammad Bahjat al-Atharī, IV, i (Baghdad, 1973), p. 35; Yāqūt, *Muʿjam al-udabāʾ* (Cairo, 1936–1938), XI, p. 197; Ibn Khallikān, *Wafayāt*, II, p. 367. There must be hundreds, if not thousands,

[1170] Tell him who blamed a mole found on the mouth
of a cute boy: This thing should not be blamed:
The mole you faulted is a turquoise stone,
his mouth is but the ring on which it's framed.

Another poet:[73]

How beautiful, the flax that sways in waves
its flowers on their stems:
Stalks of green peridot, the ends of which
are tipped with turquoise gems.

Another:

When I complained to him he merely smiled,
revealing in his fragrant mouth these nine:[74]
[1171] Camomile, pearls, cornelian, hailstones, silver,
with water-bubbles, lightning, flowers, wine.

Here ends this original *maqāma*. Praise be to God for His grace.[75]

Arabic epigrams on moles on pretty faces, male or female. See e.g. Th. Bauer, *Liebe und Liebesdichtung in der arabischen Welt des 9. und 10. Jahrhunderts. Eine literatur- und mentalitätsgeschichtliche Studie des arabischen Ġazal* (Wiesbaden, 1998), pp. 248–254. The choice of turquoise to represent the mole is rather unexpected.

[73] Unidentified, as is the following epigram.

[74] The Arabic has in fact 'ten', because the second verse lists *durr* as well as *lu'lu'*. If there is a difference (the former specifically referring to 'large pearls'), it is difficult to translate.

[75] These phrases may have been written by the copyist, who was also responsible for the addition that concludes D's text: 'It was written in the beginning of Ramadan of the year 967', which corresponds roughly to June AD 1559.

'WELCHE GÄRTEN UNS UMFANGEN . . .'

THREE POEMS BY FRIEDRICH RÜCKERT, TRANSLATED FROM A GOTHA MANUSCRIPT OF THE *SĪRAT AL-MUJĀHIDĪN* (*AL-AMĪRA DHĀT AL-HIMMA*)

Claudia Ott

Introduction

When Friedrich Rückert (1788–1866) visited the Oriental Manuscripts Collection at Gotha in 1828, he found among the manuscripts a huge number belonging to the *sīra shaʿbiyya* genre. From these manuscripts Rückert chose some poems for a translation. This was not only a poet's decision, but most probably also involved the importance of the poems for the composition of the Arabian epics.

Friedrich Rückert's translations not only express the meaning of the Arabic poems, but also reflect their poetic form: Rückert translated the Arabic rhyme as well as the metre, thereby transforming Arabic into German poetry. These outstanding translations are published here for the first time from Rückert's unpublished manuscripts, with the kind permission of the University Library of Münster, Germany.

The Arabian epic is mostly studied from a narratological or historical point of view. Friedrich Rückert, the poet, has shown us its beauty. It would be a mistake to try to translate Rückert's German translations into English. I therefore ask for the reader's understanding for the fact that this article is in German. It may also show my respect for Remke Kruk's close and fruitful relations with German scholars and universities.

Die Forschungsbibliothek Gotha (Schloss Friedenstein) beherbergt eine der größten orientalischen Handschriftensammlungen Europas. Ihre wichtigsten arabischen Bestände wurden in den Jahren 1804 bis 1807 durch Ulrich Jasper Seetzen (1767–1811) in Aleppo, Damaskus und Kairo erworben. Unter Seetzens Ankäufen befinden sich bemerkenswert viele Handschriften aus der Gattung der arabischen Epik (*sīra shaʿbiyya*). Der Grund hierfür dürfte einerseits in einem besonders großen Interesse an *Tausendundeine Nacht* und verwandter arabischer volkstümlicher Literatur liegen, andererseites an den damals schon stark gestiegenen Preisen für Handschriften 'kanonischer' literarischer bzw. wissenschaftlicher arabischer Werke, von denen ja um 1800 viele schon an andere Europäer verkauft worden waren. So notierte Seetzen am 25. Mai 1807 in sein Tagebuch:

> Mein Ankauf von Manuscripten findet hier mehr Schwierigkeiten, als ich glaubte. Zwar sind in dem Wakâl el Chalily 11 Buchhändler, also mehr, als in Halep. Allein sie versichern mir, dass die Franzosen alle Manuscripte mitgenommen, und dass die noch vorhandenen wegen der Seltenheit der Manuscripte sehr theuer wären. Und dies finde ich auch.[1]

Noch im selben Monat, also im Mai 1807, erwarb Seetzen in Kairo ein vollständiges Exemplar der *Sīrat al-Mujāhidīn* (*al-Amīra Dhāt al-Himma*) in 13 Bänden, im Kolophon datiert auf das Jahr 1139/1727.[2] Das Exemplar wurde mit einer von fünf Sendungen seiner 'Ägypten- und Orientexpedition' über Triest nach Gotha geschickt, wo es auf eine weitere Bearbeitung durch Seetzen warten sollte. Seetzen aber sollte nicht mehr nach Gotha zurückkehren. Er starb unter ungeklärten Umständen im September 1811 im Jemen.[3]

Im Jahre 1828 reiste der Dichter und Orientalist Friedrich Rückert (1788–1866) nach Gotha, um die dort vorhandenen Handschriftenschätze kennen zu lernen. Er hatte die Reise mehrfach angekündigt und jahrelang immer wieder verschoben, und als er sie endlich unternahm, behinderte schlechtes Wetter seine Pläne:

> Ich ging (. . .) den Bibliothekar Möller aufzusuchen, als den Mann, mit dem ich es eigentlich zu thun habe, und ward unterwegs dahin von einem so unverschämten Gusse überfallen, daß ich trotz des Regenschirms, unter welchen ich aus etwas unschicklichem Mitleide meinen Wegweiser mit aufnahm, so durchnäßt ward, daß ich nach zwei Worten der Begrüßung wieder abzog, weil ich am ganzen Leibe schauerte.[4]

Vielleicht war es den Wetterverhältnissen zuzuschreiben, daß Rückert denn auch kürzer als geplant in Gotha blieb. Bereits in seinem ersten

---

[1] U.J. Seetzen, *Reisen durch Syrien, Palästina, die Transjordan-Länder, Arabia Petraea und Unter-Ägypten*, ed. Fr. Kruse [et al.] (Berlin, 1855), III, p. 165.

[2] Forschungsbibliothek Gotha, MS orient. A 2497–2508. Ausführliche Beschreibung in C. Ott, *Metamorphosen des Epos. Sīrat al-Muǧāhidīn (Sīrat al-Amīra Ḏāt al-Himma) zwischen Mündlichkeit und Schriftlichkeit.* CNWS Publications; 119 (Leiden, 2003), Katalog IV, pp. 252–253.

[3] Vgl. N. Nebes, 'Ulrich Jasper Seetzen im Jemen,' in *Ulrich Jasper Seetzen (1767–1811). Leben und Werk. Die arabischen Länder und die Nahostforschung im napoleonischen Zeitalter*, Vorträge des Kolloquiums vom 23. und 24. September 1994 in der Forschungs- und Landesbibliothek Gotha, Schloß Friedenstein. Veröffentlichungen der Forschungs- und Landesbibliothek Gotha; 33 (Gotha, 1995), pp. 38–52.

[4] Zitiert nach H. Bobzin, 'Friedrich Rückert als Benutzer Gothaischer Handschriften,' in *Ulrich Jasper Seetzen* (wie Anm. 3), p. 104.

Brief aus Gotha an seine Frau heißt es: ‚Auf der Bibliothek habe ich die Hauptsache so in Gang gebracht, daß ich, wenn ich anders will, schon in wenigen Tagen mit meiner gesuchten Beute abziehen kann.'[5]

Was Rückert mit seiner 'gesuchten Beute' meinte, ist nicht ganz klar. Er hat sich etwa 20 arabische und persische Handschriften angesehen. Später ließ er sich einige Handschriften aus Gotha in sein Haus nach Neuses bei Coburg schicken, wo er sich intensiver mit ihnen beschäftigen konnte. Die meisten von Rückerts Abschriften und Übersetzungen aus diesen Handschriften sind bis heute unveröffentlicht und befinden sich in Rückerts wissenschaftlichem Nachlaß in der Universitäts- und Landesbibliothek Münster. So auch seine Notizen aus der *Sīrat al-Mujāhidīn* (*al-Amīra Dhāt al-Himma*). Friedrich Rückerts Abschriften und Übersetzungen umfassen 16 Blatt und sind überschrieben mit 'Aus Siret Elmugahedin'.[6] Sie dürften etwa um das Jahr 1843 entstanden sein.[7] Die Abschriften stammen sämtlich aus dem 1. Band des 13-bändigen Gothaer Exemplars, und sie enthalten ausnahmslos Gedichte. Drei dieser Gedichte sollen hier vorgestellt werden.

1. *ṣifat majlis* – Ein Frühlingsgelage[8]
Arabisches Metrum *ramal* (⌣̲ ∪ — —|⌣̲ ∪ — —)

Welche Gärten uns umfangen
Zwischen Rieseln, Blüh'n und Prangen!
Anthemis sind duft'ge Zähne,
Anemonen glühn'de Wangen;
Volle Wolken sind die Becher,
Draus des Weines Blitze sprangen;
Rosenwasser thaut, und Weihrauch
Ist als Nebel aufgegangen.

---

[5] Ibid.

[6] ULB Münster, Rückert-Nachlaß, Kapsel V/5/i.

[7] Vgl. H. Bobzin, 'Friedrich Rückerts wissenschaftlicher Nachlaß, mit einem vorläufigen Inventar,' in J. Erdmann (ed.) *Friedrich Rückert 1788–1866. Dichter und Gelehrter* (Coburg, 1988), p. 379.

[8] Quellenangabe: 'Aus Siret Elmugahedin, T. I. f. 204,a.' Rückert fügt dem Gedicht zwei Anmerkungen bei, in denen er Verschreibungen der Quelle emendiert hat. Die Überschriften – auch die arabischen – hat Friedrich Rückert den Gedichten vorangestellt, vielleicht zu Publikationszwecken. Aus dem Erzählzusammenhang arabischer Epik sind keine Überschriften bekannt.

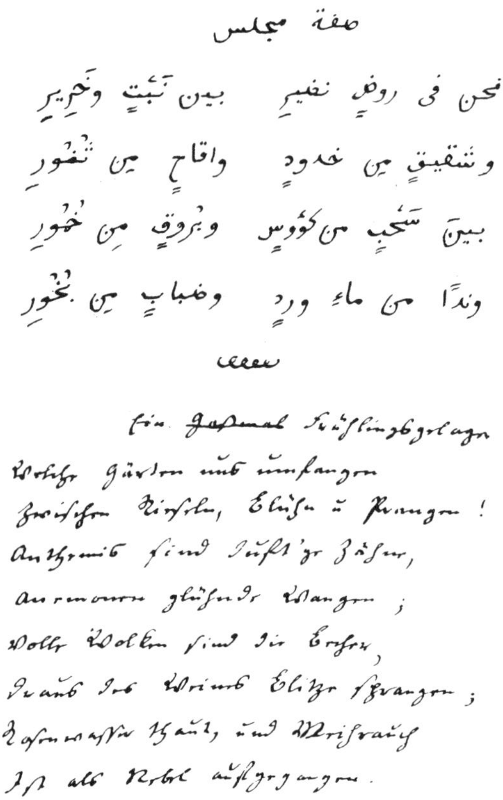

2. *bayt fī ṣifat ghulām rāʾiq*
Arabisches Metrum *ṭawīl* (∪ — ∪̲|∪ — ∪̲ — |∪ — ∪̲|∪ — ∪ —)

> (Ein Jüngling, als ob ihn Gott zum Vorschmack vom Paradies
> Gesendet, daß selig alle sey'n, die ihn schauen.)[9]
> Und lebt' er zu Josef's Zeit, so gienge durch ihn der Schnitt
> In Herzen der Männer, statt in Finger der Frauen.

Anmerkung Rückerts: 'Anspielung auf die Geschichte Joseph's im Koran: Potiphar's Weib ladet ihre Freundinnen ein, und um ihnen einen fühlbaren Beweis von der Schönheit ihres Sklaven zu geben, gibt sie jeder einen Apfel, und ein Messer um ihn zu schälen, in die Hand, dann ruft sie jenen herein, bei dessen Anblick alle, statt in die Äpfel, sich in ihre Finger schneiden.'[10]

---

[9] Diese beiden Halbverse hat Rückert dem Vers vorangestellt.

[10] Die üblichen Kommentare zu Koran 12:31f. setzen als Frucht statt Äpfeln Orangen bzw. die Zitrusfrucht *utrunja* ein.

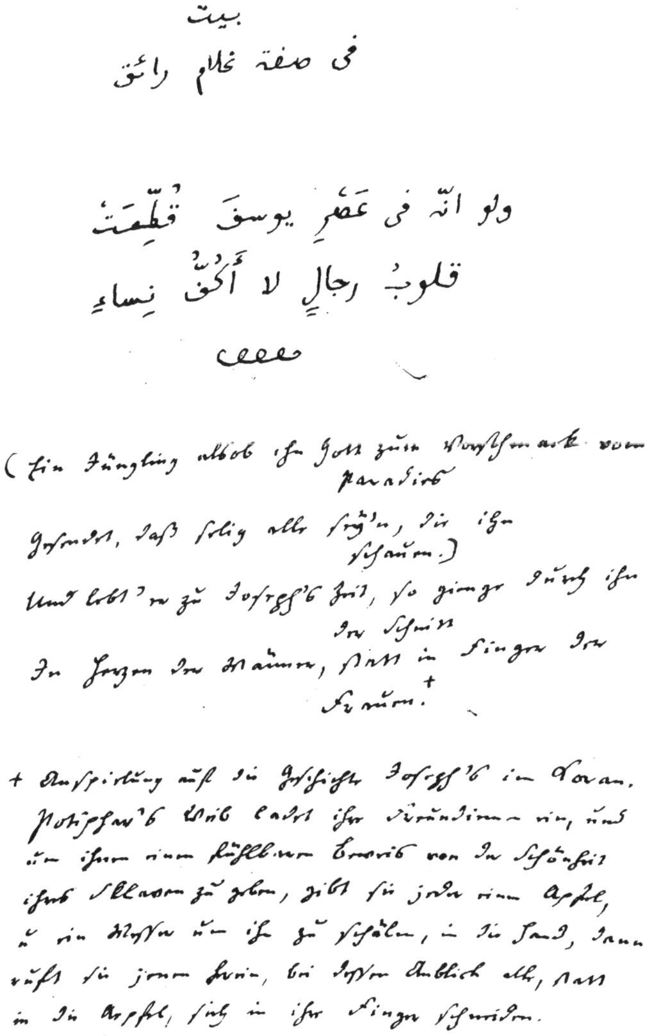
بيت
في صفة غلام رائق

ولو انه في عصر يوسف قطعت
قلوب رجال لا اكف نساء

(Ein Jüngling also lob ich Gott zum Vorschmack vom Paradies
gesendet, daß selig alle seyen, die ihn schauen.)
Und lebt' er zu Josephs Zeit, so gienge durch ihn der Schnitt
In Herzen der Männer, statt in Finger der Frauen.+

+ Anspielung auf die Geschichte Josephs im Koran. Potiphar's Weib ladet ihre Freundinnen ein, und um ihnen einen fühlbaren Beweis von der Schönheit ihres Sklaven zu geben, gibt sie jeder einen Apfel, u. ein Messer um ihn zu schälen, in die Hand, dann ruft sie jenen herein, bei dessen Anblick alle, statt in die Aepfel, sich in ihre Finger schneiden.

3. *fī madḥ al-dhamīm* – Lob des Unglücks
Arabisches Metrum *wāfir* (∪—∪∪—|∪—∪∪—|∪— —)

Vergelt' es meinem Unglück Gott mit Gutem!
Denn es verhängt mir Freuden gleichwie Leiden.
Ich lob's, aus Lust nicht an ihm selber, sondern
Weil es mich Freund und Feind lehrt unterscheiden.

Bemerkenswert sind in Rückerts Übersetzungen vor allem die formalen Annäherungen an die arabische Gedichtkomposition. Rückert folgt nämlich getreu dem arabischen Reim und Metrum. Schon der

Monoreim verleiht der deutschen Fassung eine besonderes, ganz typisch arabische Melodie. Friedrich Rückert ist aber noch einen Schritt weiter gegangen und übertrug auch das arabische Metrum ins Deutsche, indem er Längen und Kürzen der quantitierenden arabischen Metrik in betonte und unbetonte Silben im Deutschen umsetzte. So wird der arabische Gedichtrhythmus auch im Deutschen hörbar.

In der *Sīrat al-Mujāhidīn* (*al-Amīra Dhāt al-Himma*) haben wir es mit der für die arabische Epik charakteristischen Kombination von Erzählprosa, Reimprosa und Poesie zu tun. Es verwundert nicht, daß der Dichter Friedrich Rückert aus dem arabischen Epos nur die Gedichte auswählte. Von deren erzählerischem Rahmen scheint Rückert nicht die geringste Notiz zu nehmen. Das ist angesichts des Umfangs der Erzähltexte ungewöhnlich und mag schlicht der Vorliebe des Dichters geschuldet sein. Vielleicht aber hat Friedrich Rückert den Kern der arabischen Epik in ihren Gedichten gesehen und so einen wesentlichen und oft unterschätzten Aspekt der *sīra shaʿbiyya* getroffen?

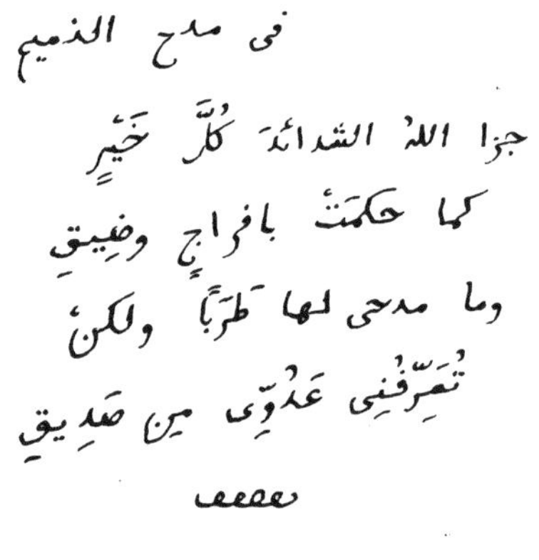
في مدح الذميم
جزا الله الشدائد كل خير
كما حكمت بافراج وضيق
وما مدحي لها طربا ولكن
تعرفني عدوي من صديق

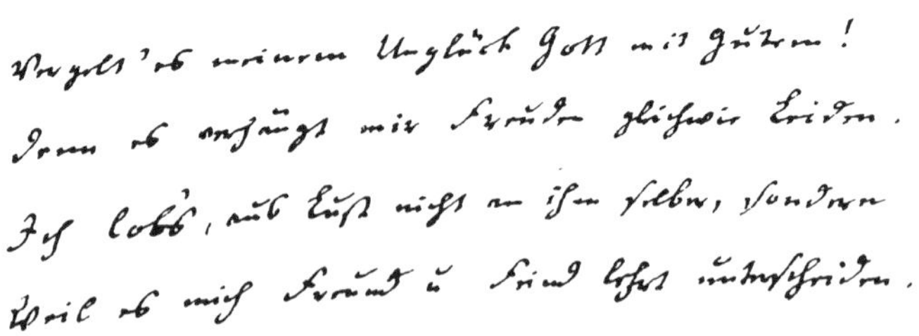
Lob des Unglücks.
Vergelt' es meinem Unglück Gott mit Gutem!
Denn es verhängt mir Freuden gleichwie Leiden.
Ich lob's, aus Lust nicht an ihm selber, sondern
Weil es mich Freund u. Feind lehrt unterscheiden.

ʿANTAR OVERSEAS

ARABIC MANUSCRIPTS IN EUROPE IN THE LATE 18TH AND EARLY 19TH CENTURY[1]

Maurits H. van den Boogert

The Austrian scholar and diplomat Joseph von Hammer-Purgstall is generally considered the first Western scholar who 'discovered' *sīrat ʿAntar*, the popular epic about the Arab Hercules, the black knight ʿAntar b. al-Shaddād.[2] When he went to Istanbul in 1799, he tried to obtain an Arabic copy of the *Arabian Nights* at the request of the Austrian diplomat, Franz von Thugut, but was unsuccessful. On the other hand, he did come across another interesting text, an Arabic manuscript containing part of the ʿAntar cycle. In 1801 he managed to obtain a complete copy of the *sīra* in Egypt, announcing his discovery in the Austrian journal *Fundgruben des Orients* a year later. In the article he argued that the ʿAntar stories were much more interesting than the *Arabian Nights Entertainments*, which were enormously popular in Europe ever since Galland's French translation had appeared at the beginning of the eighteenth century.[3] Hammer knew that he was not the first European scholar who had noticed the popular Arabic epic, however, because Sir William Jones' cursory reference to the ʿAntar cycle is mentioned in his article. In the *Poeseos Asiaticae Commentariorum*, published in London in 1774, Jones expressed his appreciation of ʿAntar, of which he had only consulted a single volume. In his recent work on the ʿAntar epic, Peter Heath, too, refers to Jones' work, but he devotes little attention to it, choosing to focus on the nineteenth-century reception of *sīrat ʿAntar* in

[1] I am indebted to David Waines, Jan Schmidt and Arnoud Vrolijk for their valuable suggestions, and to Alastair Hamilton for his useful comments and for correcting my English. It was Remke Kruk's enthusiasm when I told her about this side-line of my Russell research that encouraged me to write this article.

[2] R. Kruk, 'Sīrat ʿAntar ibn Shaddād', in R.A. Allen & D.S. Richards (eds.), *Arabic literature in the post-classical period.* Cambridge history of Arabic literature (Cambridge, 2006), pp. 292–306. For manuscript references, see Carl Brockelmann, *Geschichte der arabischen Litteratur*. Repr. of the 2nd ed. (Leiden, 1996), G II, 62 and S II, 64.

[3] P. Heath, *The thirsty sword. Sirat ʿAntar and the Arabic popular epic* (Salt Lake City, 1996), pp. 3–6.

continental Europe instead.[4] Jones' own collection of manuscripts did not include the text, so whose copy did he consult?

This article aims to make a modest contribution to our knowledge of the manuscript tradition of *sīrat ʿAntar* by examining which Arabic codices of the text were available in Europe prior to and just after Hammer-Purgstall 'discovered' it, in an attempt to identify which copy Jones might have consulted, and how it reached England.

The partial translation of *sīrat ʿAntar* by the orientalist and diplomat, Terrick Hamilton (1781–1876), sheds light on the dissemination of Arabic manuscripts of the 'Bedoueen Romance' in Europe in the late eighteenth and early nineteenth centuries.[5] Hamilton had been secretary to the British embassy in Istanbul, where he had obtained the two copies of the *sīra* upon which his translation of 1820 was based. At least one was procured for him by John Barker, the British consul in Aleppo, who assisted the traveller, William John Bankes, later in the 1820s, too.[6] It was undoubtedly also to Barker that Hamilton owed the information that the ʿAntar stories were still popular 'in coffee-houses, [where] it is read aloud by some particular person, who keeps a sheet in his hand, to which he occasionally refers to refresh his memory'. Furthermore Hamilton reports that school children in Aleppo were often obliged to copy ʿAntar manuscripts out 'and thus acquire the habit of speaking elegantly and correctly; and it may be attributed to this cause, that the copies of Antar are generally found written most execrably ill, and abounding in errors of every kind'.[7] Although many manuscripts of this text indeed appear to have been copied rather carelessly, of course this explanation cannot be accepted at face value.

The copy Barker had obtained for Hamilton had a 'smaller form than any yet sent to Europe', Hamilton claimed, adding a list of the owners of ʿAntar manuscripts he was aware of.[8] Although the author does not actually give the size of these manuscripts, the list is a useful starting point for our assessment of the interest in this Arabic popular epic in early modern Europe.

---

[4] *Ibid.*, pp. 3–21.

[5] T. Hamilton, *Antar. A Bedoueen romance*. 2 vols. (London, 1820).

[6] Cf. A. Sartre-Fauriat, *Les voyages dans le Ḥawrān (Syrie du Sud) de William John Bankes (1816 et 1818)* (Bordeaux [etc.], 2004), pp. 18, 19, 213.

[7] Hamilton, *Antar*, p. xviii.

[8] *Ibid.*, p. xxii.

The first manuscript owner on Hamilton's list was Claudius James Rich, who was appointed the East India Company's resident in Baghdad in 1810. Rich was a notable collector of Oriental manuscripts, who was not only interested in religious subjects, but also in contemporary Middle Eastern communities. It is not clear when Rich acquired his copy, which he might already have bought during his studies at Istanbul and Izmir, or during his stay in Egypt, where he had been an assistant to the consul-general. Rich was in contact with Terrick Hamilton during this period and provided him with a great deal of useful information about the ʿAntar epic.[9] Five years after Rich's death of cholera at Shiraz in 1820, the British Museum purchased his collection of 801 volumes of oriental manuscripts by Act of Parliament for the sum of £7,000. Its 390 volumes of Arabic works included his copy of *sīrat ʿAntar*.[10]

The next three copies known to Hamilton now appear to be lost. The first was owned by Andriy Yakovych Italinsky (1743–1827). Having finished his medical studies at St Petersburg in 1767 and obtained his M.D. in 1774, Italinsky entered a career in the diplomatic service in 1781. Around 1800 he was the Russian ambassador at Istanbul, where he worked together with his British colleague against Napoleon. Italinsky also maintained a correspondence with Arthur Paget, the British representative at the court of the King of the Two Sicilies.[11] According to Hammer-Purgstall, the Russian diplomat's library was in Rome the year after Italinsky's death, but it is not clear what happened to it later.[12]

It is equally unclear what happened to the Arabic manuscript of the ʿAntar cycle possessed by 'Lord Aberdeen', George Hamilton Gordon, fourth Earl of Aberdeen. He was only born in 1784, ten years after Jones published his views on the ʿAntar cycle, so Aberdeen's manuscript cannot have been the early copy we are looking for, but the fact that he had one at all is an indication of the impact of Hammer-Purgstall's recommendation.[13]

---

[9] T. Hamilton, *Antar. A Bedoueen romance*. 2 vols. (London, 1820), I, p. xlviii.

[10] H. Goodacre, U. Sims-Williams [et al.], *Arabic language collections in the British Library* (London, 1984), p. 4.

[11] Paget to Italinksy, 7 Janvier 1801, and Italinsky to Paget, 16 janvier 1801; A.B. Paget, *The Paget papers. Diplomatic and other correspondence of the Right Hon. Sir A. Paget, G.C.B., 1794–1807* (New York, 1896), I, pp. 301–302, 300, respectively.

[12] J. von Hammer-Purgstall, 'Lettera IV – Lettere sui manoscritti orientali e particolarmente arabi che si trovano nelle diverse bibliotheche d'Italia,' *Bibliotheca Italiana ossia Giornale di letteratura, scienze ed arti compilata da vari Letterati* 49 (1828), pp. 15–22.

[13] M.E. Chamberlain, 'Gordon, George Hamilton –, fourth earl of Aberdeen (1784–1860)' in *Oxford DNB*, XXII, pp. 900–909.

Another manuscript of *sīrat ʿAntar* on Hamilton's list was owned at the time by one 'M. Aidé at Constantinople'. This was Ilyās Antūn ʿĀʾida, a Greek Catholic native of Aleppo, whose family ties with the British consulate dated back to the 1730s. ʿĀʾida's grandfather, Shukrī, had been an important silk merchant with strong connections to the British mercantile community in Aleppo. Initially Shukrī ʿĀʾida was an honorary *dragoman* (interpreter) to the Dutch consulate in the 1740s, but after a dispute with the consul, ʿĀʾida switched to British protection in 1753. The transfer was arranged by Shukrī's son, Jirjis, who had become an honorary interpreter to the British consulate in Aleppo in 1747, and had soon embarked on active service. By the time his father relinquished the protection of the Dutch for that of the English, Jirjis ʿĀʾida held the office of first dragoman, which effectively made him the British community's most senior councillor in all matters Ottoman. When Jirjis ʿĀʾida died in 1775 his son, Ilyās Antūn, was considered too young to succeed him, but as soon as he had reached the right age, he, too, was appointed First Dragoman to the British consulate. The youngest ʿĀʾida soon moved to Istanbul with his family, however, hoping to find better doctors there for the treatment of an ulcer his wife was suffering from. Despite his move to the Ottoman capital, ʿĀʾida retained his appointment to the consulate in Aleppo, which was rapidly loosing its importance. Only after the Levant Company decided to close the consulate altogether from 11 April 1791, did Ilyās Antūn ʿĀʾida attempt to obtain an appointment as honorary dragoman to the British embassy in Istanbul, at which he was eventually successful. At the beginning of the nineteenth century he still actively conducted international trade in the Ottoman capital, where it was only natural that he met Terrick Hamilton, whom he evidently told about the Arabic manuscript of *sīrat ʿAntar* he owned.[14]

After these four manuscripts, Hamilton further listed the copy Hammer had donated to the Imperial Library at Vienna, and a manuscript in the library of the University of Cambridge, to which I will return below. Finally, Terrick Hamilton mentioned 'some few volumes in the possession of Mr. Hamilton'. This is undoubtedly a reference to Colonel George

[14] National Archives, London (formerly: PRO), State Papers (SP) 110/29, f. 264ff; SP 105/121, pp. 269, 323, 368–370, 375–376; SP 105/189, pp. 568–569ff; SP 110/66, pp. 34–35, 111; Başbakanlık Osmanlı Arşivi (Archive of the Prime Minister's Office, Ottoman), Istanbul: Ecnebi Defterleri (Foreigners' Registers) 35/1: 133/508: record of Ilyās Antūn ʿĀʾida's appointment as First Dragoman, dated 21 Ramadan 1196 AH/31 August 1782 AD; Archives Générales des Capucins, Rome, AD 106, I, pp. 97, 111.

William Hamilton (1807–1868), who served in various military and administrative posts in India, acquiring a sizeable collection of Oriental manuscripts. Part of his library was obtained by the British Museum in the year of his death, where his ʿAntar manuscript now appears to be kept.

The Cambridge manuscript is probably the one in twelve volumes (MSS 537–548) acquired by the Swiss traveller, John Lewis Burckhardt, who encouraged Hamilton to continue with his translation, and bought a large copy of the Arabic text for him in Cairo.[15] Burckhardt's own manuscript is dated 1195/1781 and only became part of the Cambridge collection after his death in 1817, so this cannot be the copy Jones consulted in the early 1770s.[16]

Heath mentions a group of seven Arabic fragments of the ʿAntar stories, which are kept in the library of St John's College, Oxford, but neither date nor provenance is known.[17] A third, eighteenth-century, manuscript in Cambridge listed by Heath does not occur in any of the catalogues. Nor does Heath tell us much about it, but it might be the Bruce manuscript, the whereabouts of which are unknown.[18] James Bruce of Kinnaird, who became famous for his travels to Abyssinia, but also stayed in the Levant for some time, brought home an Arabic manuscript of ʿAntar in 1774. After his death, twenty years later, his collection of manuscripts passed to his son, who in his turn left it to his wife. She initially deposited the collection at the Chelsea hospital in London, but the manuscripts were later dispersed. Some went to the Bodleian Library, others to Cambridge, the remainder of the collection being sold at auction on 30 May 1842. In the issue of March 1825 of *The Classical Journal* Britain is congratulated for having acquired part of Bruce's collection, comprising almost 100 manuscripts. One Coptic manuscript is mentioned,

---

[15] Hamilton, *Antar*, I, pp. xlvi–xlvii (footnote).

[16] E.G. Browne, *A hand-list of the Muḥammadan manuscripts preserved in the library of the University of Cambridge* (Cambridge, 1900), p. 100; Cf. T. Preston, *Catalogus Bibliothecae Burckhardtianae cum appendice librorum aliorum orientalium in bibliotheca Academiae Cantabrigiensis* (Cambridge, 1853), p. 14, where the same manuscript is dated 1190 AH/1776 AD.

[17] The fragments in question do not appear to be mentioned in E. Savage-Smith's recent catalogue, *A descriptive catalogue of Oriental manuscripts at St John's College, Oxford* (Oxford, 2005).

[18] Heath, *The Thirsty Sword*, Appendix II, p. 235; Cf. E.G. Browne, *A supplementary hand-list of the Muḥammadan manuscripts including all those written in the Arabic character, preserved in the libraries of the University and Colleges of Cambridge* (Cambridge, 1922), pp. 315–316. The appendix below provides a supplement to Heath's catalogue.

as well as one in Persian, and 24 in Ethiopic. 'Among the Arabic MSS is . . . the Romance of Antar', the journal informs us.[19] Although it is highly likely that Bruce met Jones in England, Bruce only arrived there after the publication of Jones' *Poeseos Asiaticae Commentariorum*.

The list by Hamilton may not help us to identify the text Jones consulted prior to 1774, but it does indicate that the number of Arabic manuscripts of *sīrat ʿAntar* in Europe around 1820 was limited. Heath's preliminary survey of manuscripts known today offers a wider range of possibilities.[20] If we assume that Jones' manuscript is still in England, the collections at Cambridge, Oxford, London, and Manchester are worth examining more closely. Those in Cambridge have already been discussed, and the fact that the date of the Bodleian's only copy is unknown, brings us no further. The British Library has no less than eleven relevant manuscripts, five of which date from the middle of the eighteenth century or earlier. The oldest copy –unfortunately a mere fragment – is even dated as early as 1437. Apart from Rich's copy, however, I have found no information about their provenance establishing a link with Jones. The same is true of two eighteenth-century fragments in the library of the School of Oriental and Asian Studies in London (see the appendix). Finally, the John Rylands Library in Manchester has six (partial) copies of *sīrat ʿAntar*, four of which are from the right period. One is particularly interesting, because it had belonged to a friend of William Jones.

Like many other manuscripts in the Manchester collection MS Arabic 608–610 was purchased from the Bibliotheca Lindesiana in 1901 by Mrs Rylands for the library founded by her late husband, John. Mingana's catalogue tells us that this copy of the Syrian recension of *sīrat ʿAntar* was originally written as a single volume, but was bound, probably in Aleppo, as three.[21] The Earl of Crawford, whose collections formed the Bibliotheca Lindesiana, had purchased it through the bookseller, Bernard Quaritch, in London in 1866. The previous owner was Nathaniel Bland, the Persian scholar at Oxford, who died the year before. Bland appears to have acquired the three volumes from the heirs of Claud Rus-

[19] 'Oriental manuscripts and antiquities', *The classical journal* 61 (March 1825), pp. 150–152, esp. 150–151; G. Robins, *Catalogue of a valuable collection of oriental literature, collected by James Bruce, of Kinnaird [. . .] sold by auction, by Mr. George Robins [. . .] 30th day of May, 1842* (London, 1842).

[20] Heath, Appendix II, pp. 232–239.

[21] A. Mingana, *Catalogue of the Arabic manuscripts in the John Rylands Library, Manchester* (Manchester, 1934), pp. 861–862.

sell, a former administrator of Vizagapatam. Russell is known as one of the first Britons to take home with him his Indian *munshī*, who appears to have taught him Persian, but the East India Company officer does not seem to have studied Arabic too.[22] Claud Russell's heirs had inherited the incomplete manuscript of *sīrat ʿAntar* from his elder brother, Patrick Russell, who had died, unmarried, in 1805. After Patrick's death, he bequeathed his oriental manuscripts, including the three ʿAntar volumes, to his nephews, stipulating that the books be 'preserved in his [Claud Russell's] library as they may hereafter prove acceptable to his son Claud, or other sons on their return from India'.[23] After Claud Russell senior's death on 17 April 1820, however, his sons sold the manuscripts left them by their uncle.

Patrick Russell was a man of many talents who left his mark in several academic fields, but acquired lasting fame in none. A brief description of his life is pertinent here, because it explains his connection with Jones. Patrick Russell was born on 7 February 1726 in Edinburgh, the second son of John Russell of Braidshaw, Writer to the Signet (the Scottish equivalent of a solicitor), and his third wife, Mary Anderson.[24] He was educated at Edinburgh High School and then at the city's university at the beginning of the Scottish Enlightenment. In the autumn of 1742 Patrick attended Alexander Monro Primus's famous anatomy course.[25] Russell probably also took several other courses at the university, but, like many students in those days, he did not graduate until several years later.

After having worked as a ship's surgeon for several years, Patrick Russell settled in the Syrian city of Aleppo in 1750, joining his elder half-brother, Alexander, who had arrived there a decade earlier. Both systematically studied the plague in Aleppo, Patrick witnessing three outbreaks of the epidemic in the early 1760s. By this time Alexander Russell had returned to England and published *The Natural History of Aleppo* in 1756. The book described the manners and customs of the

---

[22] S. Digby, 'An eighteenth-century narrative of a journey from Bengal to England. Munshī Ismāʿīl's *New History*,' in C. Shackle (ed.), *Urdu and Muslim South Asia. Studies in Honour of Ralph Russell* (London, 1989). I am indebted to Mohammad Seddon for this reference, and to David Waines for referring me to Seddon.

[23] National Archives, London: Prob. 11/1430, f. 236v–238v, especially 237v.

[24] G.S. Boulger, rev. Mark Harrison, 'Russell, Patrick, 1727–1805', in *Oxford DNB*, XLVIII, pp. 328–329.

[25] Edinburgh University Library, Dc.5.95: 'Record book of students, scholars, prentices, etc. 1720–1749'.

inhabitants of the city and its surrounding as well as many aspects of its flora and fauna. It was favourably reviewed by Samuel Johnson in *The Gentleman's Magazine*, in which sizeable excerpts from it were published. The book instantly won Alexander Russell a reputation as a scholar and undoubtedly contributed to his election as Fellow of the Royal Society in 1756. Patrick, who had also contributed to his brother's book, stayed in Syria until 1771, three years after Alexander had died in London. Like his brother before him, Patrick visited the lazarettos of Livorno and Naples on his journey home. He also made a detour to Rome, however, visiting the Vatican Library. Russell settled in London in 1772, and he was elected Fellow of the Royal Society five years later.

In the 1780s Patrick accompanied his younger brother, Claud, whose health was frail, on his second journey to India. Once they had arrived, however, Claud's full recovery allowed Patrick to pursue his interests in Natural History. He was eventually appointed Natural Historian to the East India Company, an office belonging to the Madras Medical Council. His endeavours to identify Indian plants that might be useful for the EIC medical service made him one of the founding fathers of Indian botany. His research on fishes and snakes – the results of which were published in the first decade of the nineteenth century, partly after his death on 2 July 1805 – was equally ground-breaking. One of the serpents he described is still known as Russell's viper (*Vipera russelli*) today.

In India Patrick Russell also worked on the second edition of his brother's book. He rewrote the list of Syrian plants, the original of which he had also made, adopting the Linnean system. He also corrected some errors in his brother's text, and added numerous footnotes and lengthy endnotes, some of which are small scholarly essays. His own research on the plague in Aleppo resulted in *A Treatise of the Plague*, which appeared in London in 1791. This is now a largely forgotten work, despite the fact that its appendix, consisting of case studies of people afflicted by the plague, offers valuable details about Syrian social life around the middle of the eighteenth century which are hard to find elsewhere.[26]

In 1794 the second edition of *The Natural History of Aleppo* was published. From the additions by Patrick Russell we learn that he was even

[26] P. Russell, *A Treatise of the Plague containing an historical journal, and medical account of the plague, at Aleppo, in the years 1760, 1761, and 1762. Also, remarks on quarantines, lazarettos, and the administration of police in times of pestilence. To which is added an appendix, containing cases of the plague; and an account of the weather, during the pestilential season* (London, 1791).

more interested in contemporary Ottoman culture than his elder half-brother. Patrick, for example, also made inquiries about Ottoman literary culture, including the Arabic popular epic.[27] Alexander Russell was certainly aware of the European interest in the *Arabian Nights*, but he did not discuss them in any detail. The second edition of *The Natural History of Aleppo*, by contrast, offers an elaborate description of the performance of story-tellers, which is worth quoting here in full, not only because few Western observers wrote about these narrators, but also because Russell owned manuscripts of the stories on which they based their repertoire:

> The recitation of Eastern fables and tales, partakes somewhat of a dramatic performance. It is not merely a simple narrative; the story is animated by the manner, and action of the speaker. A variety of other story books, besides the Arabian nights entertainment, (which, under that title, are little known at Aleppo) furnish materials for the story teller, who, by combining the incidents of different tales, and varying the catastrophe of such as he has related before, gives them an air of novelty even to persons who at first imagine they are listening to tales with which they are acquainted. He recites walking to and fro, in the middle of the coffee room, stopping only now and then when the expression requires some emphatical attitude. He is commonly heard with great attention, and, not unfrequently, in the midst of some interesting adventure, when the expectation of his audience is raised to the highest pitch, he breaks off abruptly, and makes his escape from the room, leaving, both his heroine and his audience in the utmost embarrassment. Those who happen to be near the door endeavour to detain him, insisting on the story being finished before he departs, but he always makes his retreat good; and the auditors, suspending their curiosity, are induced to return at the same hour next day, to hear the sequel. He no sooner has made his exit, than the company, in separate parties, fall to disputing about the event of the unfinished adventure. The controversy by degrees becomes serious, and opposite opinions are maintained with no less warmth, than if the fate of the city depended on the decision.[28]

Russell's collection of Arabic manuscripts provides some indication of which texts he was interested in.[29] First, he purchased a copy in two

[27] On Russell's interests in contemporary Ottoman literary culture, see M.H. van den Boogert, 'Patrick Russell and the Republic of Letters in Aleppo', in A. Hamilton, M.H. van den Boogert [et al.] (eds.), *The Republic of Letters and the Levant* (Leiden, 2005), pp. 223–264, esp. 244–248.

[28] A. Russell, *The natural history of Aleppo and parts adjacent*. 2nd ed. by P. Russell (London, 1794), I, pp. 148–150.

[29] For a discussion of the medical manuscripts Russell acquired, see van den Boogert, pp. 238–243.

volumes of the first 281 'canonical' Nights, one of the earliest sizeable manuscripts to arrive in Britain. The acquisition of this manuscript is sometimes attributed to his famous elder brother, but this is incorrect. Furthermore, Patrick bought five miscellaneous volumes, which included several 'Additional Nights', as well as translations into Arabic of Aesop's fables. He also translated a number of these 'fables and tales', some of which were published by William Beloe, the translator of Herodotus. In the preface to the third volume of his miscellanies, Beloe described their cooperation as follows: 'My friend, Dr. Russell, brought with him a small volume from Aleppo, from which he at different times recited to me so much, that I became impatient to hear more. My importunity finally prevailed, and at various intervals his kindness induced him to dictate, in the best manner he could, from the Arabic, whilst I performed the humble office of scribe'.[30] Beloe did more than merely record the nineteen tales he published, undoubtedly polishing their style considerably. Nevertheless, this was the first reliable translation of some 'Additional Nights', as Burton would later call them, including that of Bāsim, the Blacksmith. Russell appears to have acquired some twenty Arabic manuscripts altogether in Aleppo, which included the three volumes from the 'Serat Antar'.

In his will Patrick Russell stipulated that all his personal papers be burned after his death. His executors appear to have followed his instructions to the letter, because little primary material by Patrick Russell has survived. From the correspondence of Jones, however, it is clear that he and Russell were regularly in contact when they were both in India in the early 1780s.[31] It is also clear that Jones borrowed one Persian and two Arabic manuscripts from Alexander Russell around 1766, copying one of the latter in his own hand.[32] Patrick Russell arrived in London in 1772. Jones was still at Oxford at this time, but he was elected Fellow of the Royal Society in the same year. Russell was only elected in 1777, but he already attended meetings of the Royal Society in 1774, and may well

---

[30] William Beloe, *Miscellanies. Consisting of poems, classical extracts, and Oriental apologues*, III (London, 1795), Preface.

[31] G. Cannon (ed.), *The letters of Sir William Jones* (Oxford, 1970), pp. 775–777: 'Jones to Russell', 22 September 1787.

[32] John Rylands Library, Manchester: MS Ar. 264–265, copied in 1766 by William Jones from the MS borrowed from Alexander Russell; Mingana, pp. 426–427.

have done so earlier, too.[33] Moreover, Russell visited Oxford on several occasions to consult Arabic manuscripts in the Bodleian. Furthermore, both studied Arabic and shared an interest in oriental 'fables and tales', so it seems likely that Jones and Russell became acquainted soon after the latter's return from Syria in 1772.

Patrick Russell's partial copy of *sīrat ʿAntar* appears to have been the first Arabic manuscript of the text that arrived in Britain, in 1772. Russell probably met Sir William Jones soon after his return from Aleppo, either in London or at Oxford. It is tempting to conclude that the copy Jones consulted must therefore have been that owned by Russell. One problem remains, however. Jones indicated that he only saw volume 14 of the ʿAntar epic, about the contents of which he says nothing.[34] Assuming the number is correct and the manuscript in question started with volume 1, this suggests that the fragment inspected by Jones is to be found somewhere in the second half of the narrative. Russell's manuscript, however, starts with the Abraham story and covers only the beginning of the ʿAntar stories. Does this mean Russell's text cannot be the one we are looking for after all? Again there is no solid evidence, but a further investigation of Russell's oriental collection might provide the answer. We know that Russell owned about 20 manuscripts in all, chiefly Arabic texts, each of which he numbered. His copy of *Al-durr al-muntakhab fī taʾrīkh mamlakat Ḥalab*, for example, was number 18 in his private library. But Russell also inscribed the Arabic words *al-kitāb al-khāmis* ('the fifth book') in his copy, despite the fact that it was by no means the fifth volume, or fifth part, of a larger work.[35] Russell may well have made a similar mistake with his ʿAntar codices.

[33] The Royal Society, Journal Book Copy XXVII (1774–1777), 19: 10 March 1774. This is the first reference to Russell's attendance in the Journal Book. He was introduced by Daniel Solander, Sir Joseph Banks' private librarian and botanist.

[34] Hamilton, *Antar*, I, pp. xx–xxi (footnote).

[35] *The natural history of Aleppo*, I, p. 247 (Endnote 1); Mingana, pp. 416–418.

Appendix
A supplement to Heath's survey of ʿAntar manuscripts[36]

Leiden University Library

- Cod. Or. 1541: CCO 458 (I, pp. 348–349); CCA 542 (I, p. 336); Voorhoeve, *Handlist*, p. 340: 60 folios. Bound. Incomplete at the end. No colophon. Undated, but with dated annotation on flyleaf: *fī niṣf* [15] *Jumādā al-awwal laylat al-jumʿa 1160*/Thursday-Friday 25–26 May 1747. The copy appears to have been made by one copyist in black ink. The writing is not very neat, but legible. Fragments of poetry in two columns. Text corresponds with that of Hamilton, pp. 24–72. The manuscript originally belonged to a 'nobleman' named Testa. This was probably Gaspard Testa, an interpreter to the Dutch embassy in Istanbul, who became *chargé d'affaires* at the Porte in 1808. In 1846 the title of baron was conferred on him. His collection of 82 oriental manuscripts was acquired by the library in 1837 and 1839. The latter transaction included this manuscript.[37]
- Cod. Or. 1681a: CCO 2562 (V, pp. 170–171); CCA 541 (I, p. 336); Voorhoeve. *Handlist*, p. 340: 85 folios. According to De Goeje and Houtsma, 'codex anno 1081 non male exaratus est'.[38] The colophon (f. 85r) indicates that the copy was finished on 4 Muḥarram of that year, 24 May 1670. Notes on f. 1r and 85v in a North African hand. Fragments of poetry in three columns. The text consists of the first part (f. 85r: *al-juzʾ al-awwal*) of *sīrat ʿAntar*, including the Abraham story (f. 72–85). Small, pocket book format. Purchased at auction in Paris in 1860.[39]

[36] This appendix offers descriptions of six fragments of ʿAntar manuscripts. The first three, kept in Leiden, are mentioned by Heath, but his references are outdated, and the descriptions incomplete. I gratefully acknowledge the help of Dr Arnoud Vrolijk in finding the current numbers of these codices. The Turkish translation kept in Leiden is not mentioned by Heath, nor are the final two manuscripts in the library of the School of Oriental and African Studies in London.

[37] On the Testa collection, see J. Schmidt, *Catalogue of Turkish manuscripts in the library of Leiden University and other collections in the Netherlands. Vol. Two; comprising the acquisitions of Turkish manuscripts in Leiden University Library between 1800 and 1970* (Leiden, 2002), pp. 80–81.

[38] M.J. de Goeje & M.Th. Houtsma, *Catalogus codicum Arabicorum bibliothecae Academiae Lugduno-Batavae*. 2nd ed. (Leiden, 1888) I, p. 336.

[39] I have not been able to consult the *Catalogue d'une belle collection de manuscripts et livres arabes… dont la vente aura lieu le 25 juin 1860* (Paris: A. Franck, 1860).

- Cod. Or. 1681b: CCO 2562 (V, pp. 170–171); CCA 541 (I, p. 336); Voorhoeve. *Handlist*, p. 340: f. 1–126 (Voorhoeve erroneously only mentions f. 72–85). Incomplete at the beginning and the end. No colophon. Text in black ink, with headers (often marking poetry) in red ink. The title on f. 1r was added later in a different (North African?) hand from the copyist. Purchased at auction in Paris in 1860.
- Or. 12.433: Turkish translation of parts of *sīrat ʿAntar* in 242 folios. Undated, but with marginal notes by several readers from 1140/1727–28.[40] The manuscript originally belonged to the German orientalist, Franz Taeschner, who probably bought it in Istanbul or Cairo in the 1920s or '30s. It was acquired by the Leiden University Library in 1970 from the antiquarian bookshop of E.J. Brill in Leiden.[41]

SOAS, London

- MS No. 47,286: Fragment of 80 folios. No colophon; Dated to the eighteenth century by Gacek, whose catalogue offers little additional information about the manuscript, except that black and red ink were used.[42] In the margin of f. 31v the date 'Rajab sanat 1210'/12 January–10 February 1796 is found. The numbering of the folios is erratic, starting with 30v, but disappearing again until [37v] which is numbered '4'. From folio 6 the numbers occur on the verso side. After f. 50 the numbering reverts to 41 and 42, then continues from 51. Folio 73 appears to be missing. Occasional reference to the transmitter: *qāla al-nāqil* (e.g. 44, 46). The text consists of fragments of books 33 and 34 of the Cairo edition of 1961, in which ʿAntar leaves ʿAbs and helps Khufāf b. Nadba. Provenance unclear.
- MS 47,287: Fragment of 91 folios. No colophon; Dated to the eighteenth century by Gacek.[43] If this is correct, then the dates in the margin of f. 16v (16 Ṣafar 1284/19 June 1867), 60v (18 S. 1284), and 85v (19 S. 1284) were probably added by a reader. Most of the manuscript

[40] I am indebted to my colleague, Dr Jan Schmidt, for bringing this manuscript to my attention. It is described in more detail in his forthcoming article, 'The Ottoman miscellaneous manuscript. Exploration of a hitherto neglected genre'.

[41] On Taeschner (1888–1967) and his collection, see J. Schmidt, 'Franz Taeschner's collection of Turkish manuscripts in the Leiden University Library,' in J. Schmidt, *The joys of philology*, II (Istanbul, 2002), pp. 237–266.

[42] A. Gacek, *Catalogue of the Arabic manuscripts in the library of the School of Oriental and African Studies, University of London* (London, 1981), p. 182.

[43] *Ibid.*

is in one hand, which covers folios 1–11, 13–23, 25–76, 78–82, 84–85. Folios 12, 24, 77, 83, and 86–90 were added by different hands, probably replacing defective folios from the original. Fragments of poetry often in two columns, but not consistently, and frequently marked by punctuation, or by the word *shiʿr* in red ink. Frequent references to the transmitter: *qāla al-rāwī* or *qāla al-Aṣmaʿī* (e.g. 17v, 20r, 21r, 22v, 25v).[44] The text appears to consist of book 22 of the Cairo edition, which covers the war between ʿAbs and ʿĀmir. Provenance unclear.

[44] Cf. Mingana, p. 839.

THE FUNCTION(S) OF POETRY IN THE *ARABIAN NIGHTS* SOME OBSERVATIONS

Wolfhart Heinrichs

At the end of his article 'Alf Layla wa-Layla' in the second editon of the *Encyclopaedia of Islam*, Enno Littmann, translator of the 'Arabian Nights' into German, makes the following statement:

> There are about 1420 poems or fragments of poetry in the 2nd Calcutta edition, according to Horovitz (in *Festschrift Sachau*, Berlin 1915, pp. 375–379). Of these a number of 170 repetitions may be deducted, so 1250 insertions of poetry remain. Horovitz has been able to prove that those insertions whose authors he could discover are to be dated from the 12th to the 14th centuries, i.e. from the Egyptian period of the history of the Nights. These poems and verses are mostly of the kind that they might be omitted without disturbing the course of the prose texts, and, therefore, have been later added to them.[1]

This statement sent Jamel Eddine Bencheikh into a fit. In his contribution to the collective volume *Mille et un contes de la nuit* he writes:

> Cet amas de sottises illustrerait remarquablement un bêtisier de l'orientalisme arabisant des années 1930. E. Littmann partage avec de nombreux savants européens une prévention marquée à l'égard de la poésie arabe, qui reste pour eux obscure, répétitive et sans intérêt. L'historicisme philologique à l'honneur outre-Rhin a fait le reste. Cette suite de contresens sur la fonction de la poésie dans les *Nuits* exige qu'on revienne sur le problème . . .[2]

One cannot but agree with the last phrase. Bencheikh himself does this in about two pages with some perceptive and some poetic observations, all of them of a general nature, i.e., he does not actually analyze individual cases.

[1] E. Littmann, 'Alf Layla wa-Layla,' in *Encyclopaedia of Islam* (New ed.), I (Leiden [etc.], 1960), pp. 358–364, see p. 364a.

[2] J.E. Bencheikh, 'La Volupté d'en mourir,' in J. Bencheikh, C. Brémont [et al.] (eds.), *Mille et un contes de la nuit* ([Paris], 1991), pp. 268–271.

Until very recently, few scholars of the *Arabian Nights* have actually dealt with the poetry they contain in a sustained way. We owe some substantive, though brief, notes to Suhayr al-Qalamāwī in her doctoral dissertation, which she wrote under the guidance of Ṭāhā Ḥusayn,[3] to Wiebke Walther in her German introductory monograph,[4] to Sandra Naddaff in her book on the aesthetics of repetition in the *Nights*,[5] and to David Pinault in his application to the *Nights* of the 'oral-poetry' concept.[6] This was followed by the only article on the topic by Walīd Munīr in the journal *Fuṣūl*.[7] In the West it has been only in 2004 that two articles specifically devoted to the *Nights*' poetry have appeared, one by Kadhim Jihad, which is included in the proceedings of the colloquium 'Mille et une nuits en partage,' edited by Aboubakr Chraïbi,[8] and the other by Geert Jan van Gelder in the *Arabian Nights Encyclopedia*.[9] Both are valuable but not very extensive.

What accounts for this strange sin of omission? It is always precarious to name reasons for the non-existence of a thing. But the history of Arabic studies in the West may yield an indication. The fact is that Arabic poetry, especially early poetry, was for a long time a staple for the education of young Arabists, not usually from a literary-historical point of view, but from a grammatical, lexical, sometimes ethnographical angle. Prose fared worse. It was studied for content, for what it could teach about the real world, possibly for influences measured by similarities of subject-matter. It is probably only in the nineteen-eighties that classical Arabic prose became a topic, in the West, of literary studies on a larger scale. Poetry profited as well, but studies on the structure of prose works had more of the character of a novelty, and they were also stimulated by

---

[3] Suhayr al-Qalamāwī, *Alf layla wa-layla* (Cairo, 1966), pp. 126–127.

[4] W. Walther, *Tausendundeine Nacht. Eine Einführung* (München [etc.], 1987), pp. 66–68.

[5] S. Naddaff, *Arabesque. Narrative structure and the aesthetics of repetition in the 1001 Nights* (Evanston, Ill., 1991), pp. 85–86.

[6] D. Pinault, *Story-telling techniques in the Arabian Nights* (Leiden, 1992), pp. 102–107, 119–129.

[7] Walīd Munīr, 'Al-Shi'r fī Alf layla wa-layla. Tamaththul al-wāqiʿ wa-'stiqṭāb al-dhākira,' *Fuṣūl* 12/4 (1994), pp. 200–222.

[8] K. Jihad, 'Poésie des *Mille et une nuits*,' in A. Chraïbi (ed.): *Les Mille et Une Nuits en partage* ([Paris], 2004), pp. 241–250.

[9] G.J. van Gelder, 'Poetry and the *Arabian Nights*,' in U. Marzolph & R. van Leeuwen (eds.), *The Arabian Nights encyclopedia* (Santa Barbara, Cal. [etc.], 2004), pp. 13–17. It deserves mention that the precursor of the *Arabian Nights encyclopedia*, Richard van Leeuwen's *De wereld van Sjahrazaad* (Amsterdam, 1999) does not contain an entry on poetry.

general literary theory, especially narratology. So a sizable number of important studies on the art of story-telling in the *Nights* have appeared, while the poetry has mostly fallen by the wayside.

What about Bencheikh's attack on Littmann? Is he right? The point can be made that both are right, from different angles. In support of Littmann one can adduce the reading customs of countless readers that were not interested in writing an interpretive article on this or that story in the *Nights* but were simple enjoying the mysterious turns of the plots. They could not normally be bothered with the poetry and skipped it (only rarely is a poem part of the plot).[10] Another point in favor of Littmann is the fact that the poems tend to vary in the different textual traditions of the *Nights*, both in the number of slots and in the actual slot-fillers.This means that the occurrence of poetry in general is an integral part of the *Nights*, while the occurrence of individual poems or fragments is not. This in turn means that the story-teller or the copyist (or someone in between) felt free to add to, or subtract from, the version he had in front of him.

Littmann in the quotation adduced above referred to Horovitz' article in the 1915 Festschrift for Eduard Sachau. Most of the poetry in the *Nights* is anonymous, i.e. unattributed, and Horovitz' intention had been to establish the correct attribution in as many cases as he could. Strangely, Littmann misrepresented his results, by claiming that 'Horovitz has been able to prove that those insertions whose authors he could discover are to be dated from the 12th to the 14th centuries, i.e. from the Egyptian period of the history of the Nights.' Horovitz did nothing of the sort. Rather he stated that early poetry was not well represented, and that most of the poets he could identify lived between the 3rd and the 8th century of the Hijra. He also stated that his work was successful in about one quarter of all cases, but he did not list the individual results, except for al-Mutanabbī and al-Ḥarīrī.[11] Instead he gave a list of poets, arranged according to centuries. As a pure statistics, omitting all names, this would look as follows:

---

[10] I have asked readers of both the Arabic original and Western versions, and most of them agree on this. However, some Arab readers raised a protest against my assertion. Of course, my findings are anecdotal and have no statistic value.

[11] J. Horovitz, 'Poetische Zitate in Tausend und eine Nacht,' in G. Weil (ed.), *Festschrift Eduard Sachau zum siebzigsten Geburtstage gewidmet von Freunden und Schülern* (Berlin, 1915), pp. 375–379, see p. 377, and notes no. 1 and 2 for al-Mutanabbī and al-Ḥarīrī.

3rd cent. – 5 poets;
4th cent. – 8 poets;
5th cent. – 7 poets;
6th cent. – 9 poets;
7th cent. –10 poets;
8th cent. – 3 poets.

One should probably not make too much of this, but one of the goals of Horovitz was to use the poetry as one of the indicators of the history of the text of individual stories as well as of the collection as a whole. The numbers show a peak in the 7th century and a sharp drop in the 8th, which might be interpreted as indicating that the text of the so-called Calcutta II edition, on which Horovitz had based his investigation, received its final shape, as far as the poetry was concerned, around the beginning of the 8th/14th century.

Bencheikh's attack on Littmann was, to some extent, unfair. There may have been among some German and other Western scholars an antipathy to, and disinterest in, Arabic poetry. But Littmann had published translations of Arabic poems, and he was the first to devote scholarly attention to Semitic folk poetry, both in Arabic dialects and in various Ethiopian languages. But beyond that, it is clear that Bencheikh has some valid points to make. I would like to subject his points separately to a critical evaluation.

(1) The reciter. Bencheikh says that the stories call up the poetry to put it in the mouths of their protagonists. This is often true; however, interestingly enough, the poetry is also used by the narrator himself, on the surface level, so to speak, not by having a protagonist recite it. This is of great literary interest and I will come back to it later on.

(2) The genres.The repertory of love, wine, and garden poetry as well as gnomic and heroic verse in the *Nights* is quite large. This is also borne out be the collection of *Alf layla* poetry by ʿAbd al-Ṣāḥib al-ʿUqābī, which Bencheikh criticizes very harshly.[12] Al-ʿUqābī's collection is arranged according to genres, but his own activity is restricted to explanation of words; no textual criticism or literary-historical treatment (including identification of authors) takes place. One wonders what the author had in mind when he compiled his book.

[12] ʿAbd al-Ṣāḥib al-ʿUqābī, *Dīwān Alf layla wa-layla* (Baghdad, 1980), criticized in Bencheikh, p. 269 (and note 1).

(3) Attribution. A few poems are attributed to specific poets, but the large majority is anonymous. Bencheikh assumes that they were left anonymous, because they were by popular authors and everybody knew who wrote what. This is based on the assumption that ALL poems were written by professional poets. Most scholars disagree, as it is relatively easy to see that there is a definite variety in quality in the poetic corpus. Haddawi in his translation of the Mahdi edition discusses the correct procedure of the translator faced with genuine, usually high-quality, poetry on the one hand, and the miserable poetic concoctions of hacks, on the other.[13] However, in many cases Bencheikh may be right and the anonymity is based on initial popularity with subsequent loss of memory. This kind of anonymity is quite ubiquitous in classical high-brow literature as well.

(4) Construction. An attentive reading reveals that there is an organic tie between prose and poetry in the structuring of the story. The two modes of writing are employed in a division of labor. As Bencheikh says: 'If the story narrates a passion, the poem represents it.'[14] It is, in particular, passages of high emotion, where poetry as a rule makes its appearance. Bencheikh is probably mostly right, but one has also to keep in mind that there is nothing obligatory about the appearance of poetry. Large sections of the *Nights* are devoid of poems. This may have to do with the genre of the story, as various scholars have suggested. A final point of Bencheikh's is the fact that the *Arabian Nights* do not differ much in this respect from other works of Arabic literature. Prosimetrum is wide-spread.

Most authors who have, however briefly, discussed the poetry in the *Arabian Nights* have done so in general terms, which also means in generalizations. If one attentively studies the texts themselves, one finds much more variation than one would expect from those statements. But let me start with a general statement myself, one which refers to the general function of poetry in a prose context. In an earlier work, I have tried to reduce the general functions of poetry within a narrative to three types; these may be labeled *shāhid, ʿaqd wa-ḥall*, and *tamaththul*.[15]

[13] H. Haddawi, *The Arabian Nights*. Translated by H.H., based on the text of the fourteenth-century Syrian manuscript edited by Muhsin Mahdi (New York [etc.], 1990), pp. xxvii–xxix.

[14] Bencheikh, p. 270.

[15] W. Heinrichs, 'Prosimetrical genres in classical Arabic literature,' in J. Harris & K. Reichl (eds.), *Prosimetrum. Cross-cultural perspectives on narrative in prose and poetry* (Cambridge, 1997), pp. 249–275, see p. 274.

The *shāhid* 'witness' indicates a poetic prooftext that underscores the correctness of a prose account. In early literature this is a common procedure: the poetry proves the prose, the prose explains the poetry. The phrase *ʿaqd wa-ḥall* refers to the 'solidification [of prose into poetry] and the dissolution [of poetry into prose].' These two procedures, discussed in books composed by the state-scribes, were based on the assumption of the equivalence of the two modes of writing. Which in turn meant that one could write about the same topic either way and, theoretically, also switch from one to the other. Note, however, that this is not normally applied to narratives. Finally, *tamaththul* refers to using poetry as a *mathal*, a 'proverb.' For just as the proverb frequently arises from a narrative, which is its etiology, and is then applied to a new situation that bears a striking resemblance to the initial situation (*mathal* = 'similarity'!), the poem is likewise taken from its initial context, which may not be known any more but is clearly alluded to by the poem itself, and inserted into a similar context, thereby indicating the constancy of human emotions. It is clear from these definitory statements that the poetry of the *Arabian Nights* belongs in the third category.

As I said before, this is another general statement, and although I do not think that it can be denied a certain truth value, what we really need to start with is a phenomenology (and, if we are lucky, a taxonomy) of the ways of existence of *Alf layla* poetry. By phenomenology I do not intend more than an exact description of how the poem sits within the narrative. For my preliminary work I have used three editions that were available to me: the Mahdi ed., the Habicht ed., and the Qiṭṭa al-ʿAdawī ed.[16] Mahdi and Habicht mostly go together, both representing the Syrian branch of the transmission, while Qiṭṭa diverges, being based on the late Egyptian branch of transmission, as represented by the earlier Bulaq ed. (Qiṭṭa, however, abbreviates at times the prose text as well as the poetic quotations.) A painstaking comparison of the versions is not intended at this point.

---

[16] M. Mahdi (ed.), *Kitāb Alf layla wa-layla min uṣūlih al-ʿarabiyya al-ūlā*, I (Leiden, 1984); M. Habicht (ed.), *Tausend und Eine Nacht Arabisch.* Nach einer Handschrift aus Tunis herausgegeben von M.H. Nach seinem Tode fortgesetzt von M. Heinrich Leberecht Fleischer (Breslau, 1825–43), Repr. *Alf layla wa-layla.* Ṭabʿa muṣawwara ʿan ṭabʿat Brislāw (Cairo, 1998); Muḥammad Qiṭṭa al-ʿAdawī (ed.), *Kitāb Alf layla wa-layla* (Cairo, 1325 AH [1907]).

Here are some examples from the beginning of the *Nights*:

(1) The story of the cuckolded demon. When the demon takes the beautiful woman out of the trunk, the narrator first breaks into rhymed prose: *fa-kharajat minhā ṣabiyya – gharrāʾu bahiyya – kaʾannahā ʾl-shamsu ʾl-muḍiyya* 'There emerged from it a girl, resplendent and radiant, as if she were the shining sun.' Then he goes directly into poetry with the introductory formula *kamā qāla ʾl-shāʿir*:

> *ashraqat fī ʾl-dujā fa-lāḥa ʾl-nahāru –*
> *wa-ʾstanārat bi-nūrihā ʾl-asḥāru . . .*
>
> She shone in the darkness, whereupon day lit up –
> and the mornings took their light from her light. etc.

(*khafīf*), 4 lines

ed. Mahdi, trsl. Haddawi: 0
ed. Habicht: 0
ed. Qiṭṭa al-ʿAdawī, pp. 4–5 (enhancement of the description of the beautiful woman).

(2) In the same story, the woman presents her 'philosophy' first in a pithy statement: *al-marʾatu minnā idhā arādat amran lam yaghlibhā shayʾ*, 'A woman of our kind, if she desires something, nothing will defeat her,' only to break into five lines of rather outspoken poetry introduced by *kamā qāla baʿḍuhum*:

> *lā taʾmananna ilā ʾl-nisā---ʾi wa-lā tathiq bi-ʿuhūdihinnah*
> *fa-riḍāʾuhunna wa-sukhṭuhun---na muʿallaqun bi-furūjihinnah*
> *yubdīna wuddan kādhiban – wa-ʾl-ghadru ḥashwu thiyābihinnah*
> *bi-ḥadīthi Yūsufa fa-ʾʿtabir – mutaḥadhdhiran min kaydihinnah*
> *a-wa-mā tarā Iblīsa akh---raja Ādaman min ajlihinnah*

(*kāmil muraffal*) (the *hāʾ al-sakt* is omitted throughout)

> Don't feel safe of women – and don't trust their promises,
> for their contentment and their anger is dependent on their pudenda.
> They show false love, and deceit fills their garments.
> Let the story of Joseph be a lesson, he being wary of their cunning.
> Don't you see how Iblis, due to them, caused Adam to be expelled.

ed. Mahdi, trsl. Haddawi: 0
ed. Habicht: 0
ed. Qiṭṭa: pp. 5–6 (same story, women's wiles)

In this case the poem is followed by a another poem, again introduced by *qāla baʿḍuhum* and in the same vein. This is a not uncommon imitation of *adab* works, where poetic snippets and other short pieces are gathered together to elucidate one particular topic. In the *Nights*, such cases are likely to be the work of copyists rather than that of storytellers. For the latter the first poem would be sufficiently powerful not to need the addition of the other.

(3) In the story of the merchant who inadvertently killed the son of a jinni by throwing date-pits around, the merchant, on the point of being beheaded by the jinni, breaks into seven lines of poetry in the *basīṭ* meter. This is far away from any kind of realism, but it has the uncanny effect of freezing the jinni with his sword-arm in midair and producing an almost unbearable *ritardando*:

> *al-dahru yawmāni dhā amnun wa-dhā ḥadharu –*
> *wa-'l-ʿayshu shaṭrāni dhā ṣafwun wa-dhā kadaru . . .*

(*basīṭ*), 7 lines

> Time is two (kinds of) days: one is safety, the other anxiety;
> and life is two halves: one is serenity, the other distress . . .

ed. Mahdi: p. 73, trsl. Haddawi: p. 18 (the merchant speaks about fate; introductory formula: *wa-anshada yaqūlu shiʿr*)
ed. Habicht: pp. 35–36 (with variants)
ed. Qiṭṭa: 0

It is interesting to note that this poem, which is found in Mahdi and Habicht, is missing in Qiṭṭa. One may suspect that a copyist, or the editor, found the situation too improbable.

(4) In the story of the fisherman and the jinni, the former has three miscatches (the last one being the dead donkey) and each piece of bad luck is commented upon by the fisherman in a poem (of 6, 4, and 7 lines, respectively). In the edition of Qiṭṭa the first and the last poem are reduced to their first lines; obviously the copyist (or the editor) found the poetic fisherman far too loquacious. The fisherman, after the jinni's death threat, again breaks into two lines of poetry, decrying the demon's ingratitude. It is noteworthy that it is only the fisherman who produces poetry; jinnis in the *Nights* are usually not very poetic.

(5) In the Story of King Yunan and the Sage Duban there are are five poems in the Mahdi (and Habicht) edition. They are all spoken by the sage (the last one by his severed head), neither the king nor his evil

vizier produce anything poetic. The first two poems, spoken when the king was not yet ill disposed to him, as he soon would be due to the machinations of his vizier, are introduced with an intriguing gesture: *ashāra ilā 'l-maliki yaqūlu hādhihi 'l-abyāt.*[17] The later poems, spoken in despair, no longer have this. In the ed. Qiṭṭa, there are additional 'commentary' poems by the narrator, or more likely, the copyist, which in the second case again amount to a little anthology on the topic.

(6) In the story of the Enchanted King the five-line poem in the Mahdi ed., in which the king, as yet unseen by the focus person, bemoans his terrible fate, is strangely distorted in the Qiṭṭa ed. The poem is in *kāmil*; nonetheless, in the Qiṭṭa ed. it becomes a *mashṭūr* poem in which the hemistichs rhyme. In addition, the line 1 in the Mahdi ed. is turned into hemistichs 4 and 5, hemistichs 1 to 4 are new, and the rest in the Mahdi ed. is missing. It looks as if the poet in the Qiṭṭa version had subjected the first verse to a *takhmīs*, but then not continued the procedure.

(7) When the focus person finally sees the enchanted king, a beautiful young man, the narrator injects a poem about the description of a handsome youth. This is the first time that the Mahdi (and Habicht) ed. has this device. It is thus not just a matter of a later revision of a copyist, as it seems to be in the Qiṭṭa version.

I will stop here with the description of individual cases. I hope that I have made my point, which is actually more in the form of an imperative: Let us look very closely at the literary junctures, where poetry is injected, because the art of the narrator becomes apparent in small details, which often cannot, or can not yet, be accounted for, but do not seem arbitrary. Let me stay for a while with the type of poem exemplified by the last example: the *tamaththul* poem that is a commentary of the narrator. Such poems occur, but not very frequently as it seems. The Qiṭṭa edition is somewhat better endowed with these, but there they often look like interventions of copyists rather than of storytellers. In any case, we could add to these narrator's recitations also second-level cases, in which narrators of stories in stories use the same technique. What these poems amount to is that they create another level of discourse alongside the narrative (see Bencheikh's formulation above: If the story narrates a

---

[17] This seems to mean 'he gesticulated to the king, saying these lines;' it is, however, also possible that one should read *'alā* rather than *ilā*, i.e., 'he gave advice to the king ...'.

passion, the poem represents it). In a discussion with Jamal Elias about illustrations of literary works, I suggested that the poems we have been talking about might be seen to take the place of illustrations. But then I noted that I had been 'scooped' by John Payne, who, already in the first half of the 19th century, had pointed out the parallel between the poems and Western 'engravings and woodcuts.' Obviously, the change of medium makes the parallel somewhat precarious, but when one considers the fact that both, narrator's poems and illustrations, stop the narration in its tracks and pick out one moment as a still, the parallel becomes much more convincing and powerful. And there is another factor: the poems in the *Nights*, with very few exceptions, follow the classical model of prosody, and many of them simply are, of course, classical poems. This means that prose and poetry in the *Nights* belong to different registers of the language. So, in a way, one does go from one medium to the other.

NARRATIVE AND PERFORMANCE
SHAHRAZĀD'S STORYTELLING AS A RITUAL ACT

Richard van Leeuwen

For many lovers of literature, the *Thousand and one nights* is not merely a collection of stories. The tales of Shahrazād are considered as a literary phenomenon, a model for the technique of storytelling, and even the quintessential representation of the act of narration. Since the appearance of the first French translation of the collection in the beginning of the 18th century, Shahrazād has increasingly grown into the high-priestess of storytelling, the symbol of the power of the word. She has acquired this status mainly as a result of two fundamental characteristics of the work. First, as a character Shahrazād embodies the perpetual nature and the self-generative capacities of storytelling. The framing story of the *Thousand and one nights* sets in motion a seemingly endless stream of stories, which emerge one out of the other and which take all shapes and guises during the course of time. She shows that stories are without limits. They engender their own proliferation and dissemination, symbolized by the survival of Shahrazād against all odds. Second, in the framing story the work relates its own origin and, more generally, the *raison d'être* of the process of narration. It refers to the causes *how* it originated – the occurrence of a devastating catastrophe provoked by the unruly conduct of women and the subsequent threat to the survival of civilization – and it refers to the reason *why* it originated – a disruption of harmony which necessitates a far-reaching transformation, a reconstruction of Shahriyār's vision of reality and the establishment of a new equilibrium between himself and his female 'other'.

The self-referential explication and characterization of the narrative process have elicited a variety of interpretations of the framing story and its impact on the collection as a whole, most of which are derived from psycho-analytical theories. It is clear that Shahriyār has fallen victim to a state of shock caused by the betrayal of his spouse, and that his illness has to be healed to protect the empire from his deranged behaviour. His mental state has been ascribed to his alleged fear of women, to the predominance of his Id over his Ego, and to his reluctance to complete the transition from the 'imaginary' phase of his personality to the 'symbolic' phase. It has been argued that Shahriyār has failed to reconcile the feminine and male components of his character, that he – in his patriarchal obsession – has failed to acknowledge the feminine quality of imagination and therefore resorts to destructive violence. Finally, it has been suggested that Shahriyār represents patriarchal ideology, which refuses to estimate the value of female influence and which ultimately results in stagnation and the ruin of society.[1]

All these approaches have their own degree of plausibility and merit and they show the complex and powerful expressiveness of the story and the figures of Shahriyār and Shahrazād. However, there is one aspect of the story which seems to be neglected in most narratological analyses, and that is that the *Thousand and one nights*, in its self-referential explication, is not a written text, but a *narrated* text, that is, the process of narration is not presented as a textual 'object', but rather as a performative act. The stories are not presented to Shahriyār in written form, as being already finalized and recorded in a definite form; they are told to him orally in the bodily presence of the narrator and come into being before his eyes. What is more, they are linked to the act of love, suggesting that as a speech act storytelling is somehow related to eroticism as a form of bodily performance. This also determines the role of Shahrazād, who is not the author of a collection of stories of her own invention, but a conveyor and a transmitter of stories, a medium through which the stories are reproduced. She is physically involved in this process of transmittance since at any moment Shahriyār can decide to have her executed and bring the chain of stories to an end. Thus, Shahrazād presents herself as the embodiment of the process of narration, including the physical conditions of sexuality and death. Or, conversely, she shows that storytelling as a means of grasping the world intellectually is an embodied experience, with its possibilities and limitations.

It is this performative dimension of the *Thousand and one nights* which suggests a link between Shahrazād's manner of narrating and the idea of ritual. Of course, even a superficial reading of the framing story of the *Thousand and one nights* supports an association of Shahrazād's method with ritual practice. Shahrazād's storytelling has a specific rhythm,

[1] For a discussion of the story of 'Shahriyār and his brother' and its interpretations, see: U. Marzolph & R. van Leeuwen, *The Arabian nights encyclopedia* (Santa Barbara [etc.], 2004).

imposing a temporal pattern which is repeated during thousand and one nights; the storytelling represents a form of initiation of Shahriyār, who has to be divested of his aberrated, obsessive vision of life and be reinvested with a new, harmonious world-view; the period of storytelling evokes the liminal phase of a rite of passage, the outcome of which is still contingent on Shahrazād's performance and Shahriyār's responsiveness. It seems useful, therefore, to examine if the ritual elements in the story of Shahrazād and Shahriyār are confined to a few parallels or reflect the suggestion of a more intimate relationship between storytelling and ritual. In this contribution I will discuss the proposition that the framing story of the *Thousand and one nights* can be read as a form of ritual practice and that this ritual aspect even provides one of its main explanatory paradigms. After a brief outline of some recent approaches to the concept of ritual, I will proceed with an analysis of the story itself, concentrating on three questions: in what ways can the storytelling of Shahrazād be considered an implementation of forms of ritual practice? What are the consequences of this association of storytelling and ritual for the idea of narrativity? And what does this association reveal about the concept of ritual as it is used by the compiler of the collection? But before we can turn to a discussion of the story, the notion and functions of ritual have to be outlined.

Ritual and transformation

Ritual practices, often associated primarily with the manifestations of religion, are generally situated on the interface between individual and society and, as a psychological phenomenon, on the interface between body and mind. It is especially the perception of the nature of the relationship between body and mind that determines the way in which ritual is defined as a philosophical concept. The Cartesian dualism between a subjective mind and a body which it controls but from which it is essentially separated, engenders a vision of ritual as a form of social and cultural inscription of the body by forms of discipline and training. In as far as ritual has a function in the process of acquiring knowledge, it is through the impregnation of this knowledge by imposing behavioral patterns on the body. Ritual can be seen as a form of text, containing semiotic codes which refer to a knowledge that exists separately, in the form of a coherent system of signs and representations. By participating in ritual practices, these codes are inscribed on the body by the performance of various kinds of prescribed, schematic and repeated acts.[2]

This approach to ritual is elaborated especially by Michel Foucault and Pierre Bourdieu, who considered ritual behaviour as part of systems of power which 'produce' subjects with specific attitudes, habits and world-views. For Foucault, ritual is part of the technologies of the body, disciplining the body of the participants with the aim of transforming a subject according to some ideological or moral ideal. Ritual, as a form of discipline, is a means to produce a moral disposition which conforms to a specific notion of ideal man or woman, first of all by training and controlling the body. This refers not exclusively to religious practice, but includes social practices which perform the same function as rituals and which aim to mould individuals into some ideal social type, which is related to a broader context of power relations.[3]

The Foucauldian concept comes close to the notion of ritual as it is developed in phenomenological approaches. Here, the Cartesian dualism is rejected in favour of an integrative view of the relationship of body and mind. It is argued that the body is our only medium to 'have' a world and to 'be' in the world. This implies that all experiences are first of all embodied experiences and that bodily experiences are the basis for all forms of cognition. Knowledge is acquired by practice, and it is through bodily activity that experiences are structured and transformed into pre-reflective and reflective forms of understanding. Knowledge is not something that exists separated from the body; it is 'necessarily embodied, intersubjective, and active'.[4] Only through the body can our experiences be perceived, ordered, and related to each other.

In this view, ritual is more than a technology to inscribe encoded meanings on subjects. Rituals are seen as an integrated process in which thought, body and emotions are simultaneously mobilized to realize a specific transformation. Rituals are not text-like semiotic systems, but a form of creative practice resulting in understanding and change.

---

[2] For an analysis of the nature and functions of ritual, see C. Bell, *Ritual. Perspectives and dimensions* (New York [etc.], 1997); for this article I especially refer to the philosophical approaches collected in K. Schilbrack (ed.), *Thinking through rituals. Philosophical perspectives* (New York [etc.], 2004); Schilbrack, p. 2; N. Crossley, 'Ritual, body-technique, and (inter)subjectivity,' in Schilbrack, pp. 34–35.

[3] Schilbrack, p. 13; A. Hollywood, 'Practice, belief, and feminist philosophy of religion,' in Schilbrack, pp. 60–61; L. McWorther, 'Rites of passing; Foucault on power, and same-sex commitment ceremonies,' in Schilbrack, pp. 76–77, 80–81.

[4] Schilbrack, pp. 9, 13; Crossley, pp. 34–35.

Rituals teach the participants to embody schemata of perception, which makes the world intelligible and constitutes it as a 'meaningful context for action'. Rituals provide a model for the ways to behave and to act, and how to rationalize these ways, thereby providing an essential mechanism for linking individual experience to collective experiences, and for the consolidation and reproduction of socially accepted world-views and structures of power. It is through rituals that the world is constituted as an object of experience, including relationships to 'others', both in individual and in social contexts; it is through rituals that the subject can develop patterns of behaviour and actions, projecting his or her acquired knowledge into an 'ordered' future.[5]

The key-word in both perceptions of ritual is 'transformation'. Apparently, rituals are perceived to contain a component of intentionality aimed at the transformation of the subject through a structured experience. Rituals provide a framework constructed to suspend the laws of time and space, to substitute a vision of reality for what is perceived as reality itself. Rituals imply a suspension of disbelief, imposing a ritual world, aiming to transform the subject's understanding of reality. Moreover, this ritual world is evoked in cooperation with others, thus creating a system of relationships which define consciousness of the existence and nature of others, enabling the subject to develop an intersubjective notion of 'self' and 'other' within an ordered whole. Thus, through ritual we can achieve an intersubjective equilibrium and open up the possibility to reconcile the individual with the collective imagination. To participate in ritual is to learn to comprehend forms of behaviour and the 'semiotic codes which constitute the meaning of acts'.[6] To conclude, ritual can be defined as 'a set of performative actions through which will, desire, intellect, and mind are transformed and reconstituted' enabling the subject to achieve a form of self-transcendence and to internalize reality in such a way that it can be interpreted as in accord with its own aims.[7]

After this brief description of the phenomenon of ritual, we can now proceed to an analysis of the ways in which the main elements and functions of ritual, as summarized above, can be traced in the framing story

---

[5] Schilbrack, pp. 11, 14; Crossley, pp. 36–39, 43, 46; K. Schilbrack, 'Ritual metaphysics,' in Schilbrack, p. 141.

[6] Crossley, p. 38.

[7] Schilbrack, p. 6; Hollywood, p. 61; M.L. Raposa, 'Ritual inquiry; the pragmatic logic of religious practice,' in Schilbrack, p. 121; Schilbrack, 'Ritual metaphysics,' in Schilbrack, p. 141.

of the *Thousand and one nights*. Most of these elements and functions have been referred to in the previous paragraphs, but some new aspects will come up, all based on the collection of studies edited by Kevin Schilbrack.

Narration as a ritual act

As has been observed above, even a superficial analysis of the story of Shahriyār and Shahrazād will justify the association of the narrative structure of the *Thousand and one nights* with a ritual act. First, the stories are narrated within a very tight temporal framework, which is based on repetition, the creation of tension through plot-deferral, and the natural alternation of night and day. Second, the narration is a performive act, in which the narrative component is linked to a pattern of embodied practices. Storytelling is a performance, which aims to change Shahriyār's behaviour, not only by the power of speech, but also through the physical presence of Shahrazād and the threat to her life. Finally, the combination of storytelling and performance is intended as a process of initiation, a structural effort to teach Shahriyār to behave properly and fit the lesson into a broader view of life.

Shahrazād's method of storytelling is first of all conceived to attract and hold Shahriyār's attention, taking him out of his everyday activities and subjecting him to a different regime, which is gradually constructed by Shahrazād. The technique of interruption is used to create a suspension which will ensure Shahriyār's continuing focus on Shahrazād as a person and as the source of a performative schema, undoing the strict schedule of sexuality and death which used to govern Shahriyār's life. It is no coincidence that the storytelling occurs at night, of course, the time in which not only normal temporal and spatial laws, connected with Shahriyār's official authority as a king, are suspended, but which also symbolizes the realm of chaos, fantasy, and unexpected happenings.[8] The combination of a narrative strategy and the nocturnal ambience simultaneously engenders attention and a suspension of disbelief in which Shahrazād is allowed to construct her narrative world which slowly gains the status of an alternative reality. The night is the domain

[8] See: R. van Leeuwen, *The Thousand and one nights. Space, travel and transformation* (London, 2007).

of the imagination, and through the narrative strategy Shahriyār's mind is opened for the influence of an imagined representation of the world, without immediately relapsing into feelings of betrayal and mistrust.

It should be noted that the stories that are told by Shahrazād are not her own invention. The narrator delves them from an existing reservoir of stories which in some ways represent the wisdom of mankind and thus have a certain educational authority. Thus, Shahrazād presents her storytelling not as an individual 'act' in order to personally convince Shahriyār to keep her alive, but rather as the enactment of a broader set of rules, examples and lessons which transcends their specific situation. Because of the adultery of his spouse, for Shahriyār the night has become a time of uncertainty, disorder and disruptive forces, which could only be tamed by the use of force. Shahrazād now replaces this impulse by an imagined world, a constructed vision, suggesting a sense of order and stability in spite of the existence of unexpected events, evil figures and the vicissitude of fate. On the one hand, therefore, the regular 'laws' of time and space are suspended, but simultaneously a new spatiotemporal order is imposed, based on the different 'functions' of night and day, and substituting a new cumulative notion of repetition for a circular, undynamic, form of repetition which leads to destruction.

The link between the imagined reality evoked by Shahrazād and Shahriyār's real life is provided by the person of Shahrazād, who is the living proof that the narrated stories have a certain reality, or at least an influence on reality. They are capable of intervening in the course of events and are part of the contingency of life. Shahrazād is the pledge which makes the 'suspension of disbelief' possible, since she is prepared to die if the 'spell' is broken. Thus, her physical presence is the argument that an alternative order exists, that it is valuable, and that it is relevant for real life. She puts her life in jeopardy to convince Shahriyār that the world which she evokes is not a mere fantasy, but refers to a set of 'truths' which has until then not been uncovered by him. But of course, her proof is built on fiction: first she creates her own framework of time and ambience, then she constructs her imagined world within this framework, as an alternative reality, and finally she has Shahriyār believe that in fact this alternative reality is part of an already existing reality that transcends their situation. It is the combination of narrative strategy and physical presence which prevents Shahriyār from seeing through Shahrazād's ruse.

The embodied nature of Shahrazād's storytelling is strengthened by its link to sexuality. Storytelling and the sexual act, as a source of pleasure and a means for procreation, are presented as supplemental to each other, each being embedded in the other and contingent on the other. Stories, too, are a source of pleasure and a means of procreation; it is through the pleasure that they give that they can overcome temporal interruptions and generate new stories to infinity. Like sexuality, storytelling is part of the realm of desire; in fact, it is suggested that it is a drive as fundamental as the sexual drive. And like sexuality, storytelling is essentially anchored in the body, in Shahriyār's body as the target and source of desire and the domain of satisfaction, and Shahrazād's body as a source of satisfaction and the target of desire. When Shahrazād's body is eliminated, the prospect of fulfilling the combined narrative/sexual drive will be destroyed.

Still, Shahrazād suggests that there is a difference between sexual and narrative desire. Whereas the first provides a means for immediate gratification, the second is a means to postpone gratification. Of course, this idea is carefully and ingeniously exploited in the story, with the labyrinthine technique of plot-deferment, the use of interruptions and embedded stories, etcetera. Storytelling is presented as one of the primary means of the management of desire, that is, the strategies used to prevent subjects from threatening the social order by striving for an immediate fulfilment of their desires and impulses. Storytelling provides protection against irrational and rash behaviour and responses, it regulates the flow of emotions so as to evade their destructive ingredients. In order to preserve their creative and reproductive capacities, desire and emotions have to be distributed over disruptive interventions, otherwise their potential for fulfilment will be destroyed.

Shahrazād, then, teaches Shahriyār that it is not the gratification of desire that is important, but rather the deferral of gratification and the perpetuation of desire. The existence/preservation of desire is the essence of life, while the gratification of desire implies a possible death or ending. Of course Shahriyār's management of desire is inseparably linked to his authority as a man and as a king. It was the disruption of his desire-gratification pattern which caused his mental crisis, undermining his personality, his self-image, and his powers as a husband and a king. In response he used these powers to restore a certain self-image by imposing a regime aimed at the permanent and immediate fulfilment of desire, which, however, gradually leads to the destruction of the reproductive capacities of society. By her storytelling Shahrazād shows him that there is no contradiction between preserving one's authority and postponing the fulfilment of desire, that in fact a sensible postponement of satisfaction opens up the possibility of reproducing authority for eternity, a possibility which is vested in the continuity of the dialectic between desire and gratification, not by the disruption of desire by gratification and its subsequent 'reinvention'. Authority is established and preserved by strategies to overcome disruptions of the management of desire by adopting alternative strategies to secure the continuation of desire.

It is clear that by her storytelling Shahrazād wants to teach Shahriyār something, that she wants to transform him, and in this transformation lies the key to the ritual process. Her narration is a deliberate construction of a process of initiation, of cognitive learning, of transition. To achieve this, she creates her own referential framework, which installs a period of suspension of the governing 'regime', and a phase of liminality for them both. After all, the outcome of the transformation is uncertain for Shahrazād, for if Shahriyār fails to respond to her efforts, she will die. While it seems that Shahrazād finds herself in a liminal position, Shahriyār, unconsciously, is in a liminal position, too, for as long as Shahrazād's strategy works. The cognitive process initiated by Shahrazād consists of the construction of a virtual labyrinth of stories which, as suggested above, represents an imagined, alternative, reality.

The juxtaposition of imagination and reality, symbolized by the alternation of day and night, is the basis of Shahrazād's method. She presents a dialectic between two realms, one in which Shahriyār fulfills the roles of kingship and one in which he is part of a world filled with all kinds of people, a narrated world, full of unexpected events and wonders. It is an alternative world which contains a systematic reservoir of examples, a system of coded signs, which shows Shahriyār how to behave in specific situations and how to respond to unexpected turns of fate. It is suggested that this system of signs and meanings, connected with people and events, is part of a broader vision of the world, into which Shahriyār is gradually introduced by listening to the chain of stories. The technique of interrupted storytelling is intended to hold his attention, but also to use repetition as a means to stimulate the formation of a habit, which enables Shahriyār to physically and mentally internalize what he has learned. By acquiring this habit he will slowly become the person that Shahrazād envisages him to be.

Apart from the exemplary nature of the events related in Shahrazād's stories, Shahriyār is also confronted with a wide variety of persons and characters. Shahrazād not only shows Shahriyār a mirror image of himself, she also shows him that human beings come in all kinds and colours. As a result of the betrayal of his wife, Shahriyār's relationships with others –

especially women – have been disturbed. He deals with others as subjects who are indistinguishable and completely at his disposal. He has banned all personal commitment to other persons and all personal attachments. Shahrazād teaches him that humankind is endlessly differentiated, that every individual has his own character, and that everyone responds to events in his own personal way. One can say that she – as a parallel to the stories of Hārūn al-Rashīd's and Jaʿfar's nightly escapades – takes Shahriyār by the hand and leads him in disguise on a nightly journey to learn about the situation, the lives, and the problems of his subjects. She strips him of his official role to force him to interact with his subjects in their domain, and to show him how to intervene in real life. As in the stories of Hārūn al-Rashīd, this can only be achieved by the detour of a make-believe, a fantasy, a deceit. In this way he is gradually able to reconcile the royal and human components in himself.

The ritual of storytelling, then, is meant to restore Shahriyār's relationship with others, to reconstruct his vision of himself *vis-à-vis* others and to re-establish interhuman links of affection and comprehension. By exploiting the links between eroticism and storytelling, it is especially Shahriyār's relationship with women which is the main core of his initiation and transformation, creating a new balance in his personality between the female and male components. But this core is linked to a broader network of relationships, smuggling the 'other' into Shahriyār's closed domain and giving him insight in their private worlds. Shahrazād offers him a panorama of a collective imagination, which governs the lives of people and human relationships and which is perhaps part of a world of fantasy, but which still determines the behaviour of people, their world-view and their vision of others. Thus, Shahrazād's stories do not only present Shahriyār with a model for proper behaviour, but also produce a new intersubjective equilibrium, which is based on a definition, knowing, and acknowledgement of the 'other'. Shahrazād replaces his obsessed world-view dominated by a distorted vision of himself as a man and as a king, by a multiple perspective in which normal human relationships are possible.

After the storytelling has ended, Shahriyār has been initiated into a new world, he has internalized a new self-image, with its physical and psychical components, and he is transformed into a new person with a new relation to the outer world. This is all achieved by the utilization of storytelling as an instrument of cognition by turning it into a ritual act. The basic method implemented by Shahrazād – the focusing of attention, the construction of a fictional framework which substitutes and interacts with reality, the management of desire, the evocation of forms of intersubjectivity – all are essential components of ritual.

The battle of rituals

Although no authentic ending of the *Thousand and one nights* has been preserved, at least to our knowledge, we can safely assume that in the end everyone lives happily ever after. More mysterious, perhaps, is the beginning of the cycle, although it is available in a more or less standardized version. The story indicates adultery as the mobilizer of the chain of stories, but it remains uncertain how we should judge the *status quo ante*, the situation which induced the wives of both Shahriyār and his brother Shahzamān to commit adultery. Some questions remain unanswered. If Shahrazād initiated Shahriyār into the world of eroticism and the 'other,' for instance, does that mean that the kings did not have a harmonious sexual relationship with their wives? Shahrazād healed the anomalous situation that the king would be without offspring as a result of his obsession, but why didn't Shahriyār beget any offspring with his adulterous wife? Had he been impotent or unaware of the sexual act? It is clear that the transformation of Shahriyār by Shahrazād's storytelling is not the restoration of a previous lost harmony, but the formation of a new harmony, which Shahriyār had never known. This may also explain why Shahriyār reacted in such a compulsive manner to the deceit of his wife: he had never known the implications of a sexual relationship, perhaps only seeing it as a part of the exertion of his authority. But even then the question remains why the marriage did not produce any successors.

It is clear that the fatal regime imposed by Shahriyār after the discovery of his wife's infidelity is caused by a severe mental trauma, which, according to Shahriyār, could only be countered by a strong reassertion of his authority. This regime can be seen as a ritual, too, since rituals are often explicitly intended to regulate the disruptive effects of traumatic experiences, such as death and disasters. Traumas threaten to throw the world back into its initial chaos, to destroy the order of life, and to disrupt the harmony between natural and civilizational forces. It is one of the functions of rituals to restore a sense of order, restore the awareness that human society is superior to chaos, and restore the balance between individual and collective experiences. In this sense Shahriyār's response is a kind of ritual, since it aims to impose a pattern on a situation which seems utterly chaotic to the king and which he associates with the essentially disruptive nature of women. It involves a pattern of repetition, of physical acts and roles, of the suspension of temporal structures, and of the buttressing of a specific view of the world and society. However, there is one component that is absent from Shahriyār's 'ritual', and that is the cognitive function of ritual and the purpose to change a person or a situation through a ritualized cognitive process. This element of ritual is eliminated in favour of the means by which the ritual is imposed: violence.

If Shahriyār's regime is understood as a form of ritual, the framing story of the *Thousand and one nights* can be seen as a battle between two forms of ritual, one dominated by violence, the other dominated by cognition and transformation. It is clear from the outset that Shahriyār's ritual can only lead to stagnation and catastrophe, since it remains imprisoned in a cycle of ritual time, without the possibility to restore the process of reproduction and continuation. Shahriyar's ritual is closed off from the social world and its spatiotemporal structures, and it is confined to its own 'invented' temporal pattern based on repetition and the perpetuation of a specific series of acts. The intention of the installation of the ritual is to exorcise an emotional trauma by removing what seems to be the cause of it and reinstating a semblance of a previous *status quo*. The women of the empire are reduced to their role as sexual objects and are subsequently eliminated. They are thus charged with the blame for Shahriyār's negative experience and are treated as scapegoats carrying the guilt and punishment of his wife. Ultimately, they are 'sacrificed' with the aim of maintaining a specific social order and structure of authority and above all to save Shahriyār's integrity as a ruler and as a man.

This new regime, intended to buttress the power of Shahriyār, is challenged by the ritual constructed by Shahrazād. The first question to ask, of course, is why Shahrazād is in need of a ritual to try and reform Shahriyār's conduct. Some of the answers have already been suggested, such as the function of ritual as a form of mental and physical training aimed at the internalization of a specific habit or attitude. Thus, the ritual installed by Shahriyār can only be transformed by a counter-ritual, which substitutes one life pattern for another. Another possible answer is related to the question of power and authority which Shahrazād hopes to address. Shahriyār's ritual is explicitly constructed to secure his absolute power over his subjects and, especially, over women. As a woman, it is therefore impossible for Shahrazād to negotiate her relationship with Shahriyār with verbal means. There are only two tracks open to her to approach Shahriyār and have him change his ways: the first is sexuality, which is considered her essential quality; the second is the construction of a counter-reality, or an un-reality, which can only be imposed on Shahriyār in an indirect way, by a ritualized cognitive process. Thus, a ritual in which sexuality is incorporated is the only strategic opportunity at Shahrazād's disposal.

The idea of a 'battle' between rituals is supported by repeated references to binary oppositions throughout the story. The story is in fact based on a set of dualisms, between night and day, between male and female, between imagination and reality. To these Shahrazād adds the dualism between a ritual based on violence and a ritual based on the capacities of the mind. Shahrazād shows the king that whereas he has come to hate women because of their irrational, destructive behaviour, it is in fact his violent ritual that is irrational and catastrophic, and that in the end is unable to guarantee his authority. The preservation of order and authority can only be achieved when violence is banned in favour of a dynamic dialectic between the imaginative mind and reality. And this dialectic can only be realized by preserving a balance and a harmonious interaction between female and male components of the self, in relationship with the other, and, ultimately, in society. Only then will the empire retrieve its dynamism and its procreative power.

Of course it is suggested in the story that the confrontation between the two rituals is more than just an effort by Shahrazād to cure Shahriyār from his psychical aberration. Shahriyār's condition is linked to the fate of the empire, of civilization. Shahriyār's ritual can be compared with other, similar, rituals which occur mostly in stories of conversion, such as the pre-Islamic annual ritual of throwing young maidens in the Nile to assure a prosperous flood, which was according to legend abolished by ʿAmr b. al-ʿĀṣ when he conquered Egypt for Islam. Or the story of the virgins sacrificed annually to a sea-monster mentioned by Ibn Baṭṭūṭa, a custom abolished when the people converted to Islam. In the story of Shahriyār and Shahrazād, too, young women fall prey to a barbaric custom based on violence and fear, and the population is redeemed by a new form of civilization, which is not based on violence, but on a constructed, total vision of life and the world. The story is the account of Shahriyār's 'conversion,' his transformation from a monstrous, violent barbarian into a civilized, well-behaved husband. Violence is replaced by a world-view based on a symbolic system, which is not only a superior way of governing human behaviour, but which in fact is the only way to secure the survival of a social community.

Conclusions

After this discussion of the ritual elements in the story of Shahriyār and Shahrazād we can now turn to the more general questions that we posed ourselves at the outset, about the relation between ritual and narrativity. First, it is clear that the story contains many elements that are considered essential components of ritual, both in form and in content. Bodily practice, time-patterns, cognition, the management of desire, etcetera, are basic constituent elements of ritual and of Shahrazād's form of narration. Second, the association of storytelling and ritual suggests that narrativity fulfills an essential role in the ordering of ritual and its functions. The constructing of a ritual world and providing it with meanings is opposed to rituals based on violence, which depletes the ritual of all forms of meaning, and which turns ritual into a form of non-ritual. A ritual is part of a symbolic system, which cannot be built on a foundation of violence, whose essence is in fact destroyed by an act of violence. Third, ritual practice is a construction, an imaginary reality which, through ritual practice, is put into a dynamic interaction with reality. It is derived from a transcending world and 'mobilized' to shape real life and human relationships. Ritual is essential for constituting the world as an object of experience, of defining relationships with others, for defining the relationship between individuals and society, for creating spatiotemporal structures, and for upholding a collective imagination. But most of all, rituals are indispensable for producing transformation and change, making the preservation of cognitive processes and of forms of authority within social structures possible.

ORIENTALISM À LA PARISIENNE

DR MARDRUS, KEES VAN DONGEN AND THE *THOUSAND AND ONE NIGHTS*[1]

Arnoud Vrolijk

At the close of the nineteenth century the Orient was by no means an unfamiliar theme to the French public, spoiled as it was by the sweltering harem scenes of writers like Pierre Loti and the works of painters like Jean-Léon Gérôme, where the sultans, odalisques and eunuchs represented a world full of sensuality without the inhibitions of *petit bourgeois* Christianity. The World Exhibitions in Paris of 1889 and 1900, with their reconstructions of souks and Oriental cafés, also contributed to bringing the exotic world of France's colonial empire in North Africa closer to the public.[2] Of the literature of these uninhibited natives only very little was available. An exception was the translation that Jean-Antoine Galland (1646–1715) had made of the *Thousand and one nights*, but this work, published almost two centuries previously, had come to be regarded as outdated. It was also known that Galland had respected the moral reticence of his day by omitting all passages that might be regarded as obscene. In 1899 the left-wing, avant-garde journal *La Revue blanche* started with the publication of a brand new, complete, literal and uncensored translation, based on the Arabic sources: *Le livre des mille nuits et une nuit*. The *Revue blanche* went bankrupt in 1903, but another publisher, René Fasquelle, took over the translation project and brought it to a successful conclusion.[3] The reaction of the public was overwhelming. At last they had a key text at their disposal in which the

---

[1] This is a revised, enlarged and translated version of an earlier article in Dutch, 'Oriëntalisme à la Parisienne. Dr. Mardrus, de *Duizend-en-één-Nacht* en Kees van Dongen,' in J. Schaeps [et al.] (eds.), *Oostersche weelde. De Oriënt in westerse kunst en cultuur, met een keuze uit de verzamelingen van de Leidse Universiteitsbibliotheek* (Leiden, 2005), pp. 181–192.

[2] P. Blanchard [et al.], *Le Paris arabe* (Paris, 2003).

[3] The first eleven volumes were published by the *Revue blanche*, the remaining five by Fasquelle. See S. Larzul, *Les traductions françaises des Mille et une nuits* (Paris, 1996), p. 142. See also pp. 139–216 for a carefully balanced study of the merits of this translation.

Arab displayed his own world in all its untainted splendour. An author like André Gide declared unreservedly:

> Dans les Mille Nuits et une Nuit, comme dans la Bible, un monde, un peuple entier s'expose et se révèle; le récit n'a plus rien de personnellement littéraire [...]. Le récit est de la voix même du peuple; c'est son livre, et c'est tous ses livres, sa littérature, sa Somme; il n'a produit rien d'autre que cela.[4]

After the publication of the first volume another critic, the novelist Henry Bordeaux, took the measure of Oriental society: the women are on the whole libidinous and given to lying, 'lubriques et menteuses.' They are subjected to the cruelty of their men, but their very sensuality makes them creative in avoiding the consequences of this cruelty. The men, one should know, are not only cruel, but also cowardly and gullible. Orientals are unabashed gourmets who are always in search of perfumed delicacies. Because of this fixation on sensual pleasure, Oriental society is tired and insipid and consequently also very sad.[5]

Dr Mardrus

The translator who provoked these sweeping statements was Joseph-Charles Mardrus (1868–1949), an Oriental Catholic of French nationality. He was born in Cairo and studied medicine at the Université Saint-Joseph in Beirut. In 1892 he left for Paris to continue his education and in 1895 he concluded his studies with a thesis on urology.[6] Dr Mardrus worked for four years as a doctor on board the ships of the Messageries Maritimes steamline company and during that period he prepared his translation of the *Thousand and one nights*. He also befriended writers like Stéphane Mallarmé and Paul Valéry. In 1899, shortly after the publication of the first volume of his translation, he established himself permanently in Paris. He frequented the literary *salons* and in 1900 he married the lesbian poet Lucie Delarue.[7] In 1899 the public was

---

[4] A. Gide, 'Dr. J.C. Mardrus. Le livre des Mille Nuits et une Nuit. Tome IV, traduction littérale et complète du texte arabe,' in *Prétextes* (Paris, 1903), p. 209.

[5] H. Bordeaux, 'Les livres et les moeurs. Les Mille et une Nuit,' *Revue hebdomadaire* 8 (1899), pp. 271–288.

[6] D. Paulvé & M. Chesnais, *Les Mille et une Nuits et les enchantements du docteur Mardrus* (Paris, 2004), pp. 39–43. Marion Chesnais is the daughter of the sister of Mardrus's second wife and the current keeper of his papers.

[7] L. Delarue-Mardrus, *Mes mémoires* (Paris, 1938), p. 114.

made familiar with his portrait through a woodcut by the Swiss artist Félix Vallotton in the *Revue blanche*,[8] and ten years later André Rouveyre portrayed him in the *Mercure de France* (Plate 5). Both portraits show him as a bespectacled little dandy with jet-black hair and a moustache. His contemporaries claimed he used to improve his features with rouge and kohl.[9]

As an immigrant who rose to fame only after he had started publishing his translation, his background remained obscure, even to his wife. An admiring biographer, Émile-François Julia, described him as an aristocrat hailing from Mingrelia, a region in the Western Caucasus which is now part of Georgia. His grandfather, a tribal leader named 'Ovannès' (Hovhannes?) according to Mardrus,[10] allegedly battled side to side with Imam Shamil, a Daghestani Muslim rebel leader who kept the Tsar's armies at bay between 1834 and 1859. After Shamil's defeat his grandfather supposedly emigrated to Egypt.[11] The veracity of this story, most probably inspired by the lively accounts of the 'enchanter' Mardrus, is doubtful at best. In the first place it is known that Mardrus's grandfather was already living in Cairo when his son Fathallah was born in 1827.[12] Besides, there is also the chronological incongruence of the fact that Mingrelia, a Christian country, had accepted the sovereignty of the Tsar as early as 1803 and it is at least remarkable that decades later a Christian Mingrelian would still be waging war against the Russians in another part of the Caucasus in close collaboration with an Islamic insurgent calling for jihad. It is equally remarkable that Mardrus visited the Caucasus in 1897 without even mentioning his family ties in his correspondence.[13] In later years he reminisced fondly about his forebears living on 'the southern slopes of the Caucasus, where Prometheus once had suffered.'[14]

Although the name Mardrus or variants like Mardros or Mardiros, from the Greek *martyros*, do indeed occur in Mingrelia and Georgia, it

---

[8] *Revue blanche* 19 (mai–août 1899), p. 16.

[9] Ph. Dumaine, 'Souvenirs sur le docteur Mardrus,' *Points et contrepoints* 62–63 (1962), p. 45.

[10] Paulvé [et al.], p. 35.

[11] E.F. Julia, *Les mille et une nuits et l'enchanteur Mardrus* (Paris, 1935), pp. 46–54.

[12] Paulvé [et al.], p. 35.

[13] See for instance his letter of 21 November 1897 to Alfred Vallette, published in the *Mercure de France* 999–1000 (1946), pp. 230–231, in which he refers only to 'a little hike in the Caucasus'.

[14] Dumaine, p. 47.

is also very much an Armenian name. More likely, Mardrus belonged to the Armenian minority in Egypt, not quite as romantic as a Mingrelian nobleman.[15]

Especially in those days it was unusual for Armenians in Egypt to be proficient in the written Arabic language. At home they spoke Armenian and the well-to-do, like for instance Mardrus's parents, sent their children to European mission schools. In their daily contacts with the Arabic-speaking majority they picked up a basic knowledge of colloquial Arabic, an effective, but by no means elegant or exhaustive education. Mardrus himself attributed his knowledge of Arabic and Arab folktales to his nurse, a Muslim woman called Aisha.

Since Mardrus had hardly received any formal training in written Arabic, many European and American Arabists reacted with mixed feelings to the translation of this outsider. In 1900 the French Orientalist Maurice Gaudefroy-Demombynes published a review in which he demonstrated that Mardrus had failed to grasp the *finesses* of the Arabic language. He reproached him for the coarseness and moral impropriety of his translation and for the fact that he had omitted or added passages of his own accord. Gaudefroy-Demombynes declined to believe that this was indeed a literal and complete translation of the text as the title-page suggested. An acrimonious debate ensued in which Mardrus changed his story several times. Among other things he argued that his translation was intended for literati and artists ('lettrés et artistes') who felt no need for the dead weight of footnotes. Later on, he promised, he would publish a sizeable tome with documents, studies, references and reviews, enough to satisfy the demanding tastes of the most exacting scholar. However, such a volume never appeared.[16] In 1905, after the last volume had appeared, the Belgian scholar Victor Chauvin informed the scholarly public about the texts that Mardrus had assembled together and their provenance.[17] As a consequence, no love was lost between Mardrus

[15] A contemporary Arab scholar like I. Cattan took it for granted that Mardrus was of Armenian descent, see 'Une traduction dite "littérale". Le livre des mille nuits et une nuit par le docteur J.-C. Mardrus,' *Revue tunisienne* (1906), p. 17. Paulvé [et al.], p. 35, still support the theory about Mardrus's 'Caucasian' background.

[16] See the review and exchange of letters between Maurice Gaudefroy-Demombynes and Mardrus in *Revue critique d'histoire et de littératures* 49 (1900), pp. 401–406, 514–517; 50 (1900), p. 321.

[17] V. Chauvin, *Bibliographie des ouvrages arabes ou relatifs aux Arabes publiés dans l'Europe chrétienne de 1810 à 1885* (Liége, 1892–1922), IV, pp. 108–109, 159–160; VII, pp. 95–96 n., IX, pp. 77–81.

and the Arabists, although an individual like Hartwig Derenbourg, professor at the École des Langues Orientales at Paris, described Mardrus's translation rather mystifyingly as 'une belle traduction [...], une adaptation qui est aussi française qu'arabe, aussi arabe que française'.[18] In exchange, Mardrus dedicated one of his volumes to Derenbourg.

Research has revealed that a good part of the erotic contents and the descriptions of extravagant delicacies had their origin in Mardrus's fertile mind.[19] The reading public had a clear image of the 'Oriental' and Mardrus offered his readership what it wanted. From the viewpoint of customer satisfaction this may have been the right choice, but it does not make a reliable translation. It must be admitted, however, that Mardrus did nothing but follow in the footsteps of generations of Arab storytellers, who had always readily changed or adapted their stories in order to accommodate their audience.[20] Keeping this in mind, one can leisurely enjoy this exuberant, in our eyes almost grotesquely erotic recasting of the Arabic text.

The success drove Mardrus to even more achievements. Greatly assisted by his imagination he went on to make a translation of the Qur'an at the request of the French Ministry of Public Education (*Le Koran qui est la guidance et le différenciateur*), and he translated the Song of Songs from Hebrew (*Le Cantique des cantiques*), the Apocalypse of St John the Divine from Syriac (unpublished) and parts of the Book of the Dead from Ancient Egyptian (*Toute-puissance de l'adepte*).

Mardrus's 'mystification of the *Thousand and one nights*,' as the Dutch translator Richard van Leeuwen calls it, has had a lasting influence on the Western image of the Orient. Until the 1960s it was the only 'complete' modern French translation,[21] and in a country like the Netherlands, where the first complete translation, by Richard van Leeuwen, was published only from 1993, the reading public had to make do with retranslations, primarily from Mardrus's French version. The

---

[18] H. Derenbourg, '[Nouvelles littéraires],' *Journal des savants* 1902, p. 399.

[19] Cattan, pp. 16–23; M.I. Gerhardt, *The art of story-telling. A literary study of the Thousand and one nights* (Leiden, 1963), pp. 93–104; S. Larzul, 'L'exotisme fantasmatique dans *Les mille nuits et une nuit* de Mardrus,' *Revue de littérature comparée* 269 (1994), pp. 39–46.

[20] I owe this latter observation to Remke Kruk (personal communication over toasted Turkish bread with feta and fresh basil and several pots of coffee).

[21] The first complete French translation after Mardrus is that of René Khawam, *Les mille et une nuits. Traduction nouvelle et complète faite directement sur les manuscrits* (Paris, 1965–1967).

most famous among these was published by the novelist Abert Helman during and after the Nazi occupation, with illustrations by Anton Pieck (Amsterdam, 1943–1954). Selections from Helman's retranslation were published as late as 1999 under titles like *Liefdeslessen* ('Love lessons') or *Vrouwenlisten* ('Women's wiles'), again stressing Mardrus's peculiar outlook on the text.

Kees van Dongen

The success of Mardrus's *Livre des mille nuits et une nuit* triggered a hype in art and fashion that would last for years. Sergei Diaghilev's *Ballets russes* adopted the theme of the *Thousand and one nights* for their ballet *Schéhérazade*, in which Nikolai Rimsky-Korsakov's 1890 composition on the travels of Sindbad the Sailor was used to illustrate the adultery scene that caused king Shahriyār to take revenge on all his future brides. The opening night in Paris on 4 June 1910, with Ida Rubinstein and Vaclav Nijinsky in the leading roles, had an overwhelming effect. Paul Poiret, the Parisian couturier, took advantage of the moment to introduce a new 'Schéhérazade' look in ladies' fashion, with Oriental textiles and turbans with peacock feathers. Poiret's designs were exclusive and expensive, but he secured the custom of the entire Parisian high society by organising extravagant parties, where everybody who was somebody wanted to be seen, preferably in one of his creations. In June 1911, upon the introduction of his new Oriental fashion line, he threw a lavish 'Schéhérazade' party, called the 'Thousand and Second Night,' where dark, bare-chested 'slaves' served food and drinks, while Poiret and his wife presided as Sultan and Sultana.[22]

It is not clear whether Mardrus himself was present, but it is certain that he wrote the text for the invitation, designed by Raoul Dufy. His wife Lucie Delarue-Mardrus was one of the guests and she reported on the event in *Femina*, the well-known ladies' journal, but she did not mention the presence of her husband.[23] One of the people who did attend

[22] P. White, Poiret (New York, 1973), pp. 86–90.

[23] L. Delarue-Mardrus, 'La mille et deuxième nuit chez le grand couturier,' *Femina* 253 (1 août 1911), p. 415. Paulvé [et al.], pp. 78–79, suggest that Mardrus was present. However, neither Mardrus's biographer Julia, nor his wife Delarue-Mardrus in her autobiography *Mes mémoires* make mention of this party or of Mardrus's presence.

was Kees van Dongen (1877–1968), the Dutch/French fauvist painter.[24] Van Dongen, born in Delfshaven near Rotterdam, had moved to Paris before the turn of the century, but only from 1904–1905 did his career start to take shape, with his colourful and erotic paintings of dancers and vaudeville artists fetching higher and higher prices. Afterwards he developed into a fashionable and expensive portraitist of France's rich and famous, most of whom have now fallen into oblivion. Van Dongen recruited his clients from the same strata of society as Poiret, and like Poiret he entertained lavishly in his studio as part of his public relations strategy. With his wild beard, overalls or knitted pullovers he continued to cultivate the image of the poor bohemian. The Dutch artist Willem Frederik Dupont (1908–1996), for instance, portrayed him in 1930 as a kind of Rasputin (Plate 6). In this manner, Van Dongen let his clients 'share' in a world of unpolished and unfettered creativity. Later in his career he attached much less importance to his wild image, and many photographs show him at work at his easel while wearing a perfectly respectable suit and hat.

Like so many of his contemporaries, Van Dongen could not help being affected by the Oriental fashion craze that, among others, held sway in the vaudeville theatres that he loved so much. In 1906 he made a painting in vibrant fauvist colours of a belly dancer called 'La belle Fatima,' and in 1908 he painted his mistress 'Anita la Bohémienne' as a veiled Oriental dancer.[25] The 'Orient' was ubiquitous in the Paris of those days, but he decided to see the East with his own eyes and in the winter of 1911 he travelled to Morocco, one year earlier than his fellow fauvist Henri Matisse. In 1913 he went to Egypt and on his return he gave an Oriental fairytale party in his studio with Paul Poiret as one of the guests.

Unlike Matisse, however, the Oriental experience has left few traces in Van Dongen's work as a painter. He painted Oriental scenes during and immediately after his stay in North Africa, but apparently the sensuality of the *Moulin rouge* and the boudoirs of Paris was much more real and attainable than that of far-away harems, and it is very likely that to him the *Alhambra* was much more a Paris theatre than a Moorish palace in Granada.

[24] R. Engers, *Het kleurrijke leven van Kees van Dongen* (Schiedam, 2002), p. 78.
[25] J. Mélas Kyriazi, *Van Dongen et le fauvisme* (Lausanne [etc.], 1971), pp. 88, 101.

La Sirène

The Orient has a much higher profile in Van Dongen's activity as a graphic artist, an aspect that has come to attract more attention since the publication of a monograph by the Utrecht art dealers Jan Juffermans and Jan Juffermans, jr.[26] Before the end of the First World War a Paris publishing house, Les Éditions de la Sirène, started publishing so-called *livres d'artiste*, important works of modern literature by authors like Cocteau, Apollinaire and Baudelaire, illustrated by avant-garde artists. Their very first edition was a small catalogue of an exhibition of works by Kees van Dongen in March 1917.[27] Not long afterwards, in December 1918, La Sirène published a single story from Mardrus's translation of the *Thousand and one nights* under the title *Hassan Badreddine El Bassraoui. Conte des 1001 nuits*.[28] Kees van Dongen undertook the overall design of this partial edition and made 110 line drawings and eight watercolours. The clichés of the drawings were made by the company of Demoulins Frères, while Jean Saudé executed the colour illustrations with the help of the pochoir technique.[29] Juffermans & Juffermans reproduced all eight watercolours in their work on Van Dongen's graphic art: most of them are highly stylised representations of women with the large, almond-shaped eyes that Van Dongen was so famous for, against a background of strong, plain colours.

In this edition the exuberant eroticism of Mardrus's translation is reflected perfectly in Van Dongen's drawings and watercolours. The young women are unexceptionally long-legged, slender, large-eyed and in various states of undress, whereas the old women and men are decrepit and ugly. Other illustrators of the *Thousand and one nights* often took the imagery of Oriental miniatures as a source of inspiration, but Van Dongen's drawings of young women, who exclusively embody the beauty ideal of the twentieth-century *Parisienne*, reveal that he made no effort to work within the context of classical Oriental art. In all like-

---

[26] J. Juffermans & J. Juffermans, jr., *Kees van Dongen. Het grafische werk* (Zwolle, 2002).

[27] P. Fouché, *La Sirène* (Paris, 1984), p. 253.

[28] Not 'el-bass Raoul' as Juffermans & Juffermans, p. 122, have it. For a complete bibliographical description of this work see Fouché, pp. 272–273. In Mardrus's complete translation (Paris, 1899–1904) the story covers Nights 19–24 (I, pp. 249–345).

[29] A variety of stencilling in which every colour is added separately by hand, a technique that Saudé developed to perfection. See his *Traité d'enluminure d'art au pochoir* (Paris, 1925).

lihood he was quite unfamiliar with it. On his travels to Morocco and Egypt he formed an impression of both countries as they were at that moment, but he refrained from giving them a historical dimension. For the rest he devised his own Orient. For example, a girl on p. 79 is wearing stockings and high heels, an extraordinarily un-Oriental and unclassical scene. At the margin of another drawing on p. 45 a dog with the name 'Kiki' on its collar is sniffing at a bitch. Those who know that Van Dongen was nicknamed Kiki among his intimate friends will realise that this picture might be regarded as a moment of self-reflection.

As already mentioned above, the eroticism of Mardrus's translation verges on the grotesque in the eyes of the modern reader, and Van Dongen appears to have rendered this perfectly in his humorous, cartoonesque illustrations. Nevertheless, one should bear in mind that Mardrus never meant his translation to be grotesque. On the contrary, it is one of the prime examples of the oversexed, highly strung and humourless Orientalist literature of the *Belle Époque*. It is Kees van Dongen's merit that he introduced a personal element of irony and humour into Orientalist art, thereby giving the erotic a refinement it had lacked so far. It is not for nothing that Van Dongen's drawings are still attractive, while most erotic Orientalist art looks hopelessly outdated.

With a print run of 310 copies on various kinds of luxury paper and prices between 500 and 1,500 francs the edition was obviously intended for the book lover. The exclusiveness of the publication was enhanced even more by adding four leaves with rather explicit plates which were announced as 'censored,' although it has never become clear by whom. Not all copies contain these 'censored' plates, and it is likely that they were added as an extra if the buyer desired so.

Kees van Dongen's interest in the activities of La Sirène is furthermore illustrated by the fact that he bought a share in the company in 1919.[30] One year later, in 1920, he published another volume at La Sirène with illustrated stories by Rudyard Kipling.[31]

It is not clear whether Dr Mardrus and Kees van Dongen actually knew each other, but there are some indirect links. When still young, Van Dongen made illustrations for the *Revue blanche*, the periodical that first published Mardrus's translation. Fashion designer Paul Poiret knew both Mardrus and Van Dongen, and Mardrus had a painting by

[30] Fouché, p. 65.

[31] *Les plus beaux contes de Kipling*, illustrés par Kees van Dongen (Paris: Editions de la Sirène, 1920); Fouché, pp. 366–367; Juffermans & Juffermans, pp. 126–129.

Van Dongen in his drawing room. Both were prominent in social life, not only in Paris, but also in a fashionable bathing resort like Deauville, where *tout Paris* made its appearance.

Eighty watercolours

After the Second World War book illustration gained an increasingly important place in Van Dongen's work, while his production of oil paintings stagnated. It is sometimes argued that his age started playing a role. For instance, his – most entertaining – biographer Rudolf Engers suggests that the artist, who turned seventy in 1947, could do his graphic work seated, while he had to remain standing for an oil painting. It is equally possible that the *grande bourgeoisie* of France found itself so impoverished after the Crisis and the War that they were no longer in a position to commission oil paintings. Van Dongen was notorious for keeping his prices high irrespective of the economic tide.[32] As a result, Van Dongen illustrated works by authors like De Montherlant, Voltaire and Baudelaire, always in limited editions of several hundred copies. However, two books belong firmly to the post-war tradition of mass production and 'art within the reach of the common man,' and both were published by Gallimard, one of the largest publishing firms in France. The first appeared in 1947: an edition in three volumes of Marcel Proust's *À la recherche du temps perdu* with 77 colour illustrations by Van Dongen and a cover design by Paul Bonet (1899–1972). The 'limited' edition of 9,250 numbered copies, however, can hardly be called exclusive. Eight years later, in 1955, Gallimard and Fasquelle[33] published a complete, three-volume set of Mardrus's *Livre des mille nuits et une nuit* in an even larger edition of 13,750 numbered copies. Like the Proust edition, the cover and cassette were designed by Paul Bonet. Typical for this example of semi-mass production is the material of the cover, paper-thin, leather-printed plastic instead of real calf, and only one type of paper was used instead of different varieties of luxury paper. The fact that all copies were numbered gives a false feeling of exclusivity that is reminiscent rather of the Franklin Mint Company than the world of the discerning bibliophile.

---

[32] Engers, pp. 118, 120.

[33] Fasquelle had bought the rights to Mardrus's translation from the *Revue blanche* in 1903. See above, n. 3.

The three volumes of this edition were illustrated by eighty colour woodcuts by Roger Boyer and Gérard Angiolini after watercolours by Kees van Dongen. Van Dongen also designed the advertisement poster.[34] From the viewpoint of the art trade these illustrations are not original artworks, so Juffermans & Juffermans mention the edition only in passing. Nevertheless, from the viewpoint of the art historian the illustrations deserve the effort of further analysis, and Richard van Leeuwen felt justified in analysing them in his handbook for the *Thousand and one nights, De wereld van Sjahrazaad* (1999).[35] Whoever compares the two editions of 1918 and 1955 will see that Van Dongen strongly elaborated on the theme of anachronism. While he limited himself in 1918 to some modern accents, the 1955 edition is rife with little, playful anachronisms that, however, never seem to be out of place. The number of ladies with stockings and stiletto heels has multiplied. (Elderly) men wear glasses, in one instance even swimming goggles, or a tarboosh. In some cases European-style suits are worn. More often than not, the furniture is in the highly decorated *Louis-Farouk* style that is still popular in Egypt. It must be admitted, however, that Van Dongen limited himself to petty mischief in the details. For instance, people do not drive around in motorcars and the horizon is not polluted with modern office buildings. What Van Dongen also left untouched are the warm, sunny colours of the Middle East, especially in the textiles that people wear. However, the designs are not those of classical Oriental fabrics, but rather of cheap Egyptian cotton prints of 1910. The complete absence of Oriental carpets, which so often dominate Orientalist art with their heavy colours and elaborate patterns, is positively soothing to the eye. In the garments one finds the strange mixture of styles that is so much part of Orientalist art: some people go dressed like an Indian maharaja, while others wear Egyptian or Moroccan traditional dress, sometimes together in one illustration. The mosques are usually modelled after the Mosque of the Prophet in Medina, but one also finds an example of a Turkish mosque. In one illustration a gentleman is depicted in Ottoman Turkish dress against a background that consists of the horseshoe arches of the 'Mezquita' or Grand Mosque of Cordoba.

Another detail is the practice of 'recycling' older paintings or drawings, a frequent and well-known phenomenon in Van Dongen's art. For

[34] Juffermans & Juffermans, p. 172, erroneously give 1919 as the date, one year later than the edition of *Hassan Badreddine El Bassraoui*.

[35] R. van Leeuwen, *De wereld van Sjahrazaad* (Amsterdam, 1999), pp. 113–114.

example, the illustration *Les femmes colombes* (vol. 2, p. 177) is drawn after an earlier painting of 1909, *Les trois grâces*, featuring Van Dongen's first wife Augusta 'Guus' Preitinger.[36] *Un riche marchand* (vol. 3, p. 81), a picture of an elderly gentleman with pebble glasses and a bottle of champagne, is based on a 1946 painting of the Aga Khan III Sultan Muhammad Shah.[37] The oil painting *La danseuse indienne* of *c.* 1907 was re-used by Van Dongen as *La danse de Morgane* (vol. 3, p. 369).[38]

It would have stood to reason if Van Dongen had re-used the illustrations of the partial edition of 1918 for the new complete edition of 1955, but apart from some exceptions this appears not to be the case. A Nubian boy on the back of a camel (1918: p. 64) returns as the illustration *Le voyage de Kamaralzamân* (1955: vol. 1, p. 705), and a jinn in the guise of a bearded old man (1918: p. 69) makes his reappearance in the illustration *Le cimetière de Bassra* (1955: vol. 1, p. 161). Of the colour illustrations only the naked figure of Shahrazād (1918: p. 5) in the framing story of the *Thousand and one nights* returns in the 1955 edition (1955: vol. 1, p. 17), but the figure sitting opposite to her is different: in 1918 it was a naked man smoking a pipe, who bore an uncommon resemblance to Kees van Dongen himself, but in 1955 it was an anonymous figure in a white *djellaba*. The same girl, seen from the back with outstretched arms, features in the advertisement poster of the 1955 edition. Reflecting the changes in art and fashion over several decades, the two editions differ greatly in their use of colour: the pochoir prints of 1918 with their flat, opaque colours look heavy and stylised, while those from 1955 are much more vibrant and lighter through the use of transparent pastel tints.

Another interesting phenomenon is the difference in approach to the human body. In the avant-garde edition of 1918 both men and women appear fully naked whenever the situation calls for it (Plate 7). In the more middle-class oriented 1955 edition, however, the female body is almost always completely exposed, which is in agreement with the 'pin-up' culture of the postwar era with its increasing commercialisation of women's bodies, while male nudity remains of a rare occurrence.[39]

---

[36] For an image of *Les trois grâces* see Engers, p. 27. It is currently at the Galerie Tamé-naga, avenue Matignon, Paris (last seen in the spring of 2005).

[37] Van Leeuwen, p. 114; Engers, p. 119.

[38] Mélas Kyriazi, p. 89; Engers, p. 47.

[39] See for instance II, opposite pp. 472, 496; III, p. 144.

In a country like the Netherlands, academic libraries have always been much more interested in texts than in beautifully illustrated books, and original texts have always taken precedence over translations. Fortunately, the Royal Library in The Hague holds a fairly large number of French bibliophile editions from the twentieth century, donated or bequeathed by Louis Jan Koopman (1887–1968), an Amsterdam entrepreneur and book collector.[40] Among them are several editions with illustrations by Kees van Dongen, such as the 1947 Proust edition. However, Koopman was exclusively interested in French literature, which implies that the two editions of the *Thousand and one nights* are lacking in his collection. Neither edition was represented in any public collection in the Netherlands until 2003, when the Leiden University library acquired a copy of the Gallimard/Fasquelle edition of 1955 from Kunsthuis "Marc Chagall" in Amsterdam. Quite recently, Leiden University also acquired the 1918 edition from Kunsthandel Juffermans in Utrecht, which means that at last both examples of Kees van Dongen's 'Orientalist' graphic art are available for research in the Netherlands.

It is perhaps appropriate to conclude with the year 2005, when Gallimard published the first volume of an entirely new translation of the *Thousand and one nights* by Jamel Eddine Bencheikh and André Miquel in its highly prestigious *Bibliothèque de la Pléiade*.[41] Gallimard inaugurated the event with an *Album Mille et une nuits*, an iconography of translations of the *Thousand and one nights*, authored by Margaret Sironval.[42] In this work she refers fleetingly to a 1928 edition by Kees van Dongen, mentioning neither title nor publisher, so one can only assume that this is the 1918 edition of *Hassan Badreddine El Bassraoui* by Les Éditions de la Sirène.[43] In the same work she also refers to the edition of 1955 and reproduces the illustration *La danse de Morgane*, but she fails to mention the name of the publisher, either in the text or in the index of publishers.[44] Would this be an act of extreme modesty on the part of Gallimard? More important, however, is her appraisal of the impact of Mardrus's translation in intellectual circles. In a chapter entitled 'La

[40] *Verzamelaars en verzamelingen. Koninklijke Bibliotheek, 1798–1998* (Zwolle, 1998), pp. 163–167.

[41] J.E. Bencheikh & André Miquel, *Les mille et une nuits*. Bibliothèque de la Pléiade; 515. 3 vols. (Paris, 2005–....).

[42] M. Sironval, *Album Mille et une nuits. Iconographie choisie et commentée*. Album de la Pléiade; 44 (Paris, 2005).

[43] Sironval, p. 66.

[44] Sironval, pp. 167–169. For the index of publishers see pp. 262–263.

vague des *Mille nuits et une nuit* à la Belle Époque' she cites the enthusiasm of 'les esprits les plus brillants de l'époque' such as André Gide, Auguste Rodin and Anatole France, none of whom knew Arabic, but she pays little attention to the ongoing discussion among Arabists about the merits of Mardrus's translation, dismissing it in one sentence as '[les] réserves de quelques universitaires sur l'exactitude de sa traduction.'[45] This, if nothing else, illustrates the still existing divide between 'lettrés et artistes' and Arabists in present-day France.

[45] Sironval, p. 111.

LANGUAGE

GUTTA CAVAT LAPIDEM

Manfred Ullmann

Sprichwörter und sprichwörtliche Redensarten sind die kollektive Schöpfung eines Volkes oder einer Sprachgemeinschaft. Sie spiegeln die materielle Kultur des jeweiligen Volkes, seine landschaftliche Umwelt, seine Denkungsart und seine Wertvorstellungen wider, sie sind Ausdruck der Erfahrungen, die das Volk in seiner Geschichte gesammelt hat. Insofern sind solche Worte und Wendungen immer nur einem bestimmten Volk und einer bestimmten Sprache eigentümlich. Es gibt aber auch Sprichwörter mit einem hohen Symbolgehalt, in denen ein Sachverhalt so treffend und plastisch zum Ausdruck gebracht ist, daß sie politische und sprachliche Grenzen überschreiten und von anderen Kulturen übernommen werden, obwohl den dortigen Menschen die Anschauung fehlt, die das Bild geprägt hat. Zu solchem Wandergut gehört der Spruch, der in diesem Aufsatz behandelt werden soll.

Er ist allgemein bekannt in der Form *gutta cavat lapidem, non vi, sed saepe cadendo.*[1] Diese Formulierung ist aber erst seit dem Spätmittelalter bezeugt.[2] So läßt auch Giordano Bruno in seinem 1582 erschienenen Lustspiel *Il Candelaio*, dessen italienischer Text mit vielen lateinischen Redewendungen durchsetzt ist, den Manfurio sagen: *Questa lectione bisogna saepius reiterarla, et in memoriam revocarla . . .*

> *Gutta cavat lapidem, non bis, sed saepe cadendo:*
> *Sic homo fit sapiens, bis non, sed saepe legendo.*[3]

Die Quellen sind in diesem Aufsatz nach den Ausgaben und mit den Abkürzungen zitiert, die im *Wörterbuch der Klassischen Arabischen Sprache (WKAS)* verwendet werden.

[1] Zu den Formulierungen im Deutschen, Französischen, Italienischen und in anderen europäischen Sprachen vg. K.F.W. Wander, *Deutsches Sprichwörter-Lexikon*, IV (Leipzig, 1867), col. 1335 (s.v. Tropfen).

[2] Zahlreiche Nachweise bei H. Walther, *Proverbia Sententiaeque Latinitatis Medii Aevi*, II (Göttingen, 1964), p. 263f., nr. 10508–10509.

[3] Giordano Bruno, *Candelaio*, Atto terzo, Scena VII. Bibliotheca Romanica; 162–164 (Strasburgo, 1912), p. 91. Vgl. dazu H. Stadler, in *Kindlers Neues Literatur-Lexikon*, hsgb. von W. Jens, III (München, 1989), p. 275f.

In der Antike hatte die Sentenz noch anders gelautet. Ovid verzichtet auf die Begründung und setzt statt dessen dem ersten Bild ein zweites an die Seite:

> *gutta cavat lapidem, consumitur anulus usu.*[4]

Aber das Sprichwort ist schon viel älter. Bion aus Smyrna[5] sagt in seinen Bukolika:

> Ἐκ θαμινᾶς ῥαθάμιγγος, ὅπως λόγος, αἰὲς ἰοίσας
> χὰ λίθος ἐς ῥωχμὸν κοιλαίνεται . . .[6]

Doch schon vor Bion hatte im 5. Jhdt. vor Chr. Choirilos von Samos dem Gedanken folgende Gestalt gegeben:

> πέτρην κοιλαίνει ῥανὶς ὕδατος ἐνδελεχείῃ.[7]

Dieser Hexameter war in aller Munde, und auch die Gelehrten haben ihn zitiert.

So sagt Galen im I. Buch seines Werkes *De locis affectis*, es sei möglich, daß gewisse Krankheiten bereits im Körper vorhanden seien, obwohl sie wegen ihrer Geringfügigkeit noch nicht in Erscheinung träten. Das beweise der Wassertropfen, der in langer Zeit den Felsen aushöhle, was in dem Gedichtvers πέτρην κοιλαίνει ῥανὶς ὕδατος ἐνδελεχείῃ treffend ausgedrückt sei. Wenn, so fährt Galen fort, erst beim tausendsten Aufprall eine spürbare Höhlung im Felsen entstehe, so folge daraus, daß jeder einzelne Tropfen den tausendsten Teil jener Höhlung verursacht habe.[8] Durch die Übersetzung des Werkes *De locis affectis* im 9. Jhdt. haben auch die Araber die Sentenz kennengelernt. Sie lautet dort: *inna qaṭra 'l-mā'i bi-idmānihī qad yathqubu 'l-ṣakhra.*[9]

---

[4] *Publii Ovidii Nasonis epistolae ex Ponto*, IV, 10,5. Zu weiteren Stellen bei Ovid, Lukrez, Tibull, Properz und anderen vgl. A. Otto, *Die Sprichwörter und sprichwörtlichen Redensarten der Römer* (Leipzig, 1890), p. 156f.

[5] Ende des 2. Jhdts. vor Chr.

[6] *Joannis Stobaei Anthologium*, III, 29,52 (Bd. III, p. 637 Wachsmuth-Hense); J.D. Reed, *Bion of Smyrna. The fragments and the Adonis* (Cambridge, 1997), p. 108 nr. IV, dazu Kommentar p. 147f.

[7] *Epicorum Graecorum fragmenta . . .*, collegit G. Kinkel, I (Lipsiae, 1877), p. 271 nr. 10. Zur Frage des Autors vgl.: τοῦτό φησι Σιμπλίκιος Χοιρίλου τοῦ ποιητοῦ εἶναι, Φιλόπονος δὲ Μόσχου τοῦ ποιητοῦ *Ioannis Philoponi in Aristotelis Physicorum Libros quinque posteriores commentaria*, ed. H. Vitelli. Commentaria in Aristotelem Graeca, XVII (Berolini, 1888), p. 826,13–15.

[8] Galenus, *De locis affectis* I 2 (Bd. VIII 27,7ff. Kühn).

[9] Ich konnte eine Kopie der Handschrift London, Wellcome Or. 14a, benutzen, die mir Herr Prof. Dr. Rainer Degen freundlichst zur Verfügung gestellt hat. Dort steht der Vers f. 11a13f.

Denselben Vers zitiert Galen noch einmal im III. Buch seines Werkes *De temperamentis*. Er spricht davon, daß bestimmte Drogen und Nahrungsmittel wie das Bibergeil, der Pfeffer, der Wein, der Honig usw. auf den Körper einwirken, von diesem aber auch verändert werden, und diese Wechselwirkung vollziehe sich in langer Zeit, so daß sie den Sinnen verborgen bleibe. Oder wenn man einen ganzen Tag und eine ganze Nacht lang mit dem schärfsten Messer das weicheste Wachs schneide, so würde das Messer notwendig etwas stumpfer werden. Ein ähnlicher Vorgang sei schon in dem Vers πέτρην κοιλαίνει ῥανὶς ὕδατος ἐνδελεχείῃ beschrieben worden.[10] Auch die Schrift *De temperamentis* ist ins Arabische übersetzt worden. Die Passage lautet dort folgendermaßen: *fa-innaka in ʿamadta ilā sikkīnin fī ghāyati ʾl-ḥiddati fa-qaṭaʿta bihā shamʿan fī ghāyati ʾl-līni ʾl-nahāra ajmaʿa wa ʾl-layla ajmaʿa lam yakun buddun min an yaẓhara fī ʾl-sikkīni ʾl-kalālu wa-ka-dhālika ayḍani ʾl-amru fī ʾl-qawli ʾlladhī qīla: inna qaṭra ʾl-māʾi idhā alaḥḥa ʿalā ʾl-ṣakhrati thaqabahā.*[11]

Noch ein drittes Mal kommt Galen auf diesen Sachverhalt zu sprechen. Ohne den Vers zu zitieren, sagt er in seinem Werk über die einfachen Heilmittel: καὶ λίθος ἔπαθεν ὑπὸ συνεχοῦς πληγῆς ὕδατος σταλαγμοῦ καὶ ξίφος ἠμβλύνθη τέμνον κηρόν.[12] *Inna ʾl-ṣakhrata yuʾaththiru fīhā qaṭru ʾl-māʾi ʾl-muttaṣilu bi-muṣādamatin lahā wa ʾl-sayfu yakillu wa-tadhhabu ḥiddatuhā idhā quṭiʿa bihi ʾl-shamʿu.*[13]

Die Griechen hatten den Hexameter des Choirilos aber auch zu einem jambischen Trimeter umgestaltet, um ihn in die 'Menandersentenzen' aufnehmen zu können. Diese waren ein Sammelbecken der Spruchweisheit, in das vieles aus der Neuen Komödie, aber auch aus anderen Quellen Eingang gefunden hat. Sie waren als Schulbuch im Gebrauch und sind immer wieder erweitert oder auch gekürzt worden, so daß jeder der griechischen Kodizes einen anderen Textbestand aufweist. Der Zufall wollte es, daß gerade der Spruch vom steten Tropfen in keinem der erhaltenen griechischen Kodizes vorkommt. Aber in der arabischen

---

[10] Galenus, *De temperamentis* III 4 (Bd. I 675, 9ff. Kühn).

[11] MS Escorial 848 (3), f. 67a8ff. Den arabischen Text hat für mich Rainer Degen abgeschrieben, der mich auch darauf hinweist, daß Ibn Rushd diese Stelle resümiert hat: *mithālu dhālika qaṭru ʾl-māʾi ʾl-wāqiʿu ʿalā ʾl-rukhāmi ḥattā yathqubahū wa ʾl-sikkīnu ʾlladhī yakillu ʿan qaṭʿi ʾl-ashyāʾi ʾl-raṭbati*, s. *Commentaria Averrois in Galenum*, ed. M. de la Concepción Vázquez de Benito (Madrid, 1984), p. 88,25f.

[12] Galenus, *De simplicium medicamentorum temperamentis ac facultatibus* III 18 (Bd. XI 597,13f. Kühn).

[13] MS Istanbul, Ahmet III 2083, f. 68a18f.

Übersetzung der Menandersentenzen, die im 9. Jhdt. von einem gewissen Iṣṭifān angefertigt wurde, ist er erhalten. Er lautet dort nach dem *K. al-Kalim al-rūḥāniyya* des Ibn Hindū: *al-qaṭru bi-dawāmihī yaḥtafiru 'l-ṣakhra*, nach dem *Mukhtaṣar ṣiwān al-ḥikma* des ʿUmar b. Sahlān al-Sāwī: *dawāmu 'l-qaṭri yaḥfiru 'l-ṣakhra*. Da ein entsprechender, mit Rho beginnender Vers in den Kodizes und somit auch in den beiden griechischen Editionen[14] fehlt, hatte der Schreiber dieser Zeilen den arabischen Spruch im Anhang zu seiner Edition unter nr. 345 aufgeführt, ohne ihn einordnen zu können.[15] Erst Jahrzehnte später hat Rudolf Führer[16] den zugrundeliegenden griechischen Vers aufgespürt. Er ist in der *Vita Gregorii theologi episcopi Nazianzeni a Gregorio presbytero conscripta* überliefert und lautet: ῥανὶς γὰρ ἐνδελεχοῦσα κοιλαίνει πέτραν.[17] Damit wurde klar, daß der Vers in die arabische Sammlung als nr. 269 a einzureihen ist.

Bieten somit die Menandersentenzen einen vierten Beleg für die Vermittlung des griechischen Sprichwortes an die Araber, so gibt es durch einen Passus im VIII. Buch der Physik des Aristoteles noch einen fünften Überlieferungsstrang. Dort heißt es: ἔστι δ' ὅμοιος ὁ λόγος τῷ περὶ τοῦ τὸν σταλαγμὸν κατατρίβειν καὶ τὰ ἐκφυόμενα τοὺς λίθους διαιρεῖν. οὐ γὰρ εἰ τοσόνδε ἐξέωσεν ἢ ἀφεῖλεν ὁ σταλαγμός, καὶ τὸ ἥμισυ ἐν ἡμίσει χρόνῳ πρότερον.[18] In der Übersetzung des Isḥāq b. Ḥunayn, die gegen Ende des 9. Jhdts. angefertigt worden ist, lautet der Satz: *wa-hādhā 'l-qawlu shabīhun bi-mā yuqālu min amri ḥatti 'l-qaṭri wa-falqi 'l-nabti li-l-ḥijārati wa-dhālika annahū laysa yajibu in kānati 'l-qaṭratu qad dafaʿat aw ḥattat miqdāran mā an yakūna niṣfu dhālika mithlan fī niṣfi dhālika 'l-zamāni.*[19] Hier ist also nur der Erosionsvorgang als solcher erwähnt, ohne daß auf die Sentenz zurückgegriffen wurde.

---

14 *Fragmenta Comicorum Graecorum*, collegit et disposuit A. Meineke, Editio minor, II (Berolini 1847), pp. 1041–1066; *Menandri sententiae*, ed. S. Jaekel. Bibliotheca Teubneriana (Lipsiae, 1964).

15 M. Ullmann, 'Die arabische Überlieferung der sogenannten Menandersentenzen', *AKM* 34,1 (1961), p. 59.

16 R. Führer, 'Zur arabischen Übersetzung der Menandersentenzen', *Beiträge zur Altertumskunde* 43 (1993), p. 49.

17 J.-P. Migne, *Patrologia Graeca*, XXXV (Paris, 1857), col. 277 C.

18 Arist. *Physica* 253b 15–18.

19 *K. al-Ṭabīʿa li-Arisṭūṭālīs*, ed. ʿAbd al-Raḥmān Badawī, II (Kairo 1385/1965), p. 825,10ff. Dazu der Kommentar des Abū 'l-Faraj b. al-Ṭayyib, ib. p. 831,22ff. Vgl. auch M. Ullmann, *Wörterbuch zu den griechisch-arabischen Übersetzungen des 9. Jahrhunderts* (Wiesbaden, 2002), p. 633.

Soweit mir bekannt, gibt es also fünf Zeugnisse dafür, daß der Spruch von den G r i e c h e n zu den Arabern gekommen ist. Es gab aber auch noch einen ganz anderen Überlieferungsweg. Das Sprichwort kursierte nämlich auch in I n d i e n,[20] und so ist es auch im *Pañcatantra* in die Elfte Erzählung vom Löwen, seinen Ministern und dem Kamel eingeflochten. Der Löwe hatte dem Kamel zugesagt, daß sein Leben bei ihm sicher sei, aber durch langes Zureden bringen der Panther, die Krähe und der Schakal den Löwen dazu, sein Versprechen zu brechen und das Kamel zu fressen. Diese verhängnisvolle Beeinflußbarkeit durch Worte ist in folgender Sentenz glossiert: 'Wird doch des Berges harter Boden von weichem Wasser untergraben und abgerieben, geschweige der Menschen weiche Herzen vom Ohrgeflüster der Zwietracht Säenden'.[21] Durch die Übersetzung des *Pañcatantra* aus dem Mittelpersischen ins Arabische ist die Sentenz den Arabern gegen Ende des 2./8. Jhdts. vermittelt worden. In *Kalīla wa-Dimna* heißt es im Kapitel 'Der Löwe und der Stier': 'Selbst wenn der Löwe keine anderen Gefühle als Güte und Barmherzigkeit hegen würde, so würde ihn das ständige Zureden umstimmen. Denn wenn viele Worte gemacht werden, kann es nicht ausbleiben, daß Sanftmut und Freundlichkeit schwinden. Weißt du denn nicht, daß das Wasser weicher ist als die Rede und daß der Stein härter ist als das Herz? Und doch hinterläßt das Wasser, wenn es lange auf den harten Stein herabstürzt, auf ihm seine Spuren'.[22]

Um die Wende vom 5. zum 6./11. zum 12. Jhdt. ist die Fabelsammlung dann von Abū Yaʿlā Muḥammad b. al-Habbāriyya in Muzdawij-Verse gekleidet worden. Dabei hat unser Spruch folgende Gestalt angenommen:

> *fa 'l-māʾu min qawli 'l-ʿaduwwi alṭafū wa 'l-qalbu min ṣummi 'l-ṣukhūri aḍʿafū*
> *thumma idhā 'l-māʾu ʿalā 'l-ṣakhri jarā mudaydatan ghādara fīhi atharā.*[23]

---

[20] Vgl. O. Böhtlingk, *Indische Sprüche. Sanskrit und Deutsch.* Zweite Aufl., I–III (St. Petersburg, 1870–1873), nr. 4245, 4685, 5854, 7508, 7509. Ich verdanke diesen Hinweis Frau Dr. Gabriele Zeller, Tübingen.

[21] Übersetzung von Th. Benfey, *Pantschatantra. Fünf Bücher indischer Fabeln, Märchen und Erzählungen*, II (Leipzig 1859), p. 85 nr. 337. Der Spruch auch bei Böhtlingk, a.a.O. nr. 4963.

[22] *Kalīla* (de Sacy) 123,10ff./(Ch.) 87,16ff. = b. Qut. *ʿUyūn* II 22,13f. = Rāġib *Maǧmaʿ* I 141,7f. = Yamanī *Muḍāhāt* 23,5f. (§ 36). Ein ähnliches Bild findet sich auch in der Geschichte von der Eule und dem Raben: 'Der König sagte: Du wirst finden, daß ein Feind gründlicher vernichtet wird, wenn man ihn auf sanfte und behutsame (Var. listige) Weise zu Fall bringt als wenn man ihn durch eine Übermacht niederschlägt. Das Feuer nämlich mit all seiner Hitze und Wut kann nicht mehr tun, als die oberirdischen Teile eines Baumes zu verbrennen. Das Wasser hingegen, so dünn und kalt es auch ist, entwurzelt die unterirdischen Teile des Baumes': *Kalīla* (de Sacy) 205,4ff./ (Ch.) 163,15ff.

[23] b. – Habbārīya *Natāʾiǧ* 66,15f.

Wir halten also fest, daß die Sentenz *gutta cavat lapidem* den Arabern auf literarischem Wege, d.h. durch die Übersetzungen von *Kalīla wa-Dimna* und die der fünf genannten griechischen Werke,vermittelt worden ist. Dementsprechend ist sie von den arabischen Literaten nicht vor dem 3./9. Jhdt. rezipiert worden. Im Gegensatz aber zu anderen Motiven gibt es für *gutta cavat lapidem* nur eine Handvoll arabischer Belege.

Von den Menandersentenzen ist so gut wie nichts in die arabischen Sprichwörtersammlungen aufgenommen worden. Der einzige Parömiograph, der unseren Spruch zitiert, ist al-Zamakhsharī. Bei ihm lautet er: *al-qaṭratu bi-dawāmihā taḥtafiru*[24] *'l-ṣakhra.*[25] Diese Formulierung läßt vermuten, daß al-Zamakhsharī den Spruch von Ibn Hindū übernommen hat. Allerdings hat er das ursprüngliche *al-qaṭru bi-dawāmihī* durch *al-qaṭratu bi-dawāmihā* ersetzt, das Nomen generis also gegen das Nomen unitatis eingetauscht. Dieser stilistische Fehler[26] ist bei einem Autor, der als Grammatiker und Lexikograph Bedeutendes geleistet hat, immerhin bemerkenswert.

Der Göttinger Codex Michaelis 299[27] enthält in der 4. Maqāla eine Sprichwörtersammlung, die von einem Christen zu Unterrichtszwecken zusammengestellt worden war und durch eine vulgäre Sprache gekennzeichnet ist.[28] Dort heißt es unter anderem: *nuqṭatun dā'imatun takhriqu fī 'l-ḥajari.*[29]

Ibn ʿAbd Rabbihī zitiert die 'Philosophen' (*al-ḥukamā'u*) mit folgendem Spruch: *yudraku bi 'l-rifqi mā lā yudraku bi 'l-ʿunfi a-lā tarā anna 'l-mā'a ʿalā līnihī yaqṭaʿu 'l-ḥajara ʿalā shiddatihī* 'Durch Freundlichkeit kann man erreichen, was durch Barschheit nicht zu erreichen ist. Siehst du denn nicht, daß das Wasser, so weich es auch ist, in den Stein bei all dessen Härte einschneidet'?[30]

Muḥammad b. Sulaymān al-Nawfalī erklärt seinem Sohn ʿAlī, daß der Kalif Abū Jaʿfar al-Manṣūr im Jahre 158/775 an seiner Völlerei gestorben sei. Die Ärzte hätten ihm geraten, nur wenig zu essen, aber

---

[24] Var. *taḥfiru.*

[25] Zam. *Mustaqṣā* I 339,12 (nr. 1459).

[26] Vgl. M. Ullmann, *Das arabische Nomen generis*. Abh. der Akademie der Wissenschaften in Göttingen. Phil.-hist. Kl., Dritte Folge; 176 (Göttingen, 1989), p. 68.

[27] *Verzeichnis der Handschriften im Preußischen Staate.* I. Hannover: 3. Göttingen (Berlin, 1894), p. 212f.

[28] Vgl. *Arabum proverbia sententiaeque proverbiales*, quae edidit G.W. Freytag, III (Bonnae, 1843), p. xiii f.

[29] Freytag *Prov.* III, p. 416 (zu nr. 2487).

[30] *ʿIqd* I 246,4f./ (Amīn) II 360,4f.

er habe es vorgezogen, statt dessen zu jeder Mahlzeit eine scharfe Latwerge (Elektuarium, *jawārishn*) zu nehmen, die zwar das Essen zu verdauen helfe, nach und nach aber die Schleimhaut des Magens und die Darmzotten angreife und verschleiße. Das sei, so sagt al-Nawfalī, wie wenn man einen Tonkrug auf einen Schemel stelle und einen neuen gebrannten Ziegel (*ājurra*) darunterlege. Dann würden die Tropfen auf Dauer den Ziegel durchlöchern.[31]

Aḥmad b. Yaḥyā, genannt Thaʿlab,[32] berichtet, daß ihm al-Faḍl b. Saʿd b. Sālim[33] folgendes erzählt habe: Ein Mann wollte die Wissenschaften erlernen, fand dies aber zu schwer, so daß er davon abließ. Da kam er an einem Wasserlauf vorbei, der von der Höhe eines Berges auf einen Felsen herabstürzte und auf diesem einen Eindruck hinterlassen hatte. Da dachte er: Wenn das Wasser, so weich es auch ist, auf den Felsen bei all dessen Kompaktheit eingewirkt hat, so will ich, bei Gott, das Studium aufnehmen. Das tat er, und nun schaffte er es.[34]

Auf das Studium der Wissenschaften zielt auch Abū Naṣr al-Fārābī, wenn er in seiner *Risāla fī mā yanbaghī an yuqaddama qabla taʿallum al-falsafa* fordert, daß der Student der Philosophie ein starkes Streben und eine große Ausdauer besitzen müsse, und daher sage man auch, daß der Wassertropfen durch sein Andauern den Stein aushöhlen könne (*inna qaṭra ʾl-māʾi bi-dawāmihī qad yathqubu ʾl-ḥajara*).[35]

Abū Ḥāmid al-Ghazālī erklärt im *K. al-Tawba* seines *Iḥyāʾ*, daß man bei einer einmal begangenen schweren Sünde eher auf Verzeihung hoffen könne als bei einem kleinen Vergehen, das sich ständig wiederhole. Ein Gleichnis dafür böten die Wassertropfen, die in ununterbrochener Folge auf einen Stein treffen und auf ihm ihre Spuren hinterlassen (*wa-mithālu dhālika qaṭarātun mina ʾl-māʾi taqaʿu ʿalā ʾl-ḥajari ʿalā tawālin fa-tuʾaththiru fīhi*). Würde diese Wassermenge auf einen Schlag über dem Stein ausgegossen werden, so würde sie keine Wirkung haben.[36]

In G e d i c h t e n ist die Sentenz in folgenden Zusammenhängen verwendet worden: Ein von al-Suyūṭī zitierter Anonymus sagt:

uṭlub wa-lā tadjar mina ʾl-maṭlabi fa-āfatu ʾl-ṭālibi an yadjarā
a-mā tarā ʾl-māʾa bi-takrārihī fī ʾl-ṣakhrati ʾl-ṣammāʾi qad aththarā

---

31 Ṭab. *Taʾrīḫ* III 1, 387,18ff.
32 Gest. 291/904; *GAS* VIII 141–147.
33 Bei al-Suyūṭī lautet der Name: al-Faḍl b. Saʿīd b. Salm.
34 Taʿlab *Maǧālis* 169,1ff./ 141,1ff. = Suyūṭī *Muzhir* II 158, 1ff./ 303,8ff.
35 *Alfārābī's philosophische Abhandlungen*, hsgb. von F. Dieterici (Leiden, 1890), p. 54,16f.
36 Ġazālī *Iḥyāʾ* IV 32,7ff.

'Studier mit Fleiß und laß es dich nicht verdrießen, denn wenn ein Student verdrossen ist, scheitert er. Siehst du nicht, daß das Wasser durch ständig wiederholtes [Tropfen] einen Eindruck auf den tauben Fels macht?!'.[37]

Der 529/1134 gestorbene Ẓāfir b. al-Qāsim b. Manṣūr al-Ḥaddād beginnt ein paränetisches Gedicht mit den Worten:

> *wāẓib tanal kulla ṣaʿbin muʿwizin ʿasirin fa 'l-māʾu aththara bi 'l-idmāni fī 'l-ḥajarī*

'Sei fleißig und beharrlich, so wirst du alles erreichen, auch wenn es schwierig, vertrackt und mißlich ist; läßt doch das Wasser auf Dauer im Stein seine Spuren zurück'.[38]

Der philosophische Dichter Abū ʿAlī Muḥammad b. al-Ḥusayn b. ʿAbd Allāh, genannt Ibn al-Shibl,[39] der viele gedankenreiche Sinngedichte verfaßt hat, hat sich unserer Sentenz zweimal bedient:

> *fa-in lāna min farṭi 'l-ʿitābi fa-rubba-mā taghalghala luṭfu 'l-māʾi fī 'l-ḥajari 'l-ṣaldī*

'Wenn einer durch ein Übermaß an Vorhaltungen schließlich nachgibt (so ist das zu verstehen), denn manchmal dringt das weiche Wasser in einen harten Stein ein'.[40]

In einem Vierzeiler hat Ibn al-Shibl die Aufforderung zur Versöhnung mit dem Hinweis auf vier Beispiele untermauert, die das Motiv 'kleine Ursache – große Wirkung' illustrieren:

> *ulṭuf bi-khaṣmika fa 'l-labību bi-luṭfihī yastallu thārah*
> *amḍā 'l-ḥadīdi adaqquhū wa 'l-māʾu yanqubu fī 'l-ḥijārah*
> *wa 'l-hajwu baytun minhu lā yuṭfī ṭawīlu 'l-madḥi nārah*
> *yakhfā 'l-kathīru mina 'l-ḥalāwati fī 'l-qalīli mina 'l-marārah*

'Begegne deinem Feinde mit Nachsicht, denn durch Freundlichkeit kann ein kluger Mann ihm den Rachezahn ziehen. Das am dünnsten geschliffene Eisen dringt am tiefsten ein, und das Wasser macht Löcher in die Steine. Lange Lobeshymnen können den Brand, den ein einziger Schmähvers entzündet hat, nicht löschen. Eine große Menge Süßigkeit verschwindet in einer Prise Bitterstoff'.[41]

---

[37] Anon., bei Suyūṭī *Muzhir* II 158,6f./ 303,13f.

[38] Ẓāfir 105,1.

[39] Gest. 473/1080, vgl. B. Scarcia Amoretti, *Encyclopaedia of Islam* (New ed.), III (Leiden, 1971), p. 937f.

[40] Qifṭī *Muḥammadūn* 278,2.

[41] ib. 280 paen.ff.

Hier ist schließlich auch Abū ʿAbd Allāh Muḥammad b. al-Ḥusayn b. ʿUmar al-Yamanī[42] zu nennen, der sein *K. Muḍāhāt amthāl Kalīla wa-Dimna* geschrieben hat, um nachzuweisen, daß all die vielen Motive der indischen Fabelsammlung schon von arabischen Dichtern vorweggenommen worden seien. Aber die Verse, die er zu diesem Zweck zitiert, sind ausnahmslos Fälschungen, und das gilt auch für den plumpen Dreizeiler, in den er den oben behandelten Spruch aus der Geschichte vom Löwen und Stier verwandelt hat. Eine Fiktion ist natürlich auch der Dichtername al-Murāzim b. ʿUrqūb al-ʿAdawī.

wa-qultu lahū wa ʾl-māʾu alyanu fa-ʿlaman mina ʾl-qawli qadhfan bi ʾl-shatīmati wa ʾl-sabbī
ka-mā ʾl-ṣakhratu ʾl-ṣammāʾu aqsā majassatan wa-akhshanu lamsan yā ʿUmayru mina ʾl-qalbī
wa-lan yalbatha ʾl-māʾu ʾl-ṭawīlu ʾnṣibābuhū ʿalā ʾl-ṣaldi an yabdū bihī atharu ʾl-ṣabbī

'Ich sagte zu ihm: Das Wasser ist doch ganz weich. So wisse denn, wie es beim Reden zugeht, wenn man mit Schmähungen und Beschimpfungen um sich wirft! Das ist wie beim tauben Fels, der hart ist, wenn man ihn berührt, und der, mein lieber ʿAmr, rauher anzufassen ist als das Herz. Aber es kann nicht fehlen, daß sich auf dem Harten die Spur des Wassers abzeichnet, wenn es lange genug heruntergetropft ist'.[43]

N i c h t hierher gehört das Verspaar des Ibn al-Rūmī:

yā shabīha ʾl-badri fī ʾl-ḥusni wa-fī buʿdi ʾl-manālī
jud fa-qad tanfajiru ʾl-ṣakhratu bi ʾl-māʾi ʾl-zulālī

Zwar k ö n n t e der zweite Vers theoretisch auch bedeuten: 'Manchmal birst der Fels durch klares Wasser', aber der Präpositionalausdruck ist hier kein freier Aktant, der die Ursache bezeichnet, sondern er steht als Objekt in Dependenz zum Verbum. Daher ist zu übersetzen: 'Ach du [Knabe], der du in deiner Schönheit und Unerreichbarkeit dem Mond ähnelst! Gib dich hin! Läßt doch selbst ein Felsen manchmal klares Wasser sprudeln!'.[44]

Für die Sentenz *gutta cavat lapidem* konnte der Verfasser nur elf Beispiele aus arabischen Quellen beibringen. Das ist angesichts der

---

[42] Gest. 400/1009; Ṣafadī *Wāfī* 2,379 paen.ff. (nr. 848); *GAL* S I 202, nr. 3 a.
[43] Yamanī *Muḍāhāt* 23,14ff. (§ 36).
[44] b. – Rumī (Naṣṣār) V 1480 = b.a. ʿAun *Tašb.* 98,5f. = ʿAsk. *Maʿānī* I 166,10f. = Ḥuṣrī *Zahr* 563,3f. = Ǧurǧānī Asrār 269, 10f. (nr. 344) = b. – Šaǧarī *Ḥam.* 264,9f./ nr. 845.

unermeßlichen arabischen Literatur nicht eben viel. Aber auch diese wenigen Beispiele zeigen, daß es den Autoren gut gelungen ist, den Spruch als Gleichnis für bestimmte Sachverhalte und Verhaltensweisen heranzuziehen. Bei Thaʿlab, al-Fārābī, Ẓāfir al-Ḥaddād und dem von al-Suyūṭī zitierten Anonymus ist es die Beharrlichkeit und Ausdauer, die, insbesondere beim Studium der Wissenschaften, zum Ziele führt.[45] Ibn al-Shibl stellt die Sentenz in den Zusammenhang des Motivs 'kleine Ursache – große Wirkung'. In *Kalīla und Dimna* ist die Beeinflußbarkeit des Menschen thematisiert, den ständige Propaganda zu bösen Taten treibt. Aber auch der wiederholte Tadel kann einen Menschen bestimmen, nachzugeben und sich dem Guten zu öffnen, wie es Ibn al-Shibl und der unter dem Pseudonym al-Murāzim b. ʿUrqūb dichtende al-Yamanī dargestellt haben. Durch freundliche Sanftmut, so sagt Ibn ʿAbd Rabbihī, kann man mehr als durch Grobheit erreichen. Besonders originell ist der Gedanke al-Ghazālī's, wonach kleine, zur Gewohnheit gewordene Verfehlungen schwerer wiegen als eine einmal begangene große Sünde. Wenn schließlich al-Nawfalī meint, daß scharfe Medikamente auf Dauer den Körper schwächen, so erinnert dies an Galens Feststellung, daß eine heimtückische Krankheit langsam und unbemerkt entstehen kann, so wie auch die Physiologie des Stoffwechsels den Sinnen verborgen bleibt. All diese Aussagen lassen sich trefflich durch die Sentenz *gutta cavat lapidem* illustrieren.

Die angeführten arabischen Beispiele stammen alle aus einer Zeit, in der die Hellenisierung der islamischen Welt bereits weite Bereiche des kulturellen Lebens erfaßt hatte. Wenn v o r dem 3./9. Jhdt. von einem Tropfen oder vom Spalten eines Steines die Rede ist, so handelt es sich um andere Motive. Dafür nur drei Beispiele: Al-Farazdaq sagt:

> *qawāriṣu taʾtīnī wa-yaḥtaqirūnahā wa-qad yamlaʾu ʾl-qaṭru ʾl-ināʾa fa-yufʿamū*

'Es sind beißende Worte, die mich getroffen haben, denen ihre Urheber aber keine Bedeutung beimessen. Doch ein einziger Tropfen kann das Gefäß voll machen und zum Überlaufen bringen'.[46]

---

[45] Auch dieser Gedanke ist schon im Griechischen nachzuweisen. Gregorios von Nazianz sagt in einer Grabrede auf seinen Vater: ἔμελλε δὲ ἄρα κοιλαίνειν τὴν πέτραν ἡ τοῦ ὕδατος ῥανὶς ἀεὶ πλήττουσα, καὶ χρόνῳ διανύσειν τὸ σπουδαζόμενον (J.-P. Migne, *Patrologia Graeca*, XXXV (Paris, 1857), col. 997 C.)

[46] Far. I p. 60,6 / (Ṣāwī) 756,5 (die Dīwāneditionen haben statt *al-ināʾa* fälschlich *al-atiyya* 'den Wildbach') = Buḥturī *Ḥam.* 697,2 = b. Qut. *ʿUyūn* II 16,13 = Muʿāfā *Ǧalīs* I

Von ʿUbayd b. Ḥuṣayn al-Numayrī, genannt al-Rāʿī, stammt der Vers:

inna ʾl-zamāna ʾlladhī tarjū hawādatahū[47] *yaʾtī ʿalā ʾl-ḥajari ʾl-qāsī fa-yanfaliqū*

'Die Zeit, von der du hoffst, sie möge Nachsicht üben, greift den harten Stein an, so daß er Risse bekommt'.[48]

Und bei ʿAdī b. al-Riqāʿ heißt es:

usirru dumūʿan[49] *law taghalghala baʿḍuhā ilā ḥajarin ṣaldin tarakna bihī ṣadʿā*

'Wenn nur einige der Tränen, die ich verhehle, in einen harten Stein eindringen würden, so würden sie an ihm einen Spalt hinterlassen'.[50] Hier geht es nicht um das a n d a u e r n d e Tröpfeln, sondern um Tränen, die so heiß sind, daß sie wie eine Feuersglut den Stein sprengen.

Es spricht also vieles dafür, daß unsere Sentenz als Wandergut erst im Laufe des 2. bis 3./8. bis 9. Jhdts. von den Indern und vor allem von den Griechen zu den Arabern gekommen ist. Unter den geomorphologischen Verhältnissen Arabiens konnte sie auch kaum entstanden sein. Die dortigen Erosionsformen sind hauptsächlich durch den Wind und die nach einem Gewitterregen anschwellenden Fluten in den Wādīs entstanden, also horizontal ausgebildet.[51] Die vertikale Richtung, die der dauernde Tropfen nimmt, setzt perennierende Wasserläufe voraus, die aus der Höhe herabrinnen. Auch die Redensart *kataba ʿalā ʾl-māʾi* 'aufs Wasser schreiben' war ja nicht in den ariden Zonen Arabiens entstanden. Sie ist dem griechischen εἰς ὕδωρ γράφειν nachgebildet und

---

226 ult. = *Aġ.* 19,15,13/ 21,306,5 = Ḫālid. *Ašbāh* I 231,1 = Mubarrad *Kāmil* 18 paen. = b. Ḫālaw. *ŠMaqṣūra* 478 paen. = Usāma b. M. *Badīʿ* 181,1/(Mhn.) 259 ult. = *Lis.* 8,337, –3/ 7,70 a 9f. = b. Yaʿīš *ŠMufaṣṣal* I 24,14 und zahlreiche weitere Stellen.

[47] Var. *turjā hawādatuhū.*

[48] Ṯaʿāl. *Ḫāṣṣ* 106 ult. = Ṯaʿāl. *Iʿǧāz* 152,11 = Aidamur *Durr* II 329 paen. = Rāʿī (Wpt.) 47,2.

[49] So zu lesen statt *humūman* in der Ed. Nūrī Ḥammūdī al-Qaysī (Bagdad, 1407/ 1987). Allerdings hat auch Yāqūt *Buldān* IV 589,16/ V 165 a paen.f. *humūman.*

[50] ʿAdī b. – R. 25,3.

[51] An eine derartige horizontale Erosion ist auch in der hermetischen Schrift über die Zurechtweisung der Seele gedacht. Dort heißt es, daß das in Wachs eingedrückte Bild auf seinen Sinngehalt und seine Wirklichkeit im Stempel hinweise und daß das im Stempel dargestellte Bild auf die Vorstellungen verweise, die in der Seele dessen herrschen, der das Bild geschaffen hat. So könne man auch aus den Spuren, die das Wasser im Sand und Lehm hinterläßt, auf die Dynamik seiner Bewegungen und seines Wogens schließen: *K. Hirmis fī Muʿādhalat al-nafs*, ed. ʿAbd al-Raḥmān Badawī. Dirāsāt islāmiyya; 19 (Kairo, 1955), p. 56,9ff.

ebenfalls durch die Menandersentenzen und möglicherweise auf noch anderen Wegen den Arabern vermittelt worden.[52]

Die beiden Wendungen vom steten Tropfen und vom Schreiben aufs Wasser stehen als hellenistisches Lehngut nicht allein. Auch die Redensart vom 'gehörnten Ehemann' ist griechischen Ursprungs. In einer Glosse zum Traumbuch des Artemidor heißt es: προειπεῖν αὐτῷ ὅτι ἡ γυνή σου πορνεύσει καὶ τὸ λεγόμενον κέρατα αὐτῷ ποιήσει.[53] Diese Glosse fehlt in der anonymen arabischen, von Tawfīq Fahd herausgegebenen Übersetzung. Die Redensarten *yaḥmilu 'l-qurūna*, *ṭāla qarnuhū*, *huwa qarnānun*, *qarnuhū tamādā ʿuluwwuhū* usw. zur Bezeichnung eines Hahnreis sind aber vom 3./9. Jhdt. an sehr oft bezeugt.

Mit dem Ausdruck ὀρνίθων γάλα, den Aristophanes (*Vespae* 508, *Aves* 1673), Mnesimachos, Menander und andere Autoren gebrauchen,[54] wird ein unverhofftes Glück bezeichnet. Im Arabischen ist daraus *labanu 'l-ṭayri* bzw. *labanu 'l-dajāji* geworden, was man, wie al-Thaʿālibī erklärt, von einer Sache sage, deren Gelingen man nicht erhoffen kann.[55]

Das Sprichwort αἲξ τὴν μάχαιραν haben schon die frühen arabischen Dichter aufgegriffen. Wir finden es bei ʿAbd al-Ḥārith b. Ḍirār, bei dem Hudhayliten Umayya b. al-ʿAskar und bei Ḥassān b. Thābit ebenso wie zur ʿAbbāsidenzeit bei Ibn al-Rūmī (ed. Naṣṣār Bd. I nr. 344,12, Bd. III nr. 826,37 und 1034,13), al-Ṣanawbarī (127,7) und anderen.[56]

Einen Nachklang des Sisyphos-Motivs bewahrt das Sprichwort *alhafu min qālibi 'l-ṣakhrati*.[57]

Von den 'Krokodilstränen' heißt es in den gängigen Zusammenfassungen[58] übereinstimmend, daß diese Sage zuerst im Mittelalter, d.h. bei Konrad von Megenberg und in den *Physiologos*-Bearbeitungen des Codex Hamilton und des *Bestiaire* von Guillaume le Clerc auftauche. Als Redensart für geheucheltes Mitleid seien die 'Krokodilstränen' aber erst durch die *Adagia* des Erasmus von Rotterdam verbreitet worden.

52 M. Ullmann, *Aufs Wasser schreiben*. Bayerische Akademie der Wissenschaften. Phil.-hist. Klasse, Sitzungsberichte; 1989/1 (München, 1989).

53 *Artemidori Daldiani Onirocriticon*, II, 12 (p. 120,16 Pack), vgl. R. Strömberg, *Greek Proverbs* (Göteborg, 1954), p. 28f.

54 Vgl. die Nachweise bei Liddell-Scott-Jones 1254 b.

55 *WKAS* II 160 b 1ff.; 2164 a 44ff.

56 Vgl. auch die Zusammenstellung bei Fr.-Chr. Muth, in *Oriens* 33 (1992), p. 260.

57 Ḥamza *Durra* 379,6 (nr. 623) etc., s. *WKAS* II 1536 a 9ff. Vgl. R. Sellheim, *Encyclopaedia of Islam* (New ed.), VI (Leiden, 1991), p. 819 b (s.v. Mat͟hal).

58 H. Bächtold-Stäubli, *Handwörterbuch des Deutschen Aberglaubens*, V (Berlin-Leipzig, 1933), Sp. 598f.; L. Röhrich, *Das große Lexikon der sprichwörtlichen Redensarten*, II (Freiburg-Basel-Wien, 1992), p. 892f.; G. Büchmann, *Geflügelte Worte*. 42. Auflage (München, 2001), p. 79.

Aber schon der 296/908 ermordete ʿAbd Allāh b. al-Muʿtazz sagt in dem Muzdawij-Gedicht, in dem er das Leben und die Taten des Kalifen al-Muʿtaḍid bi-ʾllāh verherrlicht, über die Leute von Kūfa, die den ʿAlī und dessen Sohn al-Ḥusayn umgebracht haben:

thumma bakaw min baʿdihī wa-nāḥū jahlan ka-dhāka yafʿalu ʾl-timsāḥū.[59]

Das muß auf eine g r i e c h i s c h e Überlieferung zurückgehen, und tatsächlich schreibt Asterios, der um 400 Bischof von Amaseia in Pontos war, in einer Bußschrift: εἰ δ' ἔτι τοῖς Ἰουδαϊκοῖς προσδέδεσαι, τί σοι βούλεται ἡ νηστεία ἢ τοὺς Νειλῴους κροκοδείλους μιμεῖσθαι, οὕς φασι ταῖς κεφαλαῖς ἐπιθρηνεῖν τῶν ἀνθρώπων ὧν ἔφαγον καὶ δακρύειν ἐπὶ τοῖς λειψάνοις τῶν φόνων, οὐ μετάνοιαν τῶν γεγενημένων λαμβάνοντας, ἀλλὰ τὸ ἄσαρκον τῆς κεφαλῆς – ἐμοὶ δοκεῖν – ὀδυρομένους ὡς εἰς βρῶσιν οὐκ ἐπιτήδειον.[60] Diesen Gedanken hat der 1193 gestorbene Metropolit Eustathios in einem Brief an die Thessalonicher wiederaufgenommen, wobei er ein Verbum κροκοδειλίζω bildet: ψευδόμεθα τὸ κλαίειν, ὅτε καὶ κροκοδειλίζομεν, ἐφ' οἷς μὴ ἔχομεν κατεσθίειν ὁλοκλήρως τοὺς ἀδελφούς. καὶ γάρ τοι καὶ κροκόδειλος ποτάμιος ἀνθρώπου περιγενόμενος, καὶ τὸν ὅλον κατασπάσας ἄχρι καὶ εἰς κεφαλήν, εἶτα μαθών, ἄσαρκον αὐτὴν εἶναι καὶ ἀχρείαν εἰς βρῶσιν, κλαίειν ἐθέλει, οὐ πρὸς οἶκτον, ἀλλ' ὅτι μικρὰ κερδαίνει τοῦ ἐγκαταλείμματος.[61]

Manches andere ist noch durch die Evangelien und den *Physiologos* vermittelt worden. Ich belasse es bei diesen Andeutungen, hoffe aber gezeigt zu haben, daß es lohnt, bei der Erforschung der arabischen Sprache, ihrer Redensarten und bildhaften Ausdrücke immer auch ein Auge auf ein mögliches griechisches Substrat zu werfen.

---

[59] b. -Muʿtazz (Smr.) I 443,403, vgl. C. Lang, in *ZDMG* 40 (1886), p. 592; 41 (1887), p. 278.

[60] *Photii Bibliotheca*, ex recensione I. Bekkeri, Tomus alter (Berolini, 1825), p. 503a 34ff. (nr. 271).

[61] *Eustathii metropolitae Thessalonicensis Opuscula*, ed. Theophil. L. Fr. Tafel (Francofurti ad Moenum, 1832), p. 165,54ff. (§ 33). Bei K. Böttcher, K.H. Berger [et al.], *Geflügelte Worte. Zitate, Sentenzen und Begriffe in ihrem geschichtlichen Zusammenhang zusammengestellt und kommentiert* (Leipzig, 1981), p. 50 nr. 222, heißt es: 'In den Apostolischen Konstitutionen, einer Sammlung kirchenrechtlicher und liturgischer Quellen, um 380 in Syrien entstanden, wird (X 17) unechte Trauer umschrieben mit Krokodilstränen weinen'. Die Διαταγαὶ τῶν ἁγίων ἀποστόλων διὰ Κλήμεντος (J.-P. Migne, *Patrologia Graeca*, I [Paris, 1857], pp. 555–1156; *Didascalia et Constitutiones Apostolorum*, ed. F.X. Funk, I–II [Paderbornae, 1905]) umfassen aber nur acht Bücher, und in keinem dieser Bücher steht im 17. Kapitel etwas über Krokodilstränen. Auch über die Indizes war die Stelle nicht zu finden.

THE GENDER OF STANDARD ARABIC

Jan Jaap de Ruiter

Morocco's languages in terms of gender

In her excellent study on the relationship between gender and language in Morocco, Fatima Sadiqi (2003) argues that Standard Arabic is a male language; that Amazigh is a female language; that Moroccan Arabic is both a male and a female language, but more a male language in rural areas; and that French is a female and male language, but more a female language in urban areas.[1] She puts this picture into perspective by stating that 'this dichotomous qualification of Moroccan languages is not rigid as women and men may use various languages according to context.' The aim of her qualification is basically 'to draw attention to the fact that the use of languages in Morocco affects, and is affected by, gender'.[2] Sadiqi's idea to label languages in terms of male and female fits in well with the ideological setting in which she wrote her study. Seeking to define the role of women in Moroccan society from the perspective of the languages used, she investigates the emancipation of women in society and, in doing so, she has investigated how the languages used are linked to the sexes. Her approach is not of an empirical nature, and she does not exclude, as we shall see below, that language choices may change through future modifications in Moroccan society, be they of a social, political or economic nature.

The present article deals with the issue of the characterization of Standard Arabic in terms of gender. It will therefore leave aside the gender qualification of the other languages that Sadiqi treats in her study: Amazigh, Moroccan Arabic and French.[3] Sadiqi qualifies Standard Arabic as a male language *par excellence*, given its role in the religious domain in Moroccan society. It is the men who were and are more com-

[1] F. Sadiqi, *Women, gender and language in Morocco*. Women and gender in the Middle East and the Islamic World; 1 (Leiden [etc.], 2003), p. 218.

[2] Sadiqi 2003, p. 218.

[3] See for a discussion of the gender characteristics of these languages J.J. de Ruiter, 'Morocco's languages and gender. Evidence from the field,' *Journal of the sociology of language* (2007), forthc.

petent in that language because men are much more dominant in the religious establishment. When religious affairs are discussed it is mostly men who do so; they preach in mosques and discuss religious issues in television programs. The religious practice should be an open and public domain, but women have strangely enough hardly any access to it. If women mastered Standard Arabic, this would not automatically grant them access to the domain of this language. Women experience their religious identity much more in visits to shrines of holy men, where Standard Arabic does not play a significant role. Furthermore, Standard Arabic has legal power since it is the language of *sharia* legislation and Moroccan family law, which was only recently reformed in favor of women. *Sharia* legislation is known for its conservative attitudes to women, again a domain that excludes women. Sadiqi subsequently observes that Standard Arabic is the language of a literary tradition that is mostly supported by men. From her personal observation, Sadiqi notices that women are not very inclined to use Standard Arabic because of the attitudes of men. 'Moroccan men in general are more favorable to women's proficiency in French than to their proficiency in Standard Arabic,' she notes in her study;[4] and then there is the paradox that 'Moroccan women's "distance" from the spheres in which Standard Arabic is used explains the general tendency to disqualify women as competent speakers in Moroccan culture.' Sadiqi concludes her thoughts on Standard Arabic as a male language with the remark that Moroccan feminists deliberately use Standard Arabic in discussions in order to highlight the fact that Islam would preach universal ideas about equality and tolerance between the sexes.[5] Sadiqi's characterization of Morocco's languages in terms of gender is very much inspired by the *language choice* of Moroccan citizens. The process of arriving at a certain choice is influenced by a multitude of factors relating to standing and prestige (Standard Arabic), intimacy and family values (Amazigh), practicality (Moroccan Arabic) and social and economic mobility (French). Sadiqi's qualification of the various languages that play a role on the Moroccan linguistic market is founded on her wide-ranging and long-time experience as a sociological and anthropological observer.

---

[4] Sadiqi 2003, p. 222.
[5] Sadiqi 2003, p. 224.

The recent history of sociolinguistic research in Morocco

In the seventies and eighties of the last century Abbassi[6] and Gravel[7] executed large language surveys among students in Moroccan educational institutions, followed by the 1992 study of El-Biad.[8] These studies produced sociolinguistic images in which the subjects showed a fairly constant preference for using French over using Standard Arabic, while showing very positive attitudes towards the latter language. In general, the students used Moroccan Arabic as the dominant language in everyday communication. The policy of Arabicization of the country, implemented more or less consistently by Moroccan governments since the beginning of the eighties, apparently failed to bring about a massive shift towards Standard Arabic in terms of proficiency and use. With the publication of the *Charte d'éducation* in 1999, Morocco seemed to be on its way to a new linguistically based form of educational politics. It made an *ouverture* towards Amazigh as a language to be taught in education and also opened the door to Moroccan Arabic as a language of instruction in elementary education.[9]

At the end of the nineties I started another major sociolinguistic survey in Morocco. I had been a regular visitor to Morocco both for professional and private reasons. I had seen a country developing itself under changing language policies with alternating dominance of Standard Arabic and French,[10] and I had seen the developments in the Amazigh movement and, to a lesser extent, developments in circles that propagate the use of Moroccan Arabic vernacular as a language of culture and a language possibly to be used as a language of instruction in the education system.[11] In earlier research on the relationship between Moroccan Arabic and Standard Arabic I attempted to reconcile the two for practical purposes, such as their integrated use in education.[12]

---

[6] A. Abbassi, *A sociolinguistic analysis of multilingualism in Morocco* (Austin, 1977).

[7] L.A. Gravel, A sociolinguistic investigation of mutilingualism in Morocco (Ann Arbor, 1979).

[8] M. El-Biad, 'The role of some population sectors in the progress of Arabization in Morocco,' *International journal of the sociology of language* 87 (1991), pp. 27–44.

[9] J.J. de Ruiter, 'Analyse (socio-)linguistiqe de la Charte nationale marocaine d'éducation et de formation,' *L'Arabisant. Journal de l'Association Française des Arabisants (AFDA)* 35 (2001), pp. 63–74.

[10] G. Grandguillaume, *Arabisation et politique linguistique au Maghreb* (Paris, 1983).

[11] A. Youssi, *Grammaire et lexique de l'arabe marocain moderne* (Casablanca, 1992); A. Youssi, F. Benjelloun, M. Dahbi & Z. Iraqui-Sinaceur (eds.), *In honour of David Cohen. Aspects of the dialects of Arabic today* (Rabat, 2002).

[12] J.J. de Ruiter, 'Árabe estándar y árabe dialectal en la enseñanza de la lengua y cultura

The survey used a sociolinguistic questionnaire, with more than 70 questions, distributed among 569 students in different educational institutions in nine Moroccan cities. These were students at the Faculties of Arts in Oujda, Fes, Rabat, Beni-Mellal and Agadir, students at *Ecoles Supérieures* in Tangier, Meknes and Rabat and students at private language schools in Marrakech and Casablanca. Out of the total of 569 students, 248 were male (43,6%) and 321 female (56,4%). The female students were slightly overrepresented in the study, which may be attributed to the fact that the research was mainly carried out at Faculties of Arts, which traditionally tend to attract more women than men. The average age of the subjects was 21 years and five months. Moroccan Arabic was the mother tongue of 185 men (74,6%) and of 242 women (75,4%); Amazigh was the mother tongue of 62 men (25%) and 78 women (24,3%). Two participants indicated that their mother tongue was French.

The questionnaire that was presented to the students consisted of questions on their proficiency in Standard and Moroccan Arabic, French and Amazigh and their use of these languages. Furthermore, the questionnaire presented questions on the subjects' attitudes towards dialectal and Standard Arabic.[13] The questionnaire did not contain any questions on the subjects' attitudes towards French and Amazigh, basically because the research as a whole focused on both varieties of Arabic. The questions had a closed character, meaning that the students were presented with preformulated answers out of which they could make a choice. The answers to the proficiency, use and attitudes questions were ranged on five-point scales. This method of working facilitated the statistical processing of the data, which produced per variable absolute numbers, percentage distributions, means and standard deviations. Readers interested in the detailed results of the study are referred to the report in De Ruiter (2006).[14] Some results of the research are presented below under 'The gender of Standard Arabic.'

---

de origen (ELCO) marroquí en Europa,' in A. Franzé (ed.), *Lengua y cultura de origen. Niños marroquíes en la escuela española*. Collección encuentros. Serie TEIM; 1 (Madrid, 1999), pp. 123–142.

[13] H. Obdeijn & J.J. de Ruiter (eds.), *Le Maroc au coeur de l'Europe. L'Enseignement de la langue et culture d'origine (ELCO) aux élèves marocains dans cinq pays européens* (Tilburg, 1998); J.J. de Ruiter, 'L'arabe standard et l'arabe dialectal dans l'Enseignement de la Langue et Culture d'Origine (ELCO) marocain en Europe,' *L'Arabisant. Journal de l'Association Française des Arabisants (AFDA)* 34 (2000), pp. 19–35.

[14] J.J. de Ruiter, *Les jeunes Marocains et leurs langues*. Espaces discursifs (Paris, 2006), p. 44.

The research question

Sadiqi (2003)[15] establishes the gender associations of the languages used by Moroccans in daily life on the basis of her rich experience as a researcher and observer of the Moroccan language market, and in that context she labels Standard Arabic a male language. My research data (De Ruiter, 2006) on the other hand are based on extensive fieldwork among young Moroccans in educational institutions who make use of the languages involved. It has never been my aim to label languages in terms of male and female. Nevertheless, the study contains a chapter in which the sociolinguistic profiles of the male and the female students are compared to each other.

The analysis shows that there are – sometimes considerable – differences between the two sexes in Morocco when it comes to their proficiency in the languages involved, their use of these languages and their attitudes towards them. On the basis of these observations, one might argue that Sadiqi's characterization of a language as being male or female, or both, may be confirmed or rejected when analyzing the results of my data on the basis of gender. For instance, Standard Arabic is a male language according to Sadiqi, and it is used considerably more frequently by the male students than by the female students in my study (see below). This would support Sadiqi's claim that this language has a male character. In short, the database of my study could serve as an excellent empirical testing ground for Sadiqi's (2003) qualification of Morocco's languages in terms of gender. The main aim of this article, therefore, is to establish to what extent Sadiqi's qualification of the languages in use in Morocco in terms of gender is confirmed by data from the field, that is, by data derived from my study on the language proficiency in the languages mentioned among 569 students, their use of these languages and their attitudes towards dialectal and Standard Arabic.

In doing so, I need to explain the conditions under which the present comparison takes place. Firstly, Sadiqi's characterization of languages in terms of gender is primarily based on the concept of language use or behavior. The language profiles in my study are also composed of information on language use, but, in addition, they also involve information on language proficiency and language attitudes. This fact of my research having a wider scope, however, should not thwart the comparison.

[15] See above, n. 1.

Basing oneself on more than one language domain may also serve to reinforce the outcomes of the comparison. If Sadiqi claims that Standard Arabic is a male language, and should my data come to the same conclusion, this would reinforce the observation that Standard Arabic is a male language. The opposite is also true. If my data present a characterization that differs from Sadiqi's, they would pose a stronger argument against Sadiqi's characterization.

Secondly, Sadiqi applies her analysis of the language situation in Morocco on the whole of its society, ranging from conservative religious leaders who use only Standard Arabic to monolingual Amazigh women who live in isolation in remote areas of the kingdom. My study is, obviously, confined to young students only. Nevertheless, one may argue that Sadiqi does not exclude students from her analysis. Her assertions are valid about students, too.

A final consideration bears on possible changes in the gender character of the languages involved. Sadiqi suggested that, in the future, changes may take place in the gender of these languages. What if Amazigh grows into a well-stabilized and formal language in Moroccan society? Will it then become more male than female? This cannot be excluded. One could postulate therefore that the characterization of the languages in terms of male and female is subject to change when the behavior of the language users changes. From this last perspective, mismatches between Sadiqi's labeling of languages in terms of gender and the outcomes of the comparison between her research and mine should not be interpreted as if her study makes claims that cannot be founded.

The gender of Standard Arabic

Bearing in mind the considerations mentioned above, I analyzed the data of De Ruiter (2006) in terms of male- and femaleness and compared them to Sadiqi's (2003) gender characterization of the languages of Morocco. My research data are of three types: data on language proficiency, language use and language attitudes. In each of these three domains, several questions were submitted to the 569 subjects, and they could opt to answer on a four- or five-point scale, resulting, among other things, in a means per variable. The means between both gender groups were statistically compared to each other, which, in some cases, yielded statistically significant differences between both groups. In general, the results of the analysis showed that the differences in proficiency, use and attitudes between both gender groups were not large, but in two out of three domains – proficiency and use – some differences turned out to be statistically significant.

The following procedure for comparing Sadiqi (2003) and De Ruiter (2006) was applied. In my research, the subjects were submitted a number of questions per domain (proficiency, use and attitudes), e.g., 'What is your level in speaking Standard Arabic?' or 'To what extent do you use Standard Arabic during meals?' which they could answer on a five-point scale. Per question or variable, the answers were analyzed in four categories: the distribution of absolute numbers that a certain answer was chosen; the percentage distribution of these answers; a means of the total of answers; and the standard deviation of the means in question.

The results of my study on proficiency, use and attitudes are presented in the three tables presented below (tables 1–3). In the first column the language that we are dealing with here, Standard Arabic, is mentioned in combination with Sadiqi's characterization of this language in terms of gender. In the second column the diverse variables of the study of De Ruiter are mentioned. The third and fourth columns indicate the means per gender group per variable. Would the male group have a means of 1,8862 for speaking Standard Arabic, and the female group 1,9128, that means that the answers of both groups are close to answer 2 'I speak Standard Arabic well'. Answer 1 is 'I speak Standard Arabic very well'. The scales are explained in parentheses following the table title. In the example, the means of the male group are closer to answer 1 than the means of the female group. The male group claims to speak the language better than the female group, no matter how small the difference between both groups. In each table, the means of the group that expresses the highest proficiency (table 1), the most frequent use (table 2) or the most positive attitude (table 3) are indicated by boldface. If a difference between two means is statistically significant, this is indicated by the abbreviation 'sign.' in the column of the group in whose favor this difference is.

In the educational institutions of Beni-Mellal, Casablanca, Fes, Marrakech, Oujda and Rabat, the questionnaire was distributed twice. In the first round, the questionnaire did not yet contain questions on language proficiency. The subjects of Agadir, Meknes and Tangier filled out the complete questionnaire. Consequently, 176 subjects (31%) out of the

Table 1: Proficiency in Standard Arabic. The means in bold indicate the highest proficiency (sign = significant difference; original scale: 1 Very well; 2 Well; 3 Average; 4 Poor; 5 Not at all).

	Variable	Means Male	Means Female
Standard Arabic (Male)	Speak	**1,8862**	1,9128
	Listen	1,3772	**1,3716**
	Read	1,4731	**1,3945**
	Write	1,8383	**1,7064** (sign)

total group of 569 subjects did not answer the questions on language proficiency. Nevertheless, there is no doubt that the proficiency results of the study, based on questionnaires still coming from subjects from the educational institutions of all nine cities involved, can be considered representative for the whole group.

The question asked in the proficiency part of the questionnaire was 'What is your level in speaking/listening (understanding)/reading and writing Standard Arabic?' The accompanying scale is explained in parentheses following the title in table 1. The means of both groups for speaking Standard Arabic are close to answer 2: 'good'. The same observation goes for writing this language. The subjects in both gender groups understand spoken Arabic very well, and they are also competent in writing the language. In the case of writing Standard Arabic, the calculation of the statistical significance of the difference between males and females shows that the female subjects are significantly better at writing Standard Arabic than the males.

The data on language proficiency led to the following characterization of Standard Arabic in terms of gender. Mastery of Standard Arabic by the female subjects is better than by the male ones, but the differences in means are not very big, except for the variable of writing, which proved to be in favor of the female subjects. This makes Standard Arabic a female language in the first place and a male one in the second place.

The main question in the domain of language use was: 'To what extent do you use Standard Arabic in the following contexts?' The questions focused on the skills of reading, writing, speaking and listening or understanding. The accompanying scale is explained in parentheses following the title of table 2. The means of both groups for the questions on reading books, newspapers and magazines in Standard Arabic all center

Table 2: Use of Standard Arabic and Moroccan Arabic in terms of gender. The means in bold indicate the most frequent use (sign = significant difference; original scale: 1 Always; 2 Very often; 3 In 50% of cases; 4 Sometimes; 5 Never).

	Variable	Means Male	Means Female
Standard Arabic (Male)	Read book	**3,0528**	3,0883
	Read newspaper	**2,8571**	2,9206
	Read magazine	3,2857	**3,0415**
	Read letter	3,3211	**3,3016**
	Write letter	**3,2886**	3,3206
	Speak during meals	**4,5000**	4,5737
	Speak with other students	**4,1633** (sign)	4,3281
	Speak in the street	**4,4024** (sign)	4,5801
	Watch/listen tv/radio	**3,2146**	3,2834
	Listen CDs	**3,4024**	3,4430

around answer 3: 'in 50% of cases'. The differences between the groups are marginal, but they show a slightly stronger use of Standard Arabic by the male group. In the case of writing letters, the means of both groups are nearly identical: in both cases slightly higher than answer 3: 'in 50% of cases', although the men write just a few more letters in Standard Arabic than the women. For both groups, the means for speaking Standard Arabic in all variables range between answer 4: 'sometimes' and 5: 'never'. In this context, too, however, men speak Standard Arabic more than women during meals at home, with people in the street and with other students. In the last two variables, the difference between men and women is significant in favor of the men. With regard to understanding or listening to spoken Arabic, both groups follow television and radio programs and listen to cassettes and CDs in that language to an equal degree, with means slightly higher than option 3: 'in 50% of cases'. The differences between both groups are negligible.

The male subjects make more use of Standard Arabic than the female students: in two variables, they use it significantly more often, and, in six variables, they show more frequent use than the female subjects. From the perspective of use, this would make the language a male language.

The subjects were presented with two questions on their opinions about the beauty and importance of Standard Arabic. The questions were formulated as follows: 'To what extent do you consider Standard Arabic beautiful/important?' The scales are presented in parentheses following the title of table 3.

In both cases the means of both groups of their judgment of the beauty of Standard Arabic are around 1,55, which is between answer, 1 'very beautiful' and 2, 'beautiful'. The means of the female subjects is nonetheless closer to answer 1 than the means of the men. Therefore they consider Standard Arabic to be more beautiful than men. For the judgment of the importance of Standard Arabic more or less the same image is produced. The female students find Standard Arabic more important than the male students. The female students find Standard Arabic more beautiful and more important than the male students. From this attitudinal perspective this language bears a more female label. Nonetheless, the differences between both groups are not very great, and labeling them as exclusively female or male would therefore go too far. The differences did not turn out to be statistically significant.

The results of the comparison

Table 4 gives an overview of the results of the comparison. Per language and domain, it indicates proficiency, use and attitudes, and the number of times that means and significant differences are in favor of male or female subjects.

Table 4 shows that Standard Arabic cannot exclusively be labeled a male language. From the perspective of proficiency, it is a female lan-

Table 3: Attitudes towards Standard Arabic in terms of gender. The means in bold indicate the most positive attitudes (scale for beauty: 1 Very beautiful; 2 Beautiful; 3 Neither beautiful nor ugly; 4 Ugly; 5 Very ugly; scale for importance: 1 Very important; 2 Important; 3 Neutral; 4 Not important; 5 Not important at all).

Language	Variable	Means Male	Means Female
Standard Arabic (Male)	Beauty	1,6058	**1,5250**
	Importance	1,6857	**1,5394**

guage; from the perspective of use, it is a male language; and from the attitudinal perspective, it is primarily a female language. One would conclude from table 4, therefore, that Standard Arabic is both a female and male language, in that order.

What are the implications of these results for Sadiqi's characterization of Standard Arabic in terms of gender? Table 5 shows the characterization of Standard Arabic in terms of gender based on Sadiqi (2003) and De Ruiter (2006) and the degree to which they match.

It is clear that the comparison of the characterization of both studies does not produce a match. A large difference comes to surface. In De Ruiter's study, Standard Arabic is much more frequently used by male students, which confirms Sadiqi's judgment of this language in terms of gender, considering that language use is her starting-point. Nonetheless, the female students in De Ruiter's study have a better command of Standard Arabic and show more positive attitudes towards it. The broader basis of De Ruiter's data lead to the designation of Standard Arabic as a language that serves both genders. From the perspective of use, it is indeed male, but if we combine the factors of proficiency, use and attitudes, the language loses its male character and becomes a language of males and females.

Table 4: Results of the comparison between Sadiqi's (2003) study and De Ruiter's (2006). The number of instances where a means shows itself in favor of a gender group and the number of instances where the difference in means is statistically significant in favor of a gender group.

Language	Domain	Means Male	Sign Male	Means Female	Sign Female
Standard Arabic (Male)	Proficiency	1	–	3	1
	Use	8	2	2	–
	Attitudes	–	–	2	–

Table 5: Results of the characterization of Standard Arabic in terms of gender in Sadiqi's study (2003) and De Ruiter (2006).

Language	Label Sadiqi	Label De Ruiter	Match?
Standard Arabic	Male	Female and male	Small match

The goals of Sadiqi's and De Ruiter's studies are totally different. Sadiqi aims to describe Morocco's *marché linguistique* in terms of gender in order to highlight the oppressed position of women in Moroccan society and in order to create platforms for the further development and emancipation of women in this society. De Ruiter does not share such ideological concerns; his aim was simply to present an up-to-date picture of Moroccan sociolinguistics at the start of the new millennium. It has never been his goal to label languages in terms of gender. Furthermore, there are the limitations mentioned above, which put the outcomes of the comparison between both studies into perspective. Nevertheless, the present article shows once again that a combination of types of research (observational and empirical studies, in this case) yields a very rich basis for, in this case, a fruitful evaluation and discussion of Morocco's languages in terms of gender.[16]

Epilogue

In Classical Greek and Arabic sciences the four elements, fire, earth, air and water play an important role in understanding Nature. The four elements share four qualities: hot, shared by fire and air; dry, shared by fire and earth; cold, shared by earth and water and wet, shared by water and air. Gender is related to these four qualities as well. Males are considered dry and hot and females are considered cold and wet. Related to the four elements and qualities are the notions of left and right, where in those same scientific traditions the notion 'left' is related to females and the notion 'right' to males. In the section on human embryology (*tawallud al-insān*) in the 'Marvels of Creation' by Zakariyyā b. Muḥammad al-Qazwīnī (1200–1283), it is stated that one of the origins of a fetus to become a male or a female lies in the flowing of the male sperm from a man to a woman. Would it flow from his right side, (that is from his right testicle) to her right side (that is, to her right side of the womb), then a male embryo is conceived; if it flows from his left side to her left side a female is conceived; if it flows from his left side to her right side a 'feminine male' is begotten, and finally, if sperm flows from his left side to her right side, a 'masculine male'. One can also read in this fascinating

[16] J.J. de Ruiter, 'Compte-rendu du livre "Women, Gender and Language in Morocco" de Fatima Sadiqi', *Awal. Cahiers d'études berbères* 30 (2004), pp. 193–195.

text that if the innate heat is perfect during the moment of conception a 'masculine male' (baby) is the result and when this innate heat is less perfect a 'masculine female' is the result. It was Remke Kruk who, when I was at the time in doubt what to do for my final thesis, encouraged me to plunge in this wonderful world of classical Arabic medicine and translate and analyze al-Qazwīnī's text on human embryology.[17] The text made it clear to me that male and female are two parameters that play different roles in, in this case, classical Arabic medical science. Coming from that background I was astonished to read Sadiqi's 21st century study in which she labels the languages of Morocco in terms of gender, with Standard Arabic being a male language. I should, though, not have been that much surprised. I should have considered that Fatima Sadiqi fits well within a classical Islamic way of thinking to distinguish between sexes, not only in many aspects of science, but also in daily life and thus also in languages. From the perspective of the same classical concept of making distinction between the sexes, in for me mysterious ways, I cannot but conclude that I highly esteem the period when I as a male (conceived from right to left for that matter) learned the beautiful language that Standard Arabic is from a female tutor (in my view conceived from left to left). If I were to apply the way of thinking and of considering languages in terms of male and female, I would brand Standard Arabic a female language (again conceived from left to left), just because I learned to master and love it through Remke Kruk.

[17] J.J. de Ruiter, *Embryology in the Aja'ib al-Makhluqat of Zakariya b. Muhammad b. Mahmud al-Qazwini* (Utrecht University, master's thesis, 1985). Id., 'Human embryology in Zakariya al-Qazwini's The Marvels of Creation,' *Tijdschrift voor de geschiedenis der geneeskunde, natuurwetenschappen, wiskunde en techniek* 3 (1986), pp. 99–117. Id., 'De wording van de mens,' in A. Vrolijk (ed.), *De taal der engelen. 1250 jaar klassiek Arabisch proza* (Amsterdam, 2002), pp. 333–346.

ART & MUSIC

ISLAMIC ART AND ARCHITECTURE THROUGH THE EYES OF TESTAS

DOCUMENTARY DRAWINGS AND GENRE SCENES

Luitgard Mols

Dutch collections house a multitude of sketches with oriental themes by the 19th-century Dutch painter Willem de Famars Testas (1834–1896) that represent his observations made during a lengthy stay in Egypt from 1858 to 1860.[1] His sketches cover two different topics. Most abundant are his impressions of scenes of daily life, landscapes, depictions of horses and camels, and portraits of people. Fewer in number but yet remarkable for their degree of detail are his sketches that record constructions and architectural ornament. His choice to depict such divergent topics is explained in his diary and letters to his parents written during this trip.[2] Herein, Testas tells in detail how he joined his distant relative the famous architect-engineer and *dessinateur* Émile Prisse d'Avennes whose aim was to compose a pictorial Atlas of the monuments of Cairo. In its final form, this monumental work entitled *L'art arabe d'après les monuments du Kaire depuis le VII siècle jusqu'à la fin du XVIII* (Paris, 1877), referred to below as the Atlas, consisted of one volume of text and three volumes dedicated to the plates.[3]

[1] The five main collections are: Rijksmuseum van Oudheden in Leiden, Rijksprentenkabinet in Amsterdam, Teylers Museum in Haarlem, Prentenkabinet in Leiden and a private collection. I would like to thank Maarten Raven of the Rijksmuseum van Oudheden in Leiden, Celeste Langendijk of the Teylers Museum in Haarlem, and the private collector who wishes to remain anonymous for their kind help in sharing with me both insights and information on the draughtsman Testas. Furthermore I would like to thank Dorothy Wessels Boer-Stallman for correcting my English where it had gone wrong.

[2] His diary and letters are preserved in Leiden, Rijksmuseum van Oudheden (diary: inv. no. 19.6.1/1; letters: inv. no. 19.6.2/1). The sections in these writings pertaining to his trip form the basis of M.J. Raven (ed.), *Willem de Famars Testas. Reisschetsen uit Egypte: 1858–860*, (Maarssen [etc.], 1988).

[3] Throughout this article the Text volume is referred to as 'Text vol.,' the plate volumes as I–III. Several new editions have recently come to light, among which are E. Prisse d'Avennes, *Islamic art in Cairo from the 7th to the 18th centuries*, intr. G. Scanlon (Cairo, 1999), and Prisse d'Avennes, *Arabic art after monuments in Cairo*, intr. C. Schmidt (Paris, 2001).

Although Testas initially partook in the expedition for personal reasons, mainly for the benefit of widening his horizon at the age of 24 after finishing art school, he records that he became the group's draughtsman when the artist who had originally been assigned had called off. The team was also joined by the photographer Jarrot, which only emphasizes Prisse's intent to record the monuments as precisely as possible.[4] This article aims to assess two different questions. Firstly, what was Testas' actual contribution to the monumental Atlas? And, secondly, did his assignments for Prisse exert an influence on his own drawings?

Testas' work for Prisse

When assessing Testas' actual input into the Atlas, his contribution seems, at first glance, to be limited, for only one plate in the Atlas has been attributed to him.[5] Extant sketches by Testas, in addition to written documents in his hand, suggest, however, that his participation was more substantial. Testas left behind a list consisting of 188 entries, each of which describes an activity carried out for Prisse.[6] The majority of assignments referred to in this list constitute drawings of pharaonic and Islamic constructions and works of art, whereas Coptic and Roman artefacts are only rendered sporadically. In all, 58 items on this list address Islamic topics.[7] The scope of the tasks executed by Testas is wide, the subject matter varying from simple ornament and objects such as lamps and doors to the entire decorative design on a wall in the interior of a mosque. When these listed items are compared to the lithos in Prisse's Atlas, it becomes clear that half of the 58 topics referred to by

[4] For more information on the photographer Jarrot and his work, see M. Boom, 'Tekenen met licht. De fotograaf A. Jarrot en beeldend kunstenaar Willem de Famars Testas in Egypte, 1858–1860,' *Bulletin van het Rijksmuseum* 42/3 (1991), pp. 185–201.

[5] The plate attributed to Testas is a drawing of a *mashrabiyya* in Prisse d'Avennes 1877, III, pl. 134. The vast majority of the plates included in the three volumes are attributed to Prisse d'Avennes. In addition to Prisse d'Avennes and Testas, four other artists are mentioned in these volumes: Girault de Prangey (11 plates), Ch. De Linas (3 plates), Cournault (9 plates), and Yahia el-Wacetty (4 plates).

[6] The handwritten list is preserved in Leiden, Rijksmuseum van Oudheden. For the complete list, see Raven 1988, pp. 194–200.

[7] Besides drawing, the techniques he refers to in his list are tracing and the making of so-called 'estampages', which Testas describes as 'Impressions on unpasted and moistened paper which one beats into bas-reliefs with a brush'. For this description, see Raven 1988, p. 90. A drawing by Testas of the application of this technique is in Haarlem, Teylers Museum (inv. no. KT1866). It is depicted in Raven 1988, p. 143.

Testas in his list have been included in the Atlas, all but one of which are attributed to Prisse. For an exact comparison, the reader is referred to Appendix 1 below.

In some specific cases, his involvement in particular assignments for the Atlas is not only sustained by a reference in his list of activities but also validated by both his diary and extant sketches.[8] This is the case, for example, with the mosque of al-Burdaynī in which, according to his diary, Testas performed solitary work:[9]

> In this time I work as a paid draughtsman and I work either sitting or standing in the mosques from the early morning to dusk, Friday excepted, at which I receive many tasks at the house. I have thus spent the entire month of January in the small mosque of Bordu, in the company of the Janitsar with whom I cannot converse and who lays himself down to rest all the time. In this manner I am getting my teeth into constructional architectural drawings, with decorative motifs that are hardly visible and which are giving me the shivers.[10]

The result of his sojourn in the mosque is further specified in his list of items carried out for Prisse, in which he mentioned six tasks that are devoted to the mosque of al-Burdaynī.[11] The description of these six tasks corresponds with six plates included in the Atlas.[12] Furthermore, Testas' involvement in the preparatory drawings for one of the lithos is sustained by an extant sketch (inv. no. TTS 095) in a private collection (Plate 8). The sketch shows his attempts to grasp the varied interior decoration of a number of hexagons. These hexagons, now, are identical to those embellishing the ceiling of a *dikka* depicted in Prisse's Atlas (II, pl. 68).

---

[8] Extant sketches in the two main archives of Prisse in France, one in the Bibliothèque Nationale in Paris and the other in Avesnes-sur-Helpe, might yield more evidence of Testas' input for the Atlas. This, however, falls outside the scope of the current article.

[9] Besides the occurrence of an alternative spelling or transliteration of Arabic names, such as the reference to the mosque of al-Burdaynī as Bordu or Bordeen, the names and dates attributed to monuments by both Testas and Prisse d'Avennes were sometimes erroneous. Scanlon in Prisse d'Avennes 1999 corrected most references. One entry in the Atlas, which has not yet been corrected, is the drawing (II, pl. 98) of the medallion door attributed to the mosque of Ylgai al-Yūsufī. The blazon, i.e. a scimitar in a 3-tiered medallion depicted at the core of the door, however, is the blazon of Amir Īnāl al-Yūsufī, who became Amīr al-Silāḥ (amir of weapons) in 779/1377–78. This door is still found today in the entrance portal of the *madrasa* of the latter amir.

[10] Raven 1988, p. 88. Testas relates that upon returning to the house they rented, his work was criticised by Prisse d'Avennes who remarked upon his slowness in depicting ornament and his linear drawing.

[11] These are nos. 51 to 56 in Testas' list.

[12] These are plates 59, 60, 61, 68, and 69 in the Atlas.

Another example, in which his involvement in the drawings of the Atlas is corroborated by literary references and by sketches extant in Dutch collections alike, is the funerary complex of Sultan Qāitbāy.[13] In his list of activities for Prisse, Testas refers to two large drawings (nos. 64 and 65) of two different sides of the interior, one of which is a view on the *minbar* and *miḥrāb*. In the Atlas, two plates (I, pls. 20 and 21) are found that indeed meet the above-mentioned description. In addition to this correspondence, Testas' involvement is suggested by extant sketches which show a variety of architectural decorative detail that also recurs in the plates in the Atlas.[14] In one of these drawings, he includes two men, one a kneeling figure wearing a turban and one a man to his left, who are positioned in front of the *miḥrāb* of the mosque (Plates 9 and 10). Evidence that this sketch served as basis for the final drawing is provided by the fact that these selfsame men also recur in front of the *miḥrāb* in the litho in the Atlas (I, pl. 21), where they are joined by two other standing figures. The same sketch informs us that Testas did not develop all decorative motifs equally, for the inscription band above the *miḥrāb* is merely suggested. The absence of calligraphy is noticeable in another sketch as well, which shows a blank 3-tiered medallion surrounded by foliate scrolls set in an arcade (Plate 11).[15] In the final litho (I, pl. 21), the epigraphy in all three tiers of the medallion has been filled in. The question presents itself whether Testas was responsible for the calligraphy in the final drawing. From his extant sketches in Dutch collections, one does not get the impression that he was an active practitioner of the Arabic script, although some drawings contain small Arabic words.[16]

---

[13] For an overview of lithos and sketches with corresponding themes, see Appendix II of this article.

[14] Three sketches in the private collection (inv. nos. TTS 087, 093 and 094) show a variety of architectural ornament that is included in the two plates of the Atlas (I, pls. 20 and 21). In sketch TTS 093 the left part of a panelled window shutter is depicted, which recurs in the *īwān* on the aforementioned plates. An oblong band of floral ornament in the same sketch reappears in the plates of the Atlas below the ceiling. An ogival window, a capital, a column, and floor mosaics appear in sketch TTS 094, all of which recur in plates 20 and 21 of the Atlas. Also present in the same sketch is a floral motif that in plate 20 of the Atlas acts as a division between the lobed cartouches in the oblong inscription band below the ceiling. The floral scrolls set in a triangular shape in sketch TTS 087 corresponds with the arcade of the *īwān* in pl. 20 in Prisse's Atlas.

[15] Dutch private collection (inv. no. TTS 088). The arcade corresponds with the arcade above the *īwān* in pl. 21 of the Atlas where the 3-tiered medallion carries the epigraphic blazon of Sultan Qāitbāy. In the litho, the titles of the sultan in the central tier of the 3-tiered blazon are only partly present, whereas his name on the upper tier is indecipherable.

[16] The Rijksprentenkabinet in Amsterdam houses a sketch (inv. no. 1981:91b) in

The significance of these documentary drawings is that they not only represent the first systematic and scientific pictorial record of Islamic monuments in Cairo but sometimes also include objects or decorative motifs that have now vanished. The 3-tiered blazon surrounded by lush foliate scrolls that cover the arcade of the *qibla īwān* in the funerary complex of Qāitbāy in Testas' drawing mentioned above has been replaced by plain stones in the current situation.[17] Another difference between the drawing and the present state of the building is visible in the lower part of the *qibla* wall, for the current *ablaq* decoration surrounding the arcades of the window shutters and the oblong inscription band that nowadays covers the entire width are absent in Testas' drawing.

The Atlas contains two other drawings by Testas that enable us to reconstruct the original design of objects and their former ornament. The first example is that of a metalwork door that today has lost almost all of its metalwork mountings, although dark imprints of geometric shape on the wooden support remind us of the door's former ornament.[18] Testas' drawing not only represents the state of the door in its former glory but also shows the actual design of the door in 1911, when part of it was altered by the party responsible for the door's restoration.[19]

The second example in which the pictorial record provides information that has otherwise disappeared is a detailed drawing of a lion against a background of openwork foliate scrolls that cover the boss of a 12-pointed metalwork star.[20] According to Testas' reference (no. 23 in

---

which the words *yā Allāh yā Muḥammad* (Oh God oh Muḥammad) are written on the wall of a house. In addition to this, four names written in a bold calligraphic script have been preserved in the Dutch private collection (inv. nos. TTS 98–101). It is unknown whether they were written by Testas.

[17] For a photograph of the current state of the *qibla īwān*, see D. Behrens-Abouseif, *Islamic architecture in Cairo. An introduction* (Leiden, 1989), p. 146, pl. 104.

[18] This door is in the name of Amir Ilmās al-Yūsufī (no. 44 in the list of Testas) and was made for his funerary mosque (729–30/1329–30). It is depicted in Prisse d'Avennes 1877, II, pl. 100. For a description of the door and its restoration, see L. Mols, *Mamluk metalwork fittings in their artistic and architectural context* (Delft, 2006), pp. 200–202.

[19] The party responsible for the door's restoration was the *Comité de conservation des monuments de l'art arabe*. See *Comité de conservation des monuments de l'art arabe* 28 (1911 [publ. 1912]), p. 27. During the restoration, the original lower oblong band filled with whirling foliate ornament was removed and replaced by an oblong band with an inscription recalling the actual restoration of the door.

[20] There are three different versions of this embossed lion extant. The final drawing in Prisse d'Avennes 1877, Texte, p. 65, fig. 3. One sketch (in pencil and ink) is in the Rijksprentenkabinet in Amsterdam (inv. no. 1981:91–k) and is depicted in Raven 1988, p. 86, no. 33. The other more simple sketch is in the Dutch private collection (inv. no. TTS 009). When these two sketches are compared to the litho depicted in the Atlas, there is a marked difference between these drawings in the degree of detail with regard

his list), it was attached to a door of the mosque of Sultan Baybars. The lion *passant* represented on this boss is the blazon of Sultan Baybars (620–676/1223–1277), which was used on several of his religious buildings, his public works, and coins as well.[21] The door to which it originally belonged, a door whose main field is entirely covered with a repetitive design of metalwork stars and hexagons, is still extant, although today there remains no trace of the embossed star enlivened by the lion.[22]

Forced by circumstances – their stay in Egypt lasted a total of two years rather than the intended six months owing to illness and delays in travelling due to the low water level of the Nile – Testas had ample time to acquire experience in technical drawing relating Islamic art and architecture. While his diary tells of the difficulties in working with Prisse, Testas was proud of his contribution to the Atlas, which he deemed the best work on Egypt in both execution and exactitude.[23] Although he said not always to enjoy his work as *dessinateur*, he expected to benefit from his meticulous work in future times, especially when drawing figures in houses and monuments.[24]

---

to the star-shape, the arabesques used in the background, and depiction of the lion. The sketch in the private collection (inv. no. TTS 009) is the simplest, with only four of the 12 points of the star depicted while the lion's body is merely outlined. In the drawing now in the Rijksprentenkabinet in Amsterdam, Testas added more shadow to the body of the lion while accentuating such features as its eyes and mouth in addition to its paws. In the drawing used for the final publication, the face of the lion is further developed and the muscular tone of the lion's body is accentuated. Testas' role in the enterprise is not acknowledged, as the 12-pointed star with the lion in the Atlas is signed by J. Penel del. [an abbreviation of the Latin *delineavit*, i.e. has drawn] and by Bernard sc. [an abbreviation for the Latin *sculpsit*, meaning has engraved or has etched]. It is probable that Penel based his final version on a more advanced drawing made by Testas.

21 For an overview of the presence of this blazon on various buildings and artefacts, see L.A. Mayer, *Saracenic heraldry. A survey* (Oxford, 1933), pp. 106–10.

22 For detailed information on this door, see Mols 2006, pp. 179–181. Today the door is installed at the entrance portal to the French Embassy at Gizeh, Egypt, but formerly it served at the *madrasa* of Sultan Baybars. A somewhat earlier but far less detailed drawing in J. Bourgoin, *Les arts arabes: architecture. Le trait général de l'art arabe* (Paris, 1873), pl. 74, shows the location of this embossed star, i.e. in the lower zone of the star design, one on each leaf. The Victoria and Albert Museum in London houses a metalwork embossed star with a lion *passant* at its centre (inv. no. 909 part/49–1884), which was acquired in 1884. The sharp points of the star and the sharp-edged engravings on its surface suggest, however, that this particular piece was newly manufactured and that it is not the piece on which Testas' drawing is based. For a drawing of this embossed star, see S. Lane-Poole, *The art of the Saracens in Egypt* (London, 1886), p. 187, fig. 83.

23 Raven 1988, pp. 158–161. Testas elucidates that he is able to judge the quality of their work, for according to him they found a great deal of mistakes in the drawings of expeditions that preceded them, which they were able to correct.

24 Raven 1988, p. 90.

Testas' private work

Is any influence of his involvement in the documentation of Islamic monuments in Cairo actually visible in his private work? As far as the choice of subject matter is concerned, Testas chose a different angle. While Prisse's assignments were mostly geared towards the depiction of religious buildings ordered by the Ayyubid, Mamluk and Ottoman elite in which attention was focused on the interior design and decoration, Testas preferred to focus on a more worldly type of architecture such as houses, cafes, barber shops, etc. There, he concentrates on the exterior of buildings, another preference different from his work for Prisse.

Besides a difference in his choice of topics, Testas allows architecture to play different roles in his own drawings. Firstly, he uses buildings as a décor against which genre scenes come to life. In the foreground, the focus is on human interaction and animal behaviour – an early preference of Testas already visible in his preserved sketchbooks in which horses and dogs abound – often set in a commercial area in Cairo. In this scenery, the architecture is mostly of a simple nature, consisting of a stretch of different façades enlivened by small portals, shops, grilles, and *mashrabiyya* windows. A favoured theme, executed in pencil and in watercolour, is that of the groom running through the bustling shopping streets to make way for a horseman (Plate 12).[25] Irrespective of the seemingly secondary role played by the shops in the background, Testas adds architectural details, most of which are reminiscent of Mamluk or Ottoman times. These vary from a simple stone-cut geometric design serving to frame a portal or a bi-chrome lintel decorated with a joggled motif above a metalwork grille to the sculpted corbels positioned under the wooden *mashrabiyya*s in stone houses. The architectural ornament in the background is closely related to sketches preserved in the private collection (inv. no. TTS 013 verso). By implementing original architectural details, he aimed to enhance the realistic intent of scenes like these. His striving to authentically render these details is undoubtedly linked to his work for Prisse. The presence of photographs depicting street scenes in his own private archive shows that Testas could fall back on

[25] Examples of this, sketches in pencil (inv. nos. KT 1701 and KT1726) are in the Teylers Museum, Haarlem. A watercolour with the same theme is housed in the same museum (inv. no. DD26). It is depicted in Raven 1988, p. 184.

more than his own sketches to present a factual picture of constructions in 19th-century Cairo.[26]

However, Testas also allowed architecture to play the major role in his drawings. In contrast to his assignments for Prisse, in which interior designs and decorative motifs are usually the main focus, in his own work Testas centres on the exterior of buildings and the effect that tall buildings exert on the immediate surroundings. One of his favoured approaches is to capture the monumentality of buildings by rendering them only partially, i.e. omitting the top, with the result that their lofty façades and towering heights are emphasized even more (see Plate 13).[27] Another recurring theme is his views of cramped streets, in which the projecting balconies with *mashrabiyya* latticework draw attention to the physical embrace of the buildings and the confinement of space. In both instances, the people depicted merely act as requisites.[28] After having just arrived in Cairo, he already observed the phenomenon that, to his disappointment, most mosques cannot be seen from afar, for the streets are too narrow. He adds that this not only obstructs capturing these buildings by way of photography but also makes even drawing almost impossible.[29] Remarkably, he managed to turn the impossible into one of his favourite Oriental topics.

While some of his watercolours were made during his trip, other paintings with oriental themes were executed years after his return from visiting Egypt. For both compositional matters and architectural detail, he fell back on his own archive of sketches drawn during his two-year stay, which is illustrated by an oil painting that depicts a courtyard of a house in Cairo (Plate 14).[30] He closely followed the detailed composition of a sketch representing the exact same subject (Plate 15).[31] In all probability, Testas already planned to develop his sketch into a painting in a later stage, given the mass of architectural detail available (even the

[26] For an enumeration of the photographs and the respective collections, see Boom 1991, p. 201, n. 55.

[27] This is the case with a watercolour painted in 1859 (private collection, inv. no. TTS 029), which presents a street scene in al-Kab.

[28] For an example, see his view of a street in Cairo in watercolour (33.2 × 24.2 cm) dated 1859, depicted in Boom 1991, p. 191, pl. 8.

[29] Raven 1988, p. 47.

[30] Rijksmuseum Amsterdam, oil painting (84 × 62 cm), inv. no. SK-A-1184.

[31] The sketch, pencil on tracing paper (515 × 337 mm) is in Haarlem, Teylers Museum (inv. no. KT1710).

vanished stucco that originally covered the stone blocks has been outlined in the sketch), in addition to the fact that the sketch is squared, which allowed him to enlarge the drawing while retaining the right proportions of the building. Not all ornament in the sketch was, however, filled in. The *mashrabiyya* windows were outlined only, which is not surprising keeping in mind that he already gained a thorough experience in depicting this type of ornament as several of his assignments for Prisse show.[32] Another architectural element that had not yet been elaborated was the entrance door, which was decorated with the silhouette of a medallion and two triangles. In his oil painting, Testas creates a fully developed drawing of a shiny yellow metalwork lobed medallion with trefoil finials encompassed by four triangular-shaped cornerpieces and framed by a metal border band. A doorknocker and lock complement the door's design.[33] The degree of detail and realism are remarkable, given the lack of any detail in the sketch on which this painting is based. In order to achieve this degree of detail, he copied another of his sketches which shows an identical medallion door with exactly the same detail (Plate 16).[34] In addition to adding decorative motifs in the final painting, he also made alterations to his original design, for example by changing the figures in front of the door. The standing woman with child is replaced by a seated heavily veiled woman. For the latter, he again relied on his own sketches of seated veiled women.[35] All this gives insight into Testas' aim and method of working: he aspired after a realistic rendering of Egyptian material culture by copying his detailed sketches of single subject matter into a larger design.

Testas' two-year stay in Egypt had a lasting influence on his draughtsmanship, for after his return to Holland he remained devoted to oriental themes. His training in documentary drawing related to his work for Prisse, in combination with the sketches he made of animals, people, landscapes, and streets, left him with an eye for detail and an archive upon which he could fall back for his later work. Neither did his fascination

---

[32] See nos. 14, 15, and 27 of Testas' list.

[33] With it, Testas provides a glimpse of a now lost tradition, i.e. the use of Mamluk or Ottoman metalwork medallion doors in domestic architecture. Today, extant Mamluk or Ottoman specimens of this type of metalwork doors are found only in buildings of a religious nature.

[34] This sketch titled 'Porte de maison avec ornaments en bronze au Kaire' belongs to the Dutch private collection (inv. no. TTS 010).

[35] Inv. nos. TTS 047 and 048 in the Dutch private collection.

with the Orient end there, for in 1868 Testas revisited Egypt in the company of other painters, among whom was the famous orientalist painter Jean-Léon Gérôme.[36]

Appendix I

Assignments concerning Islamic artefacts and architectural ornament by Testas for Prisse, giving the numbers and tasks in the list left by Testas. Where applicable, the corresponding drawings in the Atlas are mentioned.

No.	Assignment*	Atlas
no. 14	Drawing of a *mashrabīya* of little known shape	
no. 15	Drawing of a courtyard with an arcade and mashrabīya	I, pl. 39
no. 18	Tracing of an arabesque in stucco above a door in an old palace	
no. 19	Impression of the former ornament	
no. 21	Impression of an arabesque ornament on a door in the Moristan (retraced in pencil afterwards)	
no. 22	Capital of a column in the mausoleum of Sultan Saleh Ayoub	
no. 23	Lion on a bronze boss on the door of a mosque of Beybars	Text vol., p. 65, fig. 3
no. 25	Door of a Harem	I, pl. 32?
no. 26	Different supports on the façades of houses in Cairo	
no. 27	Balconies of wood on a mosque	

[36] The diary that he kept during this trip is translated from the French by M. Raven (ed.), *Willem de Famars Testas. De schilderskaravaan, 1868* (Leiden, 1992). The original manuscript is kept in Leiden, Rijksmuseum van Oudheden (inv. no. 19.6.1/2).

Appendix 1 (*cont.*)

No.	Assignment*	Atlas
no. 28	Mausoleum in ruins at Karameydan	Text vol., pl. 18
no. 29	Retouche of the photographs of Gama Osman Kichyé	
no. 30	Six coloured tracings of faience slabs, one 3 metres in size, in the Gama Ibrahîm Aga	
no. 31	Pedestal of a column of the miḥrāb in the mausoleum of Saleh Ayoub	I, pl. 43, no. 7
no. 32	Drawing of the chair (minbar) in the mosque of Daoudiëh	
no. 33	Impression of the doors in the Gama Daoudiëh	
no. 37	Capital of the column in the Gama el Daoudiëh	I, pl. 43, no. 10
no. 39	Marble support for the jugs in the mosque of ?	
no. 40	Marble support for the jugs, Gama Cheikoun	
no. 41	Capital of the column in the mosque of Cheikoûn	
no. 42	Door of an Arabic house (design)	II, pl. 100
no. 43	Capital of the column in the miḥrāb of Gâma El Mâs	
no. 44	Impressions of the door of the mosque El Mâs	
no. 45	Door of a bath (watercolour)	
no. 46	View on a road with a small minaret (close to the Sikke Gedid street)	
no. 47	Courtyard in the mosque of Cheikoûn (design) watercolour	
no. 48	Drawing of the bas-reliefs in wood with animals in the Moristân	II, pl. 83
no. 49	Tracing of a design in faience representing a view on Mecca (fountain El-Kihaija)	II, pl. 83

Appendix 1 (*cont.*)

No.	Assignment*	Atlas
no. 50	Tracings of designs in faience around the ogival windows in the interior of the fountain el Kihaija	
no. 51	Joists of the ceiling of Gama Bordeen (coloured design)	II, pl. 69
no. 52	Details of a small ceiling of the Dekka of Gama Bordeen (coloured design)	II, pl. 68
no. 53	Coloured tracings of the mosaics along one of the sides in the interior of Gama Bordeen	I, pls. 60 and 61
no. 54	Drawing of the entire side of the interior wall with windows (coloured) of Gama Bordeen	I, pl. 59
no. 55	Drawing of a richly [decorated] alcove with a ceiling decorated in gold (same mosque)	
no. 56	Drawing of an inscription and borders in the mosque of Bordeen	
no. 59	Windows framed by arabesque ornament in the Gama Mohammed Nesfi Kouçoûn	I, pl. 46
no. 60	Impressions of the ornament in wood of the minbar (chair) [in the Gama Nesfi Kouçoûn]	II, pls. 85 and 87
no. 61	Large bronze lamp suspended in this mosque	I, pl. 18
no. 62	Drawing of a wooden grille, [...] of the first page of the Koran of the same mosque	
no. 64	Large drawing of an interior side with a view on the mihrab [and] minbar, etc. Gama Kaït Bey within the city	I, pl. 21
no. 65	Large drawing of another interior side, mosque Kaït Bey	I, pl. 20
no. 66	Impression of the minbar (chair), etc. [mosque Kaït Bey]	II, pl. 89?
no. 67	Arabesques traced in ink for Mr. Pr. (arabesques belonging to Mr. de la Porte, the French Consul)	

Appendix 1 (*cont.*)

No.	Assignment*	Atlas
no. 68	Tracings of faience (coloured) in the Gama Cheikoun	II, pls. 108, 109
no. 71	Arabic stove. Arabic tischt (brass basin)	III, pl. 172
no. 72	Marble tomb with a leaning figure (design) Gama?	
no. 73	Tracings of Kufic characters in mosaics and arabesques Gama Gourieh	
no. 74	Tracings in the mosque of Saleh	
no. 75	The façade decorated with six columns (int. courtyard) Gama Saleh	I, pl. 5 (top)
no. 76	Three tracings and different sketches in the small mosque of Sisarieh (in the citadel)	
no. 77	Interior view of the great mosque of Boulak: Sasaniëh	I, pl. 30 (top)
no. 80	Drawing of the mausoleum of Kalaoûn in the Morîstan	I, pl. 16
no. 81	Four coloured tracings in the mausoleum of Bersabé Acheraf (mausoleums of the caliphs)	
no. 82	Ornament in the mausoleum of Barkouk [mausoleums of the caliphs]	
no. 83	Column of the miḥrāb in the mausoleum of Kaït Bey [mausoleums of the caliphs]	I, pl. 43, no. 3
no. 84	Capital of the column in the old mosque of Omar (Old Cairo)	
no. 85	Capital in the Gama Barkouk (close to the Moristān)	I, pl. 43
no. 87	Capital of a column decorated with heads, Gama Saleh	

* Identification of the names of buildings referred to in Testas' list:
Barkouk (gama): the *madrasa* and *khānqāh* of Sultan Barqūq (786–88/1384–86)
Bersabé Acheraf (mausoleum): the *madrasa* and mausoleum of Sultan al-Ashraf Barsbāy (828–31/1425–27)

Beybars (mosque): the *madrasa* of Sultan al-Ẓāhir Baybars (660–62/1262–63)
Bordeen (gama/mosque): the mosque of al-Burdaynī (1024–38/1616–29)
Cheikoûn (mosque): the mosque of Amir Shaykhū (750/1349)
Daoudiëh (mosque, gama): the mosque of Malika Ṣafiyya (1018/1610)
El Mâs (gama/mosque): the funerary mosque of Amir Ulmās al-Nāṣirī (729–30/1329–30)
El-Kihaija (fountain) : the *sabīl* of ʿAbd al-Raḥmān Katkhudā (1156/1744)
Gourieh (Gama): the *madrasa* of Sultan al-Ghūrī (907–08/1501–03)
Ibrahîm Aga (Gama): the restoration of the Aq Sunqur mosque by Ibrahīm Aghā (1062/1652)
Kaït Bey (mosque/mausoleum): the *madrasa* and mausoleum of Sultan Qāitbāy (874–79/1470–74)
Kalaoûn (mausoleum/ moristan): the *bīmāristān* and mausoleum of Sultan al-Manṣūr Qalāʾūn (683–84/1284–86)
Mohammed Nesfi Kouçoûn (Gama): the mosque of Amir Qawṣūn al-Nāṣirī (729–30/1329–30)
Moristân: the *bīmāristān* of Sultan al-Manṣūr Qalāʾūn (683–84/1284–86)
Omar (mosque): the mosque of ʿAmr b. al-ʿĀṣ (18–20/640–41)
Osman Kichyé (Gama): the mosque of Amir ʿUthmān Katkhudā (1146/1734)
Saleh Ayoub (mausoleum): the mausoleum of Ṣāliḥ Najm al-Dīn (647–48/1249–50)
Sasaniëh (great mosque of Boulak): the mosque of Sinān Pāsha (978/1571)

Appendix II

Lithos and sketches with corresponding themes by Testas in Dutch collections

Lithos in Prisse's Atlas	Corresponding sketches by Testas
I, pl. 18	Teylers Museum, inv.no. KT 1867
I, pl. 20	Private coll. inv. nos. TTS 087, 093, 094
I, pl. 21	Private coll. inv. nos. TTS 006, 088, 093, 094
I, pl. 39	Teylers Museum, inv. no. KT 1707
I, pl. 43	Private coll. inv. no. TTS 008
I, pl. 59	Private coll. inv. no. TTS 089
II, pl. 68	Private coll. inv. no. TTS 095
Text vol. p. 65, fig. 3	Private coll. inv. no. TTS 009; Rijksprentenkabinet, inv. no. 1981:91-k

AFFE, LAUTE, NACHTIGALL

TIERE UND MUSIK IM ISLAM

Eckhard Neubauer

Der Umayyaden-Kalif Yazīd b. Muʿāwiya (reg. 60/680–64/683) war der erste Herrscher im Islam, der die Musik hoffähig machte. Seine Zuneigung zu dieser Kunst, aber auch zu seinem in Seide gekleideten Schoßäffchen Abū Qays prägt das Charakterbild, das der Historiker al-Masʿūdī (gest. 345/956) von ihm gezeichnet hat.[1] War dies ein eher zufälliges Zusammentreffen von Tier, Mensch und Musik, so kennt die Kultur des Islam zahlreiche tiefer gehende Verbindungen zwischen dem Reich der Tiere und dem der Töne. Sie reichen vom Schöpfungsmythos bis in die musikalische Praxis.

Einige Motive aus diesem Themenkreis werden hier anklingen, darunter die Entstehung der Musik aus dem Wesen der Tiere, deren 'Musikalität' und allerlei musikalische Anleihen aus dem Tierreich. Ausgespart bleiben Aspekte wie die Verwendung tierischer Materialien im Instrumentenbau (Häute, Felle, Roßhaar, Darm, Seide, Bein, Horn, Permutt, Elfenbein), die Verbindung von Musik, Tanz und Tieropfer in Heilzeremonien *(zār)* und ähnliches.

In der literarischen Tradition der Araber ist die 'Erfindung' der Musik und ihrer Instrumente in der Regel mit den Namen griechischer Gelehrter oder Gestalten aus dem Alten Testament verbunden.[2] In der ältesten Version, mitgeteilt von Hishām Ibn al-Kalbī (gest. um 204/819) in seinem 'Buch über den Beginn des Gesanges und der Lauten',[3] ist es Lamech, der Vater von Jubal, der im Schmerz um seinen jung gestorbenen Sohn dessen Leiche in einen Baum hängt ('seine Gestalt wird mir nicht aus den Augen gehen, bis sie zerfällt oder ich sterbe'). Die vom verwesenden Körper übrig bleibenden Teile, Schenkel, Bein, Fuß und

---

[1] Al-Masʿūdī, *Murūj al-dhahab*, ed. und franz. Übers. C. Barbier de Meynard & Pavet de Courteille, 9 Bde. (Paris, 1861–77), V, S. 156.

[2] s. A. Shiloah, 'The *ʿūd* and the origin of music,' in J. Blau [et al.] (eds.), *Studia orientalia memoriae D.H. Baneth dedicata* (Jerusalem, 1979), S. 395–407; vgl. T. Grame, 'The symbolism of the *ʿūd*,' *Asian music* 3/1 (1972), S. 25–34.

[3] *Kitāb Ibtidāʾ al-ghināʾ wa ʾl-ʿīdān*, s. F. Sezgin, *Geschichte des arabischen Schrifttums*, I (Leiden, 1967), S. 270, Nr. 16.

Zehen, bringen ihn zur Erfindung der Laute (*ʿūd*):[4] Der 'Körper' des Instruments entsteht als anthropomorphes Gebilde durch Sublimierung des Menschenopfers.[5]

Dagegen referiert der Historiker Maḥmūd b. Aḥmad al-ʿAynī (gest. 855/1451),[6] es sei der Autor des 'Buches der sinnreichen Erfindungen' (*Kitāb al-Ḥiyal*)[7] gewesen, der die Laute 'zuerst erfunden und derselben die Gestalt eines Kalbes gegeben' habe. 'Er gestaltete nämlich das Melavi [= *malāwī*, die Wirbel] wie die Zehen des Kalbs, die Saiten stellten die Adern vor.'[8] Hier beginnt die Geburt der Saiteninstrumente mit der Transzendierung eines Totemtiers. Auch im persischen und türkischen 'Papageienbuch' (*Ṭūṭīnāme*) ist es ein Tier, das Pate steht: Ein indischer Weiser, in der türkischen Version 'Instrumenten-Spieler' (pers. *sāz-pardāz*) genannt, beobachtet, wie ein Affe, von Baum zu Baum springend, an einem spitzen Ast hängenbleibt, der 'ihn dergestalt aufschlitzte, daß seine Gedärme (*rūde*) herausfielen.' Diese, zwischen zwei Ästen hängend, werden 'trocken und straff' und lassen im Winde 'anziehende Töne erschallen.' Der Weise bindet sie an ein 'gewölbtes Holz' (*chūbī kham*), [über ein anderes] spannt er eine Saite (*tār*) und befestigt außerdem am Holz einen halben Kürbis. So entstand, nach der persischen Version des 'Papageienbuches,' die Harfe (*chang*) und der Lautentyp *rubāb*. Auch die einsaitige indische Stockzither (*kingrā; kinnarī vīnā*) verbirgt sich hier und wird auch in Miniaturen dem Text beigegeben. 'Andere gaben,' in der türkischen Version, 'je nach ihrer Einsicht, dem Dinge verschiedene Gestalten, und so entstanden die mannigfaltigsten Instrumente.'[9] Und

[4] A. Shiloah (wie Anm. 2), S. 401, nach J. Robson, *Ancient Arabian musical instruments as described by al-Mufaḍḍal ibn Salama (9th century)* (Glasgow, 1938), S. 9, vgl. Ibn Khurradādhbih, *Mukhtār min Kitāb al-Lahw wa 'l-malāhī*, ed. I. ʿAbduh Khalīfa (Beirut, 1961), S. 14. Zum kulturhistorischen Hintergrund s. M. Schneider, 'Le rôle de la musique dans la mythologie et les rites des civilisations non européennes,' *Histoire de la musique*. Encyclopédie de la Pléiade; 1 (Paris, 1960), S. 131–214, hier S. 172–179; M. Vogel, *Onos lyras. Der Esel mit der Leier* (Düsseldorf, 1973), S. 414 ff. u.ö.

[5] Ausführlich bei Aḥmad b. Mūsā al-Sakhāwī (um 840/1436), wiedergegeben von Badr al-Dīn al-Mālikī (um 1060/1650), s. M. Schmidt-Relenberg, *Hören, Tanzen und Ekstase nach Badraddīn ibn Sālim al-Mālikī. Qūt al-arwāḥ fī aḥkām as-samāʿ al-mubāḥ. Text, Übersetzung und Kommentar* (Diss. Kiel, 1986), S. 71 ff. (Übers.), S. arab. 89 ff. (arab. Text).

[6] Im ersten Band seines *ʿIqd al-jumān fī tārīkh ahl al-zamān*, mir zur Zeit nicht zugänglich.

[7] Vielleicht ist Heron oder Philon gemeint, s. F. Sezgin, *Geschichte des arabischen Schrifttums*, V (Leiden, 1974), S. 149, 154.

[8] So übersetzt von Josef von Hammer, 'Beitrag zur Geschichte der orientalischen Musik,' *Fundgruben des Orients* 4 (1814), S. 383–384, hier S. 384.

[9] *Ṭūṭī-nāma – Das Papageienbuch. Vollständige Faksimile-Ausgabe im Originalformat*

al-Kindī (gest. nach 256/870) bestätigt: 'Man baute viele Saiteninstrumente (*ālāt watariyya*) in Anlehnung an die Zusammensetzung (*taʾlīf*) der Körper von Lebewesen/Tieren (*al-ajsād al-ḥayawāniyya*).'[10]

Im 'Papageienbuch' folgt dann an gleicher Stelle die Legende über den Vogel *qaqnus* oder *quqnis* (griech. *kyknos*, der Schwan), der den Menschen die Melodie (und die Blasinstrumente) gebracht hat: 'Was die Melodie der Musik anbetrifft, so hat sie einen anderen Ursprung. Es gibt in Indien einen Vogel, Kyknos genannt, in dessen Schnabel sich eine unzählige Menge von Löchern befindet. Ein jedes dieser Löcher läßt, wenn es intoniert wird, einen eigentümlichen seltsamen Laut hören, weshalb die Weisen Indiens die Melodie und den Tonwechsel daher entnommen haben.' In der persischen Fassung hat der Schnabel sieben Löcher, die vermutlich den Tönen der heptatonischen Skala entsprechen und üblicherweise den Planeten zugeordnet werden. Je siebzig verschiedene Töne können sie hervorbringen. Nach Farīd al-Dīn ʿAṭṭārs (6./12. Jh.) *Manṭiq al-ṭayr* hat der Schnabel des *qaqnus* etwa hundert Löcher, deren jedes anders tönt. 'Ein Philosoph' entwickelte daraus 'die Wissenschaft der Musik' (*ʿilm-i mūsīqī*).[11] Der Philosoph war Plato.[12] Ein vergleichbarer legendärer Vogel mit Löchern im Schnabel steht schon im Tierbuch von ʿUbayd Allāh b. Jibrīl b. Bukhtīshūʿ (gest. 450/1058), wo er, je nach Überlieferung, *suryānās* (aus *suryānay*, 'Schalmei'), *sīrinās* ('Sirene') oder *arghūn* ('Orgel')[13] heißt. Ibn Bukhtīshūʿ beruft sich auf die 'Zoologen' (*aṣḥāb al-ḥayawān*), die 'Musiktheoretiker' (*aṣḥāb ʿilm al-mūsīqī*) und die 'Interpreten der alten Schriften' (*aṣḥāb tafāsīr al-kutub al-ʿatīqa*), wenn er schreibt, daß die Klänge, die der Vogel hervorbringt,

(Graz, 1976), Faks.-Bd., f. 109a; *Tuti-Nameh "Das Papageienbuch". Aus der türkischen Fassung übertragen* von G. Rosen (Leipzig, 1962), S. 187–188. Die Legende lebt bis heute in kirgisischer Überlieferung, s. L. Aubert, 'Regards sur les musiques d'Asie centrale. La vièle-cheval et le luth-singe,' *Bulletin annuel* (Musée d'ethnographie de la ville de Genève) 28 (1985), S. 27–51, hier S. 38.

[10] *Muʾallafāt al-Kindī al-mūsīqiyya*, ed. Z. Yūsuf (Bagdad, 1962), S. 71, Z. -6.

[11] Farīd al-Dīn ʿAṭṭār, *Manṭiq al-ṭayr* (Teheran, 1374/1995), S. 129, nach A. Contadini, 'Musical beasts. The Swan-Phoenix in the Ibn Bakhtīshūʿ bestiaries,' in Bernard O'Kane (ed.), *The iconography of Islamic art. Studies in honour of Robert Hillenbrand* (Edinburgh, 2005), S. 93–101, hier S. 97. Ikonographische Belege aus islamischer Zeit zeigen mythische Vögel zusammen mit Musikern, s. E. Baer, *Sphinxes and harpies in medieval Islamic art* (Jerusalem, 1965), Abb. 13, 67, 72.

[12] *Bahjat al-rūḥ*, ed. H.L. Rabino de Borgomale (Teheran, 1346/1967), S. 27–28.

[13] Eine Parallele zum *sīmorgh*: 'Alle Wissenschaften stammen aus dem Flöten dieses Simurghs und sind von ihm abgeleitet; wundersame Instrumente wie die Orgel und anderes sind aus seinem Ruf und Gesang entstanden,' A. Schimmel, *Die drei Versprechen des Sperlings* (München, 1997), S. 180 nach Suhrawardī.

den Hörer paralysieren, daß die Phrygier danach ihren Aulos (*zamr*) entwickelt hätten, den sie zur Heilung von Kranken verwendeten, und daß die Seefahrer den Vogel 'Schalmeibläser' (*zāmir*) nennen.[14] Zakariyyāʾ b. Muḥammad al-Qazwīnī (gest. 682/1283) versetzt den Vogel *sīrānas* nach Afghanistan. Er trägt zwölf Löcher in seiner 'Nasenröhre' (*qaṣabat anfihī*). Beim Atmen bringt er den Ton einer Schalmei (*mizmār*) hervor, die denn auch von ihm ihren Ursprung haben soll. Mit dem süßen Klang lockt er die anderen Tiere an, lähmt sie, um sie zu töten, oder vertreibt sie mit gellendem Kreischen. Daneben kennt al-Qazwīnī ein Tier namens *shādahwār* 'aus dem fernsten Anatolien', das ein Horn (*qarn*) mit 42 hohlen 'Sprossen' (*shuʿba*) trägt, in denen der Wind 'angenehmste Töne' produziert. Für sich allein kann das Horn den Menschen entweder begeistern oder in tiefe Trauer stürzen, je nachdem, wie es gehalten wird und wie der Wind einfällt.[15] In später persischer, auf den Mystiker Najm al-Dīn Kubrā (gest. 618/1221) zurückgeführter Überlieferung hat der Schnabel des *qaqnus* 1002 Löcher verschiedener Größe, die eine sehr traurige Melodie hervorbringen. Der Vogel selbst trägt hier Züge des legendären Phoenix.[16] Nun hat es in der Tat im iranischen Raum um das 5./11. Jahrhundert ein Instrument gegeben, das, auch nach einem mythischen Vogel, *ʿanqāʾ* hieß. Es scheint eine Art Psalterium gewesen zu sein.[17]

Die Vision von einer umfassenden Harmonie zwischen Makro- und Mikrokosmos und zwischen den Lebewesen und ihren Äußerungen ging von der Spätantike auf die arabisch-islamische Kultur über. So fand al-Kindī sowohl tierische Laute als auch menschlichen Gesang im Musiksystem des syrisch-byzantinischen Oktoechos: 'Sämtliche Stile aller Völker haben Teil an den acht byzantinischen Modi…, es gibt nichts unter allem, was man hören kann, das nicht zu einem von ihnen gehörte, sei es die Stimme eines Menschen oder eines anderen Lebewesens, wie das Wiehern des Pferdes, das Schreien des Esels oder das Krähen des Hahns. Alles, was an Formen des Schreis einem jeden Lebewesen eigen ist, ist danach bekannt, zu welchem Modus (*laḥn*) der

[14] A. Contadini, 'Musical beasts' (wie Anm. 11), S. 93, 94–95.

[15] *ʿAjāʾib al-makhlūqāt*, ed. F. Wüstenfeld, *Zakarija Ben Muhammed…el-Cazwini's Kosmographie*, I, *Die Wunder der Schöpfung* (Göttingen, 1849), S. 397–398, zitiert von A. Contadini, a.a.O. S. 94.

[16] *Bahjat al-rūḥ* (wie Anm. 12), S. 25–27.

[17] s. L.I. al Faruqi, *An annotated glossary of Arabic musical terms* (Westport [etc.], 1981), S. 14–15. Zum Vogel s. E. Baer (wie Anm. 11), S. 38–42 und öfter.

acht es gehört, und es ist nicht möglich, daß es sich außerhalb eines von ihnen [bewege].'[18]

Diese Vorstellung führte dazu, daß man die Modi (*naghma, parde, maqām*) unmittelbar von Tierstimmen ableitete. In einer indo-persischen Handschrift, die möglicherweise aus dem 10./16. Jahrhundert stammt,[19] werden die zwölf grundlegenden Modi (*maqāmāt*) folgendermaßen zugeordnet: Der Hauptmakam *rāst* stammt vom Elefanten, *esfahān* vom Schaf, *ʿerāq* von Stier oder Kuh (*gāw*), *kūchek* (wörtl. 'klein') vom Schreien des Kleinkindes, *bozorg* (wörtl. 'groß') vom Rebhuhn, *ḥejāz* unklar, *būsalīk* vom Löwen, *ʿoshshāq* (wörtl. 'die Liebenden') unklar, *navā* von der Nachtigall, *ḥoseynī* vom Pferd, *rahāvī* von der Krähe und *zangūle* vom Kamel. Ergänzend und teilweise abweichend enthält eine persische Quelle aus dem 11./17. Jahrhundert[20] eine Aufzählung von 48 Modi, die aus Tierstimmen, Naturlauten und ähnlichem hergeleitet werden. Hier kommt *rāst* nicht vom Elefanten, sondern von der Trompete des Engels Esrāfīl beim Jüngsten Gericht, *esfahān* vom Falken, *kūchak* vom Hasen statt vom Kleinkind, *bozorg* von der Waldtaube (*mūsīje*), *ḥejāz* von der Schwalbe, *būsalīk* von der Maus statt vom Löwen, *ʿushshāq* von Hund und Hahn, *rahāvī* von der Taube (*fākhte*) und *zangūle* vom Kalb. Die übrigen Modi sind von den Stimmen von Fledermaus, Katze, Hund, Pferd, Esel, Maultier, Kamel, Gazelle, Hirsch, Affe, von den mythischen Vögeln *homāy* und *rokh*, von Huhn, Taube, Krähe, Falke, Reiher, Pfau, Vogel Strauß, Papagei und sogar vom Fisch abgeleitet. Eine ähnliche Aufzählung und die auffallende Zuordnung von Hund und Hahn zum gleichen *maqām* findet sich im 13./19. Jahrhundert in einer persischen Handschrift, in der *segāh* vom Hund (*sag*) abgeleitet wird und gleichzeitig den Hahnenschrei repräsentiert.[21] Dieser wird auch als Gebetsruf verstanden (s. unten).

Einige Tiere gelten an sich als 'musikalisch'. Dazu gehört neben den Singvögeln und dem Kamel vor allem die Biene,[22] deren Summen dem Ton der göttlichen Offenbarung und des Lobpreises am göttlichen

---

[18] *Risālat al-Kindī fī ʾl-Luḥūn wa ʾl-nagham*, ed. Z. Yūsuf (Bagdad, 1965), S. 26–27; franz. Übers. A. Shiloah, 'Un ancien traité sur le *ʿūd* d'Abū Yūsuf al-Kindī,' *Israel Oriental Studies* 4 (1974), S. 179–205, hier S. 203.

[19] K. Sain, 'An old manuscript on Iranian music,' *Indo-Iranica* 8 (1955), S. 29–36, hier S. 30.

[20] *Bahjat al-rūḥ* (wie Anm. 12), S. 79–80.

[21] MS Teheran, Malik 2830, f. 56b, mit Dank an Amir Hossein Pourjavady.

[22] Ibn ʿAbd Rabbih, *al-ʿIqd al-farīd*, ed. A. Amīn [et al.], 7 Bde. (Kairo, 1940–1953), hier VI (1949), S. 5.

Thron ähnelt,[23] die 'durch Gesang in Erregung gerät', deren Tanz an den Derwischtanz erinnert[24] und 'deren Junge durch angenehme Klänge zum Schlüpfen bewegt werden'.[25] In seinem 'Buch der Tiere'[26] faßt ʿAmr b. Baḥr al-Jāḥiẓ (gest. 255/869) zusammen: 'Die Gattungen (*ajnās*), die durch Gesang gekennzeichnet sind (*tūṣafu bi-l-ghināʾ*), sind diejenigen der Tauben (*ḥamām*) und der Mücken (*baʿūḍ*),[27] und diejenigen der Fliegen (*dhibbān*), zu denen die Wespe (*dabr*), die Biene (*naḥl*), die Kamelfliege (*shaʿrāʾ*), die Dasselfliege (*qamaʿ*) und die Bremse (*nuʿar*) gehören.' Streng genommen gehört das Summen der Insekten und das Gurren der Tauben sprachlich nicht zum 'Gesang' (*ghināʾ*), sondern in den Bereich der Kantillation (*tarannum*),[28] doch mit dem Satz 'die Araber nennen [auch] das Summen (*ṭanīn*) der Fliegen und Mücken Gesang (*ghināʾ*)' und einem Belegvers des Dichters al-Akhṭal (gest. um 92/710) bestätigt al-Jāḥiẓ den poetischen Sprachgebrauch.[29] Unter den Singvögeln sind es vor allem die Nachtigall (*bulbul*)[30] und der Wiedehopf (*hudhud*),[31] die für ihre schöne Stimme und Musikalität gerühmt werden. Delphin (*dulfīn*) und Krokodil (*timsāḥ*) tauchen, nach al-Kindī, aus dem Wasser auf und nähern sich dem Schiff, wenn sie den Ton einer Schalmei (*zamr*) oder eines Hornes (*būq*) hören. Auch für ihn gehört die Biene, dazu 'edle Pferde' (*ʿitāq al-khayl*), Pfauen (*ṭawāwīs*, sing. *ṭāwūs*), die Nachtigall (*hazār-dastān*), die Drossel (*sumānī*) und Gazellen (*ghizlān*, sing. *ghazāl*) zu den Tieren, die für Musik empfänglich sind: wenn der Pfau ein Saiteninstrument hört, schlägt er ein Rad.[32] Ibn Khurradādhbih (gest. 300/913) fügt die Mauersegler (*khaṭāṭīf*, sing. *khuṭṭāf*) und die Ringeltauben (*fawākhit*, sing. *fākhita*) hinzu.[33]

---

[23] s. A. Schimmel (wie Anm. 13), S. 66.

[24] Ebd. S. 66.

[25] s. Anon. (8./14. Jh.), *Kitāb al-Mīzān fī ʿilm al-adwār wa ʾl-awzān*, MS Gotha, A 85, f. 44a; Pseudo-Ṣafadī, *Risāla fī ʿilm al-mūsīqā* (Kairo, 1411/1991), S. 116; vgl. M. Schmidt-Relenberg (wie Anm. 5), S. 37, Z. -14 bis -13.

[26] ʿAmr b. Baḥr al-Jāḥiẓ, *Kitāb al-Ḥayawān*, ed. ʿA. M. Hārūn (Kairo, 1366/1947), III, S. 390.

[27] Zur Mücke vgl. A. Schimmel (wie Anm. 13), S. 44.

[28] s. E. Neubauer, 'Tarannum und terennüm in Poesie und Musik,' *Mélanges de l'Université Saint-Joseph* 48 (1973–74), S. 139–153, hier S. 139–140.

[29] *Kitāb al-Ḥayawān* (wie Anm. 26), III, S. 315.

[30] Ebd. I, S. 194; III, S. 339.

[31] Ebd. III, S. 524.

[32] *Muʾallafāt al-Kindī al-mūsīqiyya* (wie Anm. 10), S. 72.

[33] *Mukhtār min Kitāb al-Lahw wa ʾl-malāhī*, ed. I. ʿA. Khalīfa (Beirut, 1961), S. 21.

Was die Reaktion von Tieren auf die menschliche Stimme angeht, so steht im arabisch-beduinischen Raum das Kamel an erster Stelle. Eine der frühesten Formen vorislamischen arabischen Gesanges, das Kameltreiber-Lied (*ḥudā'*), ist ihm gewidmet, und zahlreich sind die Berichte über Reaktionen von Kamelen auf menschlichen Gesang.[34] Adolph von Wrede[35] schreibt 1843: 'Die Sitte, dem Kameele vorzusingen, herrscht im ganzen Orient, und die Kameele hören den Gesang gern und nehmen auch, sobald gesungen wird, einen raschern Schritt an.' Die 'Musikliebe' des Kamels führte schon früh zu legendären Schilderungen, zunächst in der erzählenden und der *samā'*-Literatur, dann auch in Musikschriften. Eine arabische Version handelt von einem Negersklaven, der wegen eines Vergehens von seinem Herren, einem Beduinen, hart bestraft wird. Der Grund: 'Er hat 400 schwer beladene Kamele in größter Hitze durch seinen Gesang so bezaubert und zur Eile angetrieben, daß sie zehn [andere Version: drei] Tagesmärsche an einem Tage zurückgelegt haben. Als ihnen die Lasten abgeladen wurden, brachen sie tot zusammen.'[36] Eine persische Legende berichtet von dem Musiker eines Herrschers, der diesem die Bedeutung seiner Kunst beweisen wollte. Er ließ ein Kamel fünf Tage lang dürsten. Als man ihm am sechsten Tag von einer Seite Wasser reichte und auf der anderen Seite der Musiker zu singen begann, wandte es sich diesem zu und trank erst, als er geendet hatte.[37] In türkischen Quellen des 9./15. Jahrhunderts wird die Geschichte dann mit dem Namen des bekannten Musikers und Musiktheoretikers Ṣafī al-Dīn al-Urmawī (gest. 693/1294) verknüpft, der dem Kalifen und dessen der Musik abgeneigten Rechtsgelehrten mit diesem Experiment erfolgreich bewies, daß die Musik eine edle und unterstützenswerte Kunst sei.[38] Der Musiker Ma'bad (gest. um 125/743) wiederum, nach seiner

---

[34] s. *Ibn Khaldûn. The Muqaddimah. An introduction to history.* Trans. from the Arabic by F. Rosenthal (New York, 1958), S. 402; A. Musil, *The manners and customs of the Rwala Bedouins* (New York, 1928), S. 358–368; J. Elsner, 'Zur Formbildung in der arabischen Musik,' in Hartmut Braun (ed.), *Probleme der Volksmusikforschung* (Bern [etc.], 1990), S. 119–134, hier S. 122–123; M. Vogel (wie Anm. 4), S. 366–368; A. Hamam, 'Bedouin music,' *The New Grove dictionary of music and musicians*, 2. Aufl. (London, 2001), III, S. 60–63 (*ḥudā'*).

[35] *Reise in Ḥadhramaut*, ed. H. von Maltzan (Braunschweig, 1870), S. 144.

[36] H. Ritter, *Das Meer der Seele* (Leiden, 1955), S. 263–264 mit Parallelüberlieferungen.

[37] Zakariyyā' b. Muḥammad al-Qazwīnī, *'Ajā'ib al-makhlūqāt*, pers. Übers., Ms. Washington, National Library of Medicine, P 2, f. 161a-b.

[38] *Seydī's book on music. A 15th century Turkish discourse.* Trans., annot. and ed. by E. Popescu-Judetz and E. Neubauer (Frankfurt, 2004), S. 24–27.

Methode des Komponierens befragt, gab zur Antwort: 'Ich besteige mein Reitkamel (*qaʿūd*), markiere das Metrum mit dem Schlagstab (*qaḍīb*) auf dem Sattel und summe (*tarannama*) das Gedicht, bis mir das Lied ausgereift ist (*istawā*).'[39] Hier trifft sich der historische Bericht über einen Komponisten der frühislamischen städtischen Musik mit der bis vor kurzem noch lebendigen Praxis, daß nächtliche Kamelreiter, ihre 'Kamelstöckchen' in der Hand,[40] altertümliche Lieder (*ṭāriq, ṭūrīq, radḥa, wanna*) anstimmen, 'um sich und das Kamel zu unterhalten.'[41]

Auch wilde Tiere können vom menschlichen Gesang angezogen werden, wie in der Orpheus-Sage und der Krishna-Legende. Wenn der Sänger Mukhāriq (gest. um 232/846) oder Prinz Ibrāhīm b. al-Mahdī (gest. 224/839) mit ihren schönen Stimmen sangen, näherten sich Tiere aus dem Wildpark des Kalifenpalastes, und selbst Tiere in freier Wildbahn entfernten sich nicht.[42]

Noch heute sagt man Musikern im Jemen nach, daß sie, wie einst Salomon, die Sprache der Vögel verstünden.[43] Beim Gesang ihres 'Konkurrenten' Isḥāq al-Mawṣilī (gest. 235/850) in Bagdad fiel eine Nachtigall tot vom Himmel.[44] Gleiches hört man noch aus den Schiraser Gärten des frühen 13./19. Jahrhunderts.[45] Zu jener Zeit ließ man in Damiette bei Abendunterhaltungen 'Nachtigallen kommen, die dazu abgerichtet sind, in Gesellschaft zu singen.'[46] Singvögel in Käfigen, heute meist Kanarienvögel, sind noch in unseren Tagen die Attraktion bei musikalischen Veranstaltungen in Afghanistan: 'Die Klänge der Musik, zusammen mit denen des Singvogels, bilden den Höhepunkt afghanischen Musikgenusses.'[47]

---

[39] Abū 'l-Faraj al-Iṣfahānī, *Kitāb al-Aghānī,* I (Kairo, 1927), S. 40 (hier und im folgenden, wenn nicht anders angegeben, nach der Edition Dār al-Kutub, 24 Bde., 1927–1974 zitiert).

[40] C.R. Raswan, *Im Lande der schwarzen Zelte* (Berlin, 1934), S. 149.

[41] s. Y. Shawqī, *Dictionary of traditional music in Oman*, revised and expanded by D. Christensen (Wilhelmshaven, 1994), S. 199, 212.

[42] Abū 'l-Faraj al-Iṣfahānī, *Kitāb al-Aghānī*, X, S. 110; XXI, ed. R. Brünnow (Leiden 1306/1889), S. 237.

[43] Ph.D. Schuyler, '*Qāt*, conversation, and song. A musical view of Yemeni social life,' *Yearbook for traditional music* 29 (1997), S. 57–72, hier S. 64–65.

[44] R.A. Nicholson, *The Kashf al-Maḥjúb. The oldest Persian treatise on Ṣúfism* by ʿAlí b. ʿUthmán al-Jullábí al-Hujwírí (Leiden und London, 1911), S. 399.

[45] W. Ouseley, *Travels in various countries of the East*, II (London, 1821), S. 220–222.

[46] M. Michaud und M. Poujoulat, *Correspondance d'Orient 1830–31*, VI (Paris, 1835), S. 358.

[47] J. Baily, 'Afghan perceptions of birdsong,' *The world of music* 39 (1997), S. 51, vgl. S. 53, 56.

Dem Frommen gilt das Krähen des Hahnes als Gebetsruf (*adhān*), der Gesang von Singvögeln als Gottesgedenken (*dhikr*).[48] Im persischen Kunstgesang wurde die Nachtigall[49] zum direkten Vorbild für eine charakteristische Gesangsmanier, einen unverwechselbaren Glottis-Schlag (*taḥrīr, takiyya*)[50] in Passagen mit Vokalisen, die als Nachahmung des 'Schlages' der Nachtigall[51] empfunden und auch so bezeichnet wird (*chehchehe, taḥrīr-e bolbolī*). Hervorragende Sänger werden mit dem Beinamen 'Nachtigall' (*bolbol*) geehrt.[52]

Die direkte, körperliche Nachahmung von Tieren findet im Tanz statt und entstammt meist den jeweils vorislamischen lokalen Kulturen. Historische Hinweise und Abbildungen zeigen Tänzer in tierischer Verkleidung oder Maske.[53] Als Erbe zentralasiatischer Turkstämme sind schamanistische Tänze, in denen das Tapsen des Bären oder der gleitende Flug des Adlers nachgeahmt werden, bis nach Anatolien gelangt. Dort finden wir außerdem Tänze benannt nach dem Widder (*koç halayı*), dem Hammel (*koyun zeybeği*), dem Kranich (*turna barı*), der viel auch in Liedern besungen wird, dem Rebhuhn (*keklik oyunu*), dem Huhn (*tavuk barı*) oder der Heuschrecke (*çekirge zeybeği*).[54] In der traditionellen persischen Tanzterminologie heißt ein Schritt 'Elefantenfuß' (*pīl-pā*), ein anderer 'Pfau' (*ṭāvūs*).[55] Die Belutschen in Ṣoḥār führen mit Gesang und Trommelspiel den *pāqit* auf, in dem Charaktere 'wie Pferde, Tiger, Löwen und Krähen' dargestellt werden.[56] Hobby-horse und Steckenpferd sind letzte Ausläufer eines im abbasidischen Bagdad noch höfischen Tanzes namens *kurraj*, bei dem den Tänzern mit Stoffen umkleidete 'Fohlen aus Holz' umgeschnallt wurden.[57]

---

[48] Ebenda, S. 54, 56.

[49] vgl. den Titel von L. Nooshin, 'The song of the nightingale. Processes of improvisation in *dastgāh segāh* (Iranian classical music),' *British journal of ethnomusicology* 7 (1998), S. 69–116.

[50] M. Caton, 'The vocal ornament *takīyah* in Persian music,' *Selected reports in ethnomusicology* 2 (1974), S. 43–53.

[51] vgl. H. Tiessen, *Musik der Natur* (Freiburg, 1953), S. 33–38.

[52] J. Baily (wie Anm. 47), S. 51.

[53] s. R. Ettinghausen, 'The dance with zoomorphic masks and other forms of entertainment seen in Islamic art,' in George Makdisi (ed.), *Arabic and Islamic studies in honor of Hamilton A.R. Gibb* (Leiden, 1965), S. 211–224 mit 25 Abb.

[54] C. Demirsipahi, *Türk halk oyunları* (Ankara, 1975), S. 13, 44 und passim.

[55] s. M. Rezvani, *Le théâtre et la danse en Iran* (Paris, 1962), S. 203.

[56] s. Y. Shawqī (wie Anm. 41), S. 134.

[57] s. Ibn Khaldûn (wie Anm. 34), S. 404; M. Gaudefroy-Demombynes, 'Sur le cheval-jupon et al-kurraj,' in *Mélanges William Marçais* (Paris, 1950), S. 155–160; A. Shiloah, 'Réflexions sur la danse artistique musulmane au moyen âge,' *Cahiers de civilisation*

Spätestens seit den frühen Abbasiden ließen sich die Kalifen, wenn sie auf die Jagd zogen, von Hofmusikern begleiten.[58] So werden auf einer Miniatur Lautenist und Falkner gemeinsam mit dem Herrscher dargestellt .[59] Auch das bei der Jagd geblasene Horn wird abgebildet, auf dem noch bei uns Halali (von arab. *ḥalāl*) geblasen wird.[60] Kleine, als 'Falkentrommel' (*ṭabl-bāz*; *davulbâz*) bezeichnete, am Sattel des Pferdes befestigte Bronzepäuklein wurden bei der Beizjagd eingesetzt.[61] Regeln für ihren Gebrauch sind in indo-iranischen Quellen überliefert.[62] Unter den Jagdtechniken und -listen heißt es: Elefanten fängt man, indem man 'Spielzeug und Musikinstrumente (*ālāt al-ṭarab*) für sie sammelt und sie damit vom Fressen abhält, so daß sie das Fliehen vernachlässigen, bis man sie ergreift und gefangen nimmt.'[63] Schon Solon soll die einschläfernde Wirkung der Musik auf Hirsche beobachtet haben,[64] auch Gazellen lassen sich von ihren Jägern in den Schlaf singen und sind dann leicht zu ergreifen.[65] Gefangene Elefanten jedoch, die aus Kummer die Nahrung verweigern, kann man mit 'rührenden Melodien' (*alḥān shajiyya*) wieder aufheitern.[66] Löwen lockt man des Nachts mit Gesang und dem Klang von Rahmentrommeln (*daff*) aus dem Dickicht.[67] Nach Ibn

---

médiévale (X^e–XIIe siècles) 5 (1962), S. 463–474, hier S. 472–474; S. Moreh, 'The hobbyhorse. A cultural contact between Asia and Africa,' in *BRISMES* (British Society for Middle Eastern Studies) *Proceedings of the 1991 international conference on Middle Eastern studies, SOAS London, 10–12 July 1991* (Exeter, 1991), S. 57–60; ders., *Live theatre and dramatic literature in the medieval Arab world* (Edinburgh, 1992), S. 27–44.

[58] s. z.B. Abū 'l-Faraj al-Iṣfahānī, *Kitāb al-Aghānī*, V, S. 394 (al-Wāthiq), ebd. S. 418 (Hārūn al-Rashīd).

[59] E.A. Polyakova und Z.I. Raḥimova, *Šarḳ miniatjurasi va adabiyëti* (Taschkent, 1987), Abb. 123.

[60] So auf einer Miniatur aus dem Houghton *Shāhnāme* von ca. 1530, s. S.C. Welch, *Wonders of the Age. Masterpieces of early Safavid painting, 1501–1576* (Cambridge, Mass., 1979) S. 77.

[61] *Museum des Institutes für Geschichte der Arabisch-Islamischen Wissenschaften. Beschreibung der Exponate*, I, *Musikinstrumente*, beschrieben von D. Franke (Frankfurt, 2000), S. 240. Eine historische Darstellung von ca. 968/1560 in S.C. Welch, *Royal Persian manuscripts* (London, 3. Aufl. 1978), S. 127.

[62] D.C. Phillott, 'Note on the drum in falconry,' *Journal and Proceedings of the Asiatic Society of Bengal*, N.S. 4 (1908, publ. 1910), S. 159–164.

[63] M. Schmidt-Relenberg (wie Anm. 5), S. 37, vgl. al-Qazwīnī (wie Anm. 37), f. 161a; M. Vogel (wie Anm. 4), S. 372.

[64] s. E. Kazemi, *Die bewegte Seele. Das spätantike Buch über das Wesen der Musik (Kitāb ʿUnṣur al-mūsīqī) von Paulos/Būlos in arabischer Übersetzung* (Frankfurt, 1999), S. 191.

[65] R.A. Nicholson (wie Anm. 44), S. 400.

[66] Ibn Khurradādhbih (wie Anm. 4), S. 21.

[67] Al-Qazwīnī (wie Anm. 37), f. 161a.

Khurradādhbih (gest. 300/913)[68] bauen die 'Leute aus den Sümpfen' (*ahl al-baṭā'iḥ*) im Südirak Gehege aus Rohr, die im Wasser stehen und Öffnungen enthalten. Sie locken die Fische mit ihren Stimmen an, füttern und hegen sie, bis die Gehege voll sind. Gesänge von Fischern (*nahma*, d.i. *naghma*, 'Ton', *mashal* etc.)[69] und von Perlentauchern[70] kennt man bis heute an den Küsten der Arabischen Halbinsel. Mit Trommeln hielten sich die Seefahrer im Indischen Ozean 'Walfische' vom Leib.[71] An Deck waren Trommeln (*zāna*, 'Instrument') als Signalinstrument und zur Begleitung der Arbeitsgesänge (*shalla*) und Seemannstänze (*midīma*) Teil der Ausrüstung des traditionellen Schiffes; beide wurden vom gleichen Meister hergestellt.[72] Zur Vogeljagd kennt man aus Afghanistan stammende höchst kunstvolle Wachtel-Locken (*tipcha bōdanā*), kleine Meisterwerke in Verarbeitung und Effekt.[73]

Das Bild, welches das Thema Jagd in Verbindung mit Musik über die Jahrhunderte hin dominierte, ist die Jagdszene des Sasanidenherrschers Bahrām Gōr aus Niẓāmī's Versepos *Haft paykar* (vor 593/1197). Die Sängerin Āzāde (oder Fetna), auf eigenem Reittier oder hinter dem Herrscher aufgesessen, nahm mit ihrer Harfe daran teil, reizte übermütig den Großkönig und verwirkte dabei ihr Leben.[74] Die zahlreichen erhaltenen Miniaturen veranschaulichen dem Instrumentenkundler die Entwicklung der persischen Harfe.[75]

Die bekannteste unter den 'musikalischen' Szenen des öffentlichen Lebens ist die herrscherliche Zeremonial- und Militärmusik mit Fanfaren und Pauken, Trommeln, Becken und Oboen, mit ihren Spielern hoch zu Roß oder auf Kamelen (*nawba*, *ṭablkhāna*, *mehtarkhāne*, Janitscharen), die von Miniaturmalern seit dem 8./14. Jahrhundert dargestellt wird. Wir finden sie von Illustrationen zu Firdawsī's (gest.

[68] Ibn Khurradādhbih (wie Anm. 4), S. 21.

[69] Y. Shawqī (wie Anm. 41), S. 61–62, 115, 130; G. Braune, *Küstenmusik in Südarabien* (Frankfurt etc., 1997), S. 132 ff.

[70] s. P.R. Olsen, *Music in Bahrain* (Hojbjerg und Bahrain, 2002), S. 85–98.

[71] Al-Masʿūdī (wie Anm. 1), I, S. 234.

[72] Y. Shawqī (wie Anm. 41), S. 117, 172 ff., 207, 214, 223.

[73] D. Franke (wie Anm. 61), S. 235.

[74] s. M.V. Fontana, 'Re Bahrām e la sua schiava nei manoscritti d'epoca timuride,' in *Atti del III Convegno internazionale sull'arte e sulla civiltà islamica* (Venezia, 1980), S. 91–113 mit 32 Abb.; H.G. Farmer, *Islam.* Musikgeschichte in Bildern; III,2 (Leipzig, [1966]), S. 30–31.

[75] Zur Entwicklung der persischen Harfe vgl. B. Lawergren, 'Harfen, 2. Persien,' in *Die Musik in Geschichte und Gegenwart, Sachteil* IV (Kassel etc., 1996), Sp. 51–55.

411/1020) *Shāhnāme*[76] bis hin zu Bildern der Orientmaler aus dem 19. und frühen 20. Jahrhundert.[77] Kampfgesänge beduinischer Reiter zu Pferd oder Kamel (*taghrūd*) blieben dagegen auf ihren Aktionskreis beschränkt.[78] Brautzüge auf Reittieren mit Musikern und ähnliche Festzüge bei Beschneidungen sind gleichfalls vorwiegend in ländlichen Gebieten zu sehen, soweit sie überhaupt noch in traditioneller Form stattfinden. Schausteller, die Bären oder Affen mit Gesang oder zum Klang eines Tamburins 'tanzen' lassen, gab es seit den Umayyaden[79] und noch bis vor kurzem.[80] Schlange, Affe und Ziege zeigten ihre Kunststücke zur Begleitung einer Flöte oder eines Tamburins in Ägypten.[81] Auch der 'Pferdetanz' (*laʿb al-khayl*, *raqṣ al-khayl*) unter Begleitung des ländlichen Ensembles aus Oboen und Trommeln (*mizmār baladī*) ist zu festlichen Anlässen noch in ägyptischen Dörfern zu bewundern.[82] Wandernde Derwische trugen ein Tierhorn umgehängt, wenn sie 'Tag und Nacht mit ihren Tamburinen unterwegs sind und Hymnen singen', wie es Clavijo[83] im Mai 1403 in der Nähe von Erzurum erlebt hat. Schließlich weiden Schafe in natura und auf Miniaturen selten allein. Ihr Hirte oder dessen Gehilfe spielt eine Flöte[84] oder singt eines der Hirtenlieder, die im heutigen Oman *ṭūrīq* heißen.[85] Die Glocken[86]

---

[76] z.B. Schlachtenszenen in B. Gray, *Persische Malerei* (Genf, 1983), S. 43, 89, 134, 135; S.C. Welch (wie Anm. 60), S. 69, S. 71.

[77] z.B. die *musiciens zambouraktchis* der persischen Artillerie um 1840, gemalt von F. Colombari, abgebildet in C. Juler, *Les orientalistes de l'école italienne* (Paris, 1987), S. 103.

[78] s. Y. Shawqī (wie Anm. 41), S. 194.

[79] Al-Masʿūdī (wie Anm. 1), V, S. 86, vgl. R. Ettinghausen (wie Anm. 53), Abb. XIX, 4 und 5.

[80] Singender Zigeuner und Tanzbär bei Bochara, s. G. Bonvalot, *En Asie centrale* (Paris, 1884), S. 270–271.

[81] M. Meyerhof, 'Schlangenbändiger in Ägypten,' *Ciba-Zeitschrift* 8/85 (1942), S. 2992–2993.

[82] P. Collaer und J. Elsner, *Nordafrika*. Musikgeschichte in Bildern; I,8 (Leipzig, 1983), S. 54b, letzter Abschnitt; M. Saleh, 'Les danses d'Égypte,' in Dj. Henni-Chebra und Ch. Poché (eds.) *Les danses dans le monde arabe ou l'héritage des almées* (Paris und Montréal, 1996), S. 113–142, hier S. 127–129; dies., 'Raqs al-khayl,' in *The Garland Encyclopedia of World Music*, VI, *The Middle East* (New York und London, 2002), S. 628; vgl. M. Vogel (wie Anm. 4), S. 368–369.

[83] *Embassy to Tamerlane*, engl. Übers. G. Le Strange (London, 1928), S. 140.

[84] s. S.C. Welch (wie Anm. 60), S. 165; E.A. Polyakova und Z.I. Raḥimova (wie Anm. 59), Abb. 110; F. Suleimanova, E. Y. Yusupov (eds.), *Nizomij "Ḫamsa" siga išlangan rasmlar* (Taschkent, 1985), Abb. 169; V. Kubičková (Text), R. Finlayson-Samsour (trans.), *Persian Miniatures*, (London, o.J.), Nr. 21.

[85] s. Y. Shawqī (wie Anm. 41), S. 203–204.

[86] Ebd. S. 261.

oder Glöckchen für die Weidetiere entstammen einer Produktion, die in den Ländern des Orients eine weit zurückreichende, im einzelnen noch unerforschte Geschichte hat.[87] Auf einer safawidischen Miniatur der Zeit um 945/1540 trägt das Leittier einer Ziegenherde eine Glocke um den Hals,[88] abgerichtete Wolfshunde tragen Glocken am Halsband (9./15. Jh.),[89] und ein Kamel auf einer Zeichnung aus dem 10./16. Jahrhundert trägt Schellenringe um den rechten Vorder- und den linken Hinterlauf.[90] Nächtliche Karawanenzüge mit der Glocke des Leittieres und den Schellen an den Beinen der übrigen Kamele boten ein weithin hörbares klingendes Spiel. Ausgediente Kamelglocken landeten im vergangenen Jahrhundert auf den Basaren.

Mannigfach sind Tierimitationen und -andeutungen im Instrumentenbau und Tiervergleiche in der schönen Literatur. Die Kastenleier der Spätantike, die im Islam als *miʿzafa* weiterlebte, wurde als Mausefalle bespöttelt.[91] Möglich ist, daß der Holzkasten mit stehenden Jocharmen, der noch im frühen 20. Jahrhundert im syrischen Bereich als Mausefalle diente, ein Erbe des Gerätes ist, das seinerzeit zu dem Vergleich Anlaß gab.[92] Der auf Miniaturen bisweilen mit einem Vogelkopf als abschließendem Ornament dargestellte,[93] breit gerundete und sich nach oben hin in der Biegung verjüngende Resonanzkasten der persischen Winkelharfe (*chang*) wurde passend mit einem Pferdehals verglichen.[94] Geschnitzte Vogelköpfe dienen bis heute als Abschluß von Wirbelkästen,[95] der 'Schnecke' unserer Violine vergleichbar, und es kommen auch vogelköpfige Saitenhalter vor.[96] Den Namen des aus dem sasanidisch-persischen

[87] E. Wiedemann, 'Über die Herstellung von Glocken bei den Muslimen,' *Mitteilungen zur Geschichte der Medizin und der Naturwissenschaften* 9 (1910), S. 475–476; Abbildungen historischer Kamelglocken s. F. Suleimanova und E.Y. Yusupov (wie Anm. 84), Abb. 9 (datiert 846/1442); *Topkapı Sarayı-Museum. Manuskripte*, ed. J.M. Rogers (Herrsching, 1986), Nr. 70 (datiert 905/1500).

[88] s. S.C. Welch (wie Anm. 60), S. 165.

[89] J.M. Rogers (wie Anm. 87), Abb. 104.

[90] S.C. Welch (wie Anm. 60), S. 190.

[91] Abū 'l-Faraj al-Iṣfahānī, *Kitāb al-Aghānī*, XII, S. 48.

[92] s. ʿA. Ḥamām, 'al-Miʿzafa fī 'l-turāth,' *Abḥāth al-Yarmūk* 10 (1994), S. 43–59, hier S. 55.

[93] H.G. Farmer (wie Anm. 74), Abb. 83 auf S. 97.

[94] s. z.B. Ḥasan Kāshānī, *Kanz al-toḥaf*, ed. T. Bīnesh, in *Seh resāle-ye fārsī dar mūsīqī* (Teheran, 1992), S. 55–128, hier S. 115, zur Pferdesymbolik allgemein s. Th.C. Grame und G. Tsuge, 'Steed symbolism on musical instruments,' *Musical quarterly* 58 (1972), S. 57–66.

[95] z.B. an *qanbūs*-Instrumenten, s. D. Franke (wie Anm. 61), S. 28, 29.

[96] Ebd. S. 60.

Bereich entlehnten Lauteninstrumentes *barbaṭ* (von griech. *bárbyton*), das mit einem kleineren Korpus und längerem Hals als der *ʿūd* versehen war und seinerseits mit der antiken Hebelwaage (*qarasṭūn*) verglichen wurde,[97] deutete man in persisch-arabischer Volksetymologie als 'Entenbauch' (*bar-i baṭ*).[98] Die im religiösen Bereich der Aleviten und Ahl-i ḥaqq gespielten und verehrten Langhalslauten (*bağlama*, *tanbūr*) tragen dagegen rein anthropomorphe Züge,[99] entsprechend 'Augen' und 'Nase', 'Gesicht', 'Hals' und 'Brust', 'Bauch' und 'Rücken' beim *ʿūd*.[100] Ein flaches, wohl der chinesischen *p'i-p'a* ähnelndes Lauteninstrument der frühen Abbasidenzeit wurde einem einheimischen Süßwasserfisch entsprechend 'Schollen-' oder 'Flunderlaute' (*ʿūd al-shabbūṭ*) genannt.[101] Der poetische Vergleich der Wirbel von Saiteninstrumenten mit 'Ohren' ist je nach dem Umfeld des Vergleiches zoomorph oder anthropomorph zu verstehen.[102] Der als 'Esel' (pers. *khar*, *kharak*, türk. *eşek*)[103] bezeichnete Steg wiederum ist eindeutig dem Tragesel zugeordnet[104] mit Ausnahme von neutürkischen Texten, in denen die osmanische Schriftform von *eşek* zu *eşik* ('Schwelle') umgedeutet wurde.[105] Es finden sich bisweilen Stege, die eine geduckte Tiergestalt andeuten.[106] Im Arabischen wird der Saitenhalter der Laute 'Pferd' (*faras*) genannt.[107] Die Dichtung liebt Vergleiche aus dem Tierreich, doch sind sie nicht nur freundlicher Art. Eine Schalmei spielende Negerin mußte es sich gefallen lassen, daß das Instrument in ihrem Mund mit dem Glied eines Maultieres verglichen wurde, ihre breiten Lippen mit der Vulva einer Eselin und die dunklen Finger auf dem Instrument mit Mistkäfern, die auf einer Schlange krabbeln.[108]

---

[97] J. Robson (wie Anm. 4), S. 6–7.

[98] Abū ʿAbd Allāh al-Khwārazmī, *Mafātīḥ al-ʿulūm*, ed. G. van Vloten (Leiden, 1895), S. 238.

[99] J. During, *Musique et mystique dans les traditions de l'Iran* (Paris und Teheran, 1989), S. 322–327; D. Franke (wie Anm. 61), S. 34.

[100] s. E. Neubauer, 'Der Bau der Laute und ihre Besaitung nach arabischen, persischen und türkischen Quellen des 9. bis 15. Jahrhunderts,' *Zeitschrift für Geschichte der arabisch-islamischen Wissenschaften* 8 (1993), S. 279–378, hier S. 282.

[101] Abū 'l-Faraj al-Iṣfahānī, *Kitāb al-Aghānī*, V, S. 202.

[102] vgl. D. Franke (wie Anm. 61), S. 42.

[103] *kharak* = *zāmila* in al-Maydānī, *al-Sāmī fī 'l-asāmī*, ed. M.M. Hindāwī ([Kairo], 1967), S. 173.

[104] vgl. D. Franke (wie Anm. 61), S. 42.

[105] vgl. ebd. S. 30, 39.

[106] s. z.B.D. Franke (wie Anm. 61), S. 148.

[107] s. z.B.M. Schmidt-Relenberg (wie Anm. 5), S. 71, Z. -11; E. Neubauer (wie Anm. 100), S. 364, Z. 13.

[108] s. M. Schmidt-Relenberg (wie Anm. 5), S. 70, Z. 8–12.

Der direkte Weg aus der Tierwelt in die Klangwelt führt über die ältesten bekannten Musikinstrumente. Ihn haben neben den Knochenflöten und den Tierhörnern die Muschelhörner genommen, die bis vor kurzem noch im Mittelmeer und im Indischen Ozean als Signalinstrumente verwendet wurden. Bezeugt sind Exemplare aus der Türkei, aus dem Irak und aus Oman,[109] wo das 'Tritonshorn' (*būq, yim*) auch zum Tanz geblasen wird.[110] Zwar sind Nachbildungen von Tieren als Gefäßflöten im islamischen Orient, wie fast überall auf der Welt, auf die Stufe von Kinderspielzeug herabgesunken,[111] doch konnte auf mittelalterlichen, in Aserbaidschan gefundenen Exemplaren anläßlich eines Kongresses 1990 in Buchara noch höchst elaborierte *mugām*-Musik geblasen werden.[112]

Anleihen, Assoziationen, Imitationen aus dem Tierreich finden sich in der Spieltechnik, der Melodik und den Namen von Modi und Instrumentalstücken. Ein vom Rhythmus geprägtes Melodiemodell ostanatolischer *aşık*-Sänger heißt 'Hoch zu Roß' (*at üstü*), die persische Entsprechung *asb-raw* war eine von 24 'Farben' (*rang*) zentralasiatischer Musik im 16. Jahrhundert.[113] Das Flattern des Vogels wird im 'Flug der Lerche' (*paresh-e jāl*) aus Afghanistan imitiert.[114] In der neueren persischen Kunstmusik (*radīf, dastgāh*) sind 'Lerche' (*chakāvak*) und 'Gesang der Lerche' (*naghme-ye chakāvak*) zwei vokal ausgeführte Abschnitte (*gūshe*) des *dastgāh* mit Namen *homāyūn*.[115] Unter den musikalischen Metren wurden mehrere von Vogelrufen inspiriert und entsprechend bezeichnet. Die nach Taubenarten benannten Metren *fākhitī* (oder *fākhte*), *warashān* und *qumrī* (*dawr-i qumriyya* und ähnlich) erscheinen in arabischen, persischen und türkischen Quellen seit dem 7./13. Jahrhundert. Ihre Länge variiert von der einfachsten Form mit 5 Zeiteinheiten bis zu zusammengesetzten Strukturen mit bis zu 80 Zeiteinheiten. An das Gurren und Rufen (Rucksen) von Tauben[116] erinnern in der Tat

---

[109] D. Franke (wie Anm. 61), S. 230.

[110] Y. Shawqī (wie Anm. 41), S. 35, 86, 148.

[111] s. H. Nixdorff, *Tönender Ton. Tongefäßflöten und Tonpfeifen aus Europa* (Berlin, 1974), Abb. 79, 80.

[112] Gespielt von Farhad Ibragimov zum Vortrag von Suraya Agajeva, *Musical instruments from excavations of medieval Baku.*

[113] A. Jung, *Quellen der traditionellen Kunstmusik der Usbeken und Tadshiken Mittelasiens* (Hamburg, 1989), S. 97.

[114] J. Baily (wie Anm. 47), S. 51–59, hier S. 56.

[115] s. H. A'lam, 'Čakāvak,' in *Encyclopaedia Iranica*, IV (London [etc.], 1990), S. 650.

[116] vgl. A. Usinger und H. Behnke, *Die Ruf-, Lock- und Reizjagd*, 4. Aufl. (Hamburg [etc.], 1975), S. 40.

Grundschläge wie, in Achtelnoten gedacht, 4+2+4+4+2+4 (*fākhitī*), 2+2+3+3+2 (*warashān*) oder 2+3 (*ḍarb qumriyya*).[117] Metren werden auch als 'schreitend' (*ravān*) oder als 'Reittier' (*rikāb*) bezeichnet.[118] 'Trabend' (*mukhabbab*) heißen bei al-Fārābī (gest. 339/950) beschleunigte Tonfolgen.[119]

Schließlich wirkten Tiere auch als Souffleure. Man kennt den Teufel (*iblīs*) als nächtlichen Einflüsterer von Melodien;[120] Tierdämonen konnten diese Funktion ebenfalls übernehmen. Dem Bagdader Hofmusiker Ibrāhīm al-Mawṣilī (gest. 188/804) erschienen im Traum eine weiße und eine schwarze Katze. Die schwarze sang ihm ein Lied vor, das ihn begeisterte, und lehrte es ihn. Diejenige unter seinen Schülerinnen (*jawārī*) aber, der er es als erste weitergab, verfiel in Wahnsinn, wie es die Katzen vorausgesagt hatten.[121] Der Begriff 'Katzenmusik' ist dem Orient fremd. Es blieb dem Abendland vorbehalten, ihn zu prägen.

---

[117] E. Neubauer, 'Glimpses of Arab music in Ottoman times,' *Zeitschrift für Geschichte der arabisch-islamischen Wissenschaften* 13 (1999–2000), S. 317–365, hier S. 346–353 passim.

[118] Ebd. S. 350.

[119] G.D. Sawa, *Music performance practice in the early ʿAbbāsid era 132–320 AH / 750–932 AD*, 2. Aufl. (Ottawa, 2004), S. 100.

[120] R. Erlich, 'Iblis-muzykant,' *Zapiski Kollegii Vostokovedov* (Leningrad) 5 (1930), S. 393–403.

[121] Abū 'l-Faraj al-Iṣfahānī, *Kitāb al-Aghānī*, V, S. 193–194. Dank an George Sawa.

REVIVAL OF THE ARABIC SUITE BY FEMALE VOICES

THE ART INTERPRETED BY AÏCHA REDOUANE AND BEIHDJA RAHAL

Anne van Oostrum

A phenomenon new to Arabic music is the interpretation of the suite by a female solo vocalist accompanied by a small ensemble. Traditionally, the male singer has always been the main figure in this genre of art music.

In Egypt and Syria, the oriental suite or *waṣla* is made up of a succession of vocal and instrumental compositions alternating with improvisations in the same musical mode or *maqām* (plur. *maqāmāt*). In the Western part of the Arab world (Morocco, Algeria and Tunisia), a type of suite called *nawba* (*nūba* in dialect) also exists which is based on the same principle.

This article focuses on two female interpreters of the Arabic suite: Aïcha Redouane who specializes in the Egyptian *waṣla* and Beihdja Rahal who performs the Algerian *nawba*. Each of these performers has her own innovative approach towards her art. During the last decade, these two artists have made successful concert tours in Europe, notably France, where their music is appreciated by an enthusiastic mixed audience of Arab and Western listeners.

In order to understand how each woman approaches this genre of art music, a brief account of the nature of Arabic music and the history of the *waṣla* and *nawba* will be given. Thereafter, the interpretations of Redouane (*waṣla*) and Rahal (*nawba*) will be discussed, paying particular attention to topics such as inspiration, education, lyrics, instrumental make-up, vocal techniques, the audience, and – last but not least – their unique role as female performers in this genre.[1]

---

[1] The social position of female performers is not discussed here. See H. Engel, *Die Stellung des Musikers im Arabisch-Islamischen Raum*. Orpheus-Schriftenreihe zur Grundfragen der Musik; 49 (Bonn, 1987).

The nature of Arabic music

Arabic music in general is based on a refined system of musical modes called *maqāmāt*. In Egyptian musical practice, for example, there are about sixty different *maqāmāt*.[2] While in the field of music *maqām* means scale or mode, it is more than a mere scale of seven tones. It is also an expression of a melodic course of the melody with fixed resting tones and a set of optional modulations. Moreover, a *maqām* can express an ethos or feeling. Professional musicians learn to master the art of improvisation within a particular *maqām*, be it instrumental (i.e., *taqsīm*),[3] or vocal (i.e., *layālī*), in order to display their virtuosity and creativity in this art.

There is no harmony in traditional Arabic music; singers and instrumentalists carry out the same melodic line, although each one embellishes it according to his or her own taste. Thus a kind of ornamented heterophony emerges. The singer is the main performer who sings several forms of poetry set to music while accompanied by a group of instrumentalists.

Each composition has an *īqāʿ* (plur. *īqāʿāt*) or rhythmic pattern or cycle consisting of heavy (*dum*) and light (*tak*) beats carried out by the percussionist. This cycle is repeated throughout the composition. However, one composition may have a few successive *īqāʿāt*.

Because it developed independently, the music in the Western part of the Arab world differs significantly from that heard in the Eastern region. There one finds other musical modes (*ṭabʿ*, plur. *ṭubūʿ*) instead of *maqām* (plur. *maqāmāt*), rhythmical cycles (*mīzān*, plur. *mawāzīn*) instead of *īqāʿ* (plur. *īqāʿāt*), instrumentation, compositional forms, styles of singing, to note a few important differences. However, in both regions, the music responds to the Arabic nature in general and includes a type of suite called *waṣla* in the East and *nawba* in the West.

An historical sketch of Arabic art music

Arabic art music dates from the early period of Islam, during the regimes of the first four caliphs (632–644 AD), when the cities Mecca

---

[2] See S. Marcus, *Arab music theory in the modern period*, 2 vols., Dissertation, University of California (Los Angeles, 1989).

[3] See A.H. van Oostrum, *The art of nāy playing in modern Egypt*. Dissertation, University of Leiden (Leiden, 2004), Ch. IV, pp. 152–213.

and Medina became the first centres of Islamic culture. The young urban elite, enriched by the spoils of war after the great conquests of parts of the neighbouring Sasanid and Byzantine empires, invited singers, musicians and dancers into their homes as entertainment. Many of these performers were of foreign origin who had learnt their craft at the Sasanid or Byzantine courts. They influenced the local music of the Arabs by introducing foreign modes, rhythms, styles of singing and instruments. In short, a new form of art music emerged.[4]

During the first dynasty of Islam, at the courts of the Umayyad Caliphs in Damascus (661–750 AD), professional performers of Mecca and Medina became court musicians. Their music developed further and blossomed during the era of the next dynasty, the Abbasids of Baghdad (749–1258 AD), who had defeated the Umayyads in 750 AD.

Meanwhile, surviving members of the Umayyad family had established an empire in Spain (756–1031 AD) where they upheld a court in Cordoba that equalled that of the Abbasids in Baghdad. The outstanding musician of Baghdad, Abū 'l-Ḥasan b. Nāfiʿ, nicknamed Ziryāb ('Blackbird'), performed and taught in Cordoba. He is accredited with the introduction of the *nawba*, the musical suite.

Ziryāb founded several conservatories in the cities of Andalusia in which young musicians were educated. They further developed his music, mixing this genre with local musical practices. A new Andalusian music arose which was distinctly different from the court music of Baghdad. This Andalusian music spread over the Western part of the Arab world (Morocco, Algeria and Tunisia) from the tenth century AD when the Berber dynasties of the Almoravides (1056–1147 AD) and Almohades (1130–1269 AD) conquered parts of Andalusia.

At the beginning of the sixteenth century, after the Christian kings had completely reconquered Spain, the last Muslim inhabitants were expelled from the Iberian Peninsula. Many refugees fled to the large cities in North Africa where they gave the Andalusian music a new impulse. They dedicated themselves to the genre, with their songs of longing for their homeland. Today there are three main forms of the *nawba*, i.e., the Moroccan, the Algerian and the Tunisian variants.

There are still unmistakably Spanish elements in the *nawba*, notably in the art of singing with its characteristic small ornaments on a long

---

[4] A standard work on the early history of Arabic music is H.G. Farmer, *A history of Arabian music to the XIIIth century* (London, 1929).

sustained tone. The vocal art of Beihdja Rahal, who performs the Algerian variant, can be seen to represent the heritage of the music from Andalusian Spain.

In the East, at the end of the 9th century AD, the political unity of the Abbasid empire slowly dissolved while local kingdoms gained power, thus in time leading to the development of local music cultures. Cities such as Cairo, Damascus, and Aleppo became cultural centres for Arabic art music.

The Egyptian waṣla

During the 19th century, Egyptian music was heavily influenced by the court music of the Ottoman empire (1281–1922 AD). From 1517 until 1917, Egypt had been a province of this reign. A form of Ottoman court music was played by Egyptian musicians in the palace of the Ottoman viceroy and in the homes of the Egyptian elite. In the course of the 19th century, Egyptian art music developed, although it still bore Ottoman traces. For example, regarding the *waṣla*, the Turkish influence can still be found in the names and compositional forms such as *samāʿī* (semai) and *bashraf* (peşrev) and musical modes such as *Ḥijāzkār* (Hicazkâr) and *Nikrīz* (Nikriz).

It was customary for a well-to-do Egyptian host to invite his male friends to his house for an evening concert called *sahra*. During such an occasion, a *takht* or small chamber music ensemble would perform a *waṣla*, accompanying the male soloist called a *muṭrib*.

The *takht* ensemble included an Arabian lute (*ʿūd*), a plucked zither (*qānūn*), and a tambourine (*riqq*). A vase drum (*darabukkā*), a reed flute (*nāy*) and violin (*kamān*) could be added to this group of instruments. The soloist was accompanied during a song's refrain by several male singers, the so-called *madhhabjiyya* (those who repeat the refrain, i.e., *madhhab*, sing.), who were often also the instrumentalists.[5] These players sat upon a sofa called *takht* in Arabic from which the ensemble derived its name.

The *waṣla* began with the instrumental overture (e.g., a *bashraf* or *samāʿī*), followed by an instrumental improvisation (*taqsīm*) and then

[5] A.J. Racy, 'Music in nineteenth-century Egypt. An historical sketch,' *Selected reports in ethnomusicology* 4 (1983), pp. 157–179.

a vocal improvisation on the words 'yā layl yā 'ayn' (*layālī*). Thereafter the first vocal composition (*muwashshaḥa* which means a strophic poem in classical Arabic) was performed, accompanied by the *takht*. A *taqsīm* played on the *qānūn* preceded a vocal improvisation on a fixed text (*mawwāl*) followed by another improvisation (*qaṣīda*) on a metered poem (ode) in classical Arabic. The *waṣla* ended with a strophic poem (*dawr*) sung in Egyptian dialect. This was the basic form of the *waṣla*, but other compositions, such as an instrumental interlude (*dūlāb*), could be added. The *sahra* could last long into the night; usually with three *waṣla*s performed in succession.[6]

The audience was all male as no women were allowed to be present. Female members of the family could have their own concerts played by female musicians (so-called *'awālim*, sing. *'ālima*) in the harem. The *'ālima*, literally 'learned woman', could play the *qānūn* and she also knew by heart various poetical forms in classical Arabic and Egyptian dialect. She was accompanied by a group of female percussionists, and in some cases by a blind male lute player. In contrast to the *ghawāzī* or singing girls who performed folk music, the *'awālim* were learned in the field of art music. Sometimes the *'ālima* sang for men, in which case she had to sit behind a curtain or lattice.

At the beginning of the twentieth century *'awālim* most often performed at wedding celebrations for the female guests.[7] The profession of *'awālim* became outdated after the 1940s, when making a career in singing was stirred by the mass media rather than through private concerts or wedding celebrations. In short, during the 19th and beginning of the 20th century, the *waṣla* accompanied by the *takht* was predominantly the domain of male singers.

The twentieth century and musical change

The rise of mass media brought important changes to musical life. The record industry, established in Egypt in 1904, recorded many singers and instrumentalists and flourished until 1932.[8] Egyptian records became

---

[6] A.J. Racy 1983, pp. 157–179.

[7] K. van Nieuwkerk, *A trade like any other. Female singers and dancers in Egypt*. Dissertation, University of Amsterdam (Amsterdam, 1991), pp. 68–75.

[8] A.J. Racy, *Musical change and commercial recording in Egypt 1904–32*. Dissertation, University of Illinois (Urbana, 1977).

popular throughout the rest of the Arab world and gave performers the opportunity to become famous abroad. This is even more true for Radio Cairo (Ṣawt al-Qāhira), founded in 1934. Its broadcasts, which could be heard throughout the Arab world and beyond thanks to its powerful transmitter, enhanced the careers of many singers.

The best known songstress during the 20th century was Umm Kulthūm (ca. 1900–1975), who established her career in Cairo by singing the songs of the *waṣla* in the houses of the elite and who was frequently recorded during the 1920s and 1930s while accompanied by her own *takht*.[9] During the 1930s and 1940s her star rose when she stopped singing the traditional *waṣla* repertory and began to concentrate on popular Arabic songs. She developed the *qaṣīda* into a concert piece for solo vocalist and Arabic orchestra. She began singing for the radio, later starred in films, and eventually performed on television.

Poets and composers wrote her songs and experimented with compositional forms, orchestration and musical modes. In particular, singer/composer Muḥammad ʿAbd al Wahhāb (1910–1991) was accredited with the introduction of foreign elements in her songs, such as Western (electronic) instruments, themes of Western classical music, and South American dance rhythms. Purists of Arabic music generally frowned upon so much modernization and regarded Egyptian music to be in a period of decline after the1940s.

From takht *to Arabic orchestra*

At the beginning of the 20th century, singers and their *takht* ensembles began to perform in coffee houses and small theatres. In this way, the music enjoyed by the elite was brought out of seclusion and popularised, thus making the singers public figures. On stage the *takht* was slowly expanded by introducing Western instruments such as the flute, clarinet, accordion and piano.[10]

[9] G. Braune, *Die Qaṣīda im Gesang von Umm Kulthūm. Die arabischen Poesie im Repertoire der grössten Sängerin unserer Zeit*, Teil 1 & 2. Beitrage zur Ethnomusikologie; 16 (Hamburg, 1987), p. 45.

[10] S. Shawan, 'Traditional Arab musical ensembles in Egypt since 1967. The continuity of tradition within a contemporary framework?,' *Ethnomusicology* (May 1984), pp. 272–276.

From the 1930s, the radio industry needed an orchestra to play new compositions during their broadcasts. For this reason, an Arabic orchestra was formed in which stringed instruments were dominant. It consisted of violins, celli, and the *takht* instruments. Sometimes Western instruments were added. This type of orchestra, with a mix of western and Arabic instruments was – and still is – called a *firqa* (band, orchestra).[11]

After the 1940s the authentic *takht* repertory, the *waṣla*, was completely forgotten by the public. However, in the 1960s, a group of musicians (including Umm Kulthūm) rescued the repertory from oblivion by reviving the *waṣla*, referred to as *turāth*, i.e. musical heritage. In 1967, they decided to establish an orchestra called the Firqat al-Musīqā al-ʿArabiyya (The Orchestra for Arabic Music) exclusively to perform the *turāth*.[12]

The instrumental make-up of this *firqa* consisted of a large section of stringed instruments, i.e. ten violins, a few celli and a contrabass, two *qānūns*, two *ʿūds*, two *nāys*, and a *riqq* and a *darabukkā*. No other Western instruments were permitted. Moreover, the soloist was replaced by a complete choir of ten men and ten women. All players and singers perform in unison the melodies of the *waṣla* compositions.

The founders of this *firqa* believed that adhering strictly to the way the orchestra should perform would prevent further decline of the authentic material, since by the 1960s few people remained who remembered the correct melody of these songs. However, this meant that no room remained for individual improvisation or embellishment of the melody which were two important characteristics of the *waṣla*.

Today the Firqat al-Musīqā al-ʿArabiyya and similar orchestras perform in Cairo's concert halls. The programme consists of a number of vocal and instrumental compositions from the *waṣla* repertory, with only a few short solo improvisations on the *nāy, qānūn, ʿūd* or *kamān*. In other words, the *waṣla* is no longer completely performed in the traditional way, with elaborate solo improvisations.[13]

Students at the conservatories in Egypt sing or play in a similar orchestra, such as the Firqat Umm Kulthūm of the Cairo Conservatory. Everybody knows by heart the traditional *takht* repertory of *samāʿī, bashraf,*

[11] Shawan, pp. 272–276.
[12] Shawan, pp. 276–285.
[13] I gained this information during fieldwork in Egypt between 1991 and 1996.

muwashshaḥāt etc., since these classics are taught during instrumental and vocal lessons. However, students are not encouraged by their teachers to embellish the melodic line or to perform improvisations. Thus the traditional repertory is well known but rather static.[14] The Egyptian mass performance of the traditional genre by a large orchestra and choir is also found in other Arab countries such as Algeria.

Aïcha Redouane and her Arabesques Vocales

The individual rendition of the *waṣla* by Aïcha Redouane as a solo vocalist accompanied by an authentic *takht* is revolutionary in this respect. She has revived the *waṣla* repertory with her style, techniques, small ornaments and her elaborate emotional improvisations.[15]

She has studied in minute detail old 78rpm recordings of performers such as Salāma al-Ḥijāzī and Yūsuf al-Manyalāwī. Her intention is not just simply to copy former masters of the art, but rather to study several renditions of one song by different singers and then create her own interpretation. In particular, vocal improvisations such as the *layālī, mawwāl* and *qaṣīda* offer the singer an opportunity to develop a unique style.

Aïcha Redouane was born in 1962 in the Middle Atlas of Morocco, in the small Berber village of Aït Attab. As a child she learnt the native Berber songs of her village as well as Arabic songs from the radio, but she also listened to Western classical music. At the age of six she travelled with her family to France and settled in the Ardèche.

After finishing secondary school she went to Grenoble to study architecture in order to become a designer of concert halls. In Grenoble she came into contact with students from the Middle East and North Africa, who lent her cassette tapes of their countries' music. When she began singing for her Arab friends, they enthusiastically encouraged her to develop her talent. After giving her first solo concert in Grenoble, she decided to stop studying architecture and dedicate herself to singing. In 1985 she went Paris in order to perfect her singing technique.

She decided on the *waṣla* repertory and spent hours training by listening to such songs recorded on old 78rpm records. The Egyptian singers

---

[14] I had the opportunity to attend courses at the Conservatory of Cairo and the Faculty of Music Education of Helwan University between 1991 and 1996.

[15] Personal communication from Aïcha Redouane in Oxford at the Conference of Arabic Music, organized by the International Folk Music Council, October 1996.

recorded at the turn of the twentieth century became her 'teachers': Yūsuf al-Manyalāwī (1850–1911), ʿAbd al-Ḥayy Ḥilmī (1857–1912), with whom she identifies herself the most, Salāma al-Ḥijāzī (d. 1917), Abū 'l-ʿIlāʾ Muḥammad (1878–1927), who was the mentor of the young Umm Kulthūm, and Sayyid al-Ṣaftī (d. 1939). She also listened to Umm Kulthūm's *adwār* (plur. of the Egyptian song called *dawr*) recorded around 1935. Another master of the art who inspired her is the Egyptian singer Ṣāliḥ ʿAbd al-Ḥayy (1896–1962). He was the last representative of the classical art music from the 19th century. He had fought a lifetime to preserve his music and rejected any attempt at modernization.[16]

Redouane had to study classical Arabic for the *qaṣīda, muwashshaḥa* and *mawwāl* but also the Egyptian dialect of the typically Egyptian *dawr*. Therefore she began studying classical Arabic at the University of Paris IV-Censier and later went to Cairo to improve her Egyptian dialect. She also learnt to play the *qānūn* in order to understand the modality of the Oriental Arab world. In fact, her education resembles the classical training of the 19th century *ʿālima*.

Riqq-player Habib Yamine, who is also an ethnomusicologist, helped Redouane grasp the typical rhythmical patterns or *īqāʿāt* of the compositions. Together with *qānūn*-player Salah el-Dine Mohamed, *ʿūd* player Brahim Meziane El-Otmani and violinist Osama Hantira they formed a *takht* ensemble, called al-Adwār.

At her concerts, Redouane and her *takht*-ensemble perform three complete *waṣla*s, just like in the old days at the *sahra*, instead of only choosing a number of compositions as done by the conductor of the Firqat al-Musīqā al-ʿArabiyya at a concert in Cairo.

Waṣla *in* maqām Ḥijāzkār

Redouane's interpretation of the *waṣla* in *maqām Ḥijāzkār* on the CD 'Arabesques Vocales' will be discussed here.[17] It begins with an instrumental overture by the Ottoman composer Jamīl Beyk al-Tambūrī

[16] C. Poché, *Egypt, vocal and instrumental art of the XIXth century, Aïcha Redouane and the group "Al-Adwar"*, CD *Egypte, Aïcha Redouane*, C 560020, Ocora, Radio France (Paris, 1993).

[17] CD *Arabesques Vocales, Aïcha Redouane*, 321015, Institut du Monde Arabe, Distrubution Harmonia Mundi (Paris, 1995).

(Tanburi Cemil Bey, 1871–1925). The short *taqsīm* on the *ʿūd* links this composition to the next piece, the *dūlāb* (instrumental interlude). The *muwashshaḥa* 'Yā Ghazāl,' composed by Muḥammad ʿUthmān (1855–1900) is the first vocal composition.

After the *muwashshaḥa* an improvisation on the *qānūn* prepares the *layālī*, vocal improvisation on the words yā layl yā ʿayn, after which Redouane sets in the *mawwāl*, or free improvisation on a fixed text. In the *mawwāl* 'Ṭūl il-layālī' (by an unknown poet) the despair of a lonely lover is expressed :

> Night after night I let you wander in my head
> My passion for you dominates my distress
> How scared I am that you may stay away from me for too long
> I could be content with your departure and long absence
> If only, oh my soul, your thoughts would be with me

The *maqām* of this *mawwāl* is *Ḥijāzkār*, which expresses a feeling of loneliness. Each *maqām* has an ethos, although this notion is vanishing in modern times.[18] It is remarkable that Redouane and her *takht* still believe in this ethos, another sign of restoring the pure tradition of the *waṣla* and its *maqāmāt*. In this *mawwāl* Redouane's voice expresses the endlessness of sleepless nights (ṭūl il-layālī) and the long absence of the beloved (ṭūl il-buʿd) by prolonging the syllables of these words.

Other special vocal techniques of Redouane are rich ornamentation and colours on sustained notes (*madda*) with vibratos (*tarjīf*), trills, glissandi, successive pauses, legati, nasalization (*ghunna*) and amplification (*tafkhīm*) of the voice.

The *dawr* or so-called 'master of the evening' (sayyid al-sahra) completes the *waṣla*. This poem in Egyptian dialect, 'Allāh yiṣūn dawlit ḥusnak' (May God protect your beauty), was set to music by the singer/composer ʿAbduh al-Ḥamūlī (1845–1901).

Based on the *waḥda sāʾira* rhythm (2/8 beat), it is structured in the form of a theme with variations.

> May God preserve your beauty forever
> But may He protect my heart against your penetrating eye,
> Sharp as a sword, who conquers me without a fight
> I can only complain to you and no one else about my love for you
> I am ill and you're my doctor,
> Please, cure me with your presence...

[18] Van Oostrum, pp. 188–189.

The *dawr* is meant to be the summit of creativity and musical emotion (*ṭarab*) provoked by listening to music at a concert. The audience in the old days showed their *ṭarab* by commenting on the music by exclamations such as 'how beautiful', and 'well done'.

In the modern rendition of complete orchestra with full choir, in which neither ornaments or improvisations nor any other sign of individuality is shown, *ṭarab* is neither provoked nor expressed during the performance. The audience in a Cairene concert hall is expected to listen quietly without comment.[19]

However, during Redouane's concerts, the young Arab urban elite express their *ṭarab* in the traditional way by exclaiming and commenting on the music. According to Redouane, it encourages the performer to make more and longer improvisations and dare to show more expression. This is an obvious feature that has been rediscovered. It seems the Arab listeners rediscover their cultural roots during her concerts and that Redouane herself deepens the traditional repertory in an individual, innovative way.

Beihdja Rahal and the nawba

Rahal, who was born in Algiers, is the representative of the Algerian *nawba*. Although she comes from a family that has performed this genre for generations, her presence is still revolutionary because this repertory used to be the domain of men.

Women, on the other hand, played and sang in the field of folk music. Until recently they were not given access to the conservatory. Today the situation has improved and the first female performers have graduated. In fact, Rahal was one of the first female students admitted to the Conservatory of Algiers. She studied singing and Arabic lute with two famous contemporary masters, Muḥammad Khaznājī and ʿAbd al-Razzāq Farkhshī. At the same time, she studied biology at the university, and after graduation taught at a secondary school. A few years later she decided to give up her job and become a full-time musician.[20]

---

[19] Personal communication from Aḥmad Shafīq Abū ʿAwf, one of the founders of this orchestra, in Cairo, 1993.

[20] CD *Nouba Raml al Maya. Beihdja Rahal*, Sunset France (Boulogne-Billancourt, France, 2003).

Although Rahal had some success in her own country as a professional singer, she decided to emigrate to France in the 1990s. It was there that her career took off. Her public is an urban intellectual elite, similar to the audience of her colleague Redouane. Although most of her listeners are of Arab origin, Western lovers of traditional or Western mediaeval music are also among her audience. Rahal not only performs the traditional *nawba*, but also renews it from within, using elaborate improvisations, subtle vocal ornaments, and special song texts.

The Algerian nawba

Texts used in the *nawba* are the strophic poems in classical Arabic, the *muwashshaḥāt*, or in dialect, the *azjāl* (sing. *zajal*). The most important collector of *nawba* texts was Muḥammad b. al-Ḥusayn al-Ḥāʾik from Tetouan in Morocco, who probably lived in the 18th century. He not only collected texts and wrote them down but also made a final classification for them on the basis of melodic (*ṭabʿ*) and rhythmical (*mīzān*) modes.[21]

Most Moroccan and Algerian collections of text assembled by musicians for their own use seem to be based on al-Ḥāʾik's collection. Although the texts are almost the same, the musical renditions of the Algerian, Moroccan and Tunisian *nawbāt* are different. The texts of the Algerian *nawba* are found in the modern collection of J. Yillis.[22]

There are eleven *nawbāt* in Algeria, named after their musical modes. Each *nawba* consists of five phases: *muṣaddar, baṭāyḥī, daraj, inṣirāf* and *(mu)khlāṣ*, while each phase of the *nawba* is characterized by a specific rhythmical cycle (*mīzān*, plur. *mawāzīn*). Each rhythmical cycle is named after its phase. The *nawba* is often interrupted by a vocal improvisation called the *istikhbār*.

Muwashshaḥāt have a simple poetical and musical structure. Therefore these songs have been and still are very popular in the Arab world. Musically speaking, *muwashshaḥāt* are compositions with a cyclical form. Rhyme scheme, musical scheme of melodic sentences and

[21] M. al-Ḥāʾik, *Majmūʿat an-nawbāt*, MS Leiden University Library, Cod. Or. 14.100, recently published as *Kunnāsh al-Ḥāʾik*, by Mālik Binnūna & ʿAbbās al-Jarārī (Rabat, 1999).

[22] J. Yillis, A. al-Ḥafnāwī, *al-Muwashshaḥāt wa ʾl-azjāl* (al-Jazāʾir, 1972–1982).

the rhythmic cycle used fit together. The genre of the *muwashshaḥāt*, invented in the 10th century in Andalusia, later developed and spread over the Arab world. This type of song (or rather poetry set to music) is found in the *waṣla* as well as in the *nawba*.

The theme of the sung *muwashshaḥāt* is usually love, and the metaphorical language is comparable to the texts of the Egyptian *waṣla*, such as the suffering of the worried lover, the beloved's eyes with their penetrating glance, and her resemblance with the moon, a star, a dove or a gazelle.

Rahal's rendition of the Nawbat Ramal Māyā

In the *istikhbār* (Track 4 of the CD Nouba Raml Maya) 'Tamallaktum ʿaqlī,' Rahal improvises and sings many small ornaments and melismas on the following text:

> You have robbed me of my sanity, my sight and hearing
> My soul, my heart, my whole being
> I am lost in your unsurpassed beauty
> I cannot find my way in this overpowering passion
> You have indeed advised me to control my secret jealousy
> Unfortunately my ample tears betray what I tried to conceal

As an extra after the final part of the *nawba*, Rahal performs a song from the light genre of female musicians in Algeria, a so-called *Qadriyya* (Track 11 of the CD) 'Qalbī wallāh man allam bihi,' a *zajal* in Algerian dialect given below. This is something remarkable in the tradition of the Algerian *nawba*.

> Nobody knows how much I suffer * only God knows my pain
> Nobody can heal my wounds * except the one who has caused them with only one glance.

In 1962 Algeria became independent and until then the *nawba* tradition had been neglected. During the French occupation, the focus of musical life had been on Western classical and French popular music.[23] In order to preserve the cultural heritage of the *nawba*, a large radio ensemble

[23] L.J. Plenckers, *De muziek van de Algerijnse Muwashshah*. Dissertation, University of Amsterdam (Amsterdam, 1989), pp. 43–44.

was formed along the same lines as the Firqat al- Mūsīqā al-ʿArabiyya in Egypt. This Algerian ensemble accompanies a choir of twelve men and twelve women.

However, Rahal seeks the intimacy of a small ensemble. The traditional *nawba* ensemble includes the *ʿūd ʿarabī* (a lute with four strings), the *quwaytra* or *quitra* (a smaller lute with a square sound box), the *rabāb* (a type of fiddle), and the *ṭār* (frame drum). Rahal is accompanied by such an ensemble, with a few adaptations: a *qānūn, nāy*, and mandolin are added and a viola replaces the *rabāb*. The voice of soloist is traditionally alternated or reinforced by a male choir. Instead, Rahal uses only two female singers as a choir.

Concluding remarks

In the 1960s, the old repertory of the *waṣla* in Egypt was rediscovered and partly preserved when a large orchestra and choir instead of a solo vocalist and a small chamber music ensemble began performing it. The intimate character of *waṣla* thus vanished, as well as the freedom of to improvise and ornament the melody freely according to one's own taste. The *nawba* of Algeria followed a similar pattern of development.

Rahal and Redouane have a revolutionary approach to the Arabic suite. They clearly have felt a need to develop their own styles instead of imitating styles of other artists. They have interpreted the genre, using refined vocal techniques, treatment of the language, the art of improvisation, and ornaments. Rahal has even added traditional songs of a lighter genre performed by female performers to the *nawba*.

Both women have scrutinised textbooks for the correct rendition of the classical Arabic poems. Redouane studied old 78rpm records dating as far back as the beginning of the 20th century, while Rahal has studied with old masters of the art who still know how to perform correctly. Their scientific approach, combined with a desire to express themselves in this field of art by developing their own style, has produced a true revival of the art. Last but not least, both Redouane and Rahal have dared enter the traditional male-dominated field of Arabic art music as solo songstresses.

LAW & RELIGION

A MEDIEVAL ISLAMIC LAW?

SOME THOUGHTS ON THE PERIODIZATION OF THE HISTORY OF ISLAMIC LAW[1]

Léon Buskens

Introduction

Remke Kruk accepted the chair of Arabic at Leiden University on 22 February 1991 with an inaugural lecture which was a plea for the use of the concept of 'Middle Ages' in the study of Muslim societies. She sketched the possibilities for further research on a period which had lasted much longer than usually understood, in some areas even well into the nineteenth or twentieth centuries. According to Remke Kruk students of Muslim cultures should turn to medievalists working on European societies for directions for further research. Her own research had been greatly enriched by her collaboration at Utrecht University with historians of European civilizations.[2]

Rather unexpectedly, the notion of 'medieval' Muslims has nowadays become quite popular in the public debate in the Netherlands. Some intellectuals and politicians argue that many Muslim immigrants hold ideas which are 'medieval,' and thereby threaten Dutch values and culture. This problem can only be solved by the coming of 'Enlightenment,' a historical stage which Muslims have not yet experienced, but which is essential for 'modernity.'

I have taken both Remke Kruk's lecture, and these crude evolutionary views in which scholarly concepts are used in a popular as well as populist way, as invitations to reflect upon the uses of the concepts of 'Middle Ages' and 'medieval' for the study of Islamic normativity. Do these notions lead to any interesting questions, to a better understanding of

[1] Amalia Zomeño was so kind as to react in an encouraging way to a first draft of this article, for which I thank her. I am also heavily indebted to my dear friend the regretted Mostapha Naji, who spent so much of his time on telling and showing me about the *turāth* of Morocco.

[2] R. Kruk, *Een kaart van de kat! Tussen Arabistiek en Mediëvistiek. Rede uitgesproken bij de aanvaarding van het ambt van gewoon hoogleraar in het Arabisch en de Arabische cultuur aan de Rijksuniversiteit te Leiden op vrijdag 22 februari 1991* (Leiden, 1991).

Islamic law? To my knowledge not much has been written about this question.[3] The following notes are intended as a proposal for further debate.

Attempts at Periodization

Some kind of periodization seems to be necessary for the study of the historical development of Islamic law. In his famous *Introduction to Islamic law* (1964), Joseph Schacht contrasts the 'classical' tradition with a 'modern' or 'modernist' period. Occasionally, he also uses the term 'Middle Ages' (*e.g.*, p. 230). Noel Coulson in his *History of Islamic law* (1964) opts for a clear tripartite division: genesis, Middle Ages, and modern times. Wael Hallaq seems to prefer the terms 'formative,' 'post-formative,' 'pre-modern,' and 'modern' in his detailed historical surveys of Islamic legal theories, although in the title of an earlier collection of essays (1995) the terms 'classical' and 'medieval' figure prominently. In many other recent works, the term 'pre-modern' has also replaced the more old-fashioned sounding 'medieval.' David Powers uses in his monograph on the practice of the law in the Islamic West between 1300–1500 AD the dynastic indication 'Marinid.' In his history of Islamic law of 2005, Knut Vikør has also little use for the concept of 'Middle Ages,' which he confines to the period between the tenth and twelfth century (p. 14), instead preferring terms such as 'classical' and 'traditional.'[4]

To a large extent the received view of Islamic legal history fits within a widespread periodization of Islamic history in general. After the lifetime of the Prophet Muhammad and the formative centuries, the Middle Ages lasted for centuries, until the coming of modernity with the French invasion of Egypt and the beginning of European colonialism in the nineteenth century. To some scholars, like Albert Hourani in

[3] R. Schulze's essay 'How medieval is Islam? Muslim intellectuals and modernity, ' in J. Hippler & A. Lueg (eds.), *The next threat. Western perceptions of Islam* (London [etc.], 1995), pp. 57–70, focuses on the concept of 'modernity,' and does not analyze the notion of 'Middle Ages.'

[4] J. Schacht, *An introduction to Islamic law*. 2nd impression (Oxford, 1965); N. Coulson, *A history of Islamic law* (Edinburgh, 1964); W.B. Hallaq, *A history of Islamic legal theories. An introduction to Sunnī* Uṣūl al-Fiqh (Cambridge, 1997); idem, *Authority, continuity and change in Islamic law* (Cambridge, 2001); idem, *Law and legal theory in classical and medieval Islam* (Aldershot, 1995); D.S. Powers, *Law, society, and culture in the Maghrib, 1300–1500* (Cambridge, 2002); K.S. Vikør, *Between God and the Sultan. A history of Islamic law* (London, 2005).

A history of the Arab peoples (1991), the Ottoman empire poses problems which lead to its exceptional status. We also find this approach in Schacht's legal history, which devotes a separate chapter to Ottoman law. Maybe, because of the centralized character of the administration of justice, the term 'pre-modern' or 'early modern' would especially fit this stage in the development of Islamic law.[5]

The way in which many scholars of Muslim history and law use the concepts of 'Middle Ages' and 'medieval' is rather vague, as Remke Kruk rightly notes. Gustave von Grunebaum is one of the scholars who upholds the notion prominently in his *Medieval Islam. A study in cultural orientation* (1953). He aims at arriving at a general characteristic of the Muslim Middle Ages, in which the dominance of religion over scientific inquiry looms large. Unfortunately, he does not go much beyond these stereotypes. For many of his contemporaries, such as the brilliant historian of Islamic law and society Robert Brunschvig, the notion of 'Middle Ages' did not seem to be problematic in itself.[6] On the contrary, it was a handy way to set off a large span of time, which some considered to be dominated by stagnation and increasing backwardness.[7] Abraham Udovitch, one of the foremost scholars of medieval Islamic law in its practical application, seems to take the categories 'Middle Ages' and 'medieval' for granted, like so many colleagues of his generation.[8]

Even Stephen Humphreys does not pay much attention to the concept of 'Middle Ages' in his handbook for historical research on early and medieval Muslim societies. For him, the period around 1500 'marks a number of important changes in Islamic life and our knowledge of it':

---

[5] On the Ottoman legal system, see: H. Gerber, *State, society, and law in Islam. Ottoman law in comparative perspective* (Albany, 1994); idem, *Islamic law and culture, 1600–1840* (Leiden, 1999); C. Imber, *Ebu's-su'ud. The Islamic legal tradition* (Edinburgh, 1997); J.E. Tucker, *In the House of the Law. Gender and Islamic law in Ottoman Syria and Palestine* (Berkeley [etc.], 1998); M.H. van den Boogert, *The capitulations and the Ottoman legal system. Qadis, consuls and* beratlıs *in the 18th century* (Leiden & Boston, 2005).

[6] R. Brunschvig, *Etudes d'islamologie. Tome second: Droit musulman* (Paris, 1976).

[7] Gustave von Grunebaum dealt with the problem of decline and stagnation in Islamic civilization in many of his writings. A number of papers on the issue, among which a study of Joseph Schacht on Islamic law in which he uses the concept of *ankylose*, were collected by: R. Brunschvig, G.E. von Grunebaum [et al.], *Classicisme et déclin culturel dans l'histoire de l'Islam. Actes du symposium international d'histoire de la civilisation musulmane (Bordeaux 25–29 juin 1956)* (Paris, 1957).

[8] A.L. Udovitch, *Partnership and profit in medieval Islam* (Princeton, 1970); idem, 'Islamic law and the social context of exchange in the medieval Middle East,' *History and anthropology* 1 (1985), pp. 445–465.

Islam was developing important centers beyond its core, Europe became increasingly powerful, and the Ottoman administration started to produce new sources for historical research.[9] One of the most sophisticated recent studies, in terms of conceptual reflection, on a Medieval Muslim society is Jonathan Berkey's monograph on popular preaching. In the conclusion the author seems to prefer the term 'Middle Period' to 'Middle Ages.' Berkey expresses valuable views on institutions, religious authority, and the relations between legal scholars and Sufism, which he considers as characteristic for this period.[10]

In what follows I would like to explore a bit further this notion of a middle period in the historical development of Islamic law. I will not focus on the contrast between the formative period, which ends around 900 AD, and the Middle Ages, since a lot has been written about this already. Instead I will concentrate on the opposition between 'medieval' and 'modern' Islamic law, which started to come into being somewhere around 1800 AD, and see to what extent this distinction might have any heuristic value.

Medieval Islamic Law

In general usage, also among Islamicists, 'Islamic law' is often equated with the classical body of normative writings of Muslim legal scholars known as *fiqh*. This classical corpus of ideas and rules developed on the basis of an orthodox understanding of the Qur'an and traditions of the Prophet, which was already established around 900 AD. Debates on this classical tradition only started on a larger scale from the end of the eighteenth century onwards. These kinds of criticisms are generally considered to be manifestations of Muslim modernity.

The most striking characteristic of the classical legal tradition was its literary form of commentaries. For medieval Muslims legal scholarship was an ongoing debate about the correct understanding of God's Will according to which all humans should live. At a certain moment an authoritative scholar would lay down a summary of the correct

[9] R.S. Humphreys, *Islamic history. A framework for inquiry*. Rev. ed. (London [etc.], 1991), pp. ix–x.

[10] J.P. Berkey, *Popular preaching and religious authority in the medieval Islamic Near East* (Seattle [etc.], 2001).

understanding of God's Will in a short text which could be learned by heart by students, a so called *matn*. As surveys of classical *fiqh* literature show, many of these basic texts, or *mutūn*, were written between the twelfth and fourteenth centuries.[11] A famous example for the Mālikī school is the *Mukhtaṣar* of the Egyptian scholar Khalīl b. Isḥāq al-Jundī (d. 1365). Scholars of later generations expounded their views in the form of detailed commentaries, sg. *sharḥ*, of these basic texts. Often, these later explanations contained quite some new views disguised as mere elucidation. These commentaries could become authoritative in their own right, embodying generally accepted teachings of particular approaches to God's Will. Every new generation of legal scholars would produce its own readings of the tradition, laid down in commentaries and supercommentaries, sg. *ḥāshiya*. In order to understand God's Law, a student had to become intimately acquainted with this intricate corpus of glosses on commentaries on scholarly understandings. For example, students in Morocco studied a commentary on Khalīl's text by ʿAbd al-Bāqī al-Zurqānī (d. 1687), with glosses by Muḥammad al-Bannānī (d. 1780), or by Muḥammad al-Rahūnī (d. 1815).[12] Wael Hallaq has stressed how commentators gradually incorporated important developments from legal practice, such as *fatāwā* and legal documents, into this literature of commentaries.[13]

Baber Johansen has shown how casuistry played a crucial role in the development of legal reasoning, and became one of the main

---

[11] Good overviews of the authoritative *fiqh* works of the classical schools can be found in Th.W. Juynboll, *Handleiding tot de kennis van de Mohammedaansche wet volgens de leer der Sjâfiʿitische school*. 4th ed. (Leiden, 1930), pp. 373–378; Schacht 1965, pp. 261–269; O. Spies & K. Pritsch, 'Klassisches islamisches Recht,' in E. Seidl [et al.], *Orientalisches Recht* (Leiden [etc.], 1964), pp. 220–343. The question of the relationship between the genesis of the the genre of *mukhtaṣar* and *taqlīd* has been analyzed by M. Fadel, 'The social logic of *taqlīd* and the rise of the *mukhtaṣar*,' *Islamic law and society* 3 (1996), pp. 193–233; see also: B.M. Wheeler, 'Identity in the margins. Unpublished Ḥanafī commentaries on the *Mukhtaṣar* of Aḥmad b. Muḥammad al-Qudūrī,' *Islamic law and society* 10 (2003), pp. 182–209.

[12] Jacques Berque has published many valuable studies on the Maghribī legal tradition. A good introduction to its intellectual history is: J. Berque, 'Ville et université. Aperçu sur l'histoire de l'Ecole de Fès,' *Revue historique de droit français et étranger*, 4ème série, 27 (1949), pp. 64–117. For an analysis of the formation of the canon and the concept of authority in the Ḥanafī school, see: B.M. Wheeler, *Applying the canon in Islam. The authorization and maintenance of interpretive reasoning in Ḥanafī scholarship* (Albany, 1996).

[13] W.B. Hallaq, 'From *fatwās* to *furūʿ*. Growth and change in Islamic substantive law,' *Islamic law and society* 1 (1994), pp. 29–65; idem, 'Model *Shurūṭ* works and the dialectic of doctrine and practice,' *Islamic law and society* 2 (1995), pp. 109–134.

characteristics of the classical tradition.[14] Most jurisconsults were thinking comfortably within the boundaries set by early scholars, who had gradually been accepted as founders of legal schools, sg. *madhhab*, such as the *imāms* Abū Ḥanīfa, Mālik b. Anas, Muḥammad b. Idrīs al-Shāfiʿī, and Aḥmad b. Ḥanbal. Belonging to a law school or *madhhab* became a matter of intellectual as well as legal identity, as George Makdisi has extensively studied.[15]

Two further key notions in the classical tradition were *ijmāʿ* and *taqlīd*. As Snouck Hurgronje kept stressing in his teachings, the authority of the consensus of the scholars, *ijmāʿ*, which could no further be questioned once reached, was the prime foundation of the classical version of Islamic law. *Taqlīd* was understood as the obligation for lesser scholars to 'follow' or 'imitate' the opinions of the founding fathers and their brilliant pupils. Again, Wael Hallaq has called for a much more nuanced understanding of this opposition between *ijtihād* and *taqlīd* in recent work.[16] The form of the commentary as a specific legal genre seems to be the literary expression of the cultural ideal of *taqlīd*. Scholars generally acquired authority by claiming to limit themselves to explaining the received views of the master scholars.

The techniques of transmission are another important characteristic of the medieval legal tradition. Franz Rosenthal has offered a brilliant overview of this tradition, which was fondly used by Remke Kruk in her introductory course on Arabic philology.[17] More recently, George Makdisi, Jonathan Berkey, Michael Chamberlain, and Houari Touati have published important monographs about the system of medieval Islamic education, in which the *madrasa* and the personal contact between the teacher and his students played such an important role.[18] Twentieth

---

[14] B. Johansen, 'Casuistry. Between legal concept and social praxis,' *Islamic law and society* 2 (1995), pp. 135–156.

[15] G. Makdisi, *The rise of colleges. Institutions of learning in Islam and the West* (Edinburgh, 1981); idem, *The rise of humanism in classical Islam and the Christian West. With special reference to Scholasticism* (Edinburgh, 1990).

[16] Hallaq 2001. See also the special issue of *Islamic law and society* 3/2 (1996) on *ijtihād and taqīd*, edited by W. Hallaq; and: Y. Rapoport, 'Legal diversity in the age of *taqlīd*. The four Chief *Qāḍīs* under the Mamluks,' *Islamic law and society* 10 (2003), pp. 210–228.

[17] F. Rosenthal, *The technique and approach of Muslim scholarship* (Rome, 1947).

[18] Makdisi 1981; Makdisi 1990; J. Berkey, *The transmission of knowledge in medieval Cairo. A social history of Islamic education* (Princeton, 1992); M. Chamberlain, *Knowledge and social practice in medieval Damascus, 1190–1350* (Cambridge, 1994); H. Touati, *Entre Dieu et les hommes. Lettrés, saints et sorciers au Maghreb (17ᵉ siècle)* (Paris, 1994); see also the proceedings edited by H. Elboudrari, *Modes de transmission de la culture religieuse en Islam* (Le Caire, 1993).

century transformations of 'traditional' Islamic education have been documented in historical ethnographies of Morocco by Dale Eickelman, and of Yemen by Brinkley Messick.[19] One of the central ideas was that knowledge was reliable and trustworthy because its possessor had 'taken' his understanding directly and personally from a teacher who was himself part of a chain of upright transmitters, which went back without interruption to the founding fathers. The most reliable way of transmission was by word of mouth, the student listening to the teacher, writing down and memorizing his words. A good pupil would acquire in his turn a licence to transmit this knowledge to younger generations, *ijāza*.

Handwritten texts functioned as an aid to the memorization of knowledge. The manuscript was corrected by the master or by another authoritative scholar, which was carefully mentioned in the colophon, placing the particular copy in a chain of transmitters and transmissions. The medieval legal tradition was a *manuscript* tradition, which meant that every copy of a text was unique, having its own individual place in the continuation of the teaching. Leiden University has from its very beginning been a place where the study of Arabic manuscripts flourished. The present *Interpres Legati Warneriani*, Jan Just Witkam, offered with his inaugural lecture an eloquent contribution to this Leiden specialty with a study on the traces of reading and transmission activities in medieval Islamic manuscripts. His work is highly appreciated by other researchers on the history of Muslim book culture, such as François Déroche and Houari Touati.[20]

---

[19] D.F. Eickelman, *Knowledge and power in Morocco. The education of a twentieth-century notable* (Princeton, 1985); B. Messick, *The Calligraphic State. Textual domination and history in a Muslim society* (Berkeley [etc.], 1993). It would be interesting to compare historical sources and ethnographic materials more extensively, also by using important studies of intellectual practices in the Sahel and Sub-Saharan Africa, such as: E.N. Saad, *Social history of Timbuktu. The role of Muslim scholars and notables, 1400–1900* (Cambridge, 1983); G. Mommersteeg, *Het domein van de Marabout. Koranleraren en magisch-religieuze specialisten in Djenné, Mali* (Utrecht, 1996); R. Santerre, *Pédagogie musulmane d'Afrique noire. L'école coranique peule du Cameroun* (Montréal, 1973).

[20] J.J. Witkam, *Van Leiden naar Damascus, en weer terug. Over vormen van islamitische lees- en leercultuur. Rede uitgesproken bij de aanvaarding van het ambt van hoogleraar in de Handschriftenkunde van de islamitische wereld aan de Universiteit Leiden op vrijdag 15 november 2002* (Leiden, 2003); F. Déroche (ed.), *Manuel de codicologie des manuscrits en écriture arabe* (Paris, 2000); idem, *Le livre manuscrit arabe. Préludes à une histoire* (Paris, 2004); H. Touati, *L'armoire à sagesse. Bibliothèques et collections en Islam* (Paris, 2003); see also the volume edited by G.N. Atiyeh, *The book in the Islamic world. The written word and communication in the Middle East* (Albany, 1995). The famous

The introduction of printing marked in many ways the end of this tradition. One of the first manuals for studying the Divine law without direct face-to-face contact with a teacher was Ibrāhīm Bājūrī's (d. 1860) famous overview of the teachings of the Shāfiʿī school. Although it still had the form of glosses on a commentary of a basic text, it was meant as a kind of 'Teach yourself' book on Islamic law. This made it an excellent foundation for early European introductions to Islamic law, such as the manuals of E. Sachau and Th.W. Juynboll.[21]

The literary form, the physical appearance as a unique manuscript, and the methods of production and transmission of legal norms, should be linked to the social position of its interpreters and the political context. Legal scholars were linked in a complex way to the detainers of political power. The *ʿulamāʾ* had, to a certain extent, an independent position, inasfar as that the political rulers needed their support in order to legitimize their rule. Economically, scholars were partly dependent on the power elite. The religious foundations, *awqāf*, from which came their income and the maintenance for their students, were often created and controlled by rulers.[22] Rulers appointed some scholars to act as judges. The judges' independence showed again in their rulings, in which they could follow to a certain extent their own discretion. Later Western scholars have distorted this discretion into the image of *Qadi-Justiz*.[23] Amalia Zomeño and David Powers offer in their monographs a view of the workings of the legal system in the Muslim West on the basis of a detailed analysis of the *fatāwā* transmitted by al-Wansharīsī (d. 1508 AD) in his famous work *al-Miʿyār*, and some of its sources.[24]

Moroccan historian Muḥammad al-Mannūnī, who was himself part of the culture of traditional learning and gave *ijāza*s to members of younger generations, such as to the late Mostapha Naji, has given an overview of book culture in the Muslim West in his fascinating study *Tārīkh al-wirāqa al-maghribiyya. Ṣināʿat al-makhṭūṭ al-maghribī min ʿuṣūr al-wasīṭ ilā al-fatra al-muʿāṣira* (al-Ribāṭ, 1991).

[21] E. Sachau, *Muhammedanisches Recht nach schafiitischer Lehre* (Stuttgart [etc.], 1897); Juynboll 1930.

[22] An excellent overview of research on the social position of the *ʿulamāʾ* is to be found in R.S. Humphreys 1991, pp. 187–208. For the later period see the volume edited by N.R. Keddie, *Scholars, saints, and sufis. Muslim religious institutions since 1500* (Berkeley [etc.], 1972).

[23] I. Schneider, 'Die Merkmale der idealtypischen *qâdî*-Justiz. Kritische Anmerkungen zu Max Webers Kategorisierung der islamischen Rechtsprechung,' *Der Islam* 70 (1993), pp. 145–159; D.S. Powers, '*Kadijustiz* or *Qāḍī*-Justice? A paternity dispute from fourteenth-century Morocco,' *Islamic law and society* 1 (1994), pp. 332–366.

[24] A. Zomeño, *Dote y matrimonio en al-Andalus y el Norte de África. Estudio sobre la jurisprudencia islámica medieval* (Madrid, 2000); Powers 2002.

The individual space which a jurist might take for legal reasoning is analysed by Khalid Masud in his famous monograph on the Granadan theoretician al-Shāṭibī (d. 1388 AD).[25]

In my view, the nature of the medieval polity determined to a large extent the shape of the classical legal tradition. The regional governments had only limited possibilities to impose and maintain a monopoly on the production of legal norms. For this restricted way of administering their subjects, a full standardization of the law was not necessary. The techniques for the development and transmission of a uniform tradition were missing. The medieval rulers left the production of legal norms to the *ʿulamāʾ* as long as they did not actively obstruct the limited working of the state. This resulted in a medieval Islamic law which was to a large degree an ongoing debate between a multiplicity of scholars on how to understand God's Will at a particular moment in a particular place. The texts which the scholars composed, commented upon, and copied, were an expression of their ideal of debate and imitation of earlier authorities.

Modern Islamic Law

The foremost embodiment of modern Islamic law is the printed law code. During the nineteenth century the Ottoman government became convinced of the usefulness of compiling uniform statements of legal rules which were valid for the entire empire. A single state-enforced text replaced the diversity of interpretations by independent scholars which only highly educated specialists could handle. The textual form of these law codes had been developed by French legal scholars to govern the European empire united by their emperor Napoleon. In continental Europe, the codification proved an effective tool for the administration of the newly developing nation states. Ottoman students went to France in order to study new technologies such as engineering, medicine, military tactics, and legal science, which their government considered useful for the modernization of the administration of the crumbling empire. The most famous product of the modern approach to legal scholarship was the civil law code of 1877 called the *Majalla*. The text became available in an Ottoman as well as an Arabic printed version. In 1875 Qadrī Pasha

[25] M.Kh. Masud, *Shāṭibī's philosophy of Islamic law* (Islamabad, 1995).

had already undertaken the codification of the Ḥanafī rules concerning family life and inheritance in Egypt.

In the colonized parts of the Muslim world, European governments also worked on the introduction of legal rules in the form of law codes. For Tunisia and Algeria, the French had compilations of Islamic law designed, which became known after the European scholars of Islamic law who had compiled them: the *Code Santillana* of 1899 and the *Code Morand* (or *Avant-projet de code du droit musulman algérien*) of 1916. In the course of the twentieth century the law code would become the common form of Islamic law. Especially independent governments and Islamist regimes embraced the codification as a symbol of national Islamic identity.[26]

It is curious to see how several prominent Western scholars declared themselves against the codification of Islamic legal norms. One of the founding fathers of the Western study of Islamic law, Christiaan Snouck Hurgronje, considered *taqnīn* to be contrary to the spirit of Islamic law. He feared that Islamic legislation would introduce norms which until then had never been respected by the people, who had lived in daily practice according to local customs.[27] Ann Elizabeth Mayer also presented the tendency to transform *fiqh* rules into law codes as un-Islamic in her influential article on Islamic law in the modern world.[28] Selecting one opinion from the extensive gamut of differing views would lead to

[26] On codification in general, see: K.S. Vikør, 'The Sharīʿa and the nation state. Who can codify the Divine law?,' in B.O. Utvik & K.S. Vikør (eds.), *The Middle East in a globalized world* (Bergen, 2000), pp. 220–250; R. Peters, 'From Jurists' law to Statute law, or what happens when the Sharīʿa is codified,' *Mediterranean politics* 7 (2002), pp. 82–95; L. Welchman, 'Islamic law. Stuck with the State?,' in A. Huxley (ed.), *Religion, law, and tradition. Comparative studies in religious law* (London, 2002), pp. 61–83; A. Layish, 'The transformation of the Sharīʿa from Jurists' law to Statutory law in the contemporary Muslim world,' *Die Welt des Islams* 44 (2004), pp. 85–113. For a case study of codification, see: R. Peters, 'The codification of Criminal law in nineteenth-century Egypt. Tradition or modernization?,' in J.M. Abun-Nasr, Ulrich Spellenberg [et al.] (eds.), *Law, society, and national identity in Africa* (Hamburg, 1990), pp. 211–225. On the formal contrasts between classical commentaries and modern law codes, see: L. Buskens, 'Islamic commentaries and French codes. The confrontation and accommodation of two forms of textualization of family law in Morocco,' in H. Driessen (ed.), *The politics of ethnographic reading and writing. Confrontations of Western and Indigenous views* (Saarbrücken [etc.], 1993), pp. 65–100.

[27] C. Snouck Hurgronje, *Verspreide geschriften*. Deel IV/2 (Bonn [etc.], 1924), pp. 259–266; see also Deel IV/1 (1924), pp. 105–106, n. 2.

[28] A.E. Mayer, 'Law and religion in the Muslim Middle East,' *The American journal of comparative law* 35 (1987), pp. 127–184.

a reduction of flexibility and the possibilities for judicial discretion. I understand the problems which these scholars associate with codification, but these are not specific to the Islamic tradition. So why should Western academics categorize this process normatively as un-Islamic?

The introduction of the law code was accompanied by the spread of related genres of legal literature, such as published court judgements, handbooks, monographs, and law journals. Western legal scholarship provided models for these new ways of writing. The modern forms were also strongly influenced by the widespread acceptance of an innovative technology of reproduction: the printing press. In several parts of the Muslim world the new technology was introduced in the form of lithograph printing, which produced results that at first glanced resembled manuscript copies.[29] Printing with movable type drastically changed the outlook of legal texts, and made large numbers of identical copies available at reduced prices. Scholars, judges and administrators living in distant areas could all refer to a uniform text, in which they could find the same rule on the same page. These copies were made easily accessible by tables of contents and indices, which replaced the 'medieval' technology of memorizing texts.[30]

The mass production of identical legal texts was linked to the rise of a new type of political organization: the centralized nation state. The rulers of a modern state claimed exclusive rights to the production of legal norms, linked to a monopoly on the use of violence and taxation. The ambitions of control of these new states made standardization of legal norms a prerequisite. A uniform and centrally controlled legal system was in its turn one of the necessary conditions for the functioning of the new administration. Colonization by Western states played an

[29] Cf. B. Messick, 'On the question of lithography,' in J. Skovgaard-Petersen (ed.), *The birth of the printing press in the Middle-East* (Copenhagen, 1997), pp. 158–176 (= *Culture & history* 16 [1997]). For studies on lithographic printing in Morocco, see: Fawzī ʿAbd al-Razzāq, *Mamlakat al-kitāb. Tārīkh al-ṭibāʿa fī 'l-Maghrib, 1865–1912* (al-Ribāṭ, 1996); and idem, *al-Maṭbūʿāt al-ḥajariyya fī 'l-Maghrib. Fihris maʿa muqaddima tārīkhiyya* (al-Ribāṭ, 1989).

[30] The standard reference for the social analysis of printing in the Muslim world is: F. Robinson, 'Technology and religious change. Islam and the impact of print,' *Modern Asian studies* 27 (1993), pp. 229–251. A valuable volume of studies was edited by J. Skovgaard-Petersen, *The birth of the printing press in the Middle-East* (Copenhagen, 1997) (= *Culture & History* 16[1997]). For a richly illustrated overview of the history of printing in the Middle East, see: E. Hanebutt-Benz, D. Glass [et al.] (eds.); *Sprachen des Nahen Ostens und die Druckrevolution. Eine interkulturelle Begegnung = Middle Eastern languages and the print revolution. A cross-cultural encounter* (Westhofen, 2002).

important role in the introduction of these modern forms of political organization, as well as the creation of independent nation states after the demise of the Ottoman empire. The political changes were closely linked to economic developments, such as processes of commercialization and the spread of capitalism.

The modern legal texts and forms of administration required a different kind of legal professionals, replacing the classical scholars of the *ʿulamāʾ* type, and the Islamic judges.[31] European-style education in law faculties became a common phenomenon in many Muslim societies, where students used textbooks in European or in vernacular languages, organized according to European models. The new kind of education gave access to modern legal professions such as lawyer, notary, or public prosecutor. These modern professions were at the service of the capitalist economy and the nation state. Brinkley Messick has analyzed the changes in legal culture in relation to transformations of society and economy in his historical ethnography of Yemen, *The calligraphic state. Textual domination and history in a Muslim society* (1993). Dale Eickelman traces these transformations by focusing on the biography of a judge in protectorate and independent Morocco in *Knowledge and power in Morocco. The education of a twentieth-century notable* (1985).

The profound reforms of legal writing, thinking, and working gave rise to debates on the reform of Islam and Islamic law, in which the concept of *ijtihād* played a key role.[32] What should be the relationship between the nation states and Islam? Was the secular European model of statecraft compatible with Islamic values? Was there still a place for the masters of the classical tradition, the *ʿulamāʾ*, in these new states? Should their education be reformed? These debates could reach larger audiences because the intellectual exchanges took place in Islamic journals such as *al-Manār*, which could be distributed at relatively low costs to intellectuals living in many different countries, thanks to the good services of printers and mail companies. Some thinkers came to the conclusion that a modern state implied a radical separation between Islam

[31] Muhammad Qasim Zaman gives in *The ulama in contemporary Islam. Custodians of change* (Princeton, 2002), an overview of changes in the social roles of *ʿulamāʾ* in colonial and contemporary societies, in which he pays special attention to their present-day re-emergence as political leaders.

[32] On the debate about *ijtihād*, see: M. Kerr, *Islamic reform. The political and legal theories of Muḥammad ʿAbduh and Rashīd Riḍā* (Los Angeles, 1966); R. Peters, '*Idjtihād* and *taqlīd* in 18th and 19th century Islam,' *Die Welt des Islams* 20 (1980), pp. 131–145.

and politics and the ideal of secularization, which might even lead to a radical banishment of Islamic law from the legal system, as happened in Turkey and the Soviet territories in the Balkans and in Central Asia.[33]

Some tentative conclusions

During recent years the concept of 'Middle Ages' has become less fashionable among historians of Islamic law. Some try to avoid using the term, employing neutral dates, or terms such as 'pre-modern' instead. Others speak of 'Middle Ages' and 'medieval' Islamic law in a rather commonsensical way, without giving much thought to the questions raised by the use of these notions. In this short essay my main aim has been to invite debate about the use of these problematical categories, and to probe to what extent these can raise interesting questions for further research and reflection. I have developed my ideas by opposing 'medieval' and 'modern' expressions of Islamic law. I am fully aware of the fact that this opposition is far too simplistic. It has not been my intention to revive such a gross binary opposition. The contrast has simply served as a means to explore the question whether it might have any use to consider a notion of a medieval Islamic legal tradition as a tool for further research.

In my view, the notion of a 'medieval' Islamic law has at least some heuristic value. The period of the Middle Ages, of which the limits vary according to the region under study, has some unity as far as the law is concerned. This unity follows from an intellectual style in which the authority of tradition, memorization, and casuistry are important elements, and from the textual genre of the commentary. These forms of intellectual reasoning took place in societies where there was a close, although often uneasy, relationship between Islam and politics. The states which were the political framework for the development, transmission and application of the law had very limited possibilities, or ambitions, to interfere deeply with the private lives of their subjects. The control of the governments over the guardians of the sacred law was at best partial.

---

[33] On processes of secularization in Muslim societies and Islamic law, see: T. Asad, *Formations of the secular. Christianity, Islam, modernity* (Stanford, 2003).

From the term 'Middle Ages' follows logically that it is a period between two other eras: the formative period of Islamic law, and the age of modernity in which the classical consensus comes under severe criticism. One of the weaknesses of the concept is that it does not indicate the characteristics of the period. Its unity seems to stem from its opposition to other, clearly defined eras. This weakness it shares with the concept of 'pre-modern,' which is also only a negative definition. The term might also suggest too much uniformity, which obscures the important changes taking place within this era. At no point I have wanted to imply that the Middle Ages of Islamic law were a stagnant period, lacking development. On the contrary, the recognition of the unity of this era through a shared style of legal reasoning and writing should be taken as an encouragement for further research on the dynamism, which has already been shown by the meticulous work done by Wael Hallaq and Baber Johansen.[34]

Another problem is that the term is borrowed from European history, with all its concomitant associations. This might lead to gross unilinear evolutionism and eurocentrism. Some people might want to conclude that the existence of 'Middle Ages' in the historical development of Islamic law also implies the need for a 'Renaissance' or an 'Enlightenment.' We need further research on the question whether we can use similar schemes for the periodization of the history of Muslim societies and for European societies. The polemics between Schulze and Radtke offer an example of the questions involved. Reinhard Schulze has published extensively about 'modernity' and the processes of modernization in Muslim societies. His views on the existence of an Islamic Enlightenment have been severely criticized by Bernd Radtke.[35] One of the important lessons to be learned, is that we need to think carefully about the connotations of the terms we use for periodization.

The notion of 'Middle Ages' also looms large in popular imagination, again with associations of backwardness, stagnation, and mystery. We can find these associations in all kind of stereotypes about Muslim

---

[34] Many of Hallaq's seminal publications have been mentioned in earlier footnotes. Several of Johansen's articles have been collected in B. Johansen, *Contingency in a sacred law. Legal and ethical norms in the Muslim* Fiqh (Leiden [etc.], 1999).

[35] R. Schulze, *A modern history of the Islamic world* (London, 2000); idem, 'Was ist die islamische Aufklärung?,' *Die Welt des Islams* 36 (1996), pp. 276–325; B. Radtke, *Autochthone islamische Aufklärung im 18. Jahrhundert* (Utrecht, 2000).

societies, both popular and scholarly, as Edward Said and his students have shown in their studies of orientalism. This is part of a larger phenomenon in which people associate other cultures with distant times, analyzed by Johannes Fabian in his book *Time and the Other* (1983). I have tried to show *en passant* that populist writers with scholarly pretensions, who choose to characterize contemporary extremist Islamist views on Islamic law as 'medieval,' are quite mistaken. Totalitarian dreams of an Islamic state which is governed according to an extremely selective and idiosyncratic interpretation of God's Will, are anything but medieval. The reduction of the polyphony of dissenting scholars' views to a single privileged opinion, as well as the use of technology such as internet, print and dvds, are all thoroughly modern, or even post-modern phenomena. The same holds true for the vision of the theocratic state which should enforce these norms. These radical thinkers have their own peculiar place among the offspring which enlightened reason has borne.

Because of all the associations clinging to the value-laden term 'Middle Ages,' we might consider adopting Jonathan Berkey's solution and use the term 'Middle Period' instead.[36] Thus, we would speak, analogous to the linguistic term 'Middle Arabic,' of 'Middle Islamic law.' But this would still leave us with the question whether it is justified to consider this 'Middle legal tradition' as a useful entity for historical research.

My essay is a plea for a more nuanced view of the Middle Ages of Islamic law. To what extent did a period with such a unity exist? Do the dynamics of the classical tradition of *fiqh* allow for such a periodization? Research done by European medievalists can be of great help to phrase new questions for further research. They have long since dealt with popular images of 'dark ages,' full of backwardness and stagnation. The span of this brief essay does not allow for any systematic comparison between the European and the Muslim legal tradition, which were both linked to manuscript culture, and to fragmented and rather weak states. George Makdisi's monumental studies of medieval education offer a model for such comparative work.[37] Thinking about the possible unity of a medieval Islamic law and research on similarities and differences between

[36] Berkey 2001, especially pp. 88ff. In his conclusions Berkey privileges the term 'Middle Period,' but in the title of this last chapter he still uses the expression 'medieval Islam.'

[37] Makdisi 1981; Makdisi 1990.

Muslim and European Middle Ages should give rise to a more nuanced and dynamic view of the classical Islamic legal tradition. At least, this is the way in which I have understood Remke Kruk's admonishment in her inaugural address.

ON WOMEN AND CAMELS
SOME COMMENTS ON A *ḤADĪTH*

Manuela Marín

In one of the writings attributed to Ibn Tūmart (d. 524/1130), the *mahdī* of the Almohads, many examples are given of the perverse behaviour and actions of the *mulaththamūn* (the veiled), also called *mujassimūn* (anthropomorphists).[1] They are, as is well known, the Almoravids, the Maghribian dynasty ultimately defeated by the Almohads, whose military might was accompanied by a powerful propaganda machine. Ibn Tūmart understood very well the strategic value of defiling the enemy, and what he said about the *mulaththamūn* is a perfect example of the use of political and religious themes as an effective and damaging weapon.

Eight actions, says Ibn Tūmart,[2] define the behaviour of the *mulaththamūn*, both men as well as women. In fact, four of these eight actions concern women in particular. Firstly, the heads of Almoravid women resemble camels' humps (*ru'ūsuhunna ka-asnimat al-bukht*), that is, explains Ibn Tūmart, these women gather their hair over their head in the shape of camels' humps. Secondly, women dress as if they are naked. Thirdly, they have strayed from the right path, and finally, they artfully attract other women to follow their ways.

As Ibn Tūmart states at the end of his accusations against the Almoravids, all these marks of perversity had already been denounced by the Prophet himself, although it is only later in the text when a *ḥadīth* condemning women whose hairdo resembles camels' humps is literally quoted on the authority of Abū Hurayra.[3]

The renewal of interest in *ḥadīth* during the Almohad period has recently been analyzed in the context of the ideological renovation and

---

[1] See D. Serrano, '¿Por qué llamaron los almohades antropomorfistas a los almorávides?,' in P. Cressier, M. Fierro [et al.] (eds.), *Los almohades. Problemas y perspectivas*, II (Madrid, 2005), pp. 815–852.

[2] Muḥammad Ibn Tūmart, *A'zz mā yuṭlab*, ed. 'Abd al-Ghanī Abū 'l-'Azm (al-Ribāṭ, 1997), p. 385.

[3] Idem, p. 387.

traditionalism which shaped the Almohad movement[4] and it is not surprising to see that all the actions listed by Ibn Tūmart (for women and for men) as unforgivable sins are Prophetic Traditions. Here I shall limit myself to an examination of the first of the points that allow the observer to identify women as Almoravids, a faction excommunicated on religious grounds and a well-known political and military opponent. Their women, according to Ibn Tūmart, embellished their heads with an extravagant coiffure, making them look like camels' humps. These, however, were not his own words, but those of – or attributed to – the Prophet himself, and it is improbable that Almoravid women actually arranged their hair in this particular fashion. Let us turn then to the *ḥadīth* itself in an attempt to understand how it was explained by scholars of different historical periods and the origins of the opinion that women's coiffures should be condemned because they resembled camels' humps.

The earliest authority I have found quoting this *ḥadīth* is the Andalusian ʿAbd al-Malik Ibn Ḥabīb (d. 238/853), a polymath who, among many other works, wrote a treatise on the proper behaviour of women, a collection of Prophetic Traditions he had gathered during his stay in Egypt and the Hijaz.[5] In Ibn Ḥabīb's book the *ḥadīth* is reproduced without a chain of transmitters on the sole authority of al-Ḥasan al-Baṣrī, the renowned preacher from Basra (d. 110/728): 'He who has not seen the denizens of Hell, let him look at these women dressed as if they were naked, strayed from the right path, and with heads resembling humps of emaciated camels, melting in the fire of Hell on the Day of Judgement!'[6]

Of the six canonical collections of *ḥadīth*, only Muslim b. al-Ḥajjāj and Aḥmad b. Ḥanbal include this tradition.[7] Muslim places it in the chapter on dress (*al-libās wa 'l-zīna*) and in a chapter on Hell (*bāb jahannam*).[8]

---

[4] See M. Fierro, 'Revolución y tradición. Algunos aspectos del mundo del saber en al-Andalus durante las épocas almorávide y almohade,' in M.L. Ávila & M. Fierro (eds.), *Biografías almohades*, II. Estudios onomástico-biográficos de al-Andalus; 10 (Madrid-Granada, 2000), pp. 131–165.

[5] M. Arcas Campoy [et al.], 'Ibn Ḥabīb al-Ilbīri,' in *Biblioteca de al-Andalus* 3 (Almería, 2004), pp. 219–227.

[6] ʿAbd al-Malik b. Ḥabīb, *Kitāb Adab al-nisāʾ al-mawsūm bi-Kitāb al-Ghāya wa 'l-nihāya*, ed. ʿAbd al-Majīd Turkī (Beirut, 1992), p. 222.

[7] A.J. Wensinck, *Concordance et indices de la tradition musulmane*, I (Leiden, 1936), p. 146.

[8] Muslim b. al-Ḥajjāj, *Ṣaḥīḥ Muslim bi-sharḥ al-Nawawī*, XIV (Cairo, n.d.), pp. 109–10; XVII, p. 190.

The text is very similar to that of ʿAbd al-Malik Ibn Ḥabīb, with a change in the last sentence, where such women are told that they will neither enter Paradise, nor smell its perfume. The chain of transmitters goes from Zuhayr b. Ḥarb to Jarīr b. Suhayl to his father and finally to Abū Hurayra. In this manner it was quoted by subsequent authors such as Ibn Tūmart and others in later times.[9]

In his commentary on Muslim's *Ṣaḥīḥ*, al-Nawawī (d. 676/1277) explains that the sentence 'their heads are like camels' humps' means that women used to enlarge the shape of their heads with veils, turbans and similar things. Another possible explanation, adds al-Nawawī, is the hairstyle adopted by some women, who gathered their plaited hair at the top of their heads.[10] In both cases they would resemble camels' humps. To this literal interpretation al-Nawawī adds another, taken from a commentary on Muslim's *Ṣaḥīḥ* by the famed Sicilian jurist al-Māzarī (d. 536/1141). This sentence could also mean, writes al-Māzarī, that women craving for men did not lower their eyes in front of them, nor did they bow their heads before them.[11]

The context in which the *ḥadīth* is placed, both by Muslim and by his commentators, is clearly of an apocalyptic nature: these women will be one of the signs of the End of Times, and their attitudes, actions and appearance will serve as an announcement of the cataclysmal reversal of values announcing the Day of Judgement.[12] In the text by Ibn Tūmart quoted above, the 'signals' used to describe the Almoravids are equally situated in an apocalyptic context, since they are followed by several Prophetic Traditions on the 'Last Hour.'[13]

The apocalyptic note is played down by other commentators, who nevertheless keep it in the background. In his Qur'anic commentary,

---

[9] ʿAbd al-Malik b. Ḥabīb could not have met Muslim, who was a child when Ibn Ḥabīb returned to al-Andalus in 210/825 or 211/826, and it is evident that his version of the *ḥadīth* was collected before traditionists established the rules governing the *ʿilm al-ḥadīth*. Thus, this *ḥadīth* represents an interesting testimony to the early spread of this particular tradition in religious circles.

[10] al-Nawawī, *Ṣaḥīḥ Muslim bi-sharḥ al-Nawawī*, XVII, p. 191.

[11] al-Māzarī's commentary on the *Ṣaḥīḥ* of Muslim is entitled *al-Muʿlim bi-fawāʾid Kitāb Muslim*, edited in Tunis and Cairo (n.d.). It is considered to be the earliest commentary on the *Ṣaḥīḥ*; see Ch. Pellat, 'Al-Māzarī,' in *Encyclopaedia of Islam* (New ed.), VI (Leiden, 1991), pp. 942–943.

[12] See W. Saleh, 'The woman as locus of apocalyptic anxiety in medieval sunni Islam,' in A. Neuwirth, B. Embaló [et al.] (eds.), *Myths, historical archetypes and symbolic figures in Arabic literature* (Beirut, 1999), pp. 123–145. On the *ḥadīth* of the camels' humps, see p. 135.

[13] Ibn Tūmart, *Aʿzz mā yuṭlab*, p. 386.

al-Qurṭubī (d. 671/1273) uses the *ḥadīth* on camels' humps when dealing with Qur'an 24:60[14] and discussing the proper appearance of women in public and private places. Women act provocatively, asserts al-Qurṭubī, when they go out covered by a dress but devoid of religious piety, with the intention of attracting men's glances. Such women are described by the Prophet as having heads like camels' humps, that is, adds al-Qurṭubī, women who have put their braids on top of their heads. Social and religious chaos (*fitna*, a concept also related to the Last Hour) will be the ineluctable result of such behaviour.[15]

It was only in 8th/14th-century Egypt when several authors started to use this *ḥadīth* to indicate a specific hairstyle in fashion among women, condemning it on religious grounds. Ibn al-Ukhuwwa (d. 729/1329), the author of a *ḥisba* treatise, when establishing rules for the proper way of behaving in public baths, makes angry remarks against inmodest women, who not only wear transparent clothes, but also don turbans similar to camels' humps. Naturally enough, he continues quoting the *ḥadīth* with its chain of transmitters.[16] A very different kind of scholar is the lexicographer Ibn Manẓūr (d. 711/1311–12), the author of the monumental dictionary *Lisān al-ʿarab*. In its entry *sanam* (hump), the word is linked to the *ḥadīth,* which is quoted with a short explanation: women so described are those who make turbans (*yataʾammamna*) for themselves with their veils (*maqāniʿ*).[17] Ibn Manẓūr does not make any comment, positive or negative, on this kind of attire.

Women actually did wear men's turbans in the Mamluk period to which Ibn al-Ukhuwwa and Ibn Manẓūr belong, as is attested by historical sources. In the 7th/13th century, the Mamluk authorities in Damascus and Cairo repeteadly forbade women the use of men's clothes and headwear, which proves the persistence of this habit among women.[18]

---

[14] 'Such women as are past child-bearing and have no hope of marriage – there is no fault in them that they put off their clothes, so be it that they flaunt no ornament;' trans. A.J. Arberry, *The Koran interpreted* (Oxford, 1964), p. 360.

[15] al-Qurṭubī, *al-Jāmiʿ li-aḥkām al-Qurʾān*, XII (Cairo, 1987), p. 311. For the use of *fitna* in another genre, see R. Kruk, 'The Bold and the Beautiful. Women and "fitna" in the Sīrat Dhāt al-Himma. The story of Nūra,' in G.R.G. Hambly (ed.), *Women in the medieval Islamic World* (London, 1998), pp. 99–116.

[16] Ibn al-Ukhuwwa, *The Maʿālim al-qurba fī aḥkām al-ḥisba*, ed. R. Levy (London, 1938), p. 157.

[17] Ibn Manẓūr, *Lisān al-ʿarab*, XV (Būlāq, 1303 AH), p. 198. The text of this entry was reproduced by Murtaḍā al-Zabīdī (d. 1205/1791), *Tāj al-ʿarūs min jawāhir al-Qāmūs*, XXXII, ed. ʿA. al-Ṭaḥāwī, (al-Kuwayt, 1987), pp. 422–23.

[18] See M. Chapoutot-Ramadi, 'Femmes dans la ville mamluke,' *JESHO* 38 (1995), pp. 145–164 and especially pp. 151–152.

In the following century, as we have just seen, the expression 'heads like camels' humps' was related to the wearing of turbans by women. A third testimonial to this custom is found in the work of the stern moralist Ibn al-Ḥājj (d. 737/1336–37), a Cairene of Maghribi origins who dealt at length with this issue in a chapter of his work, *al-Madkhal*, on the proper dress for women.[19]

The chapter begins with the usual condemation of wearing short and tight dresses, but before long Ibn al-Ḥājj takes a new turn in his discourse, asserting that wearing turbans should be forbidden and, naturally enough, he cites the *ḥadīth* on 'heads like camels' humps.' To support his argument he also adduces the interpretations by al-Nawawī and al-Qurṭubī mentioned above, but his own reflections are more interesting than his quotations. Ibn al-Ḥājj observes how harmful the turban is for the physical and religious well-being of women and offers several reasons justifying its prohibition. Firstly, but less importantly in his view, women suffer from headaches because they wear the turban down to their eyebrows. More serious, however, is the fact that women who wear the turban in this manner are hiding part of their beauty from their husbands, a thing Ibn al-Ḥājj bluntly considers as sinful. Moreover, when a woman so attired needs to wash her face and head in order to prepare herself for prayer, she will probably catch a cold, because the turban had kept her forehead warm. In conclusion, such a woman is in danger of neglecting several religious duties (*farḍ*): washing her face and rubbing her head (thus neglecting her prayer) and hiding from her husband the beauty given to her by God.[20]

It is somehow intriguing to see how Ibn al-Ḥājj, in this ingenious display of reasoning against women wearing turbans, does not use an argument widely employed by other commentators and writers dealing with the subject of women's clothes: the prohibition of women adopting men's dress and vice versa, which is also based upon Prophetic Traditions.[21] Writing in the same period as Ibn al-Ḥājj, Ibn Taymiyya (d. 728/1328) advises fathers and husbands to forbid their womenfolk to wear men's

[19] Ibn al-Ḥājj, *al-Madkhal*, I (n.pl., 1981), pp. 241–243.

[20] Idem, p. 243. See also H. Lutfi, 'Manners and customs of fourteenth-century Cairene women. Female anarchy versus male sharʿi order in Muslim prescriptive treatises,' in N.R. Keddie & B. Baron (eds.), *Women in Middle Eastern history* (New Haven, 1991), pp. 99–121, and especially p. 110.

[21] Ibn Ḥabīb, *Kitāb Adab al-nisāʾ*, pp. 206–207.

clothes, such as turbans and men's shoes.[22] Although it is not possible here to elaborate further on this point, the question itself deserves mention, since the need to establish clear boundaries between the sexes is of crucial importance in Islamic ideology.

The texts examined until now lead to several conclusions. The *ḥadīth* on 'camels' humps' had an early circulation, with apocalyptic undertones, and was accepted as sound by the first compilers of canonical collections. Later commentators used it to fight what they believed to be provocative attitudes of women in the public space, with or without reference to specific hairstyles. During the Mamluk period, however, the *ḥadīth* seems to have been used against a particular headwear, namely men's turbans adopted by women.[23] Another text, from a commentator who lived in Mosul, where he died in 606/1219, adds a new link to the social context in which this *ḥadīth* was transmitted and interpreted. Majd al-Dīn Abū 'l-Saʿādāt Ibn al-Athīr, in his work on the *ḥadīth* lexicon, gives the usual interpretation: women whose heads are like camels' humps are those who make turbans from their veils in order to increase the height of their heads. This, adds Ibn al-Athīr, is the distinguishing mark of female singers (*mughanniyāt*).[24]

We do not know if Ibn al-Athīr was merely repeating an earlier statement or if he was reflecting on actual conditions in his time. But wearing extravagant clothes was, in fact, a trade mark of songstresses,[25] and a coiffure resembling camels' humps would be appropriate for a female singer. Implicitly, the comment by Ibn al-Athīr sets a definite standard of behaviour for good Muslim women, who should avoid doing their hair in the same style as the disreputable female singers. Similarly, the Andalusian Ibn ʿAbdūn, at the beginning of the 6th/12th century, sought to forbid 'honest' women to adorn themselves with an attire similar to that of prostitutes.[26]

---

[22] Ibn Taymiyya, *al-Fatāwā al-kubrā*, IV (Beirut, 1988), p. 366.

[23] See Y.K. Stillman, *Arab dress. From the dawn of Islam to modern times. A short story* (Leiden, 2000), pp. 75–83.

[24] Ibn al-Athīr, *al-Nihāya fī gharīb al-ḥadīth wa 'l-athar*, ed. M.M. al-Ṭanāḥī & Ṭ.A. al-Zāwī (Cairo, 1963), II, p. 409.

[25] See E. Mesa, 'La indumentaria y el aspecto externo de los cantantes según el *Kitāb al-Agānī*,' in C. de la Puente (ed.), *Identidades marginales*. Estudios onomástico-biográficos de al-Andalus; 13 (Madrid, 2003), pp. 347–369.

[26] Ibn ʿAbdūn, *Risāla fī 'l-qaḍā' wa 'l-ḥisba*, ed. E. Lévi-Provençal (Cairo, 1955), pp. 50–51.

To enlarge the volume of their braids and make the coiffure so strongly disapproved of by the *ḥadīth* commentators and moralists, women could also use false hair and wigs. Again, this habit was condemned severely,[27] but only if the wig was made from the hair of another person. There was a general agreement on the permissibility of using one's own hair or strands of wool for this purpose, and the authority of some of the Prophet's wives is quoted in this respect. Thus, the mother of Bakr b. al-Ashajj told her son that she was visiting ʿĀʾisha while the latter was getting ready for her wedding, and she saw that the hairdresser was adding wool to ʿĀʾisha's hair.[28] But when another woman asked the same ʿĀʾisha if her daughter, who was ill and becoming bald, could wear a wig she was given a flat refusal.[29] It is obvious to the reader of the story that this was a 'false' wig, made from another person's hair. According to another wife of the Prophet, Umm Salama, braids could also be enhanced with strips of fabric.[30] In later times scholarly opinion followed this tradition. Ibn Rushd (the grandfather of Averroes), a jurist of great prestige, did not hesitate to give his attention to this matter, stating that a woman could use her own hair to enhance her coiffure.[31] Ibn al-Jawzī (d. 597/1200) gathered *ḥadīth* material to the same purport, and he reproduced an anecdote in which the first Umayyad caliph, Muʿāwiya b. Abī Sufyān, preached against the use of false hair by women.[32]

Wool and strips of fabric have always been a customary addition to women's hair in some Islamic societies. In nineteenth-century Egypt, E.W. Lane reported that women added to their numerous braids 'three black silk cords, with little ornaments of gold, etc., attached to them.'[33] More detailed is the picture drawn by A.-M. Goichon for the Algerian Mzab in the second decade of the 20th century, with references to the

---

[27] J. Sadan, 'Shaʿr,' in *Encyclopaedia of Islam* (New ed.), IX (Leiden, 1997), pp. 311–312.

[28] Ibn Ḥabīb, *Kitāb Adab al-nisāʾ*, pp. 225–226.

[29] Ibn Saʿd, *Kitāb al-Ṭabaqāt al-kabīr*, VIII, ed. C. Brockelmann (Leiden, 1904), p. 352.

[30] Ibn Ḥabīb, *Kitab Adab al-nisāʾ*, p. 226.

[31] Ibn Rushd, *al-Bayān wa ʾl-taḥṣīl*, IV, ed. S. Aʿrāb (Beirut, 1988), p. 384.

[32] Ibn al-Jawzī, *Aḥkām al-nisāʾ*, ed. ʿAbd al-Majīd Ṭuʿma al-Ḥalabī (Beirut, 1997), pp. 124–128. This story was also recorded by Ibn Ḥabīb, *Kitāb Adab al-nisāʾ*, p. 223. In this earliest version, Muʿāwiya warns his listeners that the Israelites were destroyed when their women began to use false hair.

[33] E.W. Lane, *An account of the manners and customs of the modern Egyptians*, I (London, 1846), p. 62.

use of wool as false hair in Morocco.[34] Both social custom and normative rules seem to concur on this matter, and it could be that the former shaped the latter. The prohibition of using another person's hair, while considering licit wool and fabrics, could be another manifestation of the Islamic aversion to alter God's creation (*taghyīr khalq Allāh).* Wool and fabrics, having but an abstract resemblance to hair, were accepted by the custodians of orthodoxy, who did not approve of the 'falsehood' represented by a wig made from real hair.

Arab lexicographers had a name for the strips of wool or textile that women added to their own hair: *al-qarāmil.* Interestingly, the root *q.r.m.l.* is also connected with camels, as we can see in al-Khalīl's pioneering work on Arab lexicography: *al-qarāmil* is 'what a woman adds to her hair,' and *al-qirmiliyya* is 'a two-humped camel (*ibl*).'[35] Very similar are the definitions given by al-Jawharī (d. 398/1007–8): *al-qirmil* is the son of *al-bukht; al-qarāmil* are two-humped camels, and also, what women tie to their hair.[36]

In the *ḥadīth* commented upon here, women's heads are compared to the humps of a *bukht*. This kind of camel was the result of crossing the *camelus bactrianus* (two-humped) with a female dromedary (one-humped). The hybrid *bukhtī* had 'either a single, rather long hump, or a single long hump with a noticeable indentation in it, [making] the fore part substantially smaller than the rear.'[37] It is easy to relate this description to a particular kind of feminine coiffure, and to see how the wrath of supervisors of public morality expressed itself against women through this meaningful comparison. By contrast, in the case of the *qarāmil* the same connection between women's hair and camels is more neutral, as Ibn al-Jawzī states, quoting Abū ʿUbayd: the '*fuqahāʾ* are indulgent

[34] A.-M. Goichon, *La vie feminine au Mzab. Étude de sociologie musulmane* (Paris, 1927), pp. 121–122. Nothing similar was recorded by A. Musil in his exhaustive study of the Rwala, *The manners and customs of the Rwala bedouins* (New York, 1928) (on women's plaits and unplaited hair, see p. 118).

[35] al-Khalīl b. Aḥmad, *Kitāb al-ʿAyn*, V, ed. Mahdī al-Makhzūmī & Ibrāhīm al-Sāmarrāʾī (Beirut, 1988), p. 265.

[36] al-Jawharī, *al-Ṣiḥāḥ*, V, ed. A.ʿA. ʿAṭṭār (Beirut, 1984), p. 1801. See also Ibn Sīda, *al-Muḥkam wa ʾl-muḥīṭ al-aʿẓam fī ʾl-lugha*, VI, ed. M. Kāmil (Cairo, 1972), p. 393 and Murtaḍā al-Zabīdī, *Tāj al-ʿarūs*, XXX, p. 248.

[37] R.W. Bulliet, *The camel and the wheel* (Harvard, 1975), p. 143. The erroneous belief in the infertility of the hybrid is reproduced by Ch. Pellat, 'Ibil,' in *Encyclopaedia of Islam* (New ed.), III (Leiden, 1986), p. 665.

towards *qarāmil* and other things attached to the hair, on condition that they are not hair.'[38]

Almohad propaganda used the *ḥadīth* material mentioned at the beginning of these pages with a conscious effort to damage the Almoravids' reputation. By representing their enemies as the perverse and corrupt harbingers of the Last Hour, Almohadism emphasized the reversal of values supposedly prevailing among Almoravids. In the same context it must be remembered how pro-Almohad chroniclers accused the Almoravids of having allowed women to get hold of political power, thus reversing the natural and religious order of things.[39] The accusations of Ibn Tūmart against women whose heads resembled camels' humps do not censure a style of head-dress or coiffure, but rather the alleged predominancy of women in society, where they should only be seen through a veil or remain hidden in their homes.

---

[38] Ibn al-Jawzī, *Aḥkām al-nisā'*, p. 128.

[39] M. Marín, *Mujeres en al-Ándalus* (Madrid, 2000), pp. 242–243.

NAÏVETÉ, VERSES OF HOLY WRIT, AND POLEMICS.

PHONEMES AND SOUNDS AS CRITERIA: BIBLICAL VERSES SUBMITTED TO MUSLIM SCHOLARS BY A CONVERTED JEW IN THE REIGN OF SULTAN BĀYAZĪD (BEYAZIT) II (1481–1512)*

Joseph Sadan

A. *A tract aimed at compelling Jews to recognize and appreciate the superiority of Islam*

Muslim polemical interest in the Holy Books of Jews and Christians is a field with a long history, one whose importance has been amply demonstrated by modern scholarship. A number of studies have focused on the polemical stance taken by Islam towards the Old Testament. Muslim polemical argumentation, which has taken numerous forms and directions, is of course quite well-known, and we shall not discuss it any further here. These polemics focus on the one hand on showing that Jews and Christians have falsified and perverted the texts of their Holy Writ, resulting in moral turpitude (for example, the ascription of morally unacceptable acts to prophets who are also revered in Islam); on the other hand, Muslim polemicists have relied on biblical verses which, so they claim, refer to the Prophet Muḥammad's advent, mission and faith. The latter kind of verses are considered free of falsification and perversion and in fact taken to be expressions of exquisite wisdom which prove the truth of their religion.[1] Modern scholarship has also dealt in a satisfactory manner with the significance of Jewish converts to

* The present paper is a revised version of a study published in Hebrew in *Peʿamim* 42 (1990), pp. 91–104. To facilitate the reader all references are given in full in the first citation and in all further citations in an abbreviated form which is indicated between quotation marks immediately after the first citation, e.g. ('Steinschneider'). I am grateful to the late Prof. H. Lazarus-Yafeh for her advice. The references to al-Qurṭubī and to Petrus Alfonsi I owe to Prof. Van Koningsveld, who also sent me photocopies of ʿAbd al-Ḥaqq in an old lithographic edition and a reproduction of the MS of al-Biqāʿī (see references).

[1] From a bibliographical perspective three types of compositions are relevant to the topic at hand: (a) Compositions (by Muslims) which refer to biblical verses that are

Islam who did their best to provide Muslims with facts and quotes from the Bible, especially for polemical purposes.[2]

quoted also in the tract discussed here (I shall not mention compositions which do not mention verses that appear in said pamphlet) – see Ibn Ḥazm al-Andalusī, *al-Fiṣal fī 'l-milal wa 'l-niḥal* (Cairo, 1348 AH) (henceforth 'Ibn Ḥazm'); Samaw'al b. Yaḥyā al-Maghribī, *Ifḥām al-yahūd = Silencing the Jews*, ed. and transl. by M. Perlmann. Proceedings of the American Academy for Jewish Research; 32 (New York, 1964) ('Samaw'al (Perlmann)'); Shihāb al-Dīn al-Qarāfī al-Ṣanhājī, *al-Ajwiba al-fākhira ʿan al-as'ila al-fājira*, ed. Bakr Zakī ʿAwaḍ (Cairo, 1987) ('Qarāfī'); Ibn Qayyim al-Jawziyya, *Ighāthat al-lahfān min maṣā'id al-shayṭān*, ed. Muḥammad Sayyid al-Kaylānī (Cairo, 1961) ('Ibn Qayyim, *Ighāthat*'); Id., *Hidāyat al-ḥayārā fī ajwibat al-yahūd wa 'l-naṣārā*, ed. A. Ḥijāzī al-Saqqā' (Cairo, 1399 AH) ('Ibn Qayyim, *Hidāyat*'); ʿAbd al-Ḥaqq al-Islāmī, *al-Sayf al-mamdūd fī al-radd ʿalā aḥbār al-yahūd*, lithographed ed. in three fascicles (I–III) (Fez, [n.d.]) (lent to me by Prof. P.S. van Koningsveld). After the publication of the 1990 original version of our present study, this source has been published: *Espada extendida para refutar a los sabios judíos*, edición, con intr., trad. y notas, E. Alfonso (Madrid, 1998) ("Abd al-Ḥaqq'); see also M. Perlmann, "Abd al-Ḥaqq al-Islāmī, a Jewish Convert,' *JQR* 31 (1940–1941), pp. 171–191 ('Perlmann, ʿAbd al-Ḥaqq'); L.J. Kassin, *A study of a fourteenth-century polemical treatise Adversus Judaeos* (unpublished Ph.D. thesis, Columbia University, 1969) ('Kassin'); M. Asín Palacios, 'Un tratado moriscos de polémica contra los judios,' *Mélanges H. Derenbourg* (Paris, 1909), pp. 343–366 ('Asín Palacios'); Ibrāhīm b. Sirāj al-Dīn al- Biqāʿī, *al-Aqwāl al-qawīma fī ḥukm al-naql ʿan al-kutub al-qadīma*, MS Escorial, Derenbourg's catalog: no. 1540 ('Biqāʿī'). (b) Some polemical or comparative Jewish compositions will also be mentioned ('rebuttals' to or clarifications of arguments made by adherents of other faiths): Judah Halevi (Yehuda ha-Levi), *Sefer ha-Kuzari*, ed. Y. Even Shemuel Kaufmann (Tel Aviv, 1973) (Heb.) ('Judah Halevi (Even Shemuel)'); Id., *Kitāb al-radd wal-dalīl fī al-dīn al-dhalīl* (*Kitāb al-Khazarī*), ed. H. (D.Z.) Baneth & H. Ben-Shamay (Jerusalem, 1977) (Heb.) ('Judah Halevi (Baneth)'); Maimonides – Moshe ben Maymon, *Igeret teman*, ed. A.S. Halkin (New York, 1952) (Heb.) ('Maimonides'); Saʿd al-Dīn Ibn Kammūnah, *Examination of the inquiries into the three faiths*, ed. M. Perlmann. University of California publications, Near Eastern studies; 4 (Berkeley [etc.], 1967) ('Ibn Kammūnah'); Joseph ben Nathan (Official), *Sefer Yosef ha-Meqanne*, ed. Y. Rosenthal (Jerusalem, 1970) (Heb.) ('Joseph ben Nathan'). (c) Scholarly studies, of which the following are the most relevant for our purposes: M. Steinschneider, *Polemische und apologetische Literatur in arabischer Sprache zwischen Muslimen, Christen und Juden*. Abhandlungen für die Kunde des Morgenlandes; 6/3 (Leipzig, 1877) ('Steinschneider'); I. Goldziher, 'Über muhammedanische Polemik gegen *Ahl al-kitāb*,' *ZDMG* 32 (1878), pp. 341–387 ('Goldziher'); M. Schreiner, 'Zur Geschichte der Polemik zwischen Juden und Muhammedaner,' *ZDMG* 42 (1888), pp. 591–674 ('Schreiner'); E. Straus (Ashtor), 'Manners of the Muslim polemics,' in *Memory book for the rabbis Beth Midrash in Vienna* (Jerusalem, 1946) (Heb.), pp. 179–182 ('Straus (Ashtor)'); M. Perlmann, 'The medieval polemics between Islam and Judaism,' in S.D. Goitein (ed.), *Religion in a Religious Age* (Cambridge MA, 1974), pp. 171–191 ('Perlmann, 'Polemics'); H. Lazarus-Yafeh, *Intertwined worlds. Medieval Islam and Bible criticism* (Princeton, 1992) ('Lazarus-Yafeh'). For a summary of references on the status of the 'Torah' (*Tawrāt* in a wider sense) in Islam, see Perlmann, 'Polemics,' pp. 135–138. For a number of additional references, see J. Sadan, 'Some literary problems concerning Judaism and Jewry in medieval Arabic sources,' in M. Sharon (ed.), *Studies in Islamic history and civilization in honour of Professor David Ayalon*, ed. M. Sharon (Jerusalem [etc.], 1986), pp. 353–398, here p. 371, n. 49.

[2] Some converts were proficient enough in Arabic to take part in these polemics;

Situated within the broad field of Muslim-Jewish polemics, the present paper deals with a specific case, of a converted Jew who lived in the Ottoman Empire during the reign of the Sultan Bāyazīd (Beyazıt) II (ruled 1481–1512). We shall focus mainly on the unusual status of this convert, which fluctuated between that of a wretched informant whom adroit Muslim clerics exploited, and a supposedly respectable writer whose tract is dedicated and presented to the Ottoman Sultan himself, although the convert was not involved in its composition, as we shall demonstrate below.

In the Fatih Camii collection, which today is located in the Süleymaniye Library in Istanbul, there is a thin tract, consisting of twenty-one pages, whose author and title are, according both the old printed Fatih catalog[3] and in the relatively new Süleymaniye card files, as follows: ʿAbd al-ʿAllām (author's name), *Risālat ilzām al-yahūd*.[4] The meaning of the title is 'Tract on compelling/forcing the Jews to accept' (i.e., to acknowledge the truth of Islam with respect to certain verses of the Torah).

The tract is written in Arabic; it deals with eight verses or short quotes, with numerous errors, deletions and confusions, from the Pentateuch (all, except for one, from the book of Deuteronomy). The contents of these verses may be characterized as follows: the Torah and prophecy in the view of the Bible. Each verse is followed by a lengthy discussion, in the spirit of medieval Islamic polemical and theological writing, and often in the spirit and style of *kalām*.[5] The main purpose of these

among them we may mention Samawʾal b. Yaḥyā al-Maghribī and ʿAbd al-Ḥaqq al-Islāmī. See Samawʾal (Perlmann), Introduction, p. 15 ff. in the English section, as well as the appendices there, pp. 127–135 in the Arabic-Hebrew section; Perlmann, 'Polemics', pp. 114–120; Perlmann, "Abd al-Ḥaqq', p. 183.

[3] See Anonymous, *Defter-i Fatih Kütübhanesi* (Catalog of the Fatih Mosque Library), printed by order of Sultan ʿAbdülḥamīd II (1876–1909), p. 172, MS cat. no. 2994 in his list.

[4] Samawʾal uses the word *ilzām* in a number of chapter headings, with the meaning of 'to force' or 'to compel' the Jews. See Samawʾal (Perlmann), for example pp. 6, 12, 20, 24 (in the Arabic part), 33, 36, 40, 42 (in the English part); Perlmann translated: 'compelling to accept', but also 'forcing'. Our tract, however, is not part of the composition written by Samawʾal and, as will be shown below, contains verses of the Torah which Samawʾal does not mention. Still, Samawʾal may well have had a direct or indirect influence on the concepts and terminology of our author, the converted Ottoman Jew.

[5] These discussions are important and deserve to be published (together with the entire text of the tract). Here our main focus is our Jewish Ottoman convert's peculiar status as an informant who 'transmits' biblical verses (whereas the discussions on these verses, as we shall see below, were neither written nor dictated by him). Therefore these polemical discussions shall remain outside the purview of the present paper. Here I shall just provide a hint at the nature of the debate by means of two examples: (a) the tract interprets the verse 'the prophet, which shall speak in my name' (consisting of a

discussions is to show that there is no contradiction between the verses in question and the appearance of Islam, that they cannot serve as arguments for Jews to adhere to their faith, and that they in fact show the need, nay the necessity, of acknowledging the superiority of Islam.

Within the broad scope of Jewish-Islamic polemics our modest tract is not devoid of interest, for both composition and author share a somewhat unusual aspect, mainly one of wretchedness. Our Jewish convert is perceived by Muslim savants in a highly uncomplimentary light. Furthermore, the skimpy selection of verses may in a way reflect a Jewish-internal tendency, emanating from within the covert's original community. We shall now flesh out these claims through the following five points:

(1) An examination of the language of the tract's declared author (especially the seemingly odd transcription of Torah verses into Arabic script) will show, as we shall see below, that he cannot be the author of the Arabic composition. His role in the process of putting the tract together was limited to providing the faulty quotes, written here in Arabic script. Perhaps he just dictated these verses, orally, (most probably) in a Turkish 'accent'. Certainly they reflect the way they were heard by the Turkish transcribers, who as a result produced rather amusing spellings, very far from what we would have expected in native speakers of 'oriental' Semitic languages (i.e., not at all in line with the 'oriental' pronunciation of Hebrew, nor with that of Arabic). The composition was thus entirely, or almost so, written by one or more Muslim religious scholars.

(2) While the Muslim 'rebuttals' or discussions following each of the biblical quotes are in line with the traditional polemical Muslim position, the somewhat confused and incomplete quotes provided by our convert and placed before each 'rebuttal' do possess a certain originality.

restructuring of elements from Deut. 18:15–22; see below for details) as referring to the various prophets who would appear in future, such as Jesus and Muḥammad' (this kind of argument is not new. See, for example, M. Perlmann, 'Affirmation of the prophecy of Muḥammad,' in *Hagut Ivrit be-Amerika*, III (Tel Aviv, 1974) (Heb.), pp. 73–97, here p. 91; (b) the biblical verse 'Since then, no prophet has risen in Israel like Moses' (Deut. 34:10. See details below) only means, according to the polemicists, that for the kind of mission which Moses undertook there was never any other prophet, but that his Torah is void (*mansūkha*, see below, notes 17, 21, 25, 40). To summarize, then, the selection of verses in the tract is original and in a sense unique, but the attendant discussions, which in my opinion were composed by Muslim theologians, although ascribed to our convert, are not novel by any means. Indeed, they would seem to continue the same line of argumentation to be found in the Muslim literature to which reference is made in note 1 above.

Certainly it is possible that some of these eight Torah fragments in varying combinations aroused the interest of earlier writers. However, the specific choice of verses in our tract does not occur in any other composition of which we are aware. Furthermore, some of these quotes cannot be found in the writings of any known Muslim polemicists. Indeed, the choice of verses as a whole (and the unique transcription of the Hebrew text) shows that what we have here is an original composition which was written in the Ottoman Empire, and not just a patchwork of older fragments.

The inaccuracies in the transcription of the verses can be divided into two types: one consists of textual confusion, deletion and omission, the other of an unusual, non-'oriental' rendering of the Hebrew text. The first type will be elaborated in section C below. The second type, however, cannot really be classified as a mistake at all, despite the impression of strangeness one gets when reading the transcribed verses. It turns out, in fact (see below, sections B and C), that when one takes into account the rules of Turkish pronunciation, the transcription makes eminent sense.

(3) Although our convert was nothing more than an informant, the entire tract, including its polemical-theological core, was ascribed to him. In this he differs from more respectable and knowledgeable Jewish converts to Islam who were better acquainted with Islamic culture and more proficient in the Arabic language, and were therefore able to attack their former faith in books which won a certain measure of popularity and respect. One prominent case in point is the twelfth-century convert Samawʾal b. Yaḥyā who, although certainly familiar with the arguments used by his new co-religionists, wrote his entire polemical discussion with his own hand. In contrast, our Ottoman Jewish convert was declared to have been the tract's author merely as an adroit polemical ploy: it is certainly wise to represent a polemical argument as originating with a (former) member of the religion being attacked.

(4) The selection of verses in the tract also only partially conforms to the traditional Muslim line (presented above), in which the emphasis is on the message of Islam and on purported mentions of Islam in the Holy Scriptures of the religions which preceded it. Although some intimations in this vein can be discerned among the eight biblical quotes in our tract,[6] most of the latter material provides support for Judaism,

[6] E.g. Gen. 49:10 (see below, section C, fourth quote).

in the sense that the verses call on believers not to abandon the ways of their forefathers. Furthermore, in one case the person who selected the verses found himself next to one of the favorite verses of Muslim polemicists, used by them to demonstrate the message of Islam through the Old Testament,[7] but the tract's authors-compilers refrain from quoting it (Deut. 18:18), although he does quote from the immediately preceding text, skips the verse in question, and goes on to quote from what follows it.[8]

---

[7] The verse (not quoted in our tract) 'I will raise them up a Prophet from among their brethren' (Deut. 18:18, and see also: Deut. 18:15, where it says 'of thy brethren') is very popular among Muslim polemicists (see Qur'an, 62:2; among exegetes of the Qur'an see, for example: Fakhr al-Dīn al-Rāzī, *Mafātīḥ al-ghayb* (Cairo, [1323 AH?]), VIII, p. 143, who provides what could be taken as a paraphrase of this verse). The Jews rebutted such Muslim arguments, and then the latter answered such rebuttals, and so on: Ibn Qutayba in an unnamed book of his, quoted by ʿAbd al-Raḥmān Ibn al-Jawzī, *al-Wafāʾ bi-aḥwāl al-Muṣṭafā*, ed. Muṣṭafā ʿAbd al-Wāḥid (Cairo, 1966), p. 63 ('Ibn al-Jawzī'). (I first became aware of this source for the verses in question by C. Brockelmann, 'Muhammedanische Weissagungen im Alten Testament,' *Zeitschrift für die alttestamentliche Wissenschaft* 15 (1895), pp. 138–142, here p. 139 ('Brockelmann')); Judah Halevi (Baneth), p. 124; Judah Halevi (Even Shemuel), p. 130; Ibn Ḥazm, part I, p. 90; Samawʾal (Perlmann), p. 29, 107 (of the Arabic part), 45, 81 (of the English part); Maimonides, pp. 46–47 (see also: *ibid.*, A. Halkin's Introduction, p. xviii); Qarāfī, pp. 419–420; Ibn Kammūnah, p. 63, lines 2–3 from the bottom, and pp. 94–96; al-Imām al- Qurṭubī (or at least ascribed to him), *al-Iʿlām bi-mā fī dīn al-naṣārā min al-fasād wa 'l-awhām*, ed. A. Ḥijāzī al-Saqqāʾ (Cairo, 1980), p. 263 ('Qurṭubī'); Ibn Qayyim, *Ighāthat*, part II, p. 353; Ibn Qayyim, *Hidāyat*, pp. 109, 115, 129, 174–175; Asín Palacios, p. 351 (the Arabic text – see Kassin, p. 330); ʿAbd al-Ḥaqq, part II, pp. 5–6 (he is aware of both Deut. 18:18 and Deut. 18:15). See also: below, Perlmann, on ʿAbd al-Ḥaqq; Biqāʿī, f. 24a, mistakenly identifies the chapter in Deuteronomy as 11 instead of the correct 18; Steinschneider, p. 326, parag. 2; Goldziher, p. 370, lines 13–14; Schreiner, pp. 601, 613; Perlmann, "Abd al-Ḥaqq,' p. 183; Perlmann, 'Polemics,' p. 118; Straus (Ashtor), pp. 191–192; Lazarus-Yafeh, pp. 104, 124–125, 150. In sources other than the polemical literature: additional materials can be gleaned from the religious-literary genre known as *Dalāʾil al-nubuwwa*, for example: Abū Nuʿaym al-Iṣbahānī (al-Iṣfahānī), *Dalāʾil al-nubuwwa* (Hayderabad, 1320 AH), pp. 16, 17, who quotes traditions and 'prophecies' about a future prophet who will come out of Arabia; however, no clear quote of the verse is involved here. On the Christian-Jewish and Jewish-Christian dispute concerning this verse, see Petrus Alfonsi, *Dialogi contra Iudaeos*, in J.P. Migne (ed.), *Patrologia cursus completus* (Paris, 1844–1999), CLVII, pp. 535–672, here col. 626 a-b ('Petrus Alfonsi'); Joseph ben Nathan, p. 61.

[8] In the polemical discussions (the 'rebuttals' of the verses, and their refutation) contained in the convert Salām's tract the subject of a prophet 'of thy brethren' or 'from among their brethren' (see previous note) is indeed raised, although the convert provides no quotes on the matter. Rather, it is mentioned in passing during a discussion of a verse quite distant from Deut. 18 (Salām ʿAbd al-ʿAllām, *Risālat ilzām al-yahūd*, MS Fatih (now in the Süleymaniye Library), No. 2994, f. 8a ('MS ʿAbd al-ʿAllām')). The polemicists (i.e., the true composers of the tract) here follow in the footsteps of pre-existing Muslim polemical literature. While their elaborations on the convert's quotes are in general quite apposite, the present case shows that they are not always entirely successful.

(5) It is true that a written Jewish polemical literature also existed, whose purpose was to protect, provide encouragement and give answers. The Jewish sages who wrote such works also quoted verses of the Torah.[9] Our converted Jew, however, did not avail himself of such precisely-formulated writings, but quoted (probably from memory) mainly confused and erroneous fragments.[10] He may thus reflect, at one or two removes, oral arguments indirectly derived from this literature as well, through Rabbis' sermons or perhaps quotes of commonly heard sayings in the community at its most basic strata, on the street, at home or in the synagogue. He thus 'discloses' the secret of these verses and their function to Muslim legal scholars, so that the latter may rebut (supposedly in his name) and debate them.

B. *The author of the tract: when he lived, his character and education, judged by his quotes*

As we mentioned above, the author of the tract is identified in the catalogs only as ʿAbd al-ʿAllām; in the text of the tract itself, however, he is called Salām ʿAbd al-ʿAllām.[11] Possibly the compilers of the catalogs thought that the word *salām* (cognate of Hebrew *shalom*, as both general and proper noun; on this possibility see few lines below forthwith), was part of a religious formula which precedes the author's name, and that the author used it to refer to himself parenthetically; however, syntactically this word cannot be joined to the formula which precedes it. Salām cannot therefore be anything by the author's 'first' name (there is of course the possibility that this was a *nom-de-plume* or an epithet given to him by the tract's true authors). The use of the name Salām can apparently go against the spirit of Islam, at least from an esthetic point of view, since it is one of the names of the Deity. A man wishing to avail himself of this name would call himself ʿAbd al-Salām (lit. Servant of God).

---

[9] See, for example, Judah Halevi (Baneth), p. 124; Id. (Even-Shmuel), p. 130; Maimonides, pp. 52–53 (and Halkin's Introduction, pp. xii–xxi); Ibn Kammūnah, p. 63 (both the quote from Genesis and the quote from Deuteronomy), 94, 96. All of the above quote verses that are similar to what is quoted in the present tract (see also below, section C).

[10] The errors may have been introduced on purpose, on the initiative of his Muslim superiors, but the character of the deviations does not support such a hypothesis; see below, toward the end of section B.

[11] See the converted Jew Salām, MS ʿAbd al-ʿAllām, at the end of f. 1b and the beginning of 2a.

While it is true that among the Jews who converted during the earliest period of Islam and provided Muslims with a wealth of Jewish and biblical materials was ʿAbd Allāh b. Salām, in this case the name was not that of the converted ʿAbd Allāh but that of his Jewish father (who was called Shalom or Shelomo/Solomon).[12] It is certainly possible that our convert was named Shalom and failed to adopt a more acceptable name after he had become a Muslim. His conversion is mentioned in the text as the result of a God's great favor, which enabled him to see the correct path.[13] The composition is dedicated to the Ottoman Sultan Bāyazīd (Beyazıt) II, who ruled in the years 1481–1512. This tells us approximately when the man lived. The text does not provide us with any other information about our convert; we can, however, extract important details from an examination of his language, especially from a reconstruction of his pronunciation.

Before analyzing the pronunciation reflected in the transcription, we should point out that the full version of the biblical quotes discussed here appears in section C below; the meanings of the words discussed here are taken from the King James Bible.

An examination of the transcription of the tract's fragmentary verses in Arabic script reveals that it does not fit the 'oriental' pronunciation tradition, either of Hebrew or of Arabic. Thus the Hebrew consonant *z* has a sound which corresponds in the pronunciation of Ottoman Turkish not only to the Arabic character ز (z), which is to be expected, but also to the characters ض (ḍ) and ظ (ẓ), which Turks do not distinguish in speech although they differentiate between them accurately in written texts (in both Ottoman Turkish, i.e. Turkish written in Arabic characters, as well as in Arabic itself). This is the case, for instance, in the following words which occur in our MS:

> *haẓo* = 'this', instead of *hazo*.
> *ẓarʿo* = 'his seed', in the sense of 'descendants' [of Israel], with *ẓ*, instead of *zarʿo*, with *z*.

No distinction exists between the Hebrew sounds *ḥ* and *kh*, both of which sound like *h* in Turkish, as in the following examples:

[12] See Steinschneider, p. 110, n. 1. Note the play on words *salām* – *islām* (which, even if unconscious, possesses an auditory-semiotic value). For another case of ʿAbd al-Salām and our convert see Perlmann, "ʿAbd al-Ḥaqq", p. 171ff.: ʿAbd al-Ḥaqq's name after converting to Islam was 'al-Islāmī' (again the same sound: *islām* – *salām*); the name ʿAbd al-Ḥaqq means 'Servant of the God of Truth'.

[13] See our converted Jew Salām, MS ʿAbd al-ʿAllām, f. 2a.

yiqḥaṭ (instead of *yiqhat*) *ʿammīm* = 'and unto him shall the gathering of the people be' (other interpretations of the expression exist); here, the word *yiqhat* undergoes two changes: *h* > *ḥ* as well as *t* > *ṭ*, since no distinction between members of these pairs is made in Turkish pronunciation. *toṣīfū* = 'you [shall not] add [unto the word which I command...]'. Here an emphatic *ṣ* has replaced *s*; again, both sound the same to Turkish ears.

Emphatic *ṭ* is transcribed with non-emphatic *t* :

Sevet instead of *sheweṭ* (*sheveṭ*), with *ṭ* > *t* (as well as: *sh* > *s* , which is irrelevant to our purposes here); the meaning of the Hebrew word is: 'tribe', 'family'.

The letters *q* (ק) and *k* (כ) alternate:

Keh[i]llat, instead of the expected *Qehillat* = 'the congregation of...'.

The phonetically-conditioned alternation between the sounds *d* and *t* is a typical feature of the Turkish language. The result can be seen in the following example:

The Hebrew word *raʿōt* = 'evils' is transliterated *raʿōd*, since the *d* at the end of this word would sound in Turkish just like *t*.[14] Notice, however, that the author(s) did not in a single case confuse the sound 'and the sound'.

While it is true that unusual orthographical features can be found in other compositions that contain transcribed Hebrew biblical verses, these can usually be explained as copying errors, random mistakes, etc. One less random example is al-Qurṭubī's book against Christianity, a work that was written in Spain and also contains transliterated Hebrew verses. In addition to scribal errors it would appear to contain also errors due to habits of pronunciation; the latter can certainly be explained in terms of the circumstances under which the book was written. This is a Muslim rebuttal to Christian polemical arguments which were sent from Toledo, then already under Christian control. The book's structure reflects this, for it contains both the Christian arguments (including Hebrew verses in transliteration) and the Muslim 'rebuttals' that were

[14] Another case in which the phonetics of Turkish affects the language of an Arabic composition (albeit to a much lesser extent than in the case of our tract) is one describing a debate between a Christian and a Rabbi in Tiberias in the eighteenth century; see the important comments in: L. Kopf, 'A controversy between a Christian and a Jew in Tiberias in the eighteenth century. An Arabic manuscript from the Genizah collection of the Jewish National Library,' *Kiryat Sefer* 39 (1974) (Heb.), pp. 273–279, here p. 279, especially section 4.

written in Muslim Spain. It is therefore quite possible that the process of transliteration was influenced systematically by a Spanish pronunciation and a Spanish ear whose linguistic habits were quite different from those of the Arab orient.[15] The situation with respect to our Ottoman tract is quite different, however, and testifies to an even greater uniformity in the process: if we bear in mind the Turkish habits of pronunciation (i.e., the Jewish convert's ear or, rather, a recitation of the verses by our convert, who perhaps did not actually write down anything, a possibility which shall be discussed immediately below), then the transliteration in fact turns out to contain no sharp deviations from the Hebrew original.

Some of these matters of pronunciation and spelling will be discussed more at length in the notes to the table of verses, below in section C (for example, the interesting cases that are reflected in notes 18 and 30). We could also have delved into the matter of the alternation between *s* and *sh* (our convert occasionally preserves the original *sh*, but more often replaces it with *s*) and asked whether these were all caused by the copyist's pronunciation or writing habits, for the two letters are very similar in shape in Arabic. We could also have raised the question of the vowels and their pronunciation, and asked whether they reflect (despite the intervention of the copyist) a certain tradition in Torah-reading. But that is a matter for another study. Here I shall only discuss those facts which lead us to the conclusion that our convert did not write the tract in question. Two possible explanations of this conclusion present themselves:

(1) Perhaps our Jewish convert did indeed write the fragmented and erroneous verses in the tract himself in Arabic script, and in the process ignored the 'oriental' pronunciation and adopted instead that of a non-Semitic language, i.e. neither Hebrew nor Arabic but one of the other languages spoken in the Ottoman Empire, in which no distinction is made, for example, between *ḥ* and *kh*, nor between these two and *h*. In all probability this language was Turkish, as assumed in the discussion

[15] See Qurṭubī. A number of verses quoted by him from a lost tract by a Christian living in Toledo are mentioned in the present article; see note 7 above and notes 25 and 28 below. Of course, one must be careful about any conclusions concerning the pronunciation of the man from Christian Toledo who wrote the epistle to Muslim Cordova. Who was the informant behind it? Was the letter written originally in Arabic or only translated after it had been written? See also: P.S. van Koningsveld, 'La apologia de Al-Kindī en la España del siglo XII. Huellas toledanas de un "animal disptax",' in *Actas del II Congresso Internacional de Estudios Mozarárabes. Estudios sobre Alfonso VI y la reconquista de Toledo* (Toledo, 1989–1990), pp. 107–129, incl. n. 4 and 5 there ('Koningsveld').

above. But this raises a question: how can a person with such a poor command of Hebrew go on to write (the polemical 'rebuttals' after each quoted verse) in excellent and grammatically flawless Arabic? The inescapable conclusion is that the entire work, except for the quotes from Hebrew, was written by one or more Muslim religious scholars, whose training and experience prepared them for writing such polemical theological compositions.[16]

(2) But even the previously-mentioned possibility, namely that the convert only wrote down the verse fragments, is highly improbable. Certainly it is much more likely that our convert pronounced the verses to the tract's real authors, who heard them with an Ottoman Turkish ear. If this is indeed the case, then our convert did not write any part of the tract; only his mouth and tongue contributed to it (nothing but verses). This would seem to be the more likely possibility.

The errors and omissions in the quoted verses show that the man, who remembered some fragments of Torah texts, was no scholar. Perhaps the reader may object that it is after all possible that it was the Muslim sages, the tract's real authors, who mixed up the verses, deleted and perverted them on purpose. Against this we argue as follows: First of all, this work was dedicated to the Sultan himself, and although this cannot constitute a safeguard against mistakes, one would think that it would lend a minimum of seriousness to the endeavor and prevent any overly blatant fabrication. Secondly, some of the errors in the quotes are due to mixing up and recombining Hebrew words and phrases, something which a Muslim scholar unfamiliar with Hebrew would not be able to do. Thirdly, unlike the choice of elements (parts of verses), the distortions cannot be interpreted as unambiguously biased in favor of Islam. Fourthly, a Muslim polemicist would certainly have wanted to produce a book that would gain a 'victory' over the opinion of the Jews, not one which if it fell into the hands of a Jew who was proficient in Arabic and familiar with the Torah would refute it easily because of inaccurate quotes. Our conclusion thus stands, that the convert was not highly educated. We do not, however, possess enough data to determine whether or not his education was in the spirit of Rabbinical Judaism.[17]

---

[16] Unlike other converts who were proficient in Arabic and played a more active role in formulating the polemical arguments which they wrote down. See above, n. 2.

[17] All eight quotes are from the Pentateuch, and two or three of them are key verses of non-rabbinical Judaism (see above, n. 5 and below, n. 25 and 40). However, they also

Our convert apparently cut quite a sorry figure; this is nowhere more apparent than in the fact that he was treated as nothing more than an informant; he was granted the status of author only as a tactical step, useful for polemical purposes, but he lacked any respectable standing among those in the know. This was the case despite the fact that, as mentioned above, the tract was dedicated to the Sultan, in an attempt to emphasize the existence of both the author and the composition.

C. The quotes in the tract and their biblical origins (according to their order in the tract)

A mixture of elements from Deut. 13:1,3,5(?),10[18]

> If there arise among you a prophet, or a dreamer[19] of dreams, and giveth thee a sign[20] or a wonder [= verse 1 ending with an element of verse 13, quoting, in the original, what the children of Belial say:] Let us go *and serve* other gods [according to verse 2: Let us go *after* other gods and the ; our converted Jew jumps from one element to a similar one and continues after a small omission], thou shalt not hearken unto the words of that prophet or [= and] that he die[21] [the last two words are taken from verse 10: '...that he die'; the huge omissions helps to shorten the biblical text].

constitute a focal point of polemics with Islam about relying on the words of the Torah as they are, unchanged (in the sense of *naskh* or voiding, as explained above, n. 5 and below, n. 21). We therefore cannot come to any conclusion on this matter, at least at the current state of research on the tract in question.

[18] MS ʿAbd al-ʿAllām, f. 4b–5a (quotation from the Bible and a discussion). This verse and all the following ones are quoted here from the King James translation of 1611, as given in the London 1930 edition by the British Foreign Bible Society. We follow the numeration of the verses in this English version which differs only rarely from the Hebrew Massorah text. Since the aim of the present section is to show what our converted Jew mixed or omitted, the English text can easily serve us when we take into consideration the Hebrew Bible as well.

[19] In the original Hebrew text: *ḥolem*, transcribed in the MS as *holem* (*ḥ* and *kh* are *h*; see section B above).

[20] *Oth* (= a sign), as *t* without *dagesh* should really be pronounced *th* (our converted Jew who confuses so many other elements vaguely remembers how *t* is recited at the synagogue).

[21] *Wa-meth* (that die) follows the same rule, but instead of the Hebrew *wa* in the sense of 'and', it should be understood as *wa*, called *waw conversive* – *waw hahippukh*, in the sense of: and will, until he..; but how could our converted Jew construct such a summary by adding this element from verse 10 (or maybe 5) to the verses 1 and 3? His memory works in a very strange way and is not refreshed by the use of the written text. Compare with Ibn Ḥazm, part I, p. 139; an older stage is discussed by Schreiner, p. 601. The meaning of such verses is especially important in the context of whether the law of

A mixture, with some omissions and distortions, of Deut. 18:15–20[22]

> The prophet which shall speak a word in my name, unto him ye shall hearken [a mixture of elements due to our converted Jew's confusion; the most important elements in the eyes of the Muslim polemicists,[23] viz. verses 15 and 18: 'The Lord, thy God, will raise up unto thee a Prophet from the midst of thee, of thy bretheren', and 'I will raise them up a Prophet from among their bretheren' are both omitted in our tract].

Deut. 34:10[24]

> And there arose not a prophet since like unto Moses in Israel [in the Bible: in Israel like unto Moses, whom the Lord knew face to face].[25]

Genesis 49:10[26]

> The sceptre shall not depart from Judah, nor lawgiver from between his feet,[27] until Shiloh come; and unto him shall the gathering of the people be.[28]

Moses must be left as it is, unchangeable, or may accept new doctrines (that of Jesus or that of Muḥammad) and see above, n. 5, 17 and below, n. 40.

[22] MS ʿAbd al-ʿAllām, f. 5a–6a (quotation from the Bible and a discussion).

[23] On its importance in Muslim polemical tracts see n. 7 above.

[24] MS ʿAbd al-ʿAllām, f. 6a–11a (quotation from the Bible and a discussion).

[25] Ibn Qutayba, in an unidentified source, quoted by Ibn al-Jawzī, p. 63, discussed by Brockelmann, p. 139; Ibn Ḥazm, part I, p. 140; Abū 'l-Fatḥ al- Shahrastānī, *al-Milal wa 'l-niḥal*, ed. Muḥammad' Sayyid al-Kaylānī (Cairo, 1967), part I, p. 211, lines 15–16 ('Shahrastānī'); Maimonides, pp. 28–29 (in Halkin's introduction, p. XVIII); Ibn Kammūnah (Perlmann), pp. 94, 96: the Muslims' claim and its refutation by the author; see also p. 22 ff. (Moses is the 'best' [*ṣafwa*] of the Israelites; a discussion concerning his qualities as prophet and legislator); Qarāfī, pp. 418, 257; Qurṭubī, p. 264; Ibn Qayyim, *Hidāyat*, p. 111; Straus (Ashtor), p. 189. The claims of the non-rabbinical Jews, relying on this verse, against Jewish orthodoxy, do not interest us here, but it is a fact that some Muslim writers know about this; see Schreiner, p. 656 and compare n. 40 below. This verse is quoted also in Christian-Jewish controversies: Petrus Alfonsi, cols. 590a, 626c–d; Joseph ben Nathan, p. 63.

[26] MS ʿAbd al-ʿAllām, f. 6a–11a (quotation from the Bible and a discussion). Concerning the use of the apparent *ī* at the end of the proper name Judah, it should be understood as *alif maqṣūra* of the Arabic orthography (which corresponds to *ā*); see also section B above.

[27] The transliterated form looks like *raglo* (his foot), but should be understood as *raglaw* (his feet; see note 31 below).

[28] See Samaw'al (Perlmann), p. 23 of the Arabic section, 41 (of the English section); Qarāfī, pp. 226, 418–419 (two different translations of the verse; the first one would probably please certain Christian polemicists); Ibn Kammūnah, p. 63 (lines 5–6 from the bottom), 64; compare with M. Perlmann, 'Ibn Maḥrama. A Christian opponent of Ibn Kammūna,' in S. Lieberman (ed.), *H.A. Wolfson jubilee volume* (Jerusalem, 1965), II, pp. 641–655, here p. 653; Qurṭubī, p. 181; Ibn Qayyim, *Ighāthat*, p. part II, p. 339; Ibn Qayyim, *Hidāyat*, p. 208; Straus (Ashtor), p. 191; Perlmann, 'Polemics', p. 117 (see also

A mixture of elements from Deut. 31:19–21[29]

> Write [did our converted Jew understand: they wrote??] ye this[30] song for you[31] [. . .] and it shall come to pass when many evils and troubles are befallen[32] them[33] that this song shall testify against them as a witness; for it shall not be forgotten out of the mouths of their seed.[34]

Deut. 30: 12–13 (with some variations)[35]

> It is not in the heaven, that thou shouldest say: Who shall go up for us to heaven [omission: and bring it unto us, that we may hear it]? Neither it beyond the sea, that thou shouldest say: Who shall go over it [go over the sea] for us [omission: and bring it unto us, that we may hear it and do it]?[36]

Deut. 4:2 (with some variations; see also Deut. 13:1)[37]

> Ye shall not add unto[38] the word which I command you,[39] neither shall ye diminish ought from it .[40]

---

p. 127), who emphasizes the Christian attitude; Lazarus-Yafeh, pp. 98, 137. Compare with Koningsveld towards the end of his article (in a paragraph dealing with Petrus Alfonsi column 624c); Joseph ben Nathan, pp. 44–45; Moshe ben Naḥman, 'The debate beween Ramban and the Pope,' in Y.D. Eisenstein (ed.), *Otzar vikukhim* (Newark, 1928) (Heb.), p. 87, col. B.

[29] MS ʿAbd al-ʿAllām, f. 14a–15b (quotation from the Bible and a discussion).

[30] This should be *hazzo* (= *hazzot* in the Massorah Hebrew bible), transcribed *haẓẓo* (as explained above, section II) and seemingly written by the copyist as *hazf* (Arabic character *fāʾ* instead of *wāw*) due to an unnecessary spot or the similarity between the two characters.

[31] For you, should be *lefanav*, but it is pronounced in classical Hebrew and rightly transcribed: *lefanaw*.

[32] Befallen = *timṣaʾena* (will find him, happen to him) transcribed as *timsaʾena* as there is no difference between *ṣ* and *s* in this tract (as explained above, section II).

[33] Them = in the Hebrew Bible: him, *oto*, divided here, for some unknown orthographic reason, to: *o to*.

[34] *Raʿod* (evils) instead of *raʿot* due to the Turkish pronunciation, and *ẓarʿo* instead of *zarʿo*, as minutely explained above, section II. See Ibn Ḥazm, part I, p. 149, who claims that Moses ordered the Israelites to keep in their memory the song of Deut. 32. Although this chapter precedes, one can assume that the connection was made in the commentaries, although not in the exact spirit of the confusion of verses made by our converted Jew. See also Samawʾal, pp. 49–50 of the Arabic section, p. 54 (of the English section); Qarāfī, p. 239, line 10 (quoting: Moses wrote this law, Torah, as a variant of: this song).

[35] MS ʿAbd al-ʿAllām, f. 15b–17a (quotation from the Bible and a discussion).

[36] Maimonides, pp. 52–53.

[37] MS ʿAbd al-ʿAllām, f. 17a–19a (quotation from the Bible and a discussion).

[38] In the Hebrew original: 'unto it': *ʿalaw*; compare with note 30 above.

[39] I command you = *meṣawwekh* instead of *meṣawwe etkhem*, due either to the Turkish pronunciation confusing *kh* and *h*, as explained above, section II, or for lapse in our converted Jew's memory.

[40] See also Judah Halevi (Baneth), p. 124; Judah Halevi (Even Shemuel), p. 130; Ibn

Deut. 33:4[41]

> Moses commanded us a law, even the inheritance of the congregation of Jacob.

To summarize, we have here a complex, and in a sense quite strange, tract. Its main thrust is to provide polemical ammunition against Judaism and for Islam, using biblical verses quoted in an Arabic script which (in any other case) would have been suitable for transcribing most Hebrew consonants and vowels. The many confusions, deletions and leaps in the text, however, tell us that the person who wrote down these quotations, a converted Jew, was not very proficient in the original Hebrew text. Furthermore, he refrained from quoting important verses which were popular among Muslim polemicists, who used them to demonstrate the superiority of their faith. Linguistic considerations, especially the orthography of the text and the reflected phonetics, show that our converted Jew did not pass on the correct pronunciation to the Muslim theologists who elaborated this tract in his name. In his writing he did not utilize the letters of the Ottoman alphabet, i.e. the Arabic characters used for writing both Arabic and Turkish. Had he done so, the Hebrew texts which he transcribed with Arabic letters could have been accurately conveyed to the Ottoman scholar or scholars who actually composed the tract and who were certainly well-acquainted with the alphabet of the Arabic language, enough to write it and probably to recite and conduct oral discourse in it as well. But our converted Jew passed on the Hebrew text using the pronunciation of Turkish, leading us to suppose that he was less than proficient in Hebrew, perhaps having only faint recollections of the Torah reading in the synagogue, and no real understanding. Apparently he did not have at his disposal a written

---

Ḥazm, part I, p. 88, argues with a Jew who told him that, in the Torah, Moses command the Jews not accept any prophet who would bring them another law, different from the law of Moses; that is in the spirit of the verses quoted here (and especially the neighboring verses omitted by the converted Jew on which the tract we are dealing with relies); Shahrastānī, part I, p. 211, who understands that Moses' law is unchangeable in the eyes of the Jews; Samaw'al (Perlmann), p. 21 of the Arabic section, p. 41 of the English section, and see especially Perlmann's comment, p. 94 of the English part, on f. 5b of the MS he has edited; 'Abd al-Ḥaqq, part II, pp. 7–8 where he develops the issue; Maimonides, pp. 52–53, 55–56; L. Nemoy, 'Ibn Kammūna's treatise on the differences between the Rabbinates and the Karaites,' *JQR* 63 (1972–1973), pp. 97–135, 222–246, here p. 115 (a controversy between rabbinical and non-rabbinical Jews; see above, n. 5, 17, 21, 25).

[41] MS 'Abd al-'Allām, f. 19a–20b (quotation from the Bible and a discussion).

copy of the Hebrew Bible when he passed on these verses. He may well not have been very literate; certainly he was not versed in the Arabic alphabet (used for writing Ottoman Turkish in his days), for if he had been he could have used it for transcribing in writing what he knew of Hebrew phonetics and so could have passed it on to the Muslim scholars. His ignorance, perhaps even illiteracy, was his undoing; unlike other converts (ʿAbd al-Ḥaqq, Samawʾal b. Yaḥyā) who were well informed about both their former and their adopted faiths, our converted Jew was naïve, uneducated and quite incompetent.

WOMEN RE-SHAPING COPTIC VISUAL CULTURE

Nelly van Doorn-Harder

Introduction

In October 2005, during a lecture at the American University in Cairo, Abuna Bishoy el-Antuny presented the new Coptic TV channel called 'Agapi' by showing a pilot of several opening features for the various programs. Those present saw pictures of Patriarch Shenouda III, great numbers of bishops, priests, monks, and deacons. As someone in the audience commented 'We mostly saw men in liturgical dress.' Naturally, a woman posed the question about where the women were and if they would ever play a significant role in this new TV channel. Abuna Bishoy answered that they had tried to convince the contemplative nuns to play a role in this pilot but that the nuns had adamantly refused. As contemplatives they had to guard their identities and did not want to be seen or heard in public. Of course, referring to the nuns only begs the question: 'why not film some other women who are not contemplative nuns?' Although I cannot say for sure, the answer might be that in the minds of the network's architects, contemplative nuns represented the closest female counterparts to the male clerics portrayed.

Judging from the images shown, we could indeed conclude that in the official Coptic Orthodox Church there is little to no space for women. Yet anyone who is prepared to take a closer look will learn that these pilots in fact presented an imagined reality of a totally male-dominated church. Yes, it is true that as is the case in all Orthodox churches – and for that matter in the Catholic Church as well – men remain the official experts of Coptic religion, officiating over the presentation of sacraments and shaping official church rules that concern both men and women. However, in many churches, including the Coptic Church, women have assumed far more important roles than meets the eye. In the Coptic Church they work as contemplative and active nuns, deaconesses, Sunday school teachers, volunteers and in their capacity as mothers they raise the new generations of Coptic believers and leaders. Even if these activities play out behind the scenes of the official church image, they are forceful and numerous enough to have created a new mindset about the role of women that, among others, is being expressed in a women-oriented religious visual culture.

In this paper I argue that during the past forty years Coptic women have managed to expand the traditionally male-oriented culture by changing the visual culture to include religious images and visual practices that are specific to women's needs, understandings and practices. Some women have become religious specialists in their own right and while honoring the official rules and systems of authority within the Church hierarchy, they have created new mental and physical spaces for women. This essay is a preliminary investigation of the culture women are shaping. It traces where women's main contributions to visual Coptic productions can be found, and how this process not only opens up spaces for women, but also strengthens the Coptic identity by contributing to the re-invention of the religious tradition.

These developments and the act of highlighting them have far-reaching consequences. By focusing on the visual culture, one shifts from a liturgical-centered discourse that is centered on rules and regulations, to what art historian David Morgan calls a 'practice-centered discourse.'[1] Even when their work was not recognized officially, women have always played prominent roles in the daily practice of religion. From the Hindu practice of *puja* to doing the church's social work, women all over the world were and are involved in wide-ranging devotional practices.

By its very nature, the practice of religion is not limited to creeds and statements of belief only; it is lived out and applied in communities shaped by their ethnic, cultural and religious environment. Their houses of worship follow particular designs; they eat specific foods, fast at certain times, and hold on to moral codes shaped in interaction with the social environment.[2] Although Coptic women – influenced by the prevailing Muslim culture – remained invisible for many centuries, they have always belonged to the agents who prepared, taught, enforced, and co-designed these building blocks of religion.

Since the 1960s, when a revival swept through their church and their services were needed to strengthen the Church body, women have become more visible. The system they operate in continues to be more

---

[1] D. Morgan, *The Sacred Gaze. Religious visual culture in theory and practice* (Berkeley [etc.], 2005), p. 32.

[2] For a discussion on the intersection of religion, belief and the visual culture, see a.o.: Morgan 2005, pp. 6–15.

or less gender-segregated, which means that women provide services mainly to women. But this segregation can be considered an asset since it has allowed them to invent, re-invent and create their own women-specific tradition and culture that can now be witnessed in scores of products made, commissioned or inspired by women.

Visual culture

Production of visual culture by and for women started within the confines of their respective living spaces – the nunneries and houses for deaconesses – and is now slowly slipping into mainstream Coptic culture. This culture ranges from hand-painted icons in churches that are officially consecrated, thus belonging to the liturgical tools, to trinkets such as mass-fabricated key chains, posters, and calendars. It also includes abbeys and monastic houses and farms designed and built under the supervision of women. These practices, rituals, and ceremonies are being translated into visual products that have become available and visible to all.

Following the observations of David Morgan, there is no hierarchy in the visual culture that thus is being constructed. This culture, in fact, is not limited to images only, but includes 'everything in society that makes images operate, including perceptual habits, cognitive frameworks, popular and elite aesthetics and such daily practices of visual organization as fashion, advertising, physiognomy and design.'[3] Those studying the visual culture of religion have found that understanding this culture helps deepen our understanding of the religion presented since people construct the worlds in which they live through visual practices. In what Morgan calls a 'visual economy' meaning of images and objects evolves in the process of circulating them among the believers who put the images to work. Seen in this way, there is a constant interaction between the world and the worldview that creates certain images and between the images and the worldview they confirm and represent.

[3] D. Morgan, 'Visual religion,' *Religion* 30 (2000), p. 42. The literature about the topic of visual culture is growing steadily, for example see: Morgan 2005; D. Morgan & S.M. Promey (eds.), *The visual culture of American religions* (Berkeley [etc.], 2001); S.M. Hoover & L.S. Clark (eds.), *Practicing religion in the age of the media. Explorations in media, religion, and culture* (New York, 2002).

In the context of everyday life we can thus detect the complexity of religious experience in an ongoing construction of hierarchies, gender roles, and realms of the sacred and the profane. Through visual constructions polarities such as hierarchies in gender roles can be overcome or in some cases even turned upside down. In this process we necessarily include the interactions between image, text and rituals that contribute to the creation of women's spaces or facilitate infusing male domains with feminine elements.

The study of religious visual culture is rigorously interdisciplinary since it includes rituals, liturgy, and other religious experiences and helps decode the role visual images play in the social construction of reality.[4] When looking at the visual culture Coptic women produce we become aware of ever-evolving projects that range from writing books about women saints to painting icons, and decorating the walls of a certain nunnery with sayings of their spiritual father.

Before elaborating on the women's culture, I now will first give some background on the role of Coptic visual culture and explain some moments in Coptic history that changed the role of women.

Coptic visual culture in historic context

Although to a lesser degree than in other Orthodox churches, religious symbols, artifacts and pictures are an inseparable part of Coptic religious life and liturgy. Coptic churches are a feast for the eye; filled with paintings and objects that are used during the service and Eucharist. The altar is hid by an iconostasis that is richly decorated with icons (Plate 17). Those material objects and many of the church rituals have a distinctive character that in some instances reaches back to the time of the pharaohs. According to the distinguished Coptic art historian Dr. Gawdat Gabra, Coptic art '...has an Egyptian character with Greek, Roman and Byzantine influences.'[5] By the fourth/fifth century it became distinctively Christian.

Many of the artifacts such as textiles decorated with religious scenes that served as clothing or burial shrouds, pottery – often found at

---

[4] Morgan 2000, p. 51.

[5] G. Gabra, 'The Coptic Church,' in G. Gabra & Z. Skalova, *Icons on the Nile Valley* (Giza, 2003), p. 23.

monastic sites –, wall paintings of Biblical representations, and icons, contributed to the formation of a distinctive Coptic tradition.[6] Copts proudly remember that some of the first known icons were painted in Egypt where they were inspired by the so-called Fayoum mummy portraits – painted panels with images of the dead – that were produced between the first and fourth century.

This rich visual tradition is enshrined in the remnants of ancient churches and monasteries, and in the modern-day buildings of churches, monasteries, nunneries and the Coptic Museum that was founded in 1908.[7] The walls in the house of a Coptic lay person, however, reflect these containers of tradition and sanctity as well, with numerous reproductions of icons and other sacred pictures.

Coptic Christianity developed when Egypt was part of the Roman Empire. In 641 AD the country was invaded by Arab Muslims, and Copts never reigned over their own country or had their own currency. Being 'strangers' in their own land, Coptic liturgical, religious and secular art became the repositories of Coptic identity, tradition and culture. Especially monasteries built in remote places preserved the early paintings that are mostly on the walls of their churches. This history is long but scattered: Between the fifth and the eighth century there was an active creation of sacred art, but with the exception of a renaissance in the twelfth/thirteenth century, there was little production of Christian visual culture until the revival of the Coptic Church in the twentieth century.

In 1956, Dr. Isaak Fanous became the director of the Art Section of the Institute of Coptic Studies (Cairo) and jumpstarted the so-called Neo-Coptic School. Referring to its Pharaonic roots, Fanous introduced a representation of Coptic religious icons and wall paintings that would re-create the 'pure' Egyptian style which, according to Fanous, had been corrupted by Byzantine and Western influences.[8] According to Dutch art historian Mat Immerzeel, holy figures and scenes from the Bible and Coptic tradition are painted in a style that reinforces 'the direct contact

[6] For examples of pieces of fabric from the sixth century, several icons and religious paintings from the fifth/sixth century, and a fourth-century portrait of a woman, see: M. Capuani, *Christian Egypt. Coptic art and monuments through two millennia* (Collegeville, MN, 2002), pp. 33, 36–40.

[7] For a short history of Coptic art, including contemporary developments, see: M. Immerzeel, 'Coptic art,' in N. van Doorn-Harder & K. Vogt (eds.), *Between Desert and City. The Coptic Orthodox Church today* (Oslo [etc.], 1997), pp. 273–288.

[8] Immerzeel, p. 281.

between believer and saint, the holy figures are always looking straight at the beholder, with wide-open, oval-shaped eyes,' and Immerzeel points out that: '... they seem to radiate a mysterious light.'[9]

Fanous' school resulted in an enormous output of Coptic icons and wall paintings that were translated into numerous expressions of visual art ranging from bookmarks to posters and coffee cups with copies of the original icons. In fact, this was a reinvention and strengthening of Coptic tradition. This process became visible and accelerated into the high speed it has today by the 1960s. The brilliant Patriarch Kyrillos VI (1959–1971) deliberately stimulated this re-invention of the tradition.[10] Reaching back to the early centuries of Christianity, he re-introduced writings, representations and memories in order to restore the Coptic collective memory of this era, thus endowing the Copts with a newly found pride and tradition. Scholars such as Peter Berger, Maurice Halbwachs and Hobsbawm and Ranger have found that this process of reintroducing memories from the past is steered by the needs of the present.[11]

Contemporary Coptic art is among the most visible results of Kyrillos' effort to strengthen the tradition and infuse the past into the present that now can be witnessed in numerous aspects of Coptic visual and non-visual culture and life. For example, it resulted in an enormous production of texts such as books about the lives and sayings of the desert fathers and mothers, intensified study of the patristic heritage that is being studied at several Coptic institutes for higher learning, and in numerous projects to clean and re-discover Coptic archeological places and art. An example of the latter are the thirteenth-century secco wall paintings at the Monastery of Saint Anthony's that were virtually rediscovered after an international team of art specialists cleaned them during the 1990s.[12]

---

[9] Immerzeel, p. 280.

[10] For more information about Patriarch Kyrillos VI see: N. van Doorn-Harder, 'Kyrillos VI (1902–1971). Planner, patriarch and saint,' in Van Doorn-Harder & Vogt, pp. 231–243; N. van Doorn-Harder, 'Practical and mystical. Patriarch Kyrillos VI (1959–1971),' *Currents in theology and mission* 33/3 (June 2006), pp. 223–232, and Brigitte Voile, *Les coptes d'Égypte sous Nasser. Sainteté, miracles, apparitions* (Paris, 2004).

[11] See, among others: P.L. Berger, *The sacred canopy. Elements of a sociological theory of religion* (New York [etc.], 1990 [1967]), M. Halbswachs, *On collective memory*, ed. and trans. by L.A. Coser (Chicago [etc.], 1992), E. Hobsbawm & T. Ranger (eds.), *The invention of tradition* (Cambridge, 1983).

[12] For the project at the Monastery of St. Antony, see the articles in: E. Bolman (ed.), *Monastic visions. Wall paintings in the monastery of St. Antony at the Red Sea* (Cairo [etc.], 2002).

Kyrillos' successor Patriarch Shenouda III (1971–) has amplified the projects begun by his predecessor. The core of these developments, however, continues to be situated in the liturgical life within churches where the communion bread is distinctively Coptic, and within the monasteries where future church leaders start their career as monks. Monks guarded the sacred paintings on their walls that in spite of being covered by centuries of soot still remained part of their spiritual and ritual life. Monks and nuns furthermore did their own share of 'living the tradition,' for example, by introducing the 'original' rule of the founder of communal monasticism Pachomius (c. 292–346 AD).[13] Another visible sign is the introduction of the *qalansuwa*, the cap worn by monks and nuns. Following Patriarch Shenouda's directions, the plain black cap was replaced by one with twelve embroidered crosses and a seam in the middle. According to Patriarch Shenouda this had been the original cap that St. Anthony (251–356), the first monk in the history of Christianity, had worn (Plate 17).[14]

The Coptic Church used to be a local church: firmly rooted in Egypt from the time of the Arab invasion it existed in virtual isolation from the rest of Christendom but for its special relationship with the Ethiopian Orthodox Church. This changed in the 1960s when Coptic migration to western countries increased and the Church launched its mission to Africa. The move outside Egypt accelerated the need for icons and wall paintings in Coptic churches and houses outside Egypt. Moving outside Egypt opened up room for a new Coptic expression of tradition and visual art. In fact, Coptic videos, DVDs and the TV channel are all expressions of reinventing and strengthening of the tradition in order to preserve it for the Copts outside Egypt. For example, one of the goals behind the TV channel is to make sure that they do not forget their heritage.

Inside Egypt, money returned from the thriving community abroad and helped support the building of new Coptic churches, retreat centers, and the restoration of monasteries. Encouraged by the newly invented expressions of art such as those by Fanous, Copts have started to follow other patterns and models as well. These initiatives are now taken by both lay people and clergy. For example, the intriguing decorations at

[13] P. van Doorn-Harder, *Contemporary Coptic nuns* (Columbia, SC, 1995), pp. 51–60.

[14] Van Doorn-Harder 1995, p. 99.

the mega-cave church on the campus of St. Samaan at the Muqattam hills that was built during the 1980s, were made by a Polish artist (Plate 18). Another illustration of this trend can be found in the church at the retreat center in Bayat (near Beni Suef). The walls of this church are decorated with scenes from the life of St. Mary, including a representation of her apparitions in Zeitoun (Cairo) in 1968 and 1986. This artwork is from the hand of a Coptic lay man who felt a special inspiration to donate his time and talents to the church in honor of St. Mary (Plate 19).

Lay women are participating in discovering alternative art forms as well. A Coptic lay woman, Mary, who works as a pediatrician, designed one of the illustrations for the 2006 English Coptic calendar sold in Coptic churches in the US, Australia and Great Britain. Although she did not have children herself, Mary had dedicated part of her income to printing a new rendition of the murder of the children by King Herod in his attempt to kill the child Jesus (Matthew 2:16–18). Reflecting on the fate of the children, she had come to realize that the traditional depiction of these murders as gruesome events ignored the final fate of the children who were taken up to heaven. Hence Mary designed a new poster portraying the children as angels surrounding the Holy Family on their flight to Egypt (Plate 20).

Women and visual art

Until Patriarch Kyrillos VI jumpstarted what is now considered to be the revival and reformation of the Coptic Orthodox Church, women had few options to participate in the official or semi-official life of the Church. Women from the upper classes were active in some of the many charitable organizations that had started in the 19th and 20th century. Convents had become the last refuge for illiterate women who for physical or psychological reasons could not be married.

This picture changed in the 1960s when Kyrillos encouraged both men and women to consider a monastic career in order to form a pool of well-educated church leaders. In 1962 he appointed twenty-one-year-old Mother Irini (d. 2006) as superior of the contemplative Convent of St. Mercurius (Abu Saifein) and gave her free range in renovating the place from a spiritual and material point of view. Mother Irini was Kyrillos' peer in reading the signs of the time. Apart from introducing new rules for the nuns, she cleaned up the abbey in Old Cairo, renovated and expanded it and then focused on raising the popularity of the convent's patron saint Abu Saifein (Plate 21). She built a church inside the convent to hold his relics and organized an annual *moulid* to commemorate his death. During this festivity she would relay the miracles that had taken place upon his intercession. As his fame grew, so did that of the convent which now receives thousands of visitors during the *moulid* on December fourth.[15] The increase in visitors resulted in an increase in gifts and endowments donated to the convent by grateful believers. Thus, enough money was gathered to build a large farm with a view on the Mediterranean in Sidi Krir between Alexandria and Marsah Matrouh. The farm is part of a complex that includes three churches, a library, a guesthouse, and a house for the nuns (Plate 22).

The renovated abbey in Cairo and the farm in Sidi Krir not only serve as spiritual centers, but also have become refuges for women seeking pastoral and spiritual advice. Moreover, they have become centers where visual art for and by women is produced and have inspired the other contemplative and active Coptic nunneries to follow in their footsteps. Nowadays increasing numbers of women are entering contemplative convents, becoming active nuns at the Convent of St. Mary in Beni Suef, or choosing to serve as deaconesses (*mukarrasāt*). The educational level of these women is rising and many hold BA or even graduate diplomas. Their academic background allowed women to start researching the history of their convents, the patron saints and women saints or saintly women. This exercise has resulted in scores of books, pamphlets and posters that serve to underline and spread the importance of the convents and the women's activities. The search for women's history and heritage is not limited to the Coptic tradition only but includes material, for example, from the Russian Orthodox Church (Plate 23).

Coptic icons

Icons, pictures of saints, hold a special place in the life of Copts. They are considered windows into heaven that can convey the life story and blessing of the saint portrayed. In a homily on Saint Mary from the fifth century, Theopilus, the Archbishop of Alexandria (385–412) describes a

[15] Van Doorn-Harder 1995, pp. 165–168.

miracle of an icon of St. Mary that oozed blood. He furthermore mentions that through the intercession of the saint icons can cure diseases and exorcise devils.[16] Icons can take on human qualities; they weep and bleed and since they are potential sources of blessing they cannot be destroyed.[17] Painting an icon is a religious exercise for which the artist prepares him or herself by fasting, praying, and studying the life of the saint to be portrayed. Especially nuns and monks have taken up this art as it suits their religious lifestyle and goal to be in continuous prayer (Plate 24).

Father Maximous El-Anthony who supervises the restoration work of the wall paintings in his monastery, provides us a glimpse of the role icons play in his daily life as a monk.[18] Inside the cell every monk or nun has several icons that help the monastic to leave the world behind and 'see the heavenly world through the windows of the icons.'[19] Icons help the monastic meditate on the life of Christ, ask for intercession, while icons of his or her favorite saint serve as a reminder to follow the saint's model.[20] The material the icon is made of, its style, or size, are indifferent since its meaning is located in the personality of the saint depicted.[21]

As all Copts do, monastics touch, kiss and hold icons. Portraits of saints can be put on anything: cups, labels on bottles and jars, spoons, key chains, T-shirts, headscarves, crosses, lamps, calendars, posters, book covers, inside a vial with holy oil, and so on. All can be effective: the Coptic community is saturated with stories of miraculous healing or rescue.[22] Some happened, for example, after the sick person put a pamphlet with the saint's picture on the affected area. On another note, students relate how saints help them study by putting his or her picture under the pillow at night.[23]

---

[16] Gabra, pp. 32–33.

[17] Gabra, p. 33.

[18] Maximous, 'Windows into heaven. Icons in monastic life today,' in Bolman, pp. 195–201.

[19] Maximous, p. 196.

[20] Ibid.

[21] Maximous, p. 197.

[22] For example, Van Doorn-Harder 1995, p. 162, relates the story of Albert Y. Youssef who put pictures of Patriarch Kyrillos VI between his important papers in order not to lose them. This story is just one of the hundreds that are told about the miraculous intercession by Kyrillos VI.

[23] See, among others: N. van Doom, 'The importance of greeting the Saints. The appreciaton of the Coptic art by laymen and clergy,' in: H. Hondelink (ed.), *Coptic art and culture* (Cairo [etc.], 1990) pp. 112–113.

Icons also play a crucial role in the liturgy; especially during the periods of Advent, Easter and the forty days following. Father Maximous explains how icons help him to enter the liturgical part of his day that begins when he enters the church. After performing his entrance prayers, he greets those present in the church: 'those in heaven and those on earth... the Lord Jesus and the Holy Virgin, the angels, the martyrs, and the saints through their icons.'[24] Greeting the icons, Father Maximous continues his explanation, is performed by kissing them or – if they are too far away – by touching them. In front of icons to whose saints he will pray for intercession, Father Maximous lights a candle.[25]

At the morning and evening services of the offering of the incense, icons of St. Mary and St. John the Baptist and others that hang above the altar are customarily honored and censed. Icons are carried around the church on feast days such as Palm Sunday, Easter, Ascension Day and Pentecost. On Good Friday an icon with the crucified Christ is 'buried' on the altar. On the night of Easter the same icon is carried in procession around the altar and the church seven times. Then the icon of the Resurrection is placed in front of the sanctuary where it stays during the time of Easter.[26]

From a young age Coptic children learn to respect icons and within the community there is consensus that icons can serve as vehicles to communicate with the Divine. The saint represented is now in heavenly glory from where he or she intercedes on behalf of the believers. With the emergence of the new school of icon painting certain themes have become popular among the Copts. Especially the Flight of the Holy Family with the pyramids in the background has become a symbolic and popular theme. Also themes from the early centuries of Christianity that had fallen into oblivion are newly re-emerging. For example, the theme of the twenty-four elders mentioned in the Apocalypse (4.4) is now being painted regularly in the area over the altar (Plate 25).[27]

Since icons can be powerful objects, naturally the artists producing them are held in great regard among the Copts.[28] Many of these artists

---

[24] Maximous, p. 199.

[25] Ibid.

[26] See: Gabra, pp. 34–35.

[27] Gabra, p. 40. Also see Otto F.A. Meinardus, *Coptic saints and pilgrimages* (Cairo, 2002), p. 22.

[28] See Morgan 2005, p. 55, for a discussion about what religious images and visual practice do to those who believe in their efficacy.

now are women, especially nuns and deaconesses (Plate 24). They also manufacture the robes of priests, altar curtains and other liturgical vestments used all over Egypt (Plates 26 and 27). This activity by itself introduces new elements into the representation of saints and symbols on these clothing and other articles. For example, the nuns of the Sisters of St. Mary explained to me that they portrayed the Virgin Mary on an altar curtain for the cathedral in Beni Suef following a special inspiration (Plate 27). For the most part their specific input goes unnoticed but it can have far-reaching consequences, and the bottom line is that women now have joined the ranks of those creating objects that hold potential blessing powers.

Women's visual culture

Mother Irini from the Convent of St. Mercurius was the first one to capitalize on the neo-Coptic trend of icon painting. Every church that was built or renovated within the convent, she had decorated with prominent icons of the patron saint and with more than average amounts of pictures with women saints. Furthermore, her nuns researched the lives of those saints and wrote pamphlets, articles and books about them. These were rapidly translated into English and other western languages to cater to the needs of immigrant Copts; especially those of the second and third generation.

Another theme that is important to the nuns at Deir Abu Saifein is the tradition that the abbey is built on a spot where the Holy Family once rested when traveling through Egypt. The tradition about the itinerary derives from the vision Patriarch Theophilus had of the Virgin Mary and was re-affirmed for the convent when in 1964 Mother Irini had a vision of Saint Mary in the garden. The Mother of Jesus revealed to her that she and her family had stayed on that specific place in the garden during her time in Egypt.[29] This vision was underscored by a similar one Patriarch Kyrillos VI had around the same time, and resulted in a specially built chapel in the garden of which the walls were decorated with scenes from the Flight to Egypt. As the convent ran out of space for new novices, the abbey was expanded and the chapel in the garden became

[29] Van Doorn-Harder 1995, p. 172.

a large chuch. The scene of the Flight, however, has been reinvented in the newly built communal hall that is decorated by a large mosaic of the Family's stay in Matariya where they entered the area that is now Cairo. As the trip through Cairo ended in the garden of the convent, a group of nuns imagined a daily scene in Matariya under the tree that according to the tradition grew on the spot where the child Jesus had been given a bath (Plate 28). The scene depicts Mary pouring water over the child, while Joseph draws water from the well assisted by an angel. Furthermore, angels surround the family while plants encircled by a halo shoot up at the places where the water runs off from the child and touches the ground.

The scene is domestic; the nuns paid attention to the details of life that are commonly overlooked or disappear in the more symbolic renditions of icons. The nuns who designed the mosaic used their fantasy to imagine what exactly happened at the place where according to the tradition the holy child had been washed and plants sprung up spontaneously due to the blessed water. They paid attention to the clothing and dressed Mary and Joseph in pharaonic looking garb. They have pondered practical questions such as where the child's clothes were drying and have hung them on Joseph's staff. The mosaic is unique – it follows Coptic themes but has been put together entirely following the inspiration of the nuns. Thus it makes two statements; the Holy Family was here and the convent's connection with St. Mary is such that the nuns feel permitted to portray her while performing an every day chore. This last statement is in fact powerful. In most scenes the mother of Jesus is depicted in rather passive positions sitting in the stable just after the birth or sitting on the donkey traveling through Egypt. This scene acknowledges that similar to women everywhere, large parts of Mary's days must have been taken up by ordinary duties. The nuns, knowingly or unknowingly, refer to findings from feminist studies that uphold that women's domestic and non-official activities contain as many potentials for being political statements as men's actions do.[30]

Over the years the nuns have also changed the icons inside the convent's churches. Apart from numerous icons of the patron saint Abu Saifein, a great number of newly painted icons of women saints now are part of the decoration. For example, St. Barbara and St. Marina are depicted (Plates 29 and 30), while the patron saint's icon is paired with

---

[30] Among others see Maila Stivens (ed.), *Why gender matters in Southeast Asian politics* (Clayton, 1991). A cruder rendition of the scene can now be found in the Muʿallaqa church as well.

a rendition of St. Mary appearing in Zeitoun (Plate 31). The statement is not only that women matter but also that the saints are still actively involved in the life and development of the convent.

The theme of the saints' involvement continues in the churches at the farm in Sidi Krir that was built with great resistance from local authorities and powerful interventions from the saints.[31] The placing of the images here is even more strategic since the iconostasis of one of the churches contains icons of Saints Irini, Marina and Dimyanah and her forty virgins (Plate 32).

The icons at Abu Saifein's convent are handpainted according to the neo-Coptic school. But down the road from the farm in Sidi Krir stands the much smaller fish farm of the convent of Mari Girgis (St. George) in Harat Zuweilah. The fact that the convent has less financial means has not withheld the nuns to decorate their chapel lavishly with posters depicting women saints. Thus we find pictures of Saint Dolagi and her four children (Plate 33), of St. Perpetua (Plate 34) and of the lesser known martyrs St. Mahra'il (Plate 35) and St. Butamina (Plate 36). Part of the nuns' display is a picture of Efrusina (Euphrosyne, 1228–1308) (Plate 37), who was the convent's superior at the beginning of the Mamluk period.[32] She was a saintly nun who had visions and healing powers. Since the death of Mother Aghapi, the head of the convent, the nuns have added to their visual culture a shrine for Aghapi and a booklet with a picture of the abbess holding the relics of St. George (Plate 38).

Since there is no hierarchy in visual culture, all representations are equally important in promoting the role of women within the Coptic Church. Women, both lay and nuns are participating in this process that includes texts and other images as well. For example, the nuns at the Convent of St. Mary in Beni Suef have sayings of Bishop Athanasius on their walls. Together with the three nuns who envisioned the convent, he guided its initiation and growth and served as the nuns' spiritual guide. His sayings are new and now placed on the convents walls mixed with those of early fathers and mothers. The nuns affirm furthermore their own role in the convent with the posters they make that, next to Bishop

---

[31] N. van Doorn-Harder, 'Discovering new roles. Coptic nuns and Church revival,' in Van Doorn-Harder & Vogt, pp. 94–95.

[32] Van Doorn-Harder 1995, pp. 133–137.

Athanasius' picture, portray their first superior the late Sister Hannah (1930–1991) and groups of the convent's nuns (Plate 39).

Conclusion

Women in churches all over the world lack access to the hierarchy and its power. For example, Francis Adeney describes in her book *Christian women in Indonesia. A narrative study of gender and religion*,[33] the struggle of Protestant women, who hold degrees in theology, to be taken seriously in their churches. She observes that in order to gain access to real power women have to rely on what are considered everyday or less important practices such as offering hospitality, creating beauty, and honoring relationships. Thus the women develop what Adeney calls 'resistance theologies that reconstruct social realities.'[34] Read carefully, this statement means that in many religious institutions, women have to rely on their own networks and power structures to shape and re-shape their own realities and eventually those of the men as well.

That the women are up against quite powerful mindsets is illustrated by the story that the art historian Karel Innemee related when I presented part of this essay at the Netherlands-Flemish Institute in Cairo on December 2nd, 2005. He told that in Dair as-Suryan his team had uncovered a picture of the Dormition of the Virgin Mary surrounded by women holding censers. This portrayal led to great confusion among the monks now living in the monastery. Left to the picture is the traditional representation in which the virgin is surrounded by the twelve apostles. The women portrayed are holding censers. In the mindset of the monks this is absolutely impossible as censers are official liturgical instruments and thus can only be held by men. According to Karel, the first reaction of the monks was that it could not be possible that this picture contained other women than the Virgin.[35] However, further research by Karel Innemee has revealed that the book of the Order of

---

[33] (Syracuse, NY, 2003).

[34] Adeney, p. 179.

[35] For the article on this wall painting see: K.C. Innemee & L. van Rompay, 'Deir al-Surian (Egypt). New discoveries of 2001–2002,' *Hugoye. Journal of Syriac Studies* 5/2 (2002) pars. 18. Photo no. 10 shows the Dormition of the Virgin surrounded by virgins. [http://syrcom.cua.edu/Hugoye/Vol5No2/HV5N2InnemeeVanRompay.html].

the priesthood that is attributed to Severus of Ashmunein, who lived towards the end of the tenth century, does not mention the censer as one of the consecrated instruments for the liturgical service.[36] In fact, Innemee found the first time consecration of the censer mentioned in the book of the *precious Pearl in explaining the rites and the dogmas of the Church* by Yuhanna Salamah in 1909. Thus the women surrounding the Virgin did not break any Church rules of their time as the censer had not yet been added to the consecrated liturgical instruments.

Whatever the influence of women's material culture on the external environment, shaping it influences the women's lives in profound ways. By uncovering models of female saints, martyrs, nuns and deaconesses, women reclaim the Biblical tradition that is filled with women religious agents. The newly discovered models provide the foundation for the newly developed monastic forms of life and fit right into the re-covering of Tradition that was practiced and encouraged by Patriarch Kyrillos VI. The newly found models confirm that women have their own hierarchies of spiritual power and show that women count, even when history has deleted them out of the picture.

Finally, these new trends are of profound importance to the generations of new Copts in the lands of immigration where young Coptic women grow up with entirely different gender models. It is inconceivable to present a church devoid of women in countries such as the USA and Australia. Thus the books produced by the nuns are quickly translated and readily available in bookstores connected to Coptic churches all over the world. The interplay between women's higher education, the globalization of the Coptic Church and the new visual economy are slowly changing patterns of expectation and mindsets of male Copts.

---

[36] Karel Innemee and Yuhanna Nessim, 'The censer as a consecrated instrument,' referring to: J. Assfalg, *Die Ordnung des Priestertums. Ein altes liturgisches Handbuch der koptischen Kirche* (Cairo, 1995), p. 15.

INDEX